AF787926

Sustainable Economy and Ecotechnology

IEREK Interdisciplinary Series for Sustainable Development

Editorial Board

Muthanna Abdulrazzaq, American University in the Emirates
Dubai, United Arab Emirates

Haitham M. Alzoubi, School of Business
Skyline University College, Sharjah, United Arab Emirates

Jacob Wood, JCU Singapore Business School, James Cook University Singapore, Singapore, Singapore

Qian Long Kweh, School of Management, Canadian University of Dubai, Dubai, United Arab Emirates

Simon Elias Bibri, Swiss Federal Institute of Technology (EPFL), Echandens, Switzerland

Fabio Fatiguso, DICATECh, Polytechnic University of Bari, Bari, Italy

Lucia Della Spina, Mediterranea University of Reggio Calabria, Reggio Calabria, Italy

Kamel Eid, College of Engineering, Qatar University, Doha, Qatar

Asma Salman, American University in the Emirates, Dubai, United Arab Emirates

Muhammad Nawaz Tunio, Business Administration, Bhit Shah, Pakistan

Mamta Gupta, Department of Management, Maharaja Surajmal Institute, New Delhi, Delhi, India

Mahmoud Yousef Ghoneim, Architecture Faculty of Fine Arts, Helwan University, Helwan, Egypt

Angeles Sánchez, Department of Applied Economics, University of Granada, Granada, Spain

Elena Cantatore, DICATECh, Politecnico di Bari, Bari, Italy

Shilpa Sindhu, Maharaja Surajmal Institute, Janakpuri, Delhi, India

Erum Shaikh, Business Administration, Shaheed Benazir Bhutto University, Sindh, Pakistan

Muhammad Qureshi, Global Supply Chain Management, Thornaby-on-Tees, UK

Saoucene Mahfoufa, Jeddah, Saudi Arabia

Pengji Wang, James Cook University Singapore, Singapore, Singapore

Somaya Askri, School of Management and HRM and Management, Canadian University Dubai, Dubai, United Arab Emirates

Series Editor

Mourad Amer, IEREK for Research, Alexandria, Egypt

The "Sustainable Economy and Ecotechnology" (SEE) book series acts as a timely initiative within the modern academic sphere, fostering a strong connection between technology, research, policy making, and a successful practical implementation for a sustainable future. In an age marked by increasingly urgent environmental challenges and by the critical importance of sustainable development. The series editor is dedicated to steering positive efforts to provide a strong platform for the dissemination of high-quality research focusing on sustainable business, technology, and ecological management and conservation. Through the contributions of experts in the field, "SEE" endeavors to create a transition toward sustainable economic models and technologically driven ecological awareness. The "SEE" series welcomes submissions of selected papers from conference proceedings, monographs, textbooks, and edited volumes, aiming to cultivate a comprehensive understanding of sustainable development by shining a spotlight on a diverse array of topics that collectively contribute to the advancement of a Sustainable Economy and Eco-technology. The scope of this series includes, but is not limited to: Sustainable business practices and management strategies and digitalization, circular economy principles and benefits, climate change impact on economy and business operation, renewable energy and its impact in driving economic growth, role of technology in ecological and environmental preservation, sustainable agriculture, sustainable economics, sustainability in digital business management, finance and investment, sustainable supply chain management: impacting marketing, finance, production, manufacturing and distribution, Implications of economic development on the environment, sustainable mobility and its economic impacts, sustainable marketing strategies and its socio-economic implications, sustainable urban and rural development impacts on economy and more topics related to business investment, economy, environment, and technology.

Series Editor: Mourad Amer **editor@see-series.com**

Anupam Prakash · Manish Kumar Dubey
Editors

Advances in Fungal Secondary Metabolites: Bioactive Attributes and Biotechnological Potential of Industrial Interest

Exploring Bioactive Compounds and Their Biotechnological Potential

Springer

Editors
Anupam Prakash
Department of Life Sciences, School of Biosciences and Technology
Galgotias University
Gautam Budha Nagar, Uttar Pradesh, India

Manish Kumar Dubey
Department of Biotechnology, University Centre for Research and Development
Chandigarh University
Mohali, Punjab, India

ISSN 3005-1274 ISSN 3005-1282 (electronic)
Sustainable Economy and Ecotechnology
ISBN 978-3-032-23506-0 ISBN 978-3-032-23507-7 (eBook)
https://doi.org/10.1007/978-3-032-23507-7

© The Editor(s) (if applicable) and The Author(s), under exclusive license to Springer Nature Switzerland AG 2026

This work is subject to copyright. All rights are solely and exclusively licensed by the Publisher, whether the whole or part of the material is concerned, specifically the rights of translation, reprinting, reuse of illustrations, recitation, broadcasting, reproduction on microfilms or in any other physical way, and transmission or information storage and retrieval, electronic adaptation, computer software, or by similar or dissimilar methodology now known or hereafter developed.
The use of general descriptive names, registered names, trademarks, service marks, etc. in this publication does not imply, even in the absence of a specific statement, that such names are exempt from the relevant protective laws and regulations and therefore free for general use.
The publisher, the authors and the editors are safe to assume that the advice and information in this book are believed to be true and accurate at the date of publication. Neither the publisher nor the authors or the editors give a warranty, expressed or implied, with respect to the material contained herein or for any errors or omissions that may have been made. The publisher remains neutral with regard to jurisdictional claims in published maps and institutional affiliations.

This Springer imprint is published by the registered company Springer Nature Switzerland AG
The registered company address is: Gewerbestrasse 11, 6330 Cham, Switzerland

If disposing of this product, please recycle the paper.

Contents

Introduction to Fungal Secondary Metabolites

Chandrabhan Prajapati, Sachchida Nand Rai, Payal Singh, and Santosh Kumar Singh

Abstract

Fungus secondary metabolites are bioactive substances generated by fungi that differ from their primary metabolites required for basic life. Fungal secondary metabolites serve crucial functions in ecological relationships, medicine, and even biotechnology, demonstrating a wide range of bioactive attributes with considerable medicinal and industrial implications. Remarkably unique and diverse secondary metabolites are essential for fungal adaptability, defence, and interaction. Further investigation into the primary and secondary metabolic processes of fungi becomes crucial, focusing on their contribution to the production of necessary medicinal and commercial chemicals. Penicillin's discovery, which transformed medicine through its use as a potent remedy for infections caused by bacteria, represents a prime example of the historical relevance of secondary metabolites of fungi. Indigenous societies highlight the relevance of fungus through their traditional uses, as ethnomycological activities reveal an early understanding of its therapeutic potential. The discovery of fungal metabolites has been transformed by technological innovations such as metabolic engineering, alongside allow NMR and LC-MS tools, allowing for improved production and accurate structural examination. Furthermore, chemical structures and biosynthesis processes, including prominent instances for fungus secondary metabolites, such as lovastatin, ergot alkaloids, as well as aflatoxins, in addition to polyketides, terpenoids, as well as alkaloid compounds, along with nonribosomal peptides, are explored. Fungal metabolite biosynthesis routes and regulatory systems are poorly understood, limiting their possibilities. Further research must be conducted to address problems with manufacturing scale and toxicity considerations. To maximize the biotechnological potential of fungus secondary metabolites, it will be imperative to address such limitations.

Keywords

Secondary metabolites · Biosynthetic · Gene clusters · Fermentation · Alkaloids · Polyketides

1 Introduction

The research and development of new drugs have relied heavily on natural chemicals from microbes, plants, animals, and even marine life. Natural sources accounted for 75.38% of the 1881 new medicines developed from 1981 to 2019. There has been a notable surge in studies into natural products, especially secondary metabolites (Conrado et al., 2022; Newman & Cragg, 2020a, b; Přichystal et al., 2016). Bacteria, fungi, and plants all have highly evolved secondary metabolisms that provide about 500,000 identified natural compounds. Plants account for 350,000, animals for 100,000, and other microorganisms for 70,000 compounds. The majority of microbiological sources are fungi (15,600), especially Actinomycetes, accounting for 13,700. The number of fungal metabolites discovered has increased, especially from filamentous Ascomycota followed by Basidiomycota, although metabolites produced by unicellular fungi together with various remaining fungal groups are lower (Bérdy, 2012; Bills & Gloer, 2016; Nett et al., 2009). It is estimated that there are roughly 1.20 million endophytes of fungal origin in the world, but only about 16% of them—190,000—have been recognized and thoroughly investigated. This demonstrates the enormous untapped opportunity of using fungal endophytes

C. Prajapati · S. K. Singh (✉)
Centre of Experimental Medicine and Surgery, Institute of Medical Sciences, Banaras Hindu University, Varanasi, India
e-mail: santosh.singh01@bhu.ac.in

S. N. Rai · P. Singh
University Centre for Research and Development (UCRD), Chandigarh University, Mohali, Punjab, India

© The Author(s), under exclusive license to Springer Nature Switzerland AG 2026
A. Prakash and M. K. Dubey (eds.), *Advances in Fungal Secondary Metabolites: Bioactive Attributes and Biotechnological Potential of Industrial Interest*, Sustainable Economy and Ecotechnology, https://doi.org/10.1007/978-3-032-23507-7_1

for drug development (Ainsworth, 2008; Hawksworth & Lücking, 2017). Biologically active secondary metabolites are produced by symbiotic fungi called fungal endophytes that live inside plant tissues. More than half of such compounds are still structurally or functionally unknown, which presents enormous opportunities for new medication research and therapeutic breakthroughs in medical research, biotechnology, and agriculture (Schulz et al., 2002). The production of several secondary metabolites by filamentous fungi is frequently associated with sporulation. Their generation is controlled by environmental as well as genetic variables (Calvo et al., 2002). Fungi produce primary and secondary metabolites, although these metabolites vary in importance, implementation, and function (Wadhwa et al., 2024). Primary metabolites comprise vital substances made in the developmental phase, including proteins, lipids, and carbohydrates. They are essential for developing sexual reproduction and fungi's fundamental metabolic functions. These metabolites can be observed among every fungus with straightforward, preserved compositions. They are used in fermentation alongside the industrial manufacturing of enzymes and are essential for generating energy, cellular repair, and biosynthesis (Dutton & Evans, 1996; Feofilova et al., 2012; Herrmann & Weaver, 1999; Wilkinson et al., 2004). Furthermore, secondary metabolites (SM) remain non-essential but ecologically important substances created during the stationary phase, frequently in response to stress or nutrient shortage. Examples include antibiotics, pigments, microbial toxins, and pharmaceutically important chemicals. These chemical compounds are usually species-specific and show high structural diversity. They play unique functions in defense, competition, and environmental adaptability, giving fungi sustainable benefits (Molelekoa et al., 2023; Silber et al., 2016; Umar et al., 2024). The primary and secondary routes of fungus metabolism generate chemicals with unique functions. The pentose phosphate route, the TCA cycle, along with glycolysis, are examples of primary metabolic processes. They produce vital substances, including ATP and NADH, in addition to proteins, amino acids, as well as lipids, known to be necessary for the generation of energy, development, and physiological maintenance. For example, the TCA cycle generates substrates regarding the synthesis of fatty acids and nucleotides, whereas glycolysis yields pyruvate energy and biosynthetic precursors. These systems remain constantly activated in fungi that promote basic biological functions (Borin et al., 2022; Masi et al., 2021). Specialized molecules that are necessary for environmental adaptability yet not required for growth are produced by secondary metabolic processes, which originated via the primary metabolic process. Some of these pathways encompass the shikimate process for aromatic chemicals, the nonribosomal peptide synthase (NRPS) process for cyclosporine, and the polyketide synthase (PKS) process for aflatoxins as well. These chemicals improve survival through facilitating interactions with the environment and competition, especially defence (Feng & Leonard, 1995; Olarte et al., 2019; Umashankar & Nygård, 2024). They need substrates including acetyl-CoA and NADPH coming from the primary metabolic process and can commonly be brought through stress or limited nutrition (Nielsen & Nielsen, 2017). SM constitutes a necessity for fungi's ability to adapt to a variety of harsh conditions. Melanins defend against UV radiation alongside oxidative stress, whereas gibberellins promote symbiosis plant development in low-nutrient soils (McGuiness et al., 2019; Singaravelan et al., 2008). In certain situations, siderophores allow fungi to gather iron, whereas mycosporines are molecules that absorb ultraviolet light in underwater fungi (Oren & Gunde-Cimerman, 2007; Winkelmann, 2007). Additional adaptations comprise trehalose, which mitigates osmotic stress, while carotenoids provide protection against light. Fungal resilience is boosted by these compounds, guaranteeing persistence under harsh environments (Avalos & Carmen Limón, 2015; Hounsa et al., 1998). SM also acts as a defensive mechanism, enabling survivability against the competition, infections, and predation. Aflatoxins resist predators, whereas penicillin and ochratoxins can hinder microbial competitors (Drott et al., 2017; Kelsic et al., 2015; Tryfinopoulou et al., 2020). Zearalenone interferes with insect development, while mycotoxins such as fumonisins weaken host defences (Antonissen et al., 2014; Niermans et al., 2019). Patulin prevents the growth of germs, while substances like griseofulvin and trichothecenes have antifungal properties (Moake et al., 2005; Xu et al., 2021). These compounds lessen the risks posed by competing organisms, enabling fungi to take control of highly competitive ecological niches. SM can also promote parasitism, competition, and symbiosis regarding environmentally friendly interactions. Cyclosporine regulates host immunological responses, whereas indole-3-acetic acid (IAA) increases plant root development, resulting in advantageous benefits (Liddicoat & Lavelle, 2019; Qiang et al., 2019). Hydrophobins allow surface attachment, whereas terpenoids attract pollinating organisms (Abbas et al., 2017; Bayry et al., 2012; Jayaram et al., 2021). Chitinases enhance fungus pathogenicity by dissolving damaged insect exoskeletons, while compounds such as strobilurins prevent microbial rivals (Jeong et al., 2023; Sauter et al., 1999). The environmental significance of fungi can be seen in the above relationships. This chapter further delves into comprehensive detail about fungal secondary metabolites, including their biosynthesis processes, ecological importance, and industrial uses. It looks toward their classification and historic significance, along with manufacturing and research innovations in technology. The chapter also emphasizes current issues and possible avenues of questioning, focusing on the function of fungal metabolites throughout the fields of agriculture, medicine, and environmental sustainability.

2 Historical Importance of Fungal Secondary Metabolites

2.1 Discovery of Penicillin and Its Impact on Medicine

A fortuitous discovery made by Alexander Fleming in 1928 revolutionized medicine. He observed that a kind of mold, which he later determined to be *Penicillium notatum*, had destroyed the nearby bacteria when he investigated *Staphylococcus aureus*. This coincidental discovery laid the foundation underlying antibiotics and demonstrated the value of inquiry in addition to paying careful attention to surprising results in research (Bennett & Chung, 2001). Despite being revolutionary, Fleming's findings had difficulties in real-world applications. Penicillin hadn't been produced in bulk while standardized until the 1940s, because of the work of Ernst Chain, Howard Florey, and their group. Their widespread availability, especially throughout World War II, was made possible by advancements in fermenting alongside extraction techniques, which helped save countless lives. Penicillin transformed how infections were treated, significantly lowering death rates and opening the door for contemporary surgical and medical breakthroughs. Its prevalence signaled the start of the antimicrobial era, which drastically changed public health as well as medical protocols (Haque et al., 2024). The initial discovery of penicillin represented a historic victory but also brought forth problems such as antibiotic resistance, which emphasizes the necessity for safe use. The sensitive equilibrium among scientific advancement and its unexpected effects can be brought to the forefront by such a legacy.

Penicillin's discovery significantly decreased infection-related fatalities, which paved the way for improvements in cancer treatment and organ transplantation, especially surgery. As a result of penicillin's success, various antibiotics were developed, revolutionizing healthcare worldwide (Banerjee et al., 2013; Chan et al., 2020). Antibiotic resistance, which continues to be a significant worldwide issue and can be seen in infections such as MRSA, was also brought about by its extensive use. Penicillin's narrative motivates continued attempts to combat antibiotic resistance while discovering alternative treatments (Laxminarayan et al., 2013). The determination of innovation remains influential in healthcare right now via personalized medicine techniques like microbiota control and bacteriophage therapies (Paule et al., 2018).

2.2 Traditional Uses of Fungi in Indigenous Cultures

Fungi were historically used by indigenous civilizations for practical, spiritual, and therapeutic purposes. The scientific investigation of fungi for cultural settings, or ethnomycology, exposes how they are used in rituals in addition to healing processes. *Amanita muscaria*, also known as fly agaric, was utilized by Siberian shamans to induce heightened levels of consciousness. In contrast, psilocybin-producing mushrooms were employed in spiritual customs by the Mazatec inhabitants of Mexico (Carboué & Lopez, 2021; Van Court et al., 2022). Additionally, fungus metabolites were employed as natural cures. Various Native American cultures deliberately exploited the antibiotic qualities of fungi covering foodstuffs to alleviate infections (Zeb & Lee, 2021). Fungi, such as *Claviceps purpurea* (ergot), played a dual function throughout European traditional medicine, serving as both a toxin and a source of psychedelic chemicals (Miedaner & Geiger, 2015). These applications are emphasized further through case reports. *Amanita muscaria* is utilized by Siberian shamans regarding spiritual trips (Feeney, 2013). The instances above demonstrate the importance of fungus in traditional belief systems and their promise for contemporary medicines.

3 Categories of Fungal Secondary Metabolites

3.1 Polyketides

Polyketides, a unique collection of secondary metabolic products, demonstrate fungi's chemical creativity (Fig. 1). These chemicals, generated via polyketide synthases (PKSs), differentiate themselves via their structural variety and wide range of biological functions. Polyketide biosynthesis involves repeated instances of acyl-CoA expansion in addition to structural change. The following modularity mechanism enables the fungus to generate chemicals with complicated chemical structures (Schuemann & Hertweck, 2009). Fungal endophytes known to evolve into symbiotic lives frequently create polyketides that strengthen host protection from environmental stressors, infections, and herbivores (Grabka et al., 2022). Medicine has been revolutionized by polyketides produced from fungus. *Griseofulvin* compound of *Penicillium griseofulvum* heals infections caused by fungi, alongside Lovastatin, regarding *Aspergillus terreus* regulates cholesterol (Aris et al., 2022; Hajjaj et al., 2001). *Pestalotiopsis microspora* and other endophytic fungi generate taxol, another polyketide found to be essential for the treatment of cancer (Gangadevi & Muthumary, 2009; Strobel et al., 1996). *Sorangium cellulosum* produces epothilones that are employed as chemotherapeutic drugs and maintain microtubules (Villegas et al., 2023). The immunosuppressive drug rapamycin originated from the fungus *Streptomyces hygroscopicus* (Vézina et al., 1975; Yoo et al., 2017). Polyketides can be employed as naturally occurring development regulators in addition to insecticides in the agricultural sector. Strobilurins of *Strobilurus tenacellus* tend to be fungicides, whereas beauvericin serves as an insect-killing polyketide found in *Beauveria bassiana* (Bartlett et al., 2002; Wang et al., 2021a). Agricultural implications because endophytes, including *Alternaria alternata*, release antibacterial tenuazonic acid alongside Colletotrichum species, generating bioherbicidal colletotrichin (Hoagland et al., 2007; Jia et al., 2023). Among the fungal endophytes were the antibiotic javanicin (*Chloridium* species), anti-oxidant compounds pestalopyrone (*Pestalotiopsis* species), in addition to antibacterial, along with cancer-preventing phomoxanthone A (*Phomopsis* species) (Kharwar et al., 2009; Rönsberg et al., 2013; Strobel & Daisy, 2003). These findings emphasize the unrealized advantages of fungi with endophytes toward the biotechnological revolution. Specific polyketides present problems, including mycotoxins that are toxic to humans as well as animals, notably as aflatoxins (*Aspergillus flavus*) along with fumonisins (*Fusarium verticillioides*), even though they have many advantages (Diener et al., 1987; Myung et al., 2009). Reducing these hazards and maximizing the opportunities for valuable applications could be achieved by comprehending their biosynthetic and ecological role.

3.2 Terpenoids

Fungi, including fungal endophytes, produce terpenoids, also known as isoprenoids, that comprise an assortment of metabolites created with isoprene units. These compounds can be described as monoterpenes (C10), diterpenes (C20), in addition to sesquiterpenes (C15) according to the number of isoprene units they contain (Luo et al., 2024) (Fig. 2). Monoterpenes, which have a total of two isoprene parts, tend to be minor, volatile molecules possessing antibacterial activity. *Candida glycerinogenes* produces geraniol, *Aspergillus niger* produces limonene, while *Trichoderma* species produce *p*-cymene. These chemicals work as antifungal alongside antibacterial remedies (García-Carnelli et al., 2014; Gualtieri et al., 2022; Zhao et al., 2023). Diterpenes, which consist of four isoprene parts, can be more structurally complicated. The endophytic fungus *Stemphylium sedicola* produces taxadiene, considered an early precursor of the anticancer medication Taxol (Mirjalili et al., 2012). Additional diterpenes, including harzianone and other products from *Trichoderma* species (Zhang & Feng, 2022). Farnesol (*Candida albicans*), an independent quorum-sensing substance, while helminthosporol (*Bipolaris* species), an antibiotic, are both examples of sesquiterpenes possessing triple units of isoprene each (Berestetskiy et al., 2020; Langford et al., 2009). For fungi to survive and interact with one another in their respective environments, terpenoids are essential for defense, communication, and controlling hormones. In this regard, the aflatrem of *Aspergillus flavus* provides a potent neurotoxin known to hinder predators (Calvo & Cary, 2015; Valdes et al., 1985). However, mellein of *Aspergillus ochraceus*, alongside trichodermin of *Trichoderma harzianum*, is both antibacterial as well as antifungal (Küçük & Kivanç, 2004; Moore et al., 1972; Mousa & Raizada, 2013). Simply indicating environmental alterations, *Streptomyces* species' geosmin promotes interactions between microbes (Becher et al., 2020). Mycorradicin, originating from fungi, promotes nutrient exchange and strengthens symbiosis via host plants (Basu et al., 2018). Gibberellins of *Fusarium* species control cell division in addition to elongation, which promotes development in plants (Hedden & Thomas, 2012). Vegetation that encounters conditions under stress, including drought, benefits from the production of abscisic acid via *Botrytis cinerea* as well (Izquierdo-Bueno et al., 2018; Vishwakarma et al., 2017). Innovative approaches to medical treatments, agriculture, and sustainability in nature can be made possible by discovering fascinating uses associated with developments in genomes alongside metabolomics. It can be expected that further research about fungal terpenoids will thoroughly comprehend their value in tackling global issues as well.

3.3 Alkaloids

Fungi produce a structurally varied class of secondary metabolites called alkaloids involving nitrogen (Fig. 3). These chemicals prove to be useful in medicine and biotechnology because of their outstanding bioactivities, including neuroactive effects and antibacterial and anticancer qualities (Lee et al., 2020; Yin et al., 2023). Intricate biosynthetic processes incorporating amino acid intermediates result in highly complicated structures of molecules while participating in the production of fungal alkaloids. Fungal alkaloids fall into categories according to their sources regarding biosynthesis, along with chemical composition (Xu et al., 2014). Indole alkaloids possess antifungal and neurotoxic characteristics. Examples of these include fumitremorgin of *Aspergillus fumigatus* as well as paxilline of *Penicillium paxillin* (Chen et al., 2022; Kato et al., 2009; Sulyok et al., 2007). In addition to its uterotonic along with vasoconstrictive properties, ergot alkaloids—which can be produced via *Claviceps purpurea*—include chemicals like ergometrine

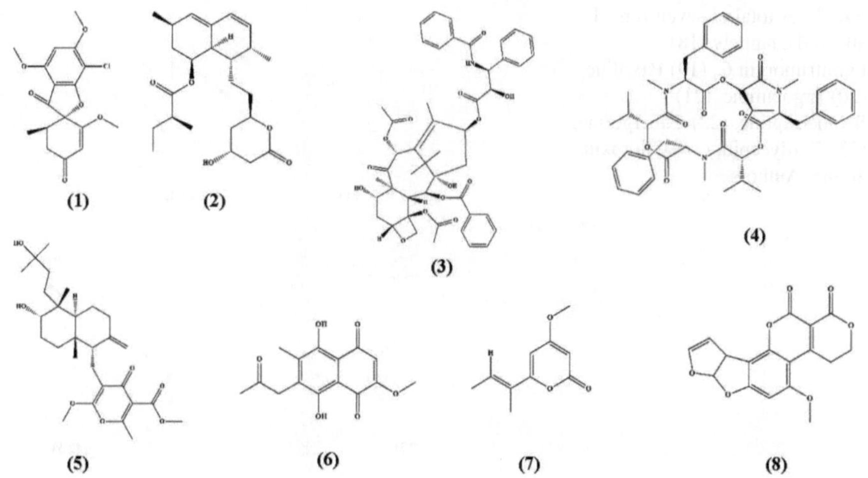

Fig. 1 A total of eight fungal polyketides, namely (1) Griseofulvin, (2) Lovastatin, (3) Paclitaxel, (4) Beauvericin, (5) Collitorichin, (6) Javanicin, (7) Pestalopyrone, (8) Aflatoxin G1. *Source* Authors

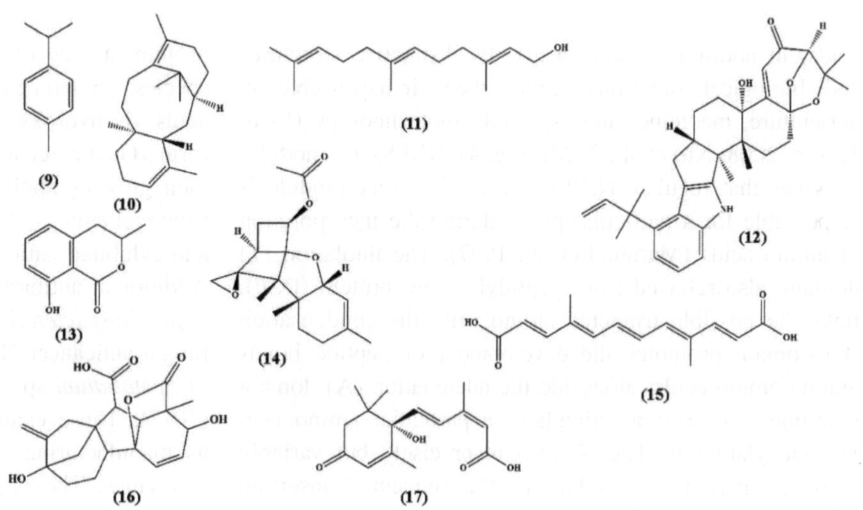

Fig. 2 A total of nine fungal terpenoids, namely (9) p-Cymene, (10) Taxadiene, (11) Farnesol, (12) Aflatrem, (13) Mellein, (14) Trichodermin, (15) Mycorradicin, (16) Gibberellate, (17) Abscisic acid. *Source* Authors

together with ergotamine that tend to be used in medicine (Schiff Jr, 2006). Eurochevalierine alkaloid of *Neosartorya pseudofischeri* has instances of anticancer properties (Goetz et al., 1985; Schnekenburger et al., 2018). Ergotamine can be prescribed to treat migraines because of its capacity to narrow the blood vessels, whereas ergometrine can be prescribed for obstetrics to safeguard against postpartum haemorrhage (Dyer et al., 2010; Perrin, 1985). Semi-synthetic materials derivatives, such as bromocriptine along with cabergoline, serve as efficient treatments for both Parkinson's disease and hyperprolactinaemia (Curran & Perry, 2004; Pela, 2004; Verhelst et al., 1999). LSD, another semi-synthetic ergot-based alkaloid, has become recognized for its hallucinogenic properties in addition to ongoing research regarding treatment for mental disorders (Lambru & Matharu, 2011). Fumitremorgin C, which originates from *Aspergillus fumigatus*, gained antitrypanosomal and antiplasmodial qualities (Diyaolu et al., 2023). Boasting anticancer properties, the cordycepin of *Cordyceps militaris* continues to be investigated for its potential to alter nucleoside analogue processes (Kunhorm et al., 2022). Gliotoxin, which can be generated by *Aspergillus fumigatus*, possesses immunosuppressive qualities to be utilized when conducting transplant therapy (Stanzani et al., 2005). The prospective benefits of alkaloids are developing as novel avenues for their manufacture, which are unlocked by developments in understanding synthetic biology alongside genomics. Fungal alkaloids continue to be an essential ingredient in creating and discovering new drugs due to their strong bioactivities and extensive structural diversity, opening the door to novel medicinal and biotechnological solutions.

3.4 Nonribosomal Peptides (NRPs)

Nonribosomal peptides (NRPs) represent a varied collection of secondary metabolites generated via huge enzyme complexes called nonribosomal peptide synthetases (NRPSs), which perform separately from ribosomes. Peptides of this kind have no limitation to just 20 standard amino acids; they frequently contain unique structural components that consist of D-amino acids, altered amino

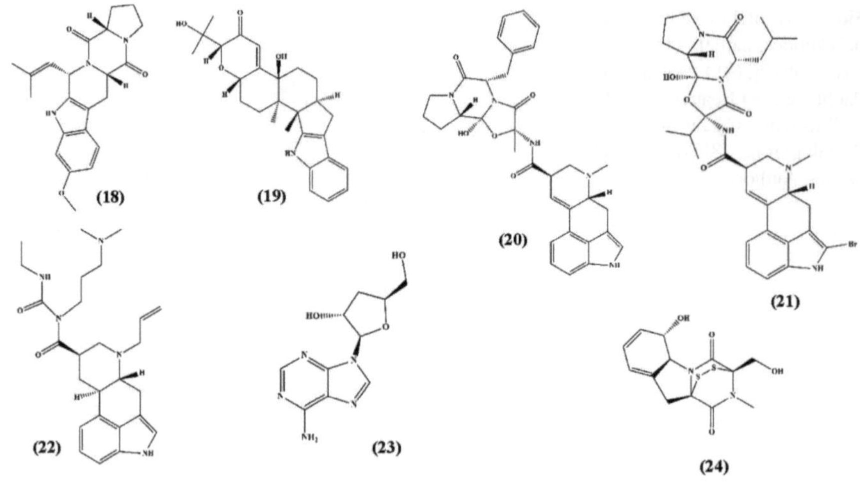

Fig. 3 A total of seven fungal alkaloids, namely (18) Fumitrimorgin C, (19) Paxilline, (20) Ergotamine, (21) Bromocriptine, (22) Cabergoline, (23) Cordycepin, (24) Gliotoxin. *Source* Authors

acids, in addition to fatty acids. Its distinctive structures and biological functions render them indispensable in agriculture, medicine, and especially biotechnology (Li & Jensen, 2008; Niu et al., 2020) (Fig. 4). NRPSs are modular enzymes that regulate NRP biosynthesis; every module is responsible for a particular phase during the incorporation of amino acids (Marahiel et al., 1997). The thiolation (T) domain, also referred to as peptidyl carrier protein (PCP), links the possible triggered amino acid; the condensation (C) domain promotes the development of peptide bonds among amino acids; alongside the adenylation (A) domain determines as well as stimulates a particular amino acid via adenylating it. The NRPS can precisely but variably construct peptides according to the sequential insertion (Zhang et al., 2023). Additionally, many NRPSs with accessory domains can modify the peptide backbone via methylation, the epimerization process, and cyclization. These changes improve the peptides' solubility, strength, and bioactivity (Duban et al., 2022). A thioesterase (TE) domain breaks down the last peptide concerning the enzyme complex around the completion of the biosynthesis process, frequently cyclizing to achieve its active state (Little & Hertweck, 2022). Gene clusters typically contain both the components required for peptide synthesis and regulation, encoding genes that regulate NRP biosynthesis. Since these genes can often be strictly controlled, the organism can manufacture NRPs when necessary while preserving energy utilization (Süssmuth & Mainz, 2017). *Tolypocladium inflatum* has become the producer of the widely recognized NRP cyclosporin. As a potent immunosuppressant, its cyclic undecapeptide transformed the transplantation of organs. It works by blocking calcineurin, a protein known to be essential for the activation of T cells. Just one NRPS enzyme regulates the production of cyclosporin through the incorporation of several odd amino acids, including methylated variants, into its molecular structure (Eberhardt et al., 2018; Wang et al., 2023). Enniatins constitute a prominent type of NRPs that are produced via *Fusarium* species. Enniatin can be made by NRPSs to alternate amino acids via hydroxy acids, leading to a unique depsipeptide form (Gautier et al., 2020). Emerging NRPs demonstrate their growing applicability throughout a variety of sectors. Emericellamides A and B originated via *Emericella* sp. and exhibited antibacterial action as well (Oh et al., 2007). Additional antibiotics like sclerotides are produced via *Aspergillus sclerotiorum* (Zheng et al., 2009, 2010). Various fungal anticancer NRPS, for instance, Scytalidamides A–B of *Scytalidium* sp., and others described by Agrawal et al. (2017). These examples show the effectiveness of NRPs as manufacturing implements, antibiotics, and anticancer medicines. The goal of contemporary NRPS engineering and genome mining studies will be to deepen the uses of these technologies while offering creative answers to key global problems.

3.5 Other Notable Fungal Secondary Metabolites

A vast array of secondary metabolites, such as volatile organic compounds (VOCs), beneficial pigments, and hazardous mycotoxins, can be generated by fungi. These substances are essential for human uses, such as industrial and medicinal applications, alongside fungal ecology (Conrado et al., 2022). Animal immune function, along with reproductive health, is affected by zearalenone along with deoxynivalenol (Bulgaru et al., 2021). Other notable mycotoxins that seriously threaten human wellness and food safety include citrinin, sterigmatocystin, and T-2 toxin (Bullerman, 1979). Melanins and carotenoids are examples of fungal pigments that support virulence, stress tolerance, and utilization in industry (Lin & Xu, 2020). Pathogenic fungi such as *Aspergillus fumigatus* can be safeguarded by melanins such as DHN melanin. In contrast, *Crypto-*

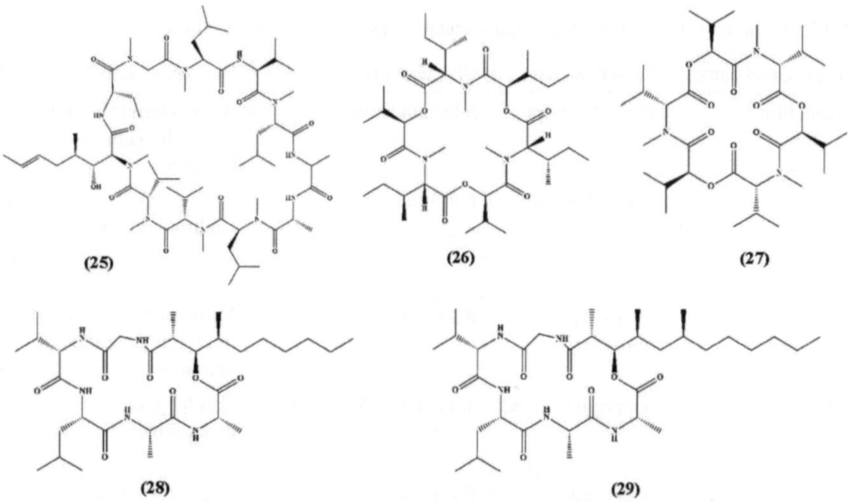

Fig. 4 A total of five fungal nonribosomal peptides namely (25) Cyclosporin A, (26) Enniatin A, (27) Enniatin B, (28) Emericellamide A, (29) Emericellamide B. *Source* Authors

coccus neoformans remains shielded against UV radiation and immunological reactions by L-DOPA melanin (Faulkner et al., 2024; Keizer et al., 2022). Beta-carotene, astaxanthin, along with lycopene are examples of carotenoids that have applications as natural food coloring alongside antioxidants (Rapoport et al., 2021). The food and textile sectors use additional pigments like monascorubrin and rubropunctatin of *Monascus* species (Kim & Ku, 2018). Promising biofuels include isoprene and ethanol, whereas earthy along with mushroom odors can be supplied by geosmin and 1-octen-3-ol (Moura et al., 2023). Aromatic substances such as limonene and phenylethanol have applications as antifungals and perfumery. Acetic acid, along with methyl isobutyl ketone, are examples of solvents used in industry that come from fungi (Jampílek & Kráľová, 2024). Various fungal metabolites collectively highlight the biotechnological prospects and ecological importance of fungus, encouraging more research regarding sectors such as agriculture, medicine, and manufacturing. A few useful secondary metabolites of fungi and their possible usefulness are given in Table 1.

4 Technological Advancements and Innovations

4.1 Role of Biotechnology in Metabolite Discovery

Metabolic engineering entails altering fungal metabolic processes to find and create new compounds. Several tactics can be utilized to accomplish this. Heterologous expression facilitates the creation and investigation of novel compounds by transferring biosynthetic gene clusters (BGCs) via less-studied fungi to simulated organisms such as *Saccharomyces cerevisiae* in addition to *Aspergillus niger* (Meng et al., 2022b). Whereas inducing pathway-specific genes promotes biosynthesis, customized gene knockouts reorient supplies regarding the creation of fungal secondary metabolites (Coyle & Panaccione, 2005; Tiwari & Dufossé, 2023; Wang et al., 2021b). Raising the accessible level of precursors, including acetyl-CoA or even malonyl-CoA, increases the generation of particular metabolites, notably polyketides (Sha et al., 2024). These strategies are further enhanced via adaptive laboratory evolution (ALE), which chooses fungus strains according to environmental stressors that enhance or alter their metabolic capacities (Portnoy et al., 2011). The precision modifying genomes made possible by CRISPR-Cas technologies make removing metabolic constraints or activating dormant BGCs easier (Leal et al., 2024). Omics innovations, like metabolomics and genetics, assist in selecting activating pathways and offering new perspectives on regulatory networks (Hautbergue et al., 2018). Under ideal circumstances, large-scale manufacturing of metabolites can be accomplished with fermentation methods, such as fed-batch and continuous processes (Allampalli & Sivaprakasam, 2024; Brewster et al., 2024). By introducing artificial promoters that closely regulate the expression of genes, promoter engineering makes it possible to produce metabolites efficiently (He et al., 2023). Furthermore, by changing chromatin states, epigenetic modification—such as the use of the enzyme histone deacetylase inhibitors—can trigger cryptic gene clusters and uncover novel metabolites (El-Hawary et al., 2018; Pfannenstiel et al., 2018). In reaction to environmental cues, metabolite production can be precisely controlled using synthetic regulatory networks, which include synthesized transcription factors (Blount et al., 2012). By modifying enzymes for greater efficiency in addition to target accuracy, enzyme engineering increases the variety of compounds that can be produced (O'Connor, 2015). Although de novo process development incorporates genes from other organisms to

Table 1 Some valuable fungal secondary metabolites with their potential uses

Compound name	Fungal source	Chemical class	Biological activity	Potential applications	References
Penicillin	*Penicillium* spp.	β-lactam antibiotic	Antibacterial (inhibits cell wall synthesis in bacteria)	Antibiotic treatment for bacterial infections	Fleming (1929), Bush (2010)
Citrinin	*Penicillium* spp.	Polyketide	Antifungal, nephrotoxic, cytotoxic	Toxicological concerns, antifungal applications	Krogh et al. (1970), Haraguchi et al. (1987), de Oliveira Filho et al. (2022)
Ergotamine	*Claviceps purpurea*	Alkaloid	Vasoconstrictive, used in migraine treatment, neurotoxic	Treatment for migraines, vasoconstriction	Merhoff and Porter (1974), Ashfaq et al. (2024)
Aflatoxin	*Aspergillus flavus*	Terpenoid (Mycotoxin)	Carcinogenic, hepatotoxic	Contaminant in food and feed, carcinogenesis	Fountain et al. (2014), Fouad et al. (2019)
Lovastatin	*Aspergillus terreus*	Polyketide	Cholesterol-lowering, HMG-CoA reductase inhibitor	Treatment of hypercholesterolemia	Jia et al. (2010), Yin et al. (2016)
Cyclosporine	*Tolypocladium inflatum*	Peptide	Immunosuppressive	Organ transplantation, autoimmune disorders	Falah et al. (2024)
Chitosan	*Mucor rouxii*	Polysaccharide	Antibacterial, antifungal, enhances wound healing, and weight loss	Pharmaceutical, agricultural, and cosmetic uses	Roller and Covill (1999), Chatterjee et al. (2005), Huq et al. (2022)
Fumonisin	*Fusarium* spp.	Mycotoxin	Carcinogenic, neurotoxic, inhibits sphingolipid metabolism	Food safety concerns, cancer research	Turner et al. (1999), Anumudu et al. (2024)

Source Authors' compilation

create entirely novel biosynthetic processes, computational techniques such as metabolic flux assessment direct the reallocation of resources within cells toward targeted processes (Martins-Santana et al., 2018; Nielsen & Nielsen, 2017). The process of identifying and manufacturing secondary metabolites became more manageable because of biotechnology's capacity to alter and enhance fungal metabolism. These methods increase the effectiveness of current procedures and open the door for novel substances alongside environmentally friendly solutions throughout various sectors.

4.2 Advances in Analytical Techniques

Advances in techniques for analysis have led to a significant advancement in the study of fungal secondary metabolites. Multi-omics measures, liquid chromatography-mass spectrometry (LC-MS), and nuclear magnetic resonance (NMR) become essential for comprehending the structures, profiles, and biosynthesis processes of these substances. Identifying and improving bioactive compounds have been rendered feasible by the broad perspectives these approaches offer (Zhong et al., 2022). NMR is a fundamental technique for determining the structure of molecules of secondary metabolic products, providing information regarding their stereochemistry, functional chains, and atomic connectivity (Zobel et al., 2016). Regarding metabolite profiling and quantification, especially in intricate biological mixes, LC-MS is a very sensitive and adaptable method (Kluger et al., 2015). For biosynthetic processes as well as connecting them with metabolite production, developing multi-omics techniques combines genomic and transcriptomic information, in addition to metabolomic information (Meng et al., 2022a; Sanches et al., 2024). These methods offer a thorough comprehension of the biosynthesis of fungal metabolites, making it easier to find new compounds and produce them sustainably.

4.3 High-Throughput Screening and Automation

Automation and high-throughput screening (HTS) transformed the search for fungal secondary metabolites because they have made it possible to quickly identify bioactive substances (Marchese et al., 2020). Screening libraries comprise carefully chosen sets of pure chemicals or fungal extracts thoroughly examined for biological properties. These libraries seemed essential in the discovery of numerous secondary metabolites with crucial uses. These illustrations highlight how important screening libraries will be to identify metabolites (Harvey et al., 2015). By optimizing fungal fermenting conditions, robotic biological reactors can be crucial (Haby et al., 2019). The rapid identification, analysis, and optimization of fungal additional metabolites by researchers through integrating HTS using robotics and AI can also accelerate their use in various industries.

5 Conclusion and Prospects

Fungal secondary metabolites hold immense potential for advancing medicine and agriculture alongside environmental sustainability. However, challenges such as the intricate regulatory networks governing biosynthesis, structural complexity, and limited production yields hinder their exploration and application. Understanding fungal biosynthesis's dynamic metabolic processes and regulatory mechanisms requires integrating advanced tools and technologies. Recent advancements in omics approaches and computational tools like machine learning promise to uncover novel bioactive compounds and expand the current understanding of the fungal metabolome. Addressing low yields and optimizing production processes will be critical for scaling up the manufacturing of therapeutically valuable metabolites. Innovative biotechnological strategies, including metabolic engineering and synthetic biology, can enhance yields, improve strain performance, and enable sustainable production. Developing high-throughput analytical techniques will further facilitate identifying and characterizing unknown metabolites, overcoming structural complexity barriers. In the future, fungal biotechnology will be pivotal in addressing global challenges, including environmental sustainability, food security, and human health. Continued research into fungal metabolic pathways and emerging technologies will unlock novel bioactive compounds, biopolymers, and eco-friendly solutions. This multidisciplinary approach will open new avenues for discovering and applying fungal-derived products, fostering innovation across various industries while promoting sustainable development.

Acknowledgments SKS lab would like to acknowledge funding from the Institution of Eminence Incentive Grant Scheme, Trans-Disciplinary Research Scheme, Department of Science and Technology, SEED grant, New Delhi, and funding from the Institute of Medical Sciences intramural grant, Banaras Hindu University, Varanasi, India. CP & SNR would like to acknowledge CSIR (09/0013(12277)/2021-EMR-I) and ICMR (3/1/3/196/Neuro/2021-NCD-I) for providing fellowship assistance, respectively. PS would like to acknowledge DBT (DBT/2023-24/BHU/2179) for providing fellowship assistance as well.

Disclosure Statement The author declares no conflict of interest.

References

Abbas, F., Ke, Y., Yu, R., Yue, Y., Amanullah, S., Jahangir, M. M., & Fan, Y. (2017). Volatile terpenoids: Multiple functions, biosynthesis, modulation and manipulation by genetic engineering. *Planta, 246*, 803–816.

Agrawal, S., Acharya, D., Adholeya, A., Barrow, C. J., & Deshmukh, S. K. (2017). Nonribosomal peptides from marine microbes and their antimicrobial and anticancer potential. *Frontiers in Pharmacology, 8*, 828.

Ainsworth, G. C. (2008). *Ainsworth & Bisby's dictionary of the fungi*. Cabi.

Allampalli, S. S. P., & Sivaprakasam, S. (2024). Unveiling the potential of specific growth rate control in fed-batch fermentation: Bridging the gap between product quantity and quality. *World Journal of Microbiology and Biotechnology, 40*(6), 1–19.

Antonissen, G., Martel, A., Pasmans, F., Ducatelle, R., Verbrugghe, E., Vandenbroucke, V., et al. (2014). The impact of Fusarium mycotoxins on human and animal host susceptibility to infectious diseases. *Toxins, 6*(2), 430–452.

Anumudu, C. K., Ekwueme, C. T., Uhegwu, C. C., Ejileugha, C., Augustine, J., Okolo, C. A., & Onyeaka, H. (2024). A review of the mycotoxin family of fumonisins, their biosynthesis, metabolism, methods of detection and effects on humans and animals. *International Journal of Molecular Sciences, 26*(1), 184.

Aris, P., Wei, Y., Mohamadzadeh, M., & Xia, X. (2022). Griseofulvin: An updated overview of old and current knowledge. *Molecules, 27*(20), 7034.

Ashfaq, M., Mushtaq, I., Mehmood, M. A., Ahmad, F., & ur Rehman, S. (2024). Ergot alkaloids from claviceps: Production and pharmacological properties. In *Fungal secondary metabolites* (pp. 241–257). Elsevier.

Avalos, J., & Carmen Limón, M. (2015). Biological roles of fungal carotenoids. *Current Genetics, 61*, 309–324.

Banerjee, A., Dahiya, M., Anand, M. T., & Kumar, S. (2013). Inhibition of proliferation of cervical and leukemic cancer cells by penicillin G. *Asian Pacific Journal of Cancer Prevention, 14*(3), 2127–2130.

Bartlett, D. W., Clough, J. M., Godwin, J. R., Hall, A. A., Hamer, M., & Parr-Dobrzanski, B. (2002). The strobilurin fungicides. *Pest Management Science: Formerly Pesticide Science, 58*(7), 649–662.

Basu, S., Rabara, R. C., & Negi, S. (2018). AMF: The future prospect for sustainable agriculture. *Physiological and Molecular Plant Pathology, 102*, 36–45.

Bayry, J., Aimanianda, V., Guijarro, J. I., Sunde, M., & Latge, J.-P. (2012). Hydrophobins—Unique fungal proteins. *PLoS Pathogens, 8*(5), e1002700.

Becher, P. G., Verschut, V., Bibb, M. J., Bush, M. J., Molnár, B. P., Barane, E., et al. (2020). Developmentally regulated volatiles geosmin and 2-methylisoborneol attract a soil arthropod to Streptomyces bacteria promoting spore dispersal. *Nature Microbiology, 5*(6), 821–829.

Bennett, J. W., & Chung, K.-T. (2001). Alexander Fleming and the discovery of penicillin.

Bérdy, J. (2012). Thoughts and facts about antibiotics: where we are now and where we are heading. *The Journal of Antibiotics, 65*(8), 385–395.

Berestetskiy, A. O., Dalinova, A. A., Dubovik, V. R., Grigoryeva, E. N., Kochura, D. M., Senderskiy, I. V., et al. (2020). Analysis and isolation of secondary metabolites of Bipolaris sorokiniana by different chromatography techniques and the spectrum of their biological activity. *Applied Biochemistry and Microbiology, 56*, 569–582.

Bills, G. F., & Gloer, J. B. (2016). Biologically active secondary metabolites from the fungi. *Microbiology Spectrum, 4*(6), 10–1128.

Blount, B. A., Weenink, T., & Ellis, T. (2012). Construction of synthetic regulatory networks in yeast. *FEBS Letters, 586*(15), 2112–2121.

Borin, G. P., & Oliveira, J. V. C. (2022). Assessing the intracellular primary metabolic profile of Trichoderma reesei and Aspergillus niger grown on different carbon sources. *Frontiers in Fungal Biology, 3*, 998361.

Brewster, A., Oudshoorn, A., van Lotringen, M., Nelisse, P., van den Berg, E., Luttik, M., & Daran, J. M. (2024). Inhibition control by continuous extractive fermentation enhances De Novo 2-phenylethanol production by yeast. *Biotechnology and Bioengineering*.

Bulgaru, C. V., Marin, D. E., Pistol, G. C., & Taranu, I. (2021). Zearalenone and the immune response. *Toxins, 13*(4), 248.

Bullerman, L. B. (1979). Significance of mycotoxins to food safety and human health. *Journal of Food Protection, 42*(1), 65–86.

Bush, K. (2010). Beta-lactam antibiotics: Penicillins. *Antibiotics and Chemotherapy*, 200–225.

Calvo, A. M., & Cary, J. W. (2015). Association of fungal secondary metabolism and sclerotial biology. *Frontiers in Microbiology, 6*, 62.

Calvo, A. M., Wilson, R. A., Bok, J. W., & Keller, N. P. (2002). Relationship between secondary metabolism and fungal development. *Microbiology and Molecular Biology Reviews, 66*(3), 447–459.

Carboué, Q., & Lopez, M. (2021). Amanita muscaria: Ecology, chemistry, myths. *Encyclopedia, 1*(3), 905–914.

Chan, S., Ng, S., Chan, H. P., Pascoe, E. M., Playford, E. G., Wong, G., et al. (2020). Perioperative antibiotics for preventing post-surgical site infections in solid organ transplant recipients. *Cochrane Database of Systematic Reviews, 8*.

Chatterjee, S., Adhya, M., Guha, A. K., & Chatterjee, B. P. (2005). Chitosan from Mucor rouxii: Production and physico-chemical characterization. *Process Biochemistry, 40*(1), 395–400.

Chen, C., Qi, J., He, Y., Lu, Y., & Wang, Y. (2022). Genomic and chemical profiling of B9, a unique Penicillium fungus derived from sponge. *Journal of Fungi, 8*(7), 686.

Conrado, R., Gomes, T. C., Roque, G. S. C., & De Souza, A. O. (2022). Overview of bioactive fungal secondary metabolites: Cytotoxic and antimicrobial compounds. *Antibiotics, 11*(11), 1604.

Coyle, C. M., & Panaccione, D. G. (2005). An ergot alkaloid biosynthesis gene and clustered hypothetical genes from Aspergillus fumigatus. *Applied and Environmental Microbiology, 71*(6), 3112–3118.

Curran, M. P., & Perry, C. M. (2004). Cabergoline: A review of its use in the treatment of Parkinson's disease. *Drugs, 64*, 2125–2141.

de Oliveira Filho, J. W. G., Andrade, T. D. J. A. D. S., de Lima, R. M. T., Silva, D. H. S., Dos Reis, A. C., Santos, J. V. D. O., de Meneses, A. A. P. M., de Carvalho, R. M., da Mata, A. M. O., de Alencar, M. V. O. B., & Dias, A. C. S. (2022). Cytogenotoxic evaluation of the acetonitrile extract, citrinin and dicitrinin-A from Penicillium citrinum. *Drug and Chemical Toxicology, 45*(2), 688–697.

Diener, U. L., Cole, R. J., Sanders, T. H., Payne, G. A., Lee, L. S., & Klich, M. A. (1987). Epidemiology of aflatoxin formation by Aspergillus flavus. *Annual Review of Phytopathology, 25*(1), 249–270.

Diyaolu, O. A., Preet, G., Fagbemi, A. A., Annang, F., Pérez-Moreno, G., Bosch-Navarrete, C., et al. (2023). Antiparasitic activities of compounds isolated from Aspergillus fumigatus strain discovered in Northcentral Nigeria. *Antibiotics, 12*(1), 109.

Drott, M. T., Lazzaro, B. P., Brown, D. L., Carbone, I., & Milgroom, M. G. (2017). Balancing selection for aflatoxin in Aspergillus flavus is maintained through interference competition with, and fungivory by insects. *Proceedings of the Royal Society B: Biological Sciences, 284*(1869), 20172408.

Duban, M., Cociancich, S., & Leclère, V. (2022). Nonribosomal peptide synthesis definitely working out of the rules. *Microorganisms, 10*(3), 577.

Dutton, M. V., & Evans, C. S. (1996). Oxalate production by fungi: Its role in pathogenicity and ecology in the soil environment. *Canadian Journal of Microbiology, 42*(9), 881–895.

Dyer, R. A., Van Dyk, D., & Dresner, A. (2010). The use of uterotonic drugs during caesarean section. *International Journal of Obstetric Anesthesia, 19*(3), 313–319.

Eberhardt, W., Nasrullah, U., & Pfeilschifter, J. (2018). Activation of renal profibrotic TGFβ controlled signaling cascades by calcineurin and mTOR inhibitors. *Cellular Signalling, 52*, 1–11.

El-Hawary, S. S., Sayed, A. M., Mohammed, R., Hassan, H. M., Zaki, M. A., Rateb, M. E., Mohammed, T. A., Amin, E., & Abdelmohsen, U. R. (2018). Epigenetic modifiers induce bioactive phenolic metabolites in the marine-derived fungus Penicillium brevicompactum. *Marine Drugs, 16*(8), 253.

Falah, F., Vasiee, A., Ramezani, M., Tabatabaee-Yazdi, F., Mortazavi, S. A., & Danesh, A. (2024). Effect of immobilization, mutation, and microbial stresses on increasing production efficiency of "Cyclosporin A". *Biomass Conversion and Biorefinery, 14*(4), 4441–4456.

Faulkner, M. E., Gong, Z., Guo, A., Laporte, J. P., Bae, J., & Bouhrara, M. (2024). Harnessing myelin water fraction as an imaging biomarker of human cerebral aging, neurodegenerative diseases, and risk factors influencing myelination: A review. *Journal of Neurochemistry, 168*(9), 2243–2263.

Feeney, K. (2013). The significance of pharmacological and biological indicators in identifying historical uses of Amanita muscaria. In *Entheogens and the development of culture: The anthropology and neurobiology of ecstatic experience* (pp. 279–318).

Feng, G. H., & Leonard, T. J. (1995). Characterization of the polyketide synthase gene (pksL1) required for aflatoxin biosynthesis in Aspergillus parasiticus. *Journal of Bacteriology, 177*(21), 6246–6254.

Feofilova, E. P., Ivashechkin, A. A., Alekhin, A. I., & Sergeeva, Y. E. (2012). Fungal spores: Dormancy, germination, chemical composition, and role in biotechnology. *Applied Biochemistry and Microbiology, 48*, 1–11.

Fleming, A. (1929). On the antibacterial action of cultures of a penicillium, with special reference to their use in the isolation of B. influenzae. *British Journal of Experimental Pathology, 10*(3), 226.

Fouad, A. M., Ruan, D., El-Senousey, H. K., Chen, W., Jiang, S., & Zheng, C. (2019). Harmful effects and control strategies of aflatoxin b1 produced by Aspergillus flavus and Aspergillus parasiticus strains on poultry. *Toxins, 11*(3), 176.

Fountain, J. C., Scully, B. T., Ni, X., Kemerait, R. C., Lee, R. D., Chen, Z.-Y., & Guo, B. (2014). Environmental influences on maize-Aspergillus flavus interactions and aflatoxin production. *Frontiers in Microbiology, 5*, 40.

Gangadevi, V., & Muthumary, J. (2009). Taxol production by Pestalotiopsis terminaliae, an endophytic fungus of Terminalia arjuna (arjun tree). *Biotechnology and Applied Biochemistry, 52*(1), 9–15.

García-Carnelli, C., Rodríguez, P., Heinzen, H., & Pilar Menéndez, P. (2014). Influence of culture conditions on the biotransformation of (+)-limonene by Aspergillus niger. *Zeitschrift für Naturforschung C, 69*(1–2), 61–67.

Gautier, C., Pinson-Gadais, L., & Richard-Forget, F. (2020). Fusarium mycotoxins enniatins: An updated review of their occurrence, the producing Fusarium species, and the abiotic determinants of their accumulation in crop harvests. *Journal of Agricultural and Food Chemistry, 68*(17), 4788–4798.

Goetz, M. A., Lopez, M., Monaghan, R. L., Chang, R. S. L., Lotti, V. J., & Chen, T. B. (1985). Asperlicin, a novel non-peptidal cholecystokinin antagonist from Aspergillus alliaceus Fermentation, isolation and biological properties. *The Journal of Antibiotics, 38*(12), 1633–1637.

Grabka, R., d'Entremont, T. W., Adams, S. J., Walker, A. K., Tanney, J. B., Abbasi, P. A., & Ali, S. (2022). Fungal endophytes and their role in agricultural plant protection against pests and pathogens. *Plants, 11*(3), 384.

Gualtieri, L., Monti, M. M., Mele, F., Russo, A., Pedata, P. A., & Ruocco, M. (2022). Volatile organic compound (VOC) profiles of different Trichoderma species and their potential application. *Journal of Fungi, 8*(10), 989.

Haby, B., Hans, S., Anane, E., Sawatzki, A., Krausch, N., Neubauer, P., & Cruz Bournazou, M. N. (2019). Integrated robotic mini bioreactor platform for automated, parallel microbial cultivation with online data handling and process control. *SLAS Technology: Translating Life Sciences Innovation, 24*(6), 569–582.

Hajjaj, H., Niederberger, P., & Duboc, P. (2001). Lovastatin biosynthesis by Aspergillus terreus in a chemically defined medium. *Applied and Environmental Microbiology, 67*(6), 2596–2602.

Haque, M. A., Nath, N. D., Johnston, T. V., Haruna, S., Ahn, J., Ovissipour, R., & Ku, S. (2024). Harnessing biotechnology for penicillin production: Opportunities and environmental considerations. *Science of the Total Environment, 174236*.

Haraguchi, H., Hashimoto, K., Shibata, K., Taniguchi, M., & Oi, S. (1987). Mechanism of antifungal action of citrinin. *Agricultural and Biological Chemistry, 51*(5), 1373–1378.

Harvey, A. L., Edrada-Ebel, R., & Quinn, R. J. (2015). The re-emergence of natural products for drug discovery in the genomics era. *Nature Reviews Drug Discovery, 14*(2), 111–129.

Hautbergue, T., Jamin, E. L., Debrauwer, L., Puel, O., & Oswald, I. P. (2018). From genomics to metabolomics, moving toward an integrated strategy for the discovery of fungal secondary metabolites. *Natural Product Reports, 35*(2), 147–173.

Hawksworth, D. L., & Lücking, R. (2017). Fungal diversity revisited: 2.2 to 3.8 million species. *Microbiology Spectrum, 5*(4), 10–1128.

He, S., Zhang, Z., & Lu, W. (2023). Natural promoters and promoter engineering strategies for metabolic regulation in Saccharomyces cerevisiae. *Journal of Industrial Microbiology and Biotechnology, 50*(1), kuac029.

Hedden, P., & Thomas, S. G. (2012). Gibberellin biosynthesis and its regulation. *Biochemical Journal, 444*(1), 11–25.

Herrmann, K. M., & Weaver, L. M. (1999). The shikimate pathway. *Annual Review of Plant Biology, 50*(1), 473–503.

Hoagland, R. E., Boyette, C. D., Weaver, M. A., & Abbas, H. K. (2007). Bioherbicides: Research and risks. *Toxin Reviews, 26*(4), 313–342.

Hounsa, C.-G., Brandt, E. V., Thevelein, J., Hohmann, S., & Prior, B. A. (1998). Role of trehalose in survival of Saccharomyces cerevisiae under osmotic stress. *Microbiology, 144*(3), 671–680.

Huq, T., Khan, A., Brown, D., Dhayagude, N., He, Z., & Ni, Y. (2022). Sources, production and commercial applications of fungal chitosan: A review. *Journal of Bioresources and Bioproducts, 7*(2), 85–98.

Izquierdo-Bueno, I., González-Rodríguez, V. E., Simon, A., Dalmais, B., Pradier, J. M., Le Pêcheur, P., et al. (2018). Biosynthesis of abscisic acid in fungi: Identification of a sesquiterpene cyclase as the key enzyme in Botrytis cinerea. *Environmental Microbiology, 20*(7), 2469–2482.

Jampílek, J., & Kráľová, K. (2024). Fungal volatile organic compounds. In *Fungal secondary metabolites* (pp. 399–426). Elsevier.

Jayaram, H., Marigowda, V., & Thara Saraswathi, K. J. (2021). Secondary metabolite production and terpenoid biosynthesis in endophytic fungi Cladosporium cladosporioides isolated from wild Cymbopogon martinii (Roxb.) Wats. *Microbiology Research, 12*(4), 812–828.

Jeong, G.-J., Khan, F., Tabassum, N., & Kim, Y.-M. (2023). Chitinases as key virulence factors in microbial pathogens: Understanding their role and potential as therapeutic targets. *International Journal of Biological Macromolecules, 126021*.

Jia, Q., Fan, Y., Duan, S., Qin, Q., Ding, Y., Yang, M., et al. (2023). Effects of *Bacillus amyloliquefaciens* XJ-BV2007 on growth of Alternaria alternata and production of tenuazonic acid. *Toxins, 15*(1), 53.

Jia, Z., Zhang, X., Zhao, Y., & Cao, X. (2010). Enhancement of lovastatin production by supplementing polyketide antibiotics to the submerged culture of *Aspergillus terreus. Applied Biochemistry and Biotechnology, 160*, 2014–2025.

Kato, N., Suzuki, H., Takagi, H., Asami, Y., Kakeya, H., Uramoto, M., et al. (2009). Identification of cytochrome P450s required for fumitremorgin biosynthesis in *Aspergillus fumigatus. Chembiochem, 10*(5), 920–928.

Keizer, E. M., Valdes, I. D., McCann, B. L., Bignell, E. M., Wösten, H. A. B., & de Cock, H. (2022). The protective role of 1, 8-dihydroxynaphthalene–melanin on conidia of the opportunistic human pathogen aspergillus fumigatus revisited: No role in protection against hydrogen peroxide and superoxides. *Msphere, 7*(1), e00874–e00821.

Kelsic, E. D., Zhao, J., Vetsigian, K., & Kishony, R. (2015). Counteraction of antibiotic production and degradation stabilizes microbial communities. *Nature, 521*(7553), 516–519.

Kharwar, R. N., Verma, V. C., Kumar, A., Gond, S. K., Harper, J. K., Hess, W. M., et al. (2009). Javanicin, an antibacterial naphthaquinone from an endophytic fungus of neem, *Chloridium* sp. *Current Microbiology, 58*, 233–238.

Kim, D., & Ku, S. (2018). Beneficial effects of Monascus sp. KCCM 10093 pigments and derivatives: A mini review. *Molecules, 23*(1), 98.

Kluger, B., Lehner, S., & Schuhmacher, R. (2015). Metabolomics and secondary metabolite profiling of filamentous fungi. *Biosynthesis and Molecular Genetics of Fungal Secondary Metabolites, 2*, 81–101.

Krogh, P., Hasselager, E., & Friis, P. (1970). Studies on fungal nephrotoxicity: 2. Isolation of two nephrotoxic compounds from *Penicillium viridicatum* Westling: Citrinin and oxalic acid. *Acta Pathologica Microbiologica Scandinavica Section B Microbiology and Immunology, 78*(4), 401–413.

Küçük, Ç., & Kivanç, M. (2004). In vitro antifungal activity of strains of *Trichoderma harzianum. Turkish Journal of Biology, 28*(2), 111–115.

Kunhorm, P., Chueaphromsri, P., Chaicharoenaudomrung, N., & Noisa, P. (2022). Enhancement of cordycepin production from Cordyceps militaris culture by epigenetic modification. *Biotechnology Letters, 44*(4), 581–593.

Lambru, G., & Matharu, M. (2011). Serotonergic agents in the management of cluster headache. *Current Pain and Headache Reports, 15*, 108–117.

Langford, M. L., Atkin, A. L., & Nickerson, K. W. (2009). Cellular interactions of farnesol, a quorum-sensing molecule produced by Candida albicans. *Future Microbiology, 4*(10), 1353–1362.

Laxminarayan, R., Duse, A., Wattal, C., Zaidi, A. K. M., Wertheim, H. F. L., Sumpradit, N., et al. (2013). Antibiotic resistance—The need for global solutions. *The Lancet Infectious Diseases, 13*(12), 1057–1098.

Leal, K., Rojas, E., Madariaga, D., Contreras, M. J., Nuñez-Montero, K., Barrientos, L., et al. (2024). Unlocking fungal potential: The CRISPR-Cas system as a strategy for secondary metabolite discovery. *Journal of Fungi, 10*(11), 748.

Lee, D., Choi, H. G., Hwang, J. H., Shim, S. H., & Kang, K. S. (2020). Neuroprotective effect of tricyclic pyridine alkaloids from Fusarium lateritium SSF2, against glutamate-induced oxidative stress and apoptosis in the HT22 hippocampal neuronal cell line. *Antioxidants, 9*(11), 1115.

Li, J., & Jensen, S. E. (2008). Nonribosomal biosynthesis of fusaricidins by Paenibacillus polymyxa PKB1 involves direct activation of a D-amino acid. *Chemistry & Biology, 15*(2), 118–127.

Liddicoat, A. M., & Lavelle, E. C. (2019). Modulation of innate immunity by cyclosporine A. *Biochemical Pharmacology, 163*, 472–480.

Lin, L., & Xu, J. (2020). Fungal pigments and their roles associated with human health. *Journal of Fungi, 6*(4), 280.

Little, R. F., & Hertweck, C. (2022). Chain release mechanisms in polyketide and non-ribosomal peptide biosynthesis. *Natural Product Reports, 39*(1), 163–205.

Luo, P., Huang, J.-H., Lv, J.-M., Wang, G.-Q., Hu, D., & Gao, H. (2024). Biosynthesis of fungal terpenoids. *Natural Product Reports*.

Marahiel, M. A., Stachelhaus, T., & Mootz, H. D. (1997). Modular peptide synthetases involved in nonribosomal peptide synthesis. *Chemical Reviews, 97*(7), 2651–2674.

Marchese, P., Mahajan, N., O'Connell, E., Fearnhead, H., Tuohy, M., Krawczyk, J., et al. (2020). A novel high-throughput screening platform identifies itaconate derivatives from marine Penicillium antarcticum as inhibitors of mesenchymal stem cell differentiation. *Marine Drugs, 18*(4), 192.

Martins-Santana, L., Nora, L. C., Sanches-Medeiros, A., Lovate, G. L., Cassiano, M. H. A., & Silva-Rocha, R. (2018). Systems and synthetic biology approaches to engineer fungi for fine chemical production. *Frontiers in Bioengineering and Biotechnology, 6*, 117.

Masi, A., Mach, R. L., & Mach-Aigner, A. R. (2021). The pentose phosphate pathway in industrially relevant fungi: Crucial insights for bioprocessing. *Applied Microbiology and Biotechnology, 105*(10), 4017–4031.

McGuiness, P. N., Reid, J. B., & Foo, E. (2019). The role of gibberellins and brassinosteroids in nodulation and arbuscular mycorrhizal associations. *Frontiers in Plant Science, 10*, 269.

Meng, L., Zhou, R., Lin, J., Wang, Q., Wang, P., Wang, W., et al. (2022a). Integrated transcriptomics and nontargeted metabolomics analysis reveal key metabolic pathways in Ganoderma lucidum in response to ethylene. *Journal of Fungi, 8*(5), 456.

Meng, X., Fang, Y., Ding, M., Zhang, Y., Jia, K., Li, Z., et al. (2022b). Developing fungal heterologous expression platforms to explore and improve the production of natural products from fungal biodiversity. *Biotechnology Advances, 54*, 107866.

Merhoff, C. C., & Porter, J. M. (1974). Ergot intoxication: historical review and description of unusual clinical manifestations. *Annals of Surgery, 180*(5), 773–779.

Miedaner, T., & Geiger, H. H. (2015). Biology, genetics, and management of ergot (Claviceps spp.) in rye, sorghum, and pearl millet. *Toxins, 7*(3), 659–678.

Mirjalili, M. H., Farzaneh, M., Bonfill, M., Rezadoost, H., & Ghassempour, A. (2012). Isolation and characterization of Stemphylium sedicola SBU-16 as a new endophytic taxol-producing fungus from Taxus baccata grown in Iran. *FEMS Microbiology Letters, 328*(2), 122–129.

Moake, M. M., Padilla-Zakour, O. I., & Worobo, R. W. (2005). Comprehensive review of patulin control methods in foods. *Comprehensive Reviews in Food Science and Food Safety, 4*(1), 8–21.

Molelekoa, T. B. J., Augustyn, W., Regnier, T., & da Silva, L. S. (2023). Chemical characterization and toxicity evaluation of fungal pigments for potential application in food, phamarceutical and agricultural industries. *Saudi Journal of Biological Sciences, 30*(5), 103630.

Moore, J. H., Davis, N. D., & Diener, U. L. (1972). Mellein and 4-hydroxymellein production by Aspergillus ochraceus Wilhelm. *Applied Microbiology, 23*(6), 1067–1072.

Moura, M. A. F. E., Gomes, D. C., & Takahashi, J. A. (2023). Applications of fungi secondary metabolites in the food industry. In *Natural secondary metabolites: From nature, through science, to industry* (pp. 739–776). Springer.

Mousa, W. K., & Raizada, M. N. (2013). The diversity of anti-microbial secondary metabolites produced by fungal endophytes: An interdisciplinary perspective. *Frontiers in Microbiology, 4*, 65.

Myung, K., Li, S., Butchko, R. A. E., Busman, M., Proctor, R. H., Abbas, H. K., & Calvo, A. M. (2009). FvVE1 regulates biosynthesis of the mycotoxins fumonisins and fusarins in Fusarium verticillioides. *Journal of Agricultural and Food Chemistry, 57*(11), 5089–5094.

Nett, M., Ikeda, H., & Moore, B. S. (2009). Genomic basis for natural product biosynthetic diversity in the actinomycetes. *Natural Product Reports, 26*(11), 1362–1384.

Newman, D. J., & Cragg, G. M. (2020a). Natural products as sources of new drugs over the nearly four decades from 01/1981 to 09/2019. *Journal of Natural Products, 83*(3), 770–803.

Newman, D. J., & Cragg, G. M. (2020b). Plant endophytes and epiphytes: Burgeoning sources of known and "unknown" cytotoxic and antibiotic agents? *Planta Medica, 86*(13/14), 891–905.

Nielsen, J. C., & Nielsen, J. (2017). Development of fungal cell factories for the production of secondary metabolites: Linking genomics and metabolism. *Synthetic and Systems Biotechnology, 2*(1), 5–12.

Niermans, K., Woyzichovski, J., Kröncke, N., Benning, R., & Maul, R. (2019). Feeding study for the mycotoxin zearalenone in yellow mealworm (Tenebrio molitor) larvae—Investigation of biological impact and metabolic conversion. *Mycotoxin Research, 35*, 231–242.

Niu, X., Thaochan, N., & Hu, Q. (2020). Diversity of linear non-ribosomal peptide in biocontrol fungi. *Journal of Fungi, 6*(2), 61.

O'Connor, S. E. (2015). Engineering of secondary metabolism. *Annual Review of Genetics, 49*(1), 71–94.

Oh, D.-C., Kauffman, C. A., Jensen, P. R., & Fenical, W. (2007). Induced production of emericellamides A and B from the marine-derived fungus Emericella sp. in competing co-culture. *Journal of Natural Products, 70*(4), 515–520.

Olarte, R. A., Menke, J., Zhang, Y., Sullivan, S., Slot, J. C., Huang, Y., et al. (2019). Chromosome rearrangements shape the diversification of secondary metabolism in the cyclosporin producing fungus Tolypocladium inflatum. *BMC Genomics, 20*, 1–23.

Oren, A., & Gunde-Cimerman, N. (2007). Mycosporines and mycosporine-like amino acids: UV protectants or multipurpose secondary metabolites? *FEMS Microbiology Letters, 269*(1), 1–10.

Paule, A., Frezza, D., & Edeas, M. (2018). Microbiota and phage therapy: Future challenges in medicine. *Medical Science, 6*(4), 86.

Pela, F. (2004). Biological activities of fungal metabolites. In *Handbook of industrial mycology* (pp. 68–111). CRC Press.

Perrin, V. L. (1985). Clinical pharmacokinetics of ergotamine in migraine and cluster headache. *Clinical Pharmacokinetics, 10*, 334–352.

Pfannenstiel, B. T., Greco, C., Sukowaty, A. T., & Keller, N. P. (2018). The epigenetic reader SntB regulates secondary metabolism, development and global histone modifications in Aspergillus flavus. *Fungal Genetics and Biology, 120*, 9–18.

Portnoy, V. A., Bezdan, D., & Zengler, K. (2011). Adaptive laboratory evolution—Harnessing the power of biology for metabolic engineering. *Current Opinion in Biotechnology, 22*(4), 590–594.

Přichystal, J., Schug, K. A., Lemr, K., Novak, J., & Havlíček, V. (2016). *Structural analysis of natural products*. ACS Publications.

Qiang, X., Ding, J., Lin, W., Li, Q., Xu, C., Zheng, Q., & Li, Y. (2019). Alleviation of the detrimental effect of water deficit on wheat (Triticum aestivum L.) growth by an indole acetic acid-producing endophytic fungus. *Plant and Soil, 439*, 373–391.

Rapoport, A., Guzhova, I., Bernetti, L., Buzzini, P., Kieliszek, M., & Kot, A. M. (2021). Carotenoids and some other pigments from fungi and yeasts. *Metabolites, 11*(2), 92.

Roller, S., & Covill, N. (1999). The antifungal properties of chitosan in laboratory media and apple juice. *International Journal of Food Microbiology, 47*(1–2), 67–77.

Rönsberg, D., Debbab, A., Mándi, A., Vasylyeva, V., Böhler, P., Stork, B., et al. (2013). Pro-apoptotic and immunostimulatory tetrahydroxanthone dimers from the endophytic fungus Phomopsis longicolla. *The Journal of Organic Chemistry, 78*(24), 12409–12425.

Sanches, P. H. G., de Melo, N. C., Porcari, A. M., & de Carvalho, L. M. (2024). Integrating molecular perspectives: Strategies for comprehensive multi-omics integrative data analysis and machine learning applications in transcriptomics, proteomics, and metabolomics. *Biology, 13*(11), 848.

Sauter, H., Steglich, W., & Anke, T. (1999). Strobilurins: Evolution of a new class of active substances. *Angewandte Chemie International Edition, 38*(10), 1328–1349.

Schiff, P. L., Jr. (2006). Ergot and its alkaloids. *American Journal of Pharmaceutical Education, 70*(5).

Schnekenburger, M., Mathieu, V., Lefranc, F., Jang, J. Y., Masi, M., Kijjoa, A., et al. (2018). The fungal metabolite eurochevalierine, a sequiterpene alkaloid, displays anti-cancer properties through selective sirtuin 1/2 inhibition. *Molecules, 23*(2), 333.

Schuemann, J., & Hertweck, C. (2009). Biosynthesis of fungal polyketides. In *Physiology and genetics: Selected basic and applied aspects* (pp. 331–351).

Schulz, B., Boyle, C., Draeger, S., Römmert, A.-K., & Krohn, K. (2002). Endophytic fungi: A source of novel biologically active secondary metabolites. *Mycological Research, 106*(9), 996–1004.

Sha, Y., Ge, M., Lu, M., Xu, Z., Zhai, R., & Jin, M. (2024). Advances in metabolic engineering for enhanced acetyl-CoA availability in yeast. *Critical Reviews in Biotechnology*, 1–19.

Silber, J., Kramer, A., Labes, A., & Tasdemir, D. (2016). From discovery to production: Biotechnology of marine fungi for the production of new antibiotics. *Marine Drugs, 14*(7), 137.

Singaravelan, N., Grishkan, I., Beharav, A., Wakamatsu, K., Ito, S., & Nevo, E. (2008). Adaptive melanin response of the soil fungus Aspergillus niger to UV radiation stress at "Evolution Canyon", Mount Carmel, Israel. *PLoS One, 3*(8), e2993.

Stanzani, M., Orciuolo, E., Lewis, R., Kontoyiannis, D. P., Martins, S. L. R., St. John, L. S., & Komanduri, K. V. (2005). Aspergillus fumigatus suppresses the human cellular immune response via gliotoxin-mediated apoptosis of monocytes. *Blood, 105*(6), 2258–2265.

Strobel, G., & Daisy, B. (2003). Bioprospecting for microbial endophytes and their natural products. *Microbiology and Molecular Biology Reviews, 67*(4), 491–502.

Strobel, G., Yang, X., Sears, J., Kramer, R., Sidhu, R. S., & Hess, W. M. (1996). Taxol from Pestalotiopsis microspora, an endophytic fungus of Taxus wallachiana. *Microbiology, 142*(2), 435–440.

Sulyok, M., Krska, R., & Schuhmacher, R. (2007). A liquid chromatography/tandem mass spectrometric multi-mycotoxin method for the quantification of 87 analytes and its application to semi-quantitative screening of moldy food samples. *Analytical and Bioanalytical Chemistry, 389*, 1505–1523.

Süssmuth, R. D., & Mainz, A. (2017). Nonribosomal peptide synthesis—Principles and prospects. *Angewandte Chemie International Edition, 56*(14), 3770–3821.

Tiwari, P., & Dufossé, L. (2023). Focus and insights into the synthetic biology-mediated chassis of economically important fungi for the production of high-value metabolites. *Microorganisms, 11*(5), 1141.

Tryfinopoulou, P., Chourdaki, A., Nychas, G.-J. E., & Panagou, E. Z. (2020). Competitive yeast action against Aspergillus carbonarius growth and ochratoxin A production. *International Journal of Food Microbiology, 317*, 108460.

Turner, P. C., Nikiema, P., & Wild, C. P. (1999). Fumonisin contamination of food: Progress in development of biomarkers to better assess human health risks. *Mutation Research/Genetic Toxicology and Environmental Mutagenesis, 443*(1–2), 81–93.

Umar, A., Darwish, D. B. E., & Alenezi, M. A. (2024). Fungal pigments: Secondary metabolites and their application. In *Fungal secondary metabolites* (pp. 173–195). Elsevier.

Umashankar, P., & Nygård, Y. (2024). Filamentous fungi as emerging cell factories for the production of aromatic compounds. *Fungal Biology and Biotechnology, 11*(1), 19.

Valdes, J. J., Cameron, J. E., & Cole, R. J. (1985). Aflatrem: A tremorgenic mycotoxin with acute neurotoxic effects. *Environmental Health Perspectives, 62*, 459–463.

Van Court, R. C., Wiseman, M. S., Meyer, K. W., Ballhorn, D. J., Amses, K. R., Slot, J. C., et al. (2022). Diversity, biology, and history of psilocybin-containing fungi: Suggestions for research and technological development. *Fungal Biology, 126*(4), 308–319.

Verhelst, J., Abs, R., Maiter, D., Van Den Bruel, A., Vandeweghe, M., Velkeniers, B., et al. (1999). Cabergoline in the treatment of hyperprolactinemia: A study in 455 patients. *The Journal of Clinical Endocrinology & Metabolism, 84*(7), 2518–2522.

Vézina, C., Kudelski, A., & Sehgal, S. N. (1975). Rapamycin (AY-22, 989), a new antifungal antibiotic I. Taxonomy of the producing streptomycete and isolation of the active principle. *The Journal of Antibiotics, 28*(10), 721–726.

Villegas, C., González-Chavarría, I., Burgos, V., Iturra-Beiza, H., Ulrich, H., & Paz, C. (2023). Epothilones as natural compounds for novel anticancer drugs development. *International Journal of Molecular Sciences, 24*(7), 6063.

Vishwakarma, K., Upadhyay, N., Kumar, N., Yadav, G., Singh, J., Mishra, R. K., et al. (2017). Abscisic acid signaling and abiotic stress tolerance in plants: A review on current knowledge and future prospects. *Frontiers in Plant Science, 8*, 161.

Wadhwa, K., Kapoor, N., Kaur, H., Abu-Seer, E. A., Tariq, M., Siddiqui, S., et al. (2024). A comprehensive review of the diversity of fungal secondary metabolites and their emerging applications in healthcare and environment. *Mycobiology*, 1–53.

Wang, H., Peng, H., Li, W., Cheng, P., & Gong, M. (2021a). The toxins of Beauveria bassiana and the strategies to improve their virulence to insects. *Frontiers in Microbiology, 12*, 705343.

Wang, J., Liu, M., Mao, C., Li, S., Zhou, J., Fan, Y., et al. (2023). Comparative proteomics reveals the mechanism of cyclosporine production and mycelial growth in Tolypocladium inflatum affected by different carbon sources. *Frontiers in Microbiology, 14*, 1259101.

Wang, W., Yu, Y., Keller, N. P., & Wang, P. (2021b). Presence, mode of action, and application of pathway specific transcription factors in Aspergillus biosynthetic gene clusters. *International Journal of Molecular Sciences, 22*(16), 8709.

Wilkinson, H. H., Ramaswamy, A., Sim, S. C., & Keller, N. P. (2004). Increased conidiation associated with progression along the sterigmatocystin biosynthetic pathway. *Mycologia, 96*(6), 1190–1198.

Winkelmann, G. (2007). Ecology of siderophores with special reference to the fungi. *Biometals, 20*, 379–392.

Xu, K., Li, X.-Q., Zhao, D.-L., & Zhang, P. (2021). Antifungal secondary metabolites produced by the fungal endophytes: Chemical diversity and potential use in the development of biopesticides. *Frontiers in Microbiology, 12*, 689527.

Xu, W., Gavia, D. J., & Tang, Y. (2014). Biosynthesis of fungal indole alkaloids. *Natural Product Reports, 31*(10), 1474–1487.

Yin, Q., Liu, X., Zhang, Z., Lei, H., & Wu, B. (2023). Chemistry and bioactivities of alkaloids isolated from marine fungi (covering 2016–2022). *Fitoterapia, 164*, 105377.

Yin, Y., Cai, M., Zhou, X., Li, Z., & Zhang, Y. (2016). Polyketides in Aspergillus terreus: Biosynthesis pathway discovery and application. *Applied Microbiology and Biotechnology, 100*, 7787–7798.

Yoo, Y. J., Kim, H., Park, S. R., & Yoon, Y. J. (2017). An overview of rapamycin: From discovery to future perspectives. *Journal of Industrial Microbiology and Biotechnology, 44*(4–5), 537–553.

Zeb, M., & Lee, C. H. (2021). Medicinal properties and bioactive compounds from wild mushrooms native to North America. *Molecules, 26*(2), 251.

Zhang, F.-L., & Feng, T. (2022). Diterpenes specially produced by fungi: Structures, biological activities, and biosynthesis (2010–2020). *Journal of Fungi, 8*(3), 244.

Zhang, L., Wang, C., Chen, K., Zhong, W., Xu, Y., & Molnár, I. (2023). Engineering the biosynthesis of fungal nonribosomal peptides. *Natural Product Reports, 40*(1), 62–88.

Zhao, C., Wang, X.-H., Lu, X.-Y., Zong, H., & Zhuge, B. (2023). Metabolic engineering of candida glycerinogenes for sustainable production of geraniol. *ACS Synthetic Biology, 12*(6), 1836–1844.

Zheng, J., Xu, Z., Wang, Y., Hong, K., Liu, P., & Zhu, W. (2010). Cyclic tripeptides from the halotolerant fungus Aspergillus sclerotiorum PT06-1. *Journal of Natural Products, 73*(6), 1133–1137.

Zheng, J., Zhu, H., Hong, K., Wang, Y., Liu, P., Wang, X., et al. (2009). Novel cyclic hexapeptides from marine-derived fungus, Aspergillus sclerotiorum PT06-1. *Organic Letters, 11*(22), 5262–5265.

Zhong, A. B., Muti, I. H., Eyles, S. J., Vachet, R. W., Sikora, K. N., Bobst, C. E., et al. (2022). Multiplatform metabolomics studies of human cancers with NMR and mass spectrometry imaging. *Frontiers in Molecular Biosciences, 9*, 785232.

Zobel, S., Boecker, S., Kulke, D., Heimbach, D., Meyer, V., & Süssmuth, R. D. (2016). Reprogramming the biosynthesis of cyclodepsipeptide synthetases to obtain new enniatins and beauvericins. *Chembiochem, 17*(4), 283–287.

Biosynthetic Pathways of Fungal Secondary Metabolites

Kallol Das, Mia Md. Tofayel Gonee Manik, Md. Rashedul Islam, Md. Mizanur Rahaman, Mohammad Muzahidur Rahman Bhuiyan, Md. Munna Aziz, and Inshad Rahman Noman

Abstract

Fungal secondary metabolites are a large class of structurally diverse compounds of significant ecological roles and high value in industry, agriculture, and medicine. Their biosynthesis includes polyketides, non-ribosomal peptides, terpenoids, and alkaloids, all produced by complex biochemical pathways encoded by biochemical gene clusters (BGCs). This chapter will review the biochemical and genetic basis for these pathways, with an emphasis on the enzymatic steps and precursor molecules leading to the synthesis of bioactive compounds. The molecular regulation of these pathways emphasizes the role of environmental signals, transcription factors, and epigenetic modifications. The development of omics technologies such as transcriptomics, metabolomics, and genomics has changed our view of route creation and control. This allowed not only the construction but also the discovery of novel pathways. The characterization of unknown pathways, the comprehension of complex regulatory networks, and metabolic flux control remain highly challenging tasks despite enormous progress. Future developments, including artificial intelligence and multi-omics integration, may lead to further development in the research of fungal secondary metabolites. The conclusion of this chapter thus underlines the necessity for an interdisciplinary approach toward the effective utilization of these metabolites when it comes to meeting global challenges related to sustainable agriculture and health. The chapter thus covers in-depth basic research on the biosynthesis of secondary metabolites by fungi and how these processes make powerful impacts on science and business.

Keywords

Biosynthetic pathways · Fungal physiology · Fungal secondary metabolites · Polyketides · Secondary metabolism

1 Introduction

Fungal secondary metabolite biosynthesis is facilitated by complex primary enzymatic pathways, which are the backbone of the formation of the metabolites. These pathways involve highly specialized enzymes that act in a coordinative manner to convert simple primary metabolites into novel, structurally more complex secondary metabolites (Cox, 2024; Winkelblech et al., 2015; Zhou et al., 2024). Prominent nodes involved in these processes include polyketide synthase (PKS), non-ribosomal peptide synthetase (NRPS, and terpene synthase (TS). Every enzyme kind is specialized to carry out certain reactions to promote the correct formation of advanced molecular arrangements. These enzymatic pathways demonstrate that fungi are metabolically creative organisms, as they can construct a virtually limitless catalog of compounds such as Acetaldehyde, Alanine, Ammonia, β-carotene, Polyprene, and more employing fundamental functional group leverages of acetyl-CoA, amino acids, and isoprene instruments (Keller et al., 2005; Younas & Younas, 2025). Synthetic routes of fungal secondary compounds are one of the fascinating and flourishing areas of research interest in any field of mycology and biochemistry, depicting the social life of fungi. These metabolites, which are a remarkable variety of structurally

K. Das (✉)
College of Agriculture, Food and Environmental Sciences, California Polytechnic State University, San Luis Obispo, CA, USA
e-mail: kalloldas91@gmail.com

M. Md. T. G. Manik · Md. R. Islam · Md. M. Rahaman · M. M. R. Bhuiyan · Md. M. Aziz
College of Business, Westcliff University, Irvine, CA, USA

I. R. Noman
Department of Computer Science, California State University, Los Angeles, CA, USA

© The Author(s), under exclusive license to Springer Nature Switzerland AG 2026

A. Prakash and M. K. Dubey (eds.), *Advances in Fungal Secondary Metabolites: Bioactive Attributes and Biotechnological Potential of Industrial Interest*, Sustainable Economy and Ecotechnology, https://doi.org/10.1007/978-3-032-23507-7_2

and functionally diverse chemical entities like polyketides, non-ribosomal peptides, terpenoids, and alkaloids, it is involved in ecological interactions, human health, and biotechnological prospects (Bills & Gloer, 2016). Many fungi synthesize secondary products for their protection against predators, competitors, and parasites. In addition, most of these compounds have been utilized based on their pharmacological potential to result in useful drugs, for instance, antibiotics and anticancer drugs (Allam et al., 2025; Hoffmeister & Keller, 2007).

Moreover, it is well understood that genes involved in the making of secondary metabolites (SMs) are under multi-tiered regulation because SM are not biosynthetic essentials but are valuable. Genetic aspects remain highly important, biosynthetic gene clusters coordinating the production of enzymes and metabolites being the most essential (Cox, 2024; Riedling et al., 2024). These clusters are regulated by transcription factors, which therefore either promote or inhibit the activity of these clusters based on signals received (Strauss & Reyes-Dominguez, 2011). More specifically, histone acetylation and DNA methylation allow fungi to even more precisely regulate metabolite production. Nutrient availability, pH, temperature, and the presence of competitors and other organisms are factors that also affect the secondary metabolism. Biosynthetic pathways for secondary metabolites can be induced only under certain conditions by certain chemical inducers, including metal ions or signaling molecules (Dubey et al., 2019). This dynamic regulation further enhances the fact that fungi are endowed with the flexibility and metabolic plasticity to observe and respond to environmental challenges. It has been specifically pointed out that only modest information has been accumulated on primary enzymatic processes and the genetic control of secondary metabolism, but without this knowledge, it would be impossible to unravel the highly varied biosynthesis of secondary metabolites in fungi (O'Connor, 2015; Suárez et al., 2024; Wang et al., 2025). In this chapter, the significance of these biosynthetic pathways in fungi will be initially discussed in terms of their ecological functions and their possible commercial utilization in medicine and biotechnology. Besides deepening our understanding of the biology of fungi, the findings open up horizons of novel developments in drug development, agriculture, and industrial uses. Mechanisms of fungal secondary biosynthesis, regulation, and modern approaches are presented herein, as understanding of them is crucial in both natural and applied sciences.

2 Key Biosynthetic Pathways

Regarding the major pathways, fungal SM biosynthesis occurs through different key biosynthetic pathways, representing unique classes of compounds such as polyketides, non-ribosomal peptides, terpenoids, and alkaloids (Fig. 1). These pathways are based on specialized enzyme systems and gene clusters for synthesizing structurally diverse and biologically active natural products. Due to the biological construct of these biosynthetic processes (Dhar et al., 2013), it can envisage the production of a diverse continuum of compounds within the field of medicine as well as in agriculture. Knowledge of these pathways gives insight into the ease of metabolism with regard to fungi and offers a new horizon for biotechnological innovation.

2.1 Polyketides

Polyketides represent one of the largest groups of fungal SMs that have a wide spectrum of biological effects, such as antibiotic, anticancer, and antifungal activities. These compounds are biosynthesized through a few coordinated enzymatic reactions, where specific enzyme clusters facilitate the sequential steps in the formation of the complicated molecular structures of these compounds (Mohan et al., 2024; Weissman, 2009).

2.1.1 Enzyme Complexes and Biosynthesis Mechanisms

Polyketides are synthesized biologically through a process that is controlled by the polyketide synthase (PKS) enzyme system. These multifunctional enzymes polymerize small monomers in a process reminiscent of fatty acid synthesis and involve the repetitive condensation of acetyl Coenzyme A and malonyl Coenzyme A (Layer et al., 2010). Based on the type of PKS, a few bike reactions occur in every cycle of chain elongation, and these are condensation, reduction, dehydration, and cyclization. Modularity of PKS enzymes is one of the key characteristics of polyketide assembly lines. Each module is associated with a particular step in the elongation cycle and independent domains of acyltransferase (AT), ketosynthase (KS), and acyl carrier protein (ACP). This domain organization enables the enzyme to determine the chain sequence, length, and the nature of other chemical groups that incorporate into the growing polyketide chain to generate bewildering structural variety.

2.1.2 Polyketide Synthases (PKS)

Polyketide synthases are broadly classified into three types. The three categories of PKS that exist are known as Type I PKS, Type II PKS, and Type III PKS, and each of these kinds has a fundamentally different organizational complexity (Dutta et al., 2014).

- **Type I PKS**—These are large, multi-domain enzymes that are formed by a single polypeptide-bound chain.

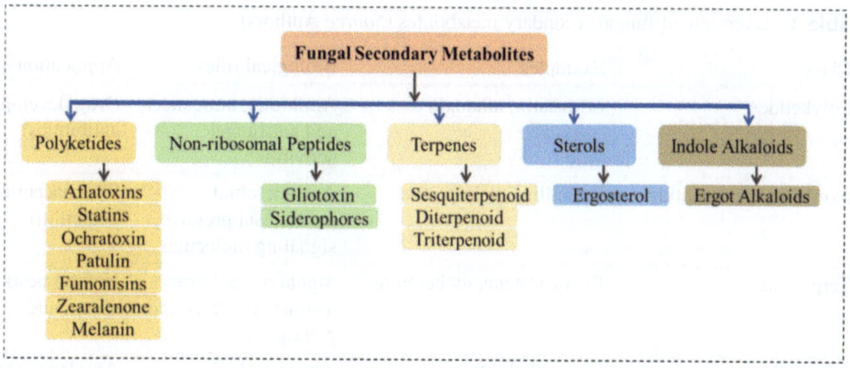

Fig. 1 Secondary metabolites (SMs) of fungi (*Source* Authors)

They work in a processive fashion, and each of them covers a cycle of elongation. The initial type of PKS, recognized in fungi, is directed at the synthesis of highly elaborative polyketides: lovastatin and aflatoxin (Katz & Donadio, 1993).

- **Type II PKS**—These are mostly present in bacteria, and the enzymes are individual, single-process enzymes. Although not a very frequent process, they also take part in the formation of certain fungal products.
- **Type III PKS**—These enzymes are not modulated and are involved in chain elongation as well as cyclization processes without having ACP domains (Shimizu et al., 2017).

The modularity of PKS enzymes affects the ability of fungi to synthesize a diverse spectrum of polyketides, reaching from simple aromatic moieties to complicated macrolactones. Moreover, customizing enzymes, including oxygenases and methyltransferases, further remodel the polyketide core skeleton and expand its chemical and biological library (Table 1).

Identification of the enzymatic components and biosynthetic processes involved in polyketide synthesis has far-reaching significance. Moreover, it brings us insights into fungal metabolism; at the same time, there is great potential in the discovery and engineering of novel polyketides that have therapeutic and industrial values. New developments in synthetic biology and genomic mining have further opened new avenues to dissect and develop the PKS pathways and accelerate the inventions in drug discovery and other biological tools (Dinglasan et al., 2025).

2.2 Non-ribosomal Peptides

Non-ribosomal peptides (NRPs) are a variety of bioactive secondary metabolites involved in the secondary metabolite system of fungi and other microorganisms that are synthesized through the non-ribosomal peptide system. Unlike ribosomal peptides, the synthesis of which involves ribosomes in collaboration with mRNA, the synthesis of NRPs is facilitated by multi-enzyme forming structures denoted as non-ribosomal peptide synthetases (NRPS). These compounds have a range of roles in other biological activities, such as acting as antimicrobial agents, having signaling properties, and being involved in other organism associations.

2.2.1 NRPS Enzymes

Nearly all NRPS are large, multi-domain enzymes that are responsible for the synthesis of NRPs. Structure of NRPS: The structural organization of NRPS is analogous with the polyketide synthases, composed of several modules that contain various modules of the peptide assembly process (Tajbakhsh et al., 2017). In the case of NRPS, each module is supposed to load a particular amino acid into a developing peptide chain.

The key components of NRPS include

- **Amino Acid Activation Domain (A)**—This domain catalyzes the activation of the amino acid by generating an aminoacyl-adenylate complex.
- From the studies done, the following domains were identified: Peptide Bond **Formation Domain (T)**—this domain is responsible for the condensation of the activated amino acid with the growing peptide chain.
- **Modifications Domains (M)**—These domains code for adducts like methyl groups, hydroxyl groups, or cyriles, which are necessary for the biological function of the product (Calcott et al., 2020).

Since NRPS is highly flexible and consists of modules allowing for the introduction of nonstandard amino acids and other kinds of modifications, there are countless structures of NRP.

Table 1 Overview of fungal secondary metabolites (*Source* Authors)

Class	Examples	Biological roles	Applications	References
Polyketides	Lovastatin, aflatoxin	Antibiotic, anticancer, and antifungal activities	Drug development (e.g., antibiotics)	Keller et al. (2005)
Non-ribosomal peptides	Penicillin, cyclosporine	Antimicrobial, immunosuppressive, signaling molecules	Pharmaceuticals (e.g., penicillin)	Süssmuth and Mainz (2017)
Terpenoids	Beta-carotene, gibberellins	Signaling, defense against predators and pathogens	Natural pesticides, bioactive compounds	Quin et al. (2014), Ivanov et al. (2025)
Alkaloids	Ergot alkaloids, psilocybin	Antimicrobial, psychoactive properties	Antidepressants, antifungal agents	Flieger et al. (2019)

2.2.2 Examples of Non-ribosomal Peptide Products

Case, one of the most famous and important NRPs, is penicillin, which is synthesized by fungi such as *Penicillium chrysogenum* and has a strong antibacterial effect. An antibiotic produced from *Tolypocladium inflatum*, cyclosporine is a potent immunosuppressive drug used in the treatment of autoimmune diseases and the prevention of organ transplant rejection. Also, the antibiotic called Gramicidin S, derived from *Bacillus brevis*, possesses remarkable bacteriocidal action for suppressing Gram-positive bacteria (Berditsch et al., 2017). Fusarium toxin has its origin in species of *Fusarium*. These mycotoxins get into the crops and are dangerous to human and animal consumption. The analysis of non-ribosomal peptides and their biosynthesis not only expands total knowledge of fungal metabolism but also gives targets toward understanding new drugs and agricultural supplements (Payne et al., 2017). It is only when the production strategies of NRPS and their products are understood that it is possible to employ different strategies to synthesize new compounds with beneficial characteristics.

2.3 Terpenoids

Terpenoids, which are also called isoprenoids, are a very extensive group of secondary metabolites that are synthesized by fungi and some other organisms. It is explained by the richness of their structure and the multiplicity arising owing to the linking of isoprene units (C5H8). It involved finding that terpenoids are involved in ecological interactions, signaling, and almost all physiological processes occurring in fungi (Ludwiczuk et al., 2017).

2.3.1 Biosynthesis of Terpenes and Terpenoids

The biosynthesis of terpenes and terpenoids involves two primary pathways: 2 pathways, of which the first is called mevalonate pathway and the second methylerythritol phosphate (MEP) pathway (Fig. 2). The mevalonate pathway takes place in the cytoplasm and participates in terpenes biosynthesis in many eukaryotes, including fungi. The process starts with the conversion of the acetyl-CoA to mevalonate, followed by the steps of a series of isoprenoid unit formation reactions catalyzed by respective enzymes (Perveen et al., 2018). Certain enzymes of interest are HMG-CoA reductase, which catalyzes the highest rate step in the pathway. Mainly present in bacteria and plastids of plants, the Methylerythritol Phosphate Pathway synthesizes isoprenoid precursors from pyruvate and glyceraldehyde-3-phosphate. It uses several unusual enzymes that are different from those used in the mevalonate pathway, which shows that terpenoid biosynthesis has evolved differently. After it has been synthesized, isoprene units are polymerized to form a range of terpenes which can be altered by oxidation, methylation, and cyclization to create a diverse range of terpenoids. Organized in this modular fashion, fungi are able to construct a seemingly endless variety of terpenoid scaffolds with multiple biological effects.

2.3.2 Role in Fungal Physiology

These compounds are very important in fungal metabolism and the environment. Most terpenoids perform a defensive role against predators, pathogens, and competitors in their natural environment. For instance, some of the terpenes cause volatility and are known to prevent microbial growth or delay the attack by herbivores. Signaling molecules for communication between fungal cells of the same species or of different species are represented by terpenoids (Schmidt-Dannert, 2015). Chemicals produced by the fungi themselves, such as secondary metabolites, can regulate their own growth, development, and even sexual behavior; among these, terpenoids play a crucial role in metabolic regulation by participating in energy metabolism pathways and contributing to cellular organization and developmental processes (Galindo-Solís & Fernández, 2022). In some

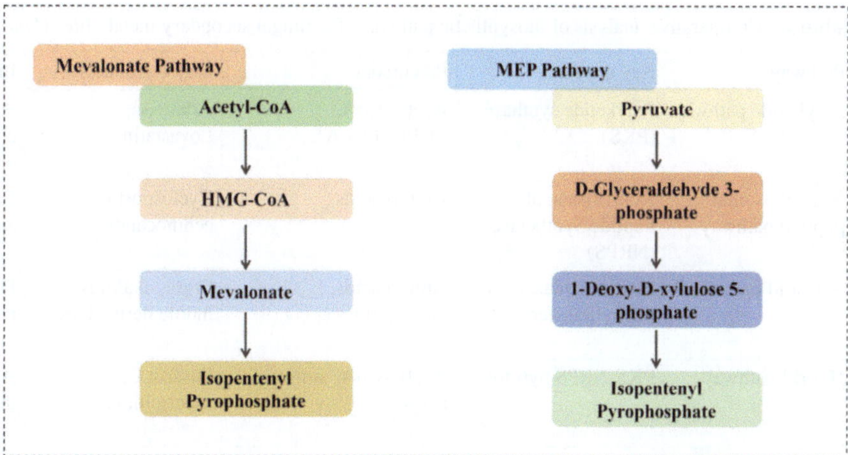

Fig. 2 A comparison diagram showing the Mevalonate and MEP pathways for terpenoid biosynthesis (*Source* Authors)

cases, the terpenoids play the role of mutualistic relationships between the fungi and plants or some other microorganisms. They can improve the transport of nutrients or act as a shield to adverse conditions within the body. The synthesis of terpenes and terpenoids is carried out within fungal cells and is a paramount feature of fungal metabolism (Panaccione et al., 2014). Such understanding of these pathways does more than help to explain fungal metabolism modalities; it also has applications in the biotechnological sector, for instance, in the synthesis of natural pesticides or bioactive pharmaceuticals from fungal-derived terpenoids.

2.4 Alkaloids

Alkaloids represent a large group of bacterio-, myco-, and phytochemicals with mostly basic properties, synthesized mainly in higher fungi, plants, and bacteria. They are described by their simple nitrogen-containing frameworks, and they have numerous organic functions, therefore playing essential roles in both biosphere and pharmaceutical sciences (Kukula-Koch & Widelski, 2017).

2.4.1 Biosynthetic Pathways of Fungal Alkaloids

The biosynthesis of fungal alkaloids involves complex pathways that typically derive from amino acids or other primary metabolites, which seems to be the rule, rather than the exception, in DNA sequence analysis (Table 2). The major biosynthetic routes include amino acid precursors: reactions of many alkaloids involve tryptophan, tyrosine, and phenylalanine amino acids (Xu et al., 2014). For example, the indole alkaloids are products of tryptophan, while the benzylisoquinoline alkaloids are products of tyrosine.

- **Enzymatic Modifications**—The biosynthesis process is sometimes complex and comprises the hydroxylation, methylation, or cyclization of an enzyme. These reactions are executed by specific enzymes, which result in the creation of different structures of alkaloids.
- **Gene Clusters**—The biosynthesis of alkaloids is often coordinated to gene clusters where all the genes required for the biosynthesis of the respective types of alkaloids are localized. This organization helps in showing cooperate expression and regulation in accordance with the environmental stimuli (Bharadwaj et al., 2021).

For example, ergot alkaloid compounds are produced by fungi such as ergot *Claviceps purpurea* and have valuable pharmacological uses, and at the same time can bring ergotism to humans and animals. Psilocybin is a psychedelic compound learned in some mushrooms which has shown promise in helping mental health (Xu et al., 2014).

2.4.2 Alkaloids in Clinical Interest

Due to the variation in their pharmacological effects, alkaloids are of great clinical importance. A considerable number of alkaloids are employed in medications, or their structures have formed the basis of certain drugs. For instance, morphine and codeine, which are derived from alkaloids present in the opium poppy, are reliably used to treat pain. Likewise, one of the anticancer drugs called vincristine is produced from the plant alkaloids of the periwinkle plant. One group of these compounds includes psilocybin and mescaline, which possess psychoactive effects; currently, they are considered useful in treating depression, anxiety, and PTSD (Khan et al., 2021). Their effectiveness in a controlled therapeutic environment is the subject of clinical research. Several fungal alkaloids contain antimicrobial properties that retard bacterial and fungal growth and thus can be considered for the construction of fresh antibiotics or antifungal drugs. Even though most alkaloids are body-friendly, some are poisonous. For example, ergot alkaloids can cause severe health problems should they be

Table 2 Comparative analysis of biosynthetic pathways for fungal secondary metabolites (*Source* Authors)

Pathway	Key enzymes	Precursors	Major products	Industrial applications	References
Polyketide pathway	Polyketide synthase (PKS)	Acetyl-CoA, Malonyl-CoA	Aflatoxin, Lovastatin	Antibiotics, cholesterol-lowering drugs	Keller et al. (2005), Riedling et al. (2024)
Non-ribosomal peptide pathway	Non-ribosomal Peptide Synthetase (NRPS)	Amino acids	Cyclosporin, echinocandin	Immunosuppressants, antifungals	Suárez et al. (2024), Cox (2024)
Alkaloid pathway	Amine oxidase, methyltransferase	Amino acids, monoamines	Ergot alkaloids, indole derivatives	Pharmaceuticals, neuroactives	Ivanov et al. (2025), Younas and Younas (2025)
Hybrid pathway	PKS-NRPS hybrids	Acetyl-CoA, amino acids	Fusarin C, aspergillicin	Agricultural fungicides, therapeutics	Zhou et al. (2024), Butkovich et al. (2025)
Terpenoid pathway	Terpene synthase (TPS)	IPP, DMAPP	Taxol, Squalene-derived metabolites	Anticancer, antimicrobial agents	Galindo-Solís and Fernández (2022), Allam et al. (2025)
Non-ribosomal peptide pathway	Non-ribosomal Peptide Synthetase (NRPS)	Amino acids	Penicillin, cyclosporine	Antimicrobials, immunosuppressants	Süssmuth and Mainz (2017)
Terpenoid pathway	Terpene synthase, HMG-CoA reductase	Acetyl-CoA, isoprene units	Beta-carotene, gibberellins	Natural pesticides, bioactive compounds	Quin et al. (2014)
Alkaloid pathway	Tyrosine decarboxylase, monooxygenase	Tryptophan, tyrosine, phenylalanine	Ergot alkaloids, psilocybin	Antidepressants, antifungal agents	Flieger et al. (2019)

consumed in the wrong way. Knowledge of their biosynthesis is vital in evaluating the probability of positive and negative impacts of their application. Thus, it is possible to conclude that the fungal alkaloids are an essential type of secondary metabolites with multiple biosynthetic routes and great clinical interest. The mentioned types of bioactive compounds result from different pharmacological activities, which highlights the further study of their mechanisms and therapeutic uses (Heinrich et al., 2021).

3 Regulation of Secondary Metabolite Production

The synthesis of secondary metabolites in fungi is a tightly controlled process that is controlled by multigene circuits, the organism and environmental factors, and chemical messengers. Profitable metabolites are often the focus of such synthetic systems (Shwab & Keller, 2008); knowledge of these regulatory mechanisms is therefore important in the improvement of these metabolites in applications such as in industry or the treatment of diseases (Table 3).

3.1 Genetic Regulation and Cluster Genes

The biosynthesis of fungal secondary metabolites is frequently regulated by gene clusters that are a set of co-localized genes that code protein products involved in the production of these metabolites (Fig. 3). Biosynthetic genes are responsible for the synthesis of enzymes that carry out the production of secondary metabolites like polyketide synthases (PKS) and non-ribosomal peptide synthetases (NRPS). Not only biosynthetic genes but also gene clusters contain a few regulatory genes that govern the synthesis of biosynthetic genes (Brakhage & Schroeckh, 2011). These regulators can also respond to one or the other signal to make certain that metabolite production will happen under favorable states (Khan et al., 2025). The nature of grouping these clustered genes provides opportunities for synchronized regulation in terms of gene expression, whereby fungi can produce secondary metabolites according to the expression of certain stimuli or signals that are within a developmental program (Razin et al., 2021; Sproul et al., 2005). These gene clusters have been convincingly identified and characterized because of another genomic sequencing technology, and their manipulation for increased metabolite accumulation has become much easier now (Fig. 3).

3.2 Transcription Factors and Epigenetics with Respect to Their Roles

Due to their importance in controlling the expression of the genes associated with secondary metabolism, transcription factors (TFs) draw reasonable attention. Many of these

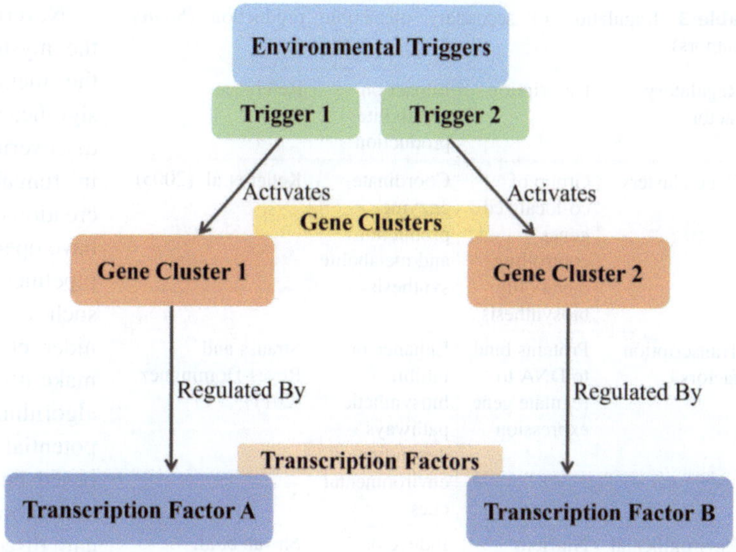

Fig. 3 A conceptual diagram showing gene clusters, transcription factors, and environmental triggers regulating metabolite production (*Source* Authors)

proteins interact with DNA at unique DNA sequences within gene promoters, regulating the transcription of target genes. Depending on the metabolic requirement and the challenges in the environment, different TFs can either enhance or inhibit a given biosynthetic network in the fungus (Hawkins et al., 2018). However, the regulation of metabolite production in cells is usually not the sole responsibility of any single TF but is achieved in cooperation with other regulatory proteins and signaling pathways. This intricate interaction guarantees that the synthesis of secondary metabolites is done in the correct amount and time. Secondary metabolism is also controlled by epigenetic modifications. Other mechanisms that lead to variation in biological processes without prompting changes at the DNA level involve changes to DNA molecules, such as methylation and histones. These epigenetic mechanisms enable fungi to adapt or modify their metabolic state depending on the above-mentioned factors such as the environment, stress, or developmental stimuli.

3.3 Environmental Triggers and Chemical Inducers

This section provides evidence on how ecological factors affect the synthesis of secondary metabolites in fungi. Various triggers can induce or enhance the biosynthesis of these compounds: Implementing essential nutrients like carbon, nitrogen, as well as other nutrients, can determine metabolic processes. For example, nutrient limitations tend to be behind the synthesis of secondary metabolites to compensate for energy loss (Wu et al., 2016). External and environmental influences, for example, temperature, pH values, and light, might influence metabolic activity and the genes controlling SMs. Secondary inducers are compounds that stimulate the synthesis of specific compounds, such as metal ions (zinc, copper), and signaling molecules (jasmonic acid). Such chemical signals may stimulate certain signal transduction that results in increased biosynthesis. Knowledge of how environmental stimuli and chemical stimuli influence secondary metabolite synthesis is critical to optimizing the fermentation process in bioprocessing environments. Through regulating these factors, the subsequent production of desired metabolites useful in pharmaceutical or agricultural use can be improved (Dubey et al., 2019). Therefore, the regulation of secondary metabolite production in fungi encompasses gene regulation, local and temporal mRNA expression (Hartwig et al., 2002), DNA methylation, histone modifications, and response to the environment. Knowledge of such regulation mechanisms not only enlightens fungal biology but also opens up for further technological developments necessary to optimize metabolite production concerning its applications.

4 Machine Learning in Fungal Secondary Metabolites

The secondary metabolites that are produced by fungi are extremely diverse. Many of these secondary metabolites have powerful biological activity and offer a great deal of promise for the development of pharmaceuticals, agricultural applications, and industrial biotechnology (Fig. 4). Machine learning (ML) has emerged as a powerful technique that has the potential to expedite the prediction and identification of secondary metabolites produced by fungi. This has the potential to supplement the conventional wet-lab methods and to direct the validation of experimental results. The ability of microorganisms, particularly fungi,

Table 3 Regulation of secondary metabolite production (*Source* Authors)

Regulatory factor	Description	Impact on metabolite production	References
Gene clusters	Group of co-localized genes controlling metabolite biosynthesis	Coordinate enzyme production and metabolite synthesis	Keller et al. (2005)
Transcription factors	Proteins bind to DNA to regulate gene expression	Enhance or inhibit biosynthetic pathways depending on environmental cues	Strauss and Reyes-Dominguez (2011)
Environmental triggers	Nutrient availability, pH, temperature, and presence of competitors	Induce or suppress metabolite production	Shwab et al. (2007)
Epigenetic modifications	DNA methylation, histone modifications	Modulate metabolic states and allow adaptation to environmental changes	Strauss and Reyes-Dominguez (2011)
Chemical inducers	Compounds like metal ions (zinc, copper) or signaling molecules (Jasmonic acid)	Stimulate biosynthesis of specific metabolites	Hoffmeister and Keller (2007)

which are believed to be the most natural occupants of a wide variety of habitats, to create active molecules is characterized by extensive interaction capacities. The physiologically active secondary metabolites that were derived from endophytic fungi exhibited a variety of bioactivities (Navarro-Muñozet al., 2020; Riedling et al., 2024). These bioactivities included antibacterial, anti-tumor, anti-inflammatory, immunomodulatory, and anti-diabetes activities (Mohamed et al., 2021). Additionally, these compounds developed functional constipation activities. Fungi are responsible for the production of approximately 9000 of the 22,500 bioactive chemicals that have been derived from microbes up until this point (Brakhage & Schroeckh, 2011). According to Demain et al. (2004), the significance of the function that fungi play in the manufacture of antibiotics and other medications for the treatment of diseases that are not caused by infectious agents cannot be overstated.

Nevertheless, the complexity of fungal genomes and the mysterious interaction that exists between genes and the metabolites that they produce is one of the most significant challenges that arise throughout the process of discovering secondary metabolites. Recent developments in fungal genomes and transcriptomics, as well as the creation of extensive databases and tools for bioinformatics, have opened the path for machine learning-driven discovery pipelines. To detect and classify BGCs, state-of-the-art tools such as MIBiG (Medema et al., 2015), PRISM (Skinnider et al., 2017), and antiSMASH (Blin et al., 2021) make use of supervised and unsupervised machine learning algorithms. These algorithms make predictions about the potential chemical structures and bioactivities of BGCs based on sequence homology, conserved protein domains, and known biosynthetic logic. Machine learning algorithms can discern patterns that are not easily visible by standard alignment-based methods. This is accomplished by integrating genomic, transcriptomic, and metabolomic data. These computational predictions can then be used to drive more targeted isolation and characterization efforts in the laboratory, which will significantly reduce the amount of time and resources that are spent screening random fungal isolates and extracts. In addition, machine learning techniques make it possible to make a de novo prediction of groups of metabolites based solely on genomic characteristics. As a result of the availability of thousands of fungal genomes, models are now able to make use of established structure–function connections and training sets that are drawn from BGCs that have been experimentally validated. According to Cimermancic et al. (2014), these techniques not only increase the accuracy of predicting recognized classes of metabolites, but they also can attract the attention of previously uncharacterized clusters that have the potential to produce novel chemical scaffolds that possess bioactive capabilities.

Despite these advancements, the incorporation of machine learning into the discovery of secondary metabolites produced by fungi is still proceeding. Deficits in annotated training data, biological complexity, and an insufficient understanding of fungal metabolic pathways are all factors that cause difficulties (Butkovich et al., 2025). Furthermore, thorough wet-lab validation is required to link newly predicted gene clusters to metabolite structures that have been proven by experimentation. Nevertheless, the accuracy of models will be improved through ongoing developments in machine learning algorithms, the creation of databases that are more comprehensive and more curated, and advancements in the integration of several omics. It is expected that these efforts will, over the course of time, quicken the pace of the discovery of secondary metabolites produced by fungi, thereby revealing new chemicals and

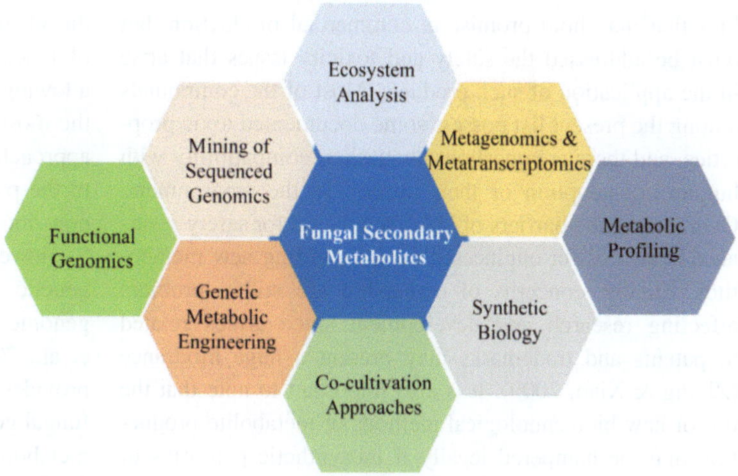

Fig. 4 Screening approaches for novel fungal secondary metabolites (*Source* Authors)

leading to a better knowledge and utilization of the chemical variety of fungi.

5 Challenges

The study and use of biosynthesis of fungal secondary metabolites present a number of challenges that may hinder their further research and practical application. This is because determining the biosynthetic pathways of the desired compound can be technically daunting since they typically include a multitude of enzymes and regulatory systems (Nawrot-Chorabik et al., 2022). This intrinsic multilayered connectivity makes it challenging to decipher comprehensively how these pathways function and hinders attempts at optimizing their functioning to yield higher metabolite profiles. Further, low yield is another problem with secondary metabolites of fungi, which challenges their potential commercial use. To reach the desired levels of productivity, one has to experiment a lot in conditions of growth and genetic engineering; furthermore, even with these means, one is not guaranteed to find what one is looking for. Secondly, strain variability is another biological issue; the two related strains within the species can differ significantly in the levels of metabolites being produced. Locating higher-yielding strains and dissecting out the genotype basis for such variations is sometimes a painful process (Fig. 5).

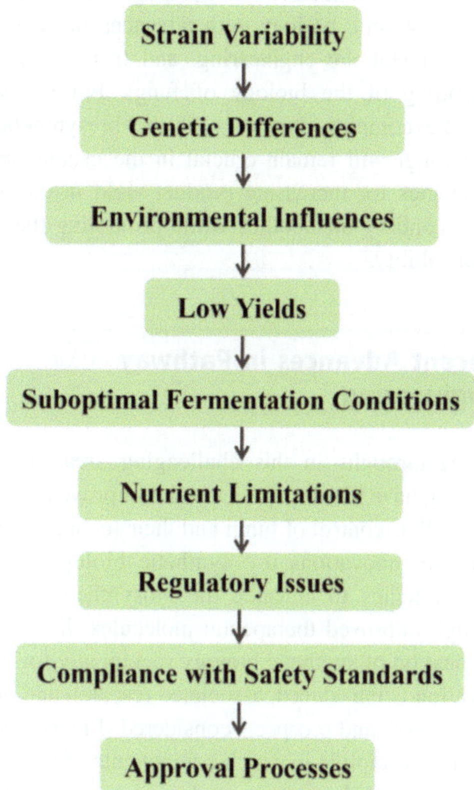

Fig. 5 The challenges in fungal secondary metabolite research (e.g., strain variability, low yields, and regulatory issues) (*Source* Authors)

However, there are challenges in the regulation of the production of secondary metabolites. The regulatory environment related to the production of the secondary metabolite also poses some challenges (Brakhage, 2013). The next area, which constitutes regulation, includes the ability to determine how specific intracellular factors, including nutrient availability and extracellular limitations, including stress factors that affect metabolite synthesis. This regulatory complexity makes it challenging to forecast outcomes touching on metabolic changes consequent to genetic alterations on metabolite productivity. In addition, optimization of one biosynthetic pathway might influence the others, for example, it may decrease the yields of other valuable metabolites or elevate the levels of byproducts. Efficient and minimal overload of the central metabolic pathways continues to be a major problem in metabolic engineering.

Furthermore, the pest-control and pathogen-eradication products are only some of the fungal secondary metabolites that may hold promise in commercial production, but must be addressed the safety and toxicity issues that arise in the application of such products. Most of the compounds among the present list possess some documented toxic properties, and there are questions about their compatibility with human consumption or their impact on the environment. Overcoming the barriers of MAC regulation for safety assurance is critical for engineered strains yielding new metabolites. Finally, concerns of ownership are still a problem affecting research and development since issues related to patents and trademarks may present a huge hindrance (Zhong & Xiao, 2009). It is also important to note that the use of new biotechnological methods of metabolite production may be hampered legally if biosynthetic pathways or genetic equipments are patented. The resolution of these complex issues can notably benefit from the cooperation that combines synthetic biology with advances in genomics as well as in metabolic engineering, and from the enhanced understanding of the biology of fungi. Further work to unravel the complex dynamics of the biosynthetic pathways of fungi will remain crucial in the establishment of new strategies for increasing product yield in the application of secondary metabolites while promoting their safety and applicability.

6 Recent Advances in Pathway Engineering

New breakthroughs in this challenging area of pathway engineering have filled major gaps and improved our knowledge as well as control of fungi and their secondary metabolites. These innovations use synthetic biology and novel genetic strategies to enhance the biosynthetic algorithms and design improved therapeutic molecules. In this article, the application of synthetic biology specifically for pathway modification is introduced, associated research and development discussed, and prospects considered (Fig. 6). Synthetic biology is a discipline in which components of living organisms are elaborated and integrated according to engineering principles (Shen et al., 2019). Thus, through the redesigning of the existing biosynthetic pathways, the new pathways that can support the synthesis of target SMs can be designed and developed. This approach enables the use of non-native genes or enzymes from other organisms, hence broadening the repertoire of possible products. It means that using synthetic biology tools, one can carefully adjust the levels of some enzymes in a pathway (Misawa, 2011). When the metabolic flux is controlled in this manner, researchers can increase the yield of the desired metabolite while decreasing the yields of other unwanted metabolites. The structures of biosynthetic pathways are composed of modules that indicate the ease of control, change, and optimization. One of the advantages is that the individual components or modules of a metabolic pathway can easily be changed by other ones, allowing the researchers to find out which specific setups are the most effective in terms of metabolite generation. These approaches of synthetic biology have given a dramatic boost to the production of diverse FCMs, especially polyketides, non-ribosomal peptides, terpenoids, and alkaloids.

However, hailed as the most significant step toward genetic engineering is the discovery of a broad-range genome editing tool called CRISPR technology (Chae et al., 2017). Comparable with other eukaryotes, CRISPR provides opportunities to create targeted changes in the fungal genome where individual genes impacting secondary metabolites can be knocked out or overexpressed (Cho et al., 2018). Such a targeted method can cause enhanced yields or production of new compounds by modulating the biological control system. One useful feature of CRISPR-based systems is that they can be programmed to produce gain and loss of function without causing genomic insertions (Molla & Yang, 2019). This flexibility makes it easy for researchers to enhance or even suppress metabolite production based on changes in the environment or experimental conditions.

Sophisticated molecular biological techniques make it possible to implement efficient methods of analyzing multiple genetic constructs at once. This efficiency increases the rate of determining the right genetic makeup for the production of better metabolites. Other genetic tools apart from CRISPR include TALENs, which affect the enhancement of secondary metabolites through offering other methods of gene regulation and modification, and RNAi techniques as well. Pathway engineering by synthetic biology with the help of genetic modulators like CRISPR has shown novel approaches to enhance the production of fungal secondary metabolites (Sinha et al., 2019). What is more, these innovations not only improve our knowledge of the existing metabolic pathways but also pave the way toward the production of new products possessing great potential in medicine, agriculture, and many other sectors. These technologies are therefore useful and can continue to aid future research development in the field of fungal biotechnology.

7 Conclusion

Research on fungal secondary metabolites and their biosynthesis is an area that has been explored and is developing very effectively in recent years, and has enormous prospects in medicine, agriculture, and other fields of biotechnology. From analyzing mechanisms in the biosynthesis of metabolites like polyketides, non-ribosomal peptides, terpenoids, and alkaloids, researchers have understood various func-

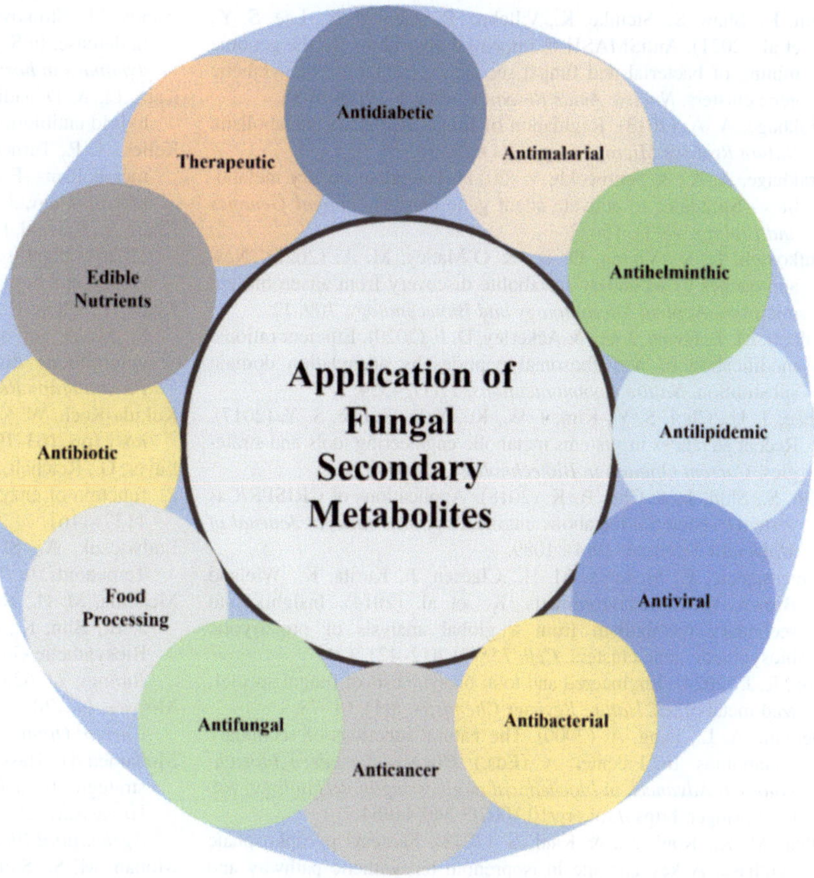

Fig. 6 Application of fungal secondary metabolites (*Source* Authors)

tional aspects of the metabolism of fungi. With new accumulated knowledge about clustered genes, transcription factors, and epigenetic regulation features, researchers have gained better insights into how fungi make changes to the environment and maximize the synthesis of metabolites. Moreover, the infusion of synthetic biology and other advanced genetic resources, such as CRISPR, has provided new prospects for pathway engineering and optimized the metabolic pathways to produce diversified secondary metabolites. Still, several threats exist even as engineers press on toward future milestones. The challenge for the comprehensive exploitation of FOSM is, however, due to the intricacy of the biosynthetic pathways, the strain variation, the regulatory issues, and the safety issues. These will require appreciable synergies between involved researchers, relevant industries, and regulatory agencies. Further work is warranted to understand the biosynthetic pathways of various fungal secondary metabolites for them to be fully exploited. Challenges can be better addressed, and coping strategies have to be developed so that the synthesis of relevant products that impact human welfare and healthier agriculture will be continuously improved. The future of fungal biotechnology has good prospects, in the form of significantly enhancing its contribution to various sectors.

Acknowledgements We would like to express our gratitude to all the co-authors for their contributions and critical reviews from the anonymous reviewers.

Funding This research received no external funding.

Conflicts of Interest The authors declare no conflict of interest.

References

Allam, A. A., Rudayni, H. A., Ahmed, N. A., Aba Alkhayl, F. F., Lamsabhi, A. M., & Kamel, E. M. (2025). Multidimensional insights into squalene epoxidase drug development: in vitro mechanisms, in silico modeling, and in vivo implications. *Expert Opinion on Therapeutic Targets, 29*(6), 385–403.

Berditsch, M., Trapp, M., Afonin, S., Weber, C., Misiewicz, J., Turkson, J., et al. (2017). Antimicrobial peptide gramicidin S is accumulated in granules of producer cells for storage of bacterial phosphagens. *Scientific Reports, 7*, 44324.

Bharadwaj, R., Kumar, S. R., Sharma, A., & Sathishkumar, R. (2021). Plant metabolic gene clusters: Evolution, organization, and their applications in synthetic biology. *Frontiers in Plant Science, 12*, Article 697318.

Bills, G. F., & Gloer, J. B. (2016). Biologically active secondary metabolites from the fungi. *Microbiology Spectrum, 4*(6), 10–1128.

Blin, K., Shaw, S., Steinke, K., Villebro, R., Ziemert, N., Lee, S. Y., et al. (2021). AntiSMASH 6: Improved algorithms for the genome mining of bacterial and fungal secondary metabolite biosynthetic gene clusters. *Nucleic Acids Research, 49*(W1), W29–W35.
Brakhage, A. A. (2013). Regulation of fungal secondary metabolism. *Nature Reviews Microbiology, 11*(1), 21–32.
Brakhage, A. A., & Schroeckh, V. (2011). Fungal secondary metabolites—Strategies to activate silent gene clusters. *Fungal Genetics and Biology, 48*(1), 15–22.
Butkovich, L. V., Vining, O. B., & O'Malley, M. A. (2025). New approaches to secondary metabolite discovery from anaerobic gut microbes. *Applied Microbiology and Biotechnology, 109*, 12.
Calcott, M. J., Owen, J. G., & Ackerley, D. F. (2020). Efficient rational modification of non-ribosomal peptides by adenylation domain substitution. *Nature Communications, 11*(1), 4554.
Chae, T. U., Choi, S. Y., Kim, J. W., Ko, Y. S., & Lee, S. Y. (2017). Recent advances in systems metabolic engineering tools and strategies. *Current Opinion in Biotechnology, 47*, 67–82.
Cho, S., Shin, J., & Cho, B. K. (2018). Applications of CRISPR/Cas system to bacterial metabolic engineering. *International Journal of Molecular Sciences, 19*(4), 1089.
Cimermancic, P., Medema, M. H., Claesen, J., Kurita, K., Wieland Brown, L. C., Mavrommatis, K., et al. (2014). Insights into secondary metabolism from a global analysis of prokaryotic biosynthetic gene clusters. *Cell, 158*(2), 412–421.
Cox, R. J. (2024). Engineered and total biosynthesis of fungal specialized metabolites. *Nature Reviews Chemistry, 8*(1), 61–78.
Demain, A. L., Fang, A. (2000). The natural functions of secondary metabolites. In: Fiechter, A. (Eds.), *History of modern biotechnology I. Advances in biochemical engineering/biotechnology*, vol. 69. Springer. https://doi.org/10.1007/3-540-44964-7_1
Dhar, M. K., Koul, A., & Kaul, S. (2013). Farnesyl pyrophosphate synthase: A key enzyme in isoprenoid biosynthetic pathway and potential molecular target for drug development. *New Biotechnology, 30*(2), 114–123.
Dinglasan, J. L. N., Otani, H., Doering, D. T., Udwary, D., & Mouncey, N. J. (2025). Microbial secondary metabolites: Advancements to accelerate discovery towards application. *Nature Reviews Microbiology, 23*, 338–354. https://doi.org/10.1038/s41579-024-01141-y
Dubey, M. K., Meena, M., Aamir, M., Zehra, A., & Upadhyay, R. S. (2019). Regulation and role of metal ions in secondary metabolite production by microorganisms. In *New and future developments in microbial biotechnology and bioengineering* (pp. 259–277). Elsevier.
Dutta, S., Whicher, J. R., Hansen, D. A., Hale, W. A., Chemler, J. A., Congdon, G. R., et al. (2014). Structure of a modular polyketide synthase. *Nature, 510*(7506), 512–517.
Flieger, M., Stodůlková, E., Wyka, S. A., Černý, J., Grobárová, V., Píchová, K., et al. (2019). Ergochromes: Heretofore neglected side of ergot toxicity. *Toxins (Basel), 11*(8), 439.
Galindo-Solís, J. M., & Fernández, F. J. (2022). Endophytic fungal terpenoids: Natural role and bioactivities. *Microorganisms, 10*(2), 339.
Hartwig, A., Asmuss, M., Ehleben, I., Herzer, U., Kostelac, D., Pelzer, A., et al. (2002). Interference by toxic metal ions with DNA repair processes and cell cycle control: molecular mechanisms. *Environmental Health Perspectives, 110*(suppl5), 797–799.
Hawkins, L. J., Al-Attar, R., & Storey, K. B. (2018). Transcriptional regulation of metabolism in disease: From transcription factors to epigenetics. *PeerJ, 6*, Article e5062.
Heinrich, M., Mah, J., & Amirkia, V. (2021). Alkaloids used as medicines: Structural phytochemistry meets biodiversity—An update and forward look. *Molecules, 26*(7), 1836.
Hoffmeister, D., & Keller, N. P. (2007). Natural products of filamentous fungi: Enzymes, genes, and their regulation. *Natural Product Reports, 24*(2), 393–416.
Ivanov, M., Stojković, D., & Soković, M. (2025). Fungal metabolites in defense. In S. S. M. Soliman & M. I. Husseiny (Eds.), *Metabolic dynamics in host-microbe interaction.* Springer.
Katz, L., & Donadio, S. (1993). Polyketide synthesis: Prospects for hybrid antibiotics. *Annual Review of Microbiology, 47*, 875–912.
Keller, N. P., Turner, G., & Bennett, J. W. (2005). Fungal secondary metabolism—From biochemistry to genomics. *Nature Reviews Microbiology, 3*(12), 937–947.
Khan, A., Kanwal, F., Ullah, S., Fahad, M., Tariq, L., Altaf, M. T., et al. (2025). Plant secondary metabolites—Central regulators against abiotic and biotic stresses. *Metabolites, 15*(4), 276.
Khan, H., Ullah, H., Khattak, S., Aschner, M., Aguilar, C. N., Halimi, S. M. A., et al. (2021). Therapeutic potential of alkaloids in autoimmune diseases: Promising candidates for clinical trials. *Phytotherapy Research, 35*(1), 50–62.
Kukula-Koch, W. A., & Widelski, J. (2017). Alkaloids. In *Pharmacognosy* (pp. 163–198). Academic Press.
Layer, G., Reichelt, J., Jahn, D., & Heinz, D. W. (2010). Structure and function of enzymes in heme biosynthesis. *Protein Science, 19*(6), 1137–1161.
Ludwiczuk, A., Skalicka-Woźniak, K., & Georgiev, M. I. (2017). Terpenoids. In *Pharmacognosy* (pp. 233–266). Academic Press.
Medema, M. H., Kottmann, R., Yilmaz, P., Cummings, M., Biggins, J. B., Blin, K., et al. (2015). The Minimum Information about a Biosynthetic Gene cluster (MIBiG) specification. *Nature Chemical Biology, 11*, 625–631.
Misawa, N. (2011). Pathway engineering for functional isoprenoids. *Current Opinion in Biotechnology, 22*(5), 627–633.
Mohamed, H., Hassane, A., Atta, O., & Song, Y. (2021). Deep learning strategies for active secondary metabolites biosynthesis from fungi: Harnessing artificial manipulation and application. *Biocatalysis and Agricultural Biotechnology, 38*, Article 102195.
Mohan, M. S., Salim, S. A., Pakhira, P., & Busi, S. (2024). Microbial production of polyketides and non-ribosomal peptides and their applications. In *Microbial products for health and nutrition* (pp. 365–390). Springer.
Molla, K. A., & Yang, Y. (2019). CRISPR/Cas-mediated base editing: Technical considerations and practical applications. *Trends in Biotechnology, 37*(10), 1121–1142.
Navarro-Muñoz, J. C., Selem-Mojica, N., Mullowney, M. W., Kautsar, S. A., Tryon, J. H., Parkinson, E. I., et al. (2020). A computational framework to explore large-scale biosynthetic diversity. *Nature Chemical Biology, 16*(1), 60–68.
Nawrot-Chorabik, K., Sułkowska, M., & Gumulak, N. (2022). Secondary metabolites produced by trees and fungi: Achievements so far and challenges remaining. *Forests, 13*(8), 1338.
O'Connor, S. E. (2015). Engineering of secondary metabolism. *Annual Review of Genetics, 49*(1), 71–94.
Panaccione, D. G., Beaulieu, W. T., & Cook, D. (2014). Bioactive alkaloids in vertically transmitted fungal endophytes. *Functional Ecology, 28*(2), 299–314.
Payne, J. A., Schoppet, M., Hansen, M. H., & Cryle, M. J. (2017). Diversity of nature's assembly lines–recent discoveries in non-ribosomal peptide synthesis. *Molecular BioSystems, 13*(1), 9–22.
Perveen, S., & Al-Taweel, A. (Eds.). (2018). Terpenes and terpenoids. BoD–Books on Demand.
Quin, M. B., Flynn, C. M., & Schmidt-Dannert, C. (2014). Traversing the fungal terpenome. *Natural Product Reports, 31*(10), 1449–1473.
Razin, S. V., Ioudinkova, E. S., Kantidze, O. L., & Iarovaia, O. V. (2021). Co-regulated genes and gene clusters. *Genes, 12*(6), 907.
Riedling, O., Walker, A. S., & Rokas, A. (2024). Predicting fungal secondary metabolite activity from biosynthetic gene cluster data using machine learning. *Microbiology Spectrum, 12*, e03400-e3423.
Schmidt-Dannert, C. (2015). Biosynthesis of terpenoid natural products in fungi. *Advances in Biochemical Engineering/Biotechnology, 148*, 19–61.

Shen, X., Wang, J., Li, C., Yuan, Q., & Yan, Y. (2019). Dynamic gene expression engineering as a tool in pathway engineering. *Current Opinion in Biotechnology, 59*, 122–129.

Shimizu, Y., Ogata, H., & Goto, S. (2017). Type III polyketide synthases: Functional classification and phylogenomics. *ChemBioChem, 18*(1), 50–65.

Shwab, E. K., Bok, J. W., Tribus, M., Galehr, J., Graessle, S., & Keller, N. P. (2007). Histone deacetylase activity regulates chemical diversity in *Aspergillus*. *Eukaryotic Cell, 6*(9), 1656–1664.

Shwab, E. K., & Keller, N. P. (2008). Regulation of secondary metabolite production in filamentous ascomycetes. *Mycological Research, 112*(2), 225–230.

Sinha, R., Sharma, B., Dangi, A. K., & Shukla, P. (2019). Recent metabolomics and gene editing approaches for synthesis of microbial secondary metabolites for drug discovery and development. *World Journal of Microbiology and Biotechnology, 35*, 1–14.

Skinnider, M. A., Merwin, N. J., Johnston, C. W., & Magarvey, N. A. (2017). PRISM 3: Expanded prediction of natural product chemical structures from microbial genomes. *Nucleic Acids Research, 45*(W1), W49–W54.

Sproul, D., Gilbert, N., & Bickmore, W. A. (2005). The role of chromatin structure in regulating the expression of clustered genes. *Nature Reviews Genetics, 6*(10), 775–781.

Strauss, J., & Reyes-Dominguez, Y. (2011). Regulation of secondary metabolism by chromatin structure and epigenetic codes. *Fungal Genetics and Biology, 48*(1), 62–69.

Suárez, I., Collado, I. G., & Garrido, C. (2024). Revealing hidden genes in *Botrytis cinerea*: New insights into genes involved in the biosynthesis of secondary metabolites. *International Journal of Molecular Sciences, 25*(11), 5900.

Süssmuth, R. D., & Mainz, A. (2017). Nonribosomal peptide synthesis—Principles and prospects. *Angewandte Chemie International Edition, 56*(14), 3770–3821.

Tajbakhsh, M., Karimi, A., Fallah, F., & Akhavan, M. M. (2017). Overview of ribosomal and non-ribosomal antimicrobial peptides produced by Gram positive bacteria. *Cellular and Molecular Biology, 63*(10), 20–32.

Wang, Z., Zhang, P., Wei, Y., Fang, Z., Zhang, L., Zhang, Y., et al. (2025). Metabolic and transcriptomic responses of Taxus mairei to nano-pollutants: Insights into AgNPs and PsNPs impact. *Medicinal Plant Biology, 4*, Article e010.

Weissman, K. J. (2009). Introduction to polyketide biosynthesis. *Methods in Enzymology, 459*, 3–16.

Winkelblech, J., Fan, A., & Li, S. M. (2015). Prenyltransferases as key enzymes in primary and secondary metabolism. *Applied Microbiology and Biotechnology, 99*, 7379–7397.

Wu, X., Cobbina, S. J., Mao, G., Xu, H., Zhang, Z., & Yang, L. (2016). A review of toxicity and mechanisms of individual and mixtures of heavy metals in the environment. *Environmental Science and Pollution Research, 23*, 8244–8259.

Xu, W., Gavia, D. J., & Tang, Y. (2014). Biosynthesis of fungal indole alkaloids. *Natural Product Reports, 31*(10), 1474–1487.

Younas, H., & Younas, F. (2025). Phytohormone production by endophytic fungi. In K. A. Abd-Elsalam & S. F. AbuQamar (Eds.), *Fungal endophytes*. (Vol. I). Springer.

Zhong, J. J., & Xiao, J. H. (2009). Secondary metabolites from higher fungi: Discovery, bioactivity, and bioproduction. Biotechnology in Chin I: From bioreaction to bioseparation and bioremediation. *Advances in Biochemical Engineering/Biotechnology, 113*, 79–150.

Zhou, J., Chen, X., & Li, S. M. (2024). Construction of an expression platform for fungal secondary metabolite biosynthesis in *Penicillium crustosum*. *Applied Microbiology and Biotechnology, 108*, 427.

Fungal Secondary Metabolites as Antibiotics

Shachi Bishnoi, Vishakha Rathi, Rita Sharma, Garima Rathi, Jagdip Singh Sohal, Anupam Prakash, and Umesh Kumar

Abstract

In addition to their primary function as antibiotics, microbial secondary metabolites are essential to modern biological technology and medicine. Throughout the idiophase, bacteria, fungi, and plants produce these biologically active substances, which are necessary for avoiding infections while promoting advanced healthcare procedures such as cancer treatments and certain surgeries. Secondary metabolic products can be described as either bacteriostatic, which prevents bacterial growth by interacting with protein synthesis or folate production, or bactericidal, which damages cell wall integrity and vital cellular functions. The ability of these systems to change is an important factor in the fight against antimicrobial resistance (AMR), which has become an increasing concern for global health, sustainability, food security, and healthcare systems. The overuse of antimicrobials in both agriculture and human health contributes to the growing challenge of AMR, so to combat this challenge, we need to think of new and better solutions. Recent developments in genomic data science, high-resolution mass spectrometry, and elucidation of biochemical pathways are helping to identify new secondary metabolite compounds that may be clinically useful in the treatment of multidrug-resistant infections. This review provides a thorough understanding of how secondary metabolites develop, what they do, and what they are contributing to our efforts to combat AMR by providing increased support for both the development of sustainable practices and the creation of novel methods for drug discovery to fight against antibiotic resistance.

Keywords

Antibiotic-resistant · Bactericidal agents · Penicillin · Secondary metabolites

1 Introduction

Antibiotics are substances produced by microbes, semi-synthetic antibiotics are modified forms of naturally occurring antibiotics using chemical processes to provide improved efficacy, stability, and safety (Koch et al., 2019). Antibiotics work in various ways to inhibit bacterial growth or destroy bacteria via interfering with cell wall synthesis and function, preventing protein synthesis, disrupting nucleic acid replication and transcription, or changing membrane integrity and function (Walsh, 2003). The term "antibiotic" was originally created by Selman Waksman in 1942, referring to compounds obtained from living organisms that inhibit the growth of other organisms (Waksman, 1947). Antibiotics that were originally discovered were only available from natural sources; with advances in science and technology, antibiotics have also been manufactured using chemical modifications to increase

S. Bishnoi
Quality Department, Peters Surgical India Pvt. Ltd, Gurugram, Haryana, India

V. Rathi
Forensic Science Laboratory, Rohini, Government of NCT Delhi, India

R. Sharma
Department of Biotechnology and Microbiology, School of Sciences, ITM University, Gwalior, Madhya Pradesh, India

G. Rathi
Delhi Public School Ghaziabad Meerut Road, Ghaziabad, Uttar Pradesh, India

J. S. Sohal · U. Kumar (✉)
University Centre for Research and Development, Chandigarh University, Mohali, Punjab, India
e-mail: umeshkumar82@gmail.com

A. Prakash
Department of Life Sciences, School of Biosciences and Technology, Galgotias University, Greater Noida, India

© The Author(s), under exclusive license to Springer Nature Switzerland AG 2026

A. Prakash and M. K. Dubey (eds.), *Advances in Fungal Secondary Metabolites: Bioactive Attributes and Biotechnological Potential of Industrial Interest*, Sustainable Economy and Ecotechnology, https://doi.org/10.1007/978-3-032-23507-7_3

their effectiveness, stability, and decrease their toxicities. Antibiotics come in many different chemical structures and can be classified as broad or narrow-spectrum based on their range of activity against different bacteria (Katzung et al. 2017).

Antibiotics are classified as either natural (i.e., derived from the natural products of organisms such as fungi and bacteria) or synthetic. The role of antibiotics in treating bacterial infections is of historic importance, as demonstrated by the work of Alexander Fleming in 1929. An understanding of these classifications is essential in determining antibiotic treatment options and in managing antimicrobial resistance (Fleming et al., 1929). Examples of major antibiotic classes include beta-lactams (penicillin and cephalosporins), aminoglycosides (streptomycin, gentamicin), tetracyclines, macrolides (erythromycin), and quinolones (ciprofloxacin) (Leekha et al., 2011). Not only do these medications treat a variety of conditions in people, but they can be used to manage medical issues in animals, as well as promote healthy growth of fish and provide protection against disease for livestock and crops (Davies & Davies et al., 2010).

The emergence of antibiotic-resistant bacteria has increasingly challenged the effectiveness of antibiotics. This has occurred because the misuse, overuse, and non-completion of prescribed courses of antibiotics continue to drive the evolution of resistant strains of bacteria such as Methicillin-resistant Staphylococcus aureus (MRSA) and Multidrug-resistant Mycobacterium tuberculosis (Ventola et al., 2015). As a result, global public health organizations advocate for the responsible use of antibiotics and the need for new classes of antimicrobial agents.

Thus, in the search for new antimicrobial agents, natural sources such as bacteria, Actinomycetes, and fungi have been studied extensively as possible sources of antimicrobial compounds. Fungi are of particular interest because they have historically contributed and continue to contribute to the discovery of antibiotics (Demain & Sanchez, 2009), and there is a continued interest in the ability of fungi to produce secondary metabolites to compete and survive in different environmental conditions.

Antibiotics act on different processes in bacteria compared to humans, thus killing the proper types of bacteria. This is accomplished through the disruption of cell wall structures, DNA and protein synthesis, and many other cellular processes. One type of antibiotic is broad-spectrum antibiotics, such as tetracycline, minocycline, azithromycin, etc., which will effectively treat multiple types of bacterial infections. Another type of antibiotic is narrow-spectrum antibiotics, which only treat certain types of bacterial infections (Demain, and Sanchez et al., 2009). With respect to combating the rising trend of antimicrobial resistance, researchers are working to identify new antibiotics, using bacteriophages as an alternative form of therapy, and developing more accurate diagnostics to aid in therapeutic decisions in treating patients. In addition to research, public health campaigns promoting the appropriate use of antibiotics are also being developed.

1.1 Fungal Antibiotics

Fungi often produce a particular type of chemical called secondary metabolites. Many of these secondary metabolites have strong antimicrobial properties. The primary function of these compounds is not to help with fungal growth, development, or reproduction; rather, they provide for ecological advantages such as competition against other microorganisms present in their environment (Bills & Gloer, 2016). The discovery of penicillin by Alexander Fleming in 1928 is perhaps the most widely known example of fungi being used as a source of antibiotics and marked the beginning of an "antibiotic era" (Fleming et al., 1929).

1.2 Penicillin and Its Derivatives

Penicillin is a type of beta-lactam antibiotic made from the fungus named Penicillium notatum (which was later renamed Penicillium chrysogenum). Penicillin inhibits bacterial cell wall production by interacting with transpeptidase, which is an enzyme that cross-links chains of peptidoglycan (Bennett & Chung et al., 2001). Penicillin was a critical turning point in the ability of humankind to control bacterial infections during World War II, and it opened a new era of controlling infections. After penicillin was successfully developed, thousands of hours of research were conducted to develop products with more activity, more stability, and the ability to resist being broken down by the β-lactamases (the degrading enzymes made by bacteria that can degrade penicillin) (Aminov, 2010).

Fungi have provided antibiotics for every medical practice since the start of modern medicine. Fungi have myriad types of bioactive compounds and produce many that can be used for the treatment of fungal disease. The successful development of penicillin opened up the field of exploration into the fungi that produced other antibiotics (Katz and Baltz et al., 2016; Clardy et al., 2009). In this chapter, we will briefly review the development of penicillin and of antibiotics derived from fungi and how fungi can be viewed as a primary source of natural product isolates.

1.3 Cephalosporins

Cephalosporins, another important group of antifungal agents, were originally sourced from the fungus Acremonium (previously Cephalosporium acremonium), and like penicillins, they are beta-lactam-type antifungals, which inhibit the synthesis of the bacterial cell wall. Compared to penicillins, cephalosporins have a broader antibacterial action and are more stable to the action of beta-lactamase enzymes (Abraham & Newton et al., 1961). With time, cephalosporins have been adapted into several generations, providing additional pharmacological benefits and therefore playing an important role in the treatment of numerous infections (Katzung, 2017).

1.4 Griseofulvin

In addition to antibacterial agents, fungi also produce compounds active against fungal infections. Griseofulvin, produced by *Penicillium griseofulvin*, is an antifungal agent that disrupts fungal cell mitosis by interfering with microtubule function (Sutton et al., 1994). It is primarily used in the treatment of dermatophyte infections such as ringworm, athlete's foot, and other skin-related fungal diseases. According to Berger and Paparella (2013), Griseofulvin was one of the first antifungal medications to be used in clinical settings. It is still used today despite the availability of more recent antifungal medications. Additional fungal antibiotics and new findings beyond penicillin, cephalosporins, and griseofulvin, fungi have yielded several other bioactive compounds with antimicrobial and pharmaceutical importance. For instance, *Penicillium brevicompactum* mycophenolic acid has immunosuppressive properties and is used to prevent graft rejection in organ transplantation (Bentley et al., 2000). *Aspergillus fumigatus*-derived fumagillin has been shown to have antiparasitic and antiangiogenic as well as anticancer properties (Olson et al., 2000).

With the rise of antibiotic resistance, there is renewed interest in exploring underexplored and extreme environments for novel fungal strains capable of producing unique antibiotics. According to Kusari et al. (2012), marine fungi and endophytic fungi, which reside within plant tissues without causing harm, have emerged as promising sources of novel secondary metabolites. Modern techniques such as genome mining, metabolomics, and synthetic biology are facilitating the discovery and characterization of previously inaccessible or silent biosynthetic gene clusters responsible for antibiotic production in fungi (Brakhage, 2013).

Fungal Antibiotics: Contributions to Health Care

The major contributions of fungal antibiotics have been monumental to the evolution of health care, and they continue to shape the way we treat both bacterial and fungal-related diseases. (Mahajan & Balachandran, 2015) Beginning with the initial discovery of penicillin, the study of the vast diversity of fungal organisms for the discovery of new bioactive compounds is still ongoing and will continue to be a crucial component of the discovery and development of new antibiotics. (Hossain et al., 2025) There is currently an increase in the incidence of antimicrobial resistance, which highlights the need for using the substantial chemical diversity of natural metabolites from fungi, along with recent advances in biotechnology, to facilitate the discovery of next-generation antibiotics and in the area of antibiotic development, below is a history of the contributions made.

2 Penicillin: Discovery and Development

2.1 Discovery of Penicillin

Penicillin's discovery by chance has changed the face of medicine. In 1928, Alexander Fleming made an incredible breakthrough in antibiotic discovery when he discovered that a fungus (Penicillium notatum or Penicillium chrysogenum) could protect the bacteria (Staphylococcus aureus) growing on his agar plate. He described the chemical secreted by the fungus as penicillin and recognized that it could stop bacterial growth. In the 1940s, Ernst Chain, Howard Florey, and their team at Oxford University took it a step further to isolate and produce penicillin in large quantities; this was a game-changer for medicine, specifically due to its application as a treatment of infected wounds and disease during WWII (Chain et al., 1945). Due to the mass use of penicillin during WWII, the number of people dying from infected wounds and diseases significantly decreased; that demonstrates how penicillin changed medicine forever. Penicillin's success also encouraged researchers to continue searching for additional antibiotics; many other antibiotics were developed to fight infectious diseases. Penicillin and other antibiotics saved countless lives due to bacterial infection.

Streptomyces griseus (also a type of bacteria), although described as a bacterium, stimulated fungal research, resulting in the discovery of streptomycin in 1945, and led to the creation of cephalosporins, broad-spectrum antibiotics effective against resistant strains, which were discovered in Cephalosporium acremonium (Abraham & Chain, 1940).

Other than penicillin and streptomycin, many important antibiotic discoveries have played important roles in the fight against bacterial infections. For example Chloramphenicol which is a broad-spectrum antibiotic was isolated from Streptomyces Venezuela in 1947 and found to have significant therapeutic value (Waksman & Woodruff, 1947)and there was erythromycin (also a broad-spectrum antibiotic) which was isolated from Saccharopolyspora erythraea in 1952 and is used to treat respiratory infection and skin infection (McGuire, 1952) as well as tetracyclines (another type of antibiotic that inhibits bacterial protein synthesis) derived from Stptomyces aureofaciens;‘s 1948 discovery (Duggar, 1948). All three of these antibiotics and their development had major impacts on medicine today.

2.2 Penicillin Production

The early stages of growing penicillin were characterized by low yields, difficulties purifying penicillin, and thus, after much screening and strain improvement, showed significantly increased levels of penicillium chrysogenum (Demain & Elander, 1999). The development of submerged fermentation technology, as well as the optimization of growth media, contributed significantly to increasing the scale of production for penicillin (Elander, 2003). Penicillin is classified as a β-lactam antibiotic produced through the secondary metabolism of fungi, particularly from strains of Penicillium chrysogenum (previously named P.notatum) (Chain et al., 1945; Fleming, 1929). The biosynthesis of penicillin involves three main steps; these are each catalyzed by different forms of enzymes, and all take place in a nitrogen-limited environment (Brakhage, 1998). Below is a list of the enzymes involved in the biosynthesis of penicillin.

- δ-(L-α-aminoadipyl)-L-cysteinyl-D-Valine synthetase (ACV synthetase)
- Acremonium chrysogenum Acyl-CoA: IPN acyltransferase (IAT)
- Phenylacetyl-CoA ligase (PCL)
- Isopenicillin N synthase (IPNS).

The assembly of a tripeptide precursor starts the biosynthesis of the penicillin antibiotics, and then cyclization occurs to produce the well-known β-lactam ring structure. The final steps in the biosynthetic pathway involve the chemical modifications of the β-lactam structure into the various penicillin antibiotics. The first step in the biosynthesis is the condensation of three amino acid precursors, L-α-aminoadipic acid, L-cysteine, and L-valine, into the tripeptide δ- (L-α-amino adipyl)-L-cysteinyl-D-valine (ACV). The major multifunctional enzyme complex referred to as δ- (L-α-amino adipoyl)-L-cysteinyl-D-valine synthetase (ACVS) initiates this process. ACV's creation is required in order to produce what will ultimately form the core structure of penicillins. Enzymes produced by Penicillium fungi also assist in making ACV by carrying out the necessary biochemical process.

- **Enzymes Involved:**
  - ACVS initiates the formation of the tripeptide ACV.
  - The transformation of ACV into isopenicillin N (the direct precursor to both penicillin G and penicillin V) is facilitated by the action of isopenicillin N synthase.
- **Formation of Isopenicillin N**

  Isopenicillin N synthase (IPNS), an iron-dependent enzyme, catalyzes the cyclization of tripeptide ACV into isopenicillin N through a series of chemical reactions involving the formation of a unique four-membered β-lactam and a five-membered thiazolidine ring. Since the β-lactam ring is essential for the antibacterial activity of many antibiotics, this process represents one of the most important stages in how these drugs work (Keller et al., 2005).
- **Conversion to Penicillin G and V**

  Isopenicillin N serves as the common precursor to the production of a number of different penicillins. As an illustration:
  - The side chain of penicillin G (also known as benzylpenicillin) incorporates phenyl acetate.
  - The side chain to penicillin V (also known as phenoxymethylpenicillin) is phenoxyacetic acid.

Side chain modifications are accomplished by acyltransferase enzymes, resulting in numerous penicillin compounds with distinct stabilities and activities. Synthesis of penicillin is regulated by many factors. External factors such as the abundance of carbon, nitrogen, and phosphate significantly control the synthesis of penicillin. According to Demain and Elander (1999), the level of penicillin production is affected by the level of nitrogen and glucose, as high level of nitrogen and glucose prevents its production. The production is also decreased by inhibiting the secondary metabolism of the fungi. Moreover, to maintain its efficient production, co-regulation of the gene clusters expressing the biosynthesis enzymes (such as *pcbAB, pcbC,* and *penDE*) is required.

Fungal secondary metabolism, illustrated by penicillin production, showcases the complex coordination of genetic regulation and biochemical pathways, highlighting its sophistication and industrial relevance (Brakhage, 2013). This procedure has been further optimized by utilizing the advancements of fermentation and genetic engineering, allowing for the large-scale manufacturing of penicillin to satisfy the demands of the global healthcare industry (Demain & Elander, 1999).

2.3 Fermentation

Fermentation techniques are widely used in biotechnology and industrial value; the two most common methods are

Solid-State Fermentation (SSF) and Submerged Fermentation (SmF) (Pankey & Sabath, 2004). Other methods, such as batch-fed fermentation and surface fermentation, do exist, albeit they are not as well-known. Because of its success in large-scale manufacturing and simplicity in nutrient control, SmF is the method of choice and widespread use in manufacturing sectors for the generation of penicillin (Demain & Elander et al., 1999).

The primary way of producing penicillin is fermentation, in which fungi such as *P. chrysogenum* are cultivated under closely controlled conditions to yield penicillin as a secondary product. From small-scale operations to highly efficient industrial systems, the method has changed over time (Fleming et al., 1945).

2.3.1 Submerged State Fermentation

- **Process**: The primary method for commercial-scale manufacturing of penicillin is through submerged fermentation (Krishna, 2005). This method involves the use of large-scale bioreactors filled with liquid media containing sufficient nutrients for growth.
- **Medium Composition**: The fermentation media will generally be based on the carbon sources of either glucose or lactose (which could include Corn Steep Liquor), combined with a nitrogen source such as ammonium salts. The growth of the fungus, as well as penicillin manufacturing, requires the addition of trace minerals (zinc and magnesium) as necessary components for successful growth and production.
- **Aeration and Agitation**: The aeration system and the agitation system of a submerged fermentation process provide the means to maintain oxygen levels as well as mix the components of the fermentation media thoroughly. Without adequate mixing and aeration, the fungus may have inadequate access to oxygen for its metabolism.

Advantages of SmF:

- Better parameter control, such as pH, temperature, oxygen, and nutrients, which are necessary for optimal production of penicillin
- Increased productivity, as SmF has been used on a large scale to produce penicillin from the fungal organism P. chrysogenum and other related fungi in relation to current methodologies of penicillin production.
- Greater efficiency of nutrient and gas mixing (throughout), allowing for proper function and development of the fungi, which allows for better penicillin production.
- Increased ability to automate and scale SmF technology for the industrial manufacture of penicillin.
- Greater efficiency of recovery/purification of penicillin during downstream processing of penicillin when using the liquid media of SMF compared to the solid media of penicillin cultivation.

Limitations of SmF:

- There's a high cost associated with production due to expensive bioreactors, constant aeration and agitation, along with the need for tightly controlled systems that contribute to high operational costs, as well as energy costs.
- Liquids are susceptible to microbial contamination, so the liquid medium must be sterilized and monitored.
- Small fluctuations in pH, temperature, oxygen level, and nutrient concentration could adversely affect the yield of penicillin.
- Shear forces from high aeration and/or agitation will damage the fungal mycelia, thereby affecting the growth and production of the fungus.
- Foam created during fermentation will require the addition of antifoaming agents, which have negative effects on oxygen transfer and will impede downstream processing.
- Continuous monitoring and control require skilled workers to do this, which adds further complications to the process.

2.3.2 *Solid-State Fermentation Technique*:

- **Process**: In SSF (Submerged State Fermentation), fungi are cultivated on solid substrates (i.e., wheat bran, rice husk, and soybean meal) under conditions with limited availability of moisture (Krishna, 2005). The method of submerged fermentation is used to produce penicillin in large fermenters that have been aerated and agitated with a capacity of between 50 000 and 200 000 L. Typical conditions during penicillin production will include a temperature maintained at 25 - 27 °C and a pH of 6.5–6.8 that is achieved by the addition of buffering agents. Further, aeration should be provided by spargers and agitators to supply sufficient oxygen for fungal growth and metabolic activity, achieved using antifoaming agents such as vegetable oils or silicone-based compounds. The process lasts 5–7 days, with penicillin secreted into the broth as a secondary metabolite during the stationary phase of fungal growth. While less common for industrial penicillin production, SSF is used in regions with limited access to liquid fermentation technology.

Advantages of solid-state fermentation are:

- Lower capital investment and operational costs
- suitable for smaller-scale production or research purposes.

Limitations of solid-state fermentation are:

- Difficulties in scaling up
- Challenges in maintaining uniform temperature and oxygen distribution.

Submerged fermentation process is highly preferred for the production of penicillin due to high productivity rate, better control of various parameters, efficient mixing with automation and scaling, which ultimately plays a significant role in downstream processing of penicillin recovery whereas, SSF is not ideal for penicillin on an industrial scale and it might be used in specific cases or for other types of products.

Fermentation optimization has played a crucial part in maximizing product yield, improving process efficiency, and hence in reducing costs in penicillin industries. Some of the major steps used to increase the penicillin production are discussed below.

- **Strain Improvement**: Genetic modification and mutagenesis have enhanced the productivity of *P. chrysogenum.* Modern strains can produce significantly higher penicillin yields compared to those used during the 1940s.
- **Environmental Control**: The best pH for maximum penicillin creation is at 6.5 to 7.5. The best range in terms of temperature will be from 25 - 28 degrees Celsius.
- **Precursor Feeds**: Precursor substrates will add to the concentration of the penicillin, thereby increasing the production of penicillin G and penicillin V.

2.4 Downstream Processing

The isolation and purification of penicillin must take place using solvent extraction and crystallization techniques to produce the pharmaceutical-grade product. The use of Submerged Fermentation (SmF's) in a liquid system allows for an easier recovery of penicillin than Solid-State Fermentation (SSF), which is more suitable for resource-constrained environments (Krishna., 2005).

- Submerged Fermentation (SmF): Refers to a type of fermentation in liquid form that yields penicillin in high volume. The growth of fungi and the production of penicillin increase with nutrient-rich media like lactose and glucose.
- Solid-State Fermentation (SSF): Primarily used in research and under-resourced regions; involves the use of solid-growing media for the growth of fungi. (Krishna., 2005).

3 Challenges in Penicillin Production

- The emergence of resistant bacterial strains like MRSA is a result of unregulated usage and overprescription of Penicillins (Ventola et al.,, 2015).
- It is difficult to develop new penicillin medications due to production costs and inefficient use against synthetic antibiotics.
- Penicillins transformed the treatment of infectious diseases and also spurred interest in the scientific community about other types of metabolites created by fungi that may possess antibacterial capabilities. This research showed that fungi may indeed be an enormous source of bioactive compounds, of which some could have antibacterial properties.

3.1 Fungi as a Major Source of Natural Antibiotics

Fungi have been known to produce many bioactive chemicals. Antibiotics like penicillin and cephalosporins come from a variety of different fungi. Fungi have changed medicine for the better by helping us treat various types of infections (Demain & Fang, 2000; Hawksworth, 2012). This shows the role fungi will play in the continued development of biotechnology and pharmaceuticals into the future.

3.2 Other Antibiotics from Fungi

After the introduction of penicillin as a successful antibiotic, other antibiotics derived from fungi or molds were found to treat certain types of infections. Examples of these other antomicrobials include the following:

- Griseofulvin is an antibiotic made from Penicillium spp. that is used to treat fungal infections in humans.
- Cephalosporins are broad-spectrum antibiotics obtained from Acremonium spp. that interfere with the production of bacterial cell walls.
- Fusidic acid is used for treating skin infections caused by Staphylococcus aureus and was originally isolated from Fusidium coccineum.

In addition to the antibiotics above, there are several other examples of pharmaceuticals derived from fungi, such as lovastatin produced by Aspergillus terreus and mycophenolic acid produced by Penicillium brevicompactum (Bentley et al., 2000; Olson et al., 2000).

3.3 Diversity and Potential of Fungi

Due to their remarkable diversity and adaptability, fungi are one of the richest sources of natural bioactive compounds. As produced by the majority of fungal species, these effectively inactive compounds are known to possess a variety of properties, including antimicrobials, antifungals, and antivirals (Al-Fakih & Almaqtri, 2019; Assaf et al., 2020; Guzmán-Guzmán et al., 2023). The considerable interest in identifying naturally occurring compounds within fungal genera such as Aspergillus, Penicillium, and Trichoderma has led to widespread research into their biological activity.

A systematic approach to screening fungal isolates from a range of environmental sources (e.g., soil, decaying organic matter, and marine ecosystems) has resulted in the identification of new compounds (Al-Fakih & Almaqtri, 2019; Assaf et al., 2020; Guzmán-Guzmán et al., 2023). Identifying fungal antibiotics is now being assisted by improved technologies, including genome sequencing and high-throughput screening.

Three main areas of recent drug discovery efforts focus on the exploration of novel sources, the refinement of discovery techniques, and the development of more effective means of overcoming resistance. Antimicrobial metabolites originating from microorganisms will continue to be a key component of the development of new antibiotic classes. However, the use of fungal antibiotics will face significant barriers, including the emergence of resistance, high production costs, and limited opportunities for growth. Future explorations of fungal biodiversity and hybridization between natural and synthetic biology present exciting and unexplored pathways for discovering and creating new drug leads.

3.4 Discovery and Development of New Antibacterial Compounds

The creation and discovery of new antibiotics are an essential part of fighting antibiotic resistance. It serves to address a number of unmet medical needs. Advances in natural product discovery, genomics, and AI are revitalizing the search for new antibacterial compounds. Collaborative efforts across academia, industry, and governments are essential to address the growing antibiotic resistance crisis.

3.4.1 Potential of Fungi in Antibiotic Discovery

Ascomycetes and Basidiomycetes, which make up filamentous fungus, have a broad and mainly unknown secondary metabolome. They are the source of numerous chemicals that are important in agriculture, industry, and medicine. Filamentous fungi have yielded several antibiotics. The first and second classes of antibiotics were cephalosporines like cephalosporine C and penicillins like penicillin V (Bérdy, 2005). Since they are both non-ribosomal peptides with a β-lactam ring (four-atom cyclic amide) that attaches to peptidoglycans and stops them from cross-linking, they are chemically extremely similar. This compromises the stability of the cell walls, particularly those of Gram-positive bacteria.

Antibiotics act by inhibiting protein production. The bacterial ribosome is composed of the 50 s and 30 s ribosomal subunits. The synthesis of the polypeptide chain begins once these two subunits link along the mRNA strand. Antibiotics must target the ribosomal 30 or 50 subunits in order to prevent the synthesis of proteins.

Antibiotics work against the ribosomal 305 subunits to prevent the synthesis of proteins:

Tetracycline antibiotics: This class of antibiotics targets the ribosomal 30S subunit's highly conserved 16S rRNA sequence. For example, tetracycline works by preventing t-RNA from attaching to the ribosome's A-site, which eventually prevents protein synthesis from occurring (Barrenechea et al., 2021).

Chloramphenicol antibiotics: block the activity of peptidyl transferase, an enzyme found on the 50S ribosomal subunit that is essential for the synthesis of proteins. This blockage stops t-RNA from binding to the A-site of the ribosome, resulting in the suppression of protein synthesis. (Uddin et al., 2021).

3.4.2 Look for Novel (Natural) Antimicrobial Substance

Newly developed antibiotics are typically saved for situations in which the current antibiotics are ineffective to prevent bacteria from eventually developing resistance to the novel compounds (Ventola, 2015). A significant bottleneck in natural product research is the rediscovery of previously identified molecules, which leads to an increased burden without producing useful results. Fortunately, High-Resolution Mass Spectrometry (HRMS) has become available and can be used instead of other established methods of analysis, such as Nuclear Magnetic Resonance (NMR) Spectroscopy and Pre-fractionation Techniques, for example, Liquid- and Solid-State Extracts, or C18 HPLC. These aforementioned technologies for molecular identification serve to reduce the incidence of copied molecules during the process of detecting new sources of bioactive chemicals, which has led to increasing success in identifying new sources of such compounds.

3.4.3 Approaches to Discovery Based on Informatics:

Those antibiotics originally developed using traditional methods of searching for new antibiotics—ie, "Grind and Find" methods—which rely on growing microorganisms and then extracting their chemicals - are increasingly being replaced by more sophisticated genomics approaches (Ziemert et al., 2016). The AntiSMASH program provides a web-based opportunity to investigate the rigid structures of many different groups of chemical compounds. Using anti-SMASH, scientists can construct an informative database of the required chemical building blocks and, using current methods of molecular modeling, construct a molecular model based upon the chemical structure of these chemical building blocks (Blin et al., 2021).

Soil metagenomic shotgun sequencing is an underdeveloped means of discovering new antibiotics, as deep sequencing remains a limiting factor for producing useful BGC data. Researchers recently developed a barcoding technique for analyzing BGC combinations between environmental samples, using degenerated polymerase chain reaction (PCR) primers via the Environmental Surveyor of Natural Product Diversity (eSNaPD) system (Reddy et al., 2014). Using eSNAPD data, researchers were able to identify calcium-dependent antibiotic BGCs associated with multiple soil metagenomes; notably, more than 70% of all BGCs sequenced through eSNAPD were unique to the sample's metagenome. This led to the identification of a novel family of antibiotics called malacidins, which did not become resistant in lab settings (Wrighton, 2018).

In order to combat multidrug-resistant *Mycobacterium tuberculosis*, a novel group of semi-synthetic rifamycins was recently produced. This is an example of a known modular BGC that assembles specific parts of existing antibiotics to produce novel variants of well-known antimicrobials. It is predicted that the number of pathways with biochemical characterizations is far smaller than the potential for secondary metabolite generation. Many bacteria have homologues of known genes involved in antibiotic manufacturing, including those encoding non-ribosomal peptide synthases and polyketide synthases, in their genomic sequences. For instance, only a small number of the 20 and 30 operons for secondary metabolites that the Actinomycetes, *Streptomyces coelicolor*, and *Streptomyces avermitilis*, respectively, harbor have been characterized. The discovery of silent or cryptic BGC that might not be conveyed under conventional laboratory cultures (Mantravadi et al., 2019).

4 Mechanisms of Action: Bactericidal Versus Bacteriostatic

Antibacterial activity can be categorized as either bacteriostatic or bactericidal. Antibiotics have a unique chemical framework, like a backbone, decorated with special groups that give them the power to fight and stop harmful microbes as shown in the Fig. 1. Several in vitro microbiological techniques, such as the serum bactericidal titter (SBT), time-kill curve, and minimum bactericidal concentration (MBC), (Table 1) are used to assess the bactericidal activity of antibacterial agents against various isolates. It's important to keep in mind that even bacteriostatic antibacterial kills bacteria, even if they don't meet certain lab requirements to be classified as bactericidal. The MBC is the minimum concentration of antimicrobial needed to provide a 3-log10 ($\geq$99.9% killing) reduction in surviving bacteria (cfu/mL) compared with the initial inoculum at 24 h of growth, while the MIC is the lowest concentration of antimicrobial that will prevent the visible growth of a microorganism at 24 h of growth in specific media. The MBC/MIC ratio can be used to determine whether an antibacterial agent is bacteriostatic or bactericidal in vitro. The impact is classified as bacteriostatic if the MBC/MIC ratio is greater than 4, and as bactericidal if it is less than 4. The kinetics of bacterial killing in vitro have been determined using time-kill curves, however they are not frequently employed to ascertain the bacteriostatic or bactericidal nature of antibacterial agents (Ishak et al., 2024; Pankey & Sabath et al., 2004).

Modern antibiotics, such as daptomycin, a lipopeptide antibiotic, also exhibit bactericidal activity by disrupting bacterial cell membranes.

5 Antifungal Secondary Metabolites

Fungi, as opportunistic pathogens, have severe implications for human health and crop productivity. These are exceptional producers of secondary metabolites that exhibit a wide range of bioactivities, including potent antifungal properties (Bodade & Lonkar, 2024). The majority of traditional antifungal drugs are fungistatic, meaning they do not kill the fungus but rather stop it from growing. This can lead to problems with relapse or incomplete eradication of the organism after treatment is finished. When an individual has a compromised immune system, achieving a cure becomes increasingly difficult, as halting fungal growth may not be enough to prevent the infection's spread. Many of the traditional antifungal drugs, such as polyenes and azoles, have been found to have significant resistance and toxicities (Cowen et al., 2015), leading to the need for new antifungal agents. One area of potential new antifungal agents is found

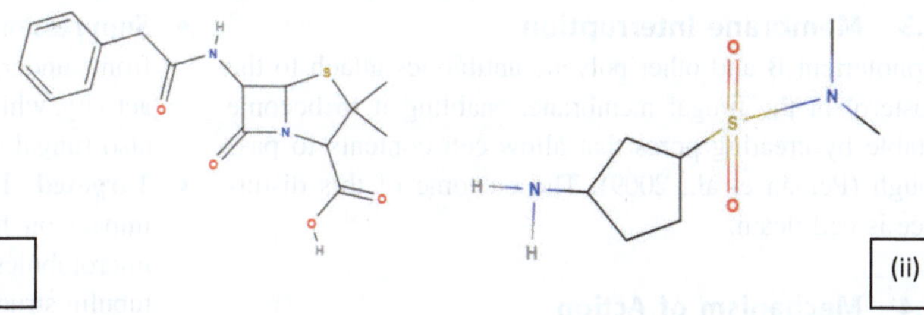

Fig. 1 Structure of (i) Penicillin and (ii) Sulfonamide. (*Source* Authors)

Table 1 Mechanisms of action and drug reactions of antibacterial agents

Antibacterial agent	Mechanism of action	Drug reaction
Bactericidal (e.g., Penicillin)	• Inhibits bacterial growth by preventing peptidoglycan cross-linking in the cell wall • Binds irreversibly to DD-transpeptidase, an enzyme essential for cross-linking, via its beta-lactam ring • This inactivation disrupts cell wall synthesis, leading to bacterial cell death	• Allergic reactions such as angioedema, pruritus, and urticaria, acute interstitial nephritis
Bacteriostatic (eg., Sulfonamides)	• Stop the early phases of folate secretion • Inhibit folic acid production and competitively block dihydropteroate synthetase (DHPS) • Disrupt the microbial production of folic acid at different stages of the biosynthesis process, which eventually results in the suppression of bacterial DNA synthesis	• Hypersensitivity responses are brought on by sulphonamides • Photosensitivity, drug fever, and rash are common mild responses

in the secondary metabolites of fungi. Secondary metabolites are non-essential life-sustaining chemicals produced by fungi that, while not essential for the life of the fungus, play a large role in many ecological functions, including deterrence of predator species, competition for space, and facilitation of spore dispersal (Hang et al., 2024). Understanding the ecological roles of these secondary metabolites will lead to the discovery of even better antifungal drugs, specifically targeting fungi resistant to currently used drugs (Pemán et al., 2009). Because of the ability of many fungi to produce antimicrobial compounds, numerous secondary metabolites are currently being researched as potential treatments for fungal infections of humans (Bhattacharya et al., 2020).

5.1 Mechanisms of Antifungal Activity

The mechanisms by which fungal secondary metabolites act to disrupt the growth and survival of fungal pathogens include a variety of methods. Some examples include

5.1.1 Breakdown of the Cell Wall

Antifungal agents work by disrupting the cell wall, which supports the integrity of the fungal cell. The cell membranes of fungi support shape, structure, and protection from environmental conditions. Fungal cells have a wall that contains polysaccharides that are not found in human cells (i.e., mannan, chitin, and glucan), and these materials are critical to maintaining the integrity of the cells and allowing for their proper growth and development in their environments (Cowen et al., 2015). For example, echinocandins inhibit the production of glucan (a primary component of the cell wall), leading to the death of the cell through lysis.

5.1.2 Mechanism of Action

Echinocandins, which include both Caspofungin, micafungin, and anidulafungin, target the synthesis of the polysaccharide β-1,3-D-glucan, which is an intricate component found within the fungal cell wall. The process used by echinocandin agents to achieve this synthesis is selective, as human cells do not produce β-1,3-D-glucan. This property is why Chosen Fungicide titles are able to limit unwanted consequences of treatment (Cowen et al., 2015).

- **Glucan Synthase Restriction**: Echinocandins target glucan synthase, which is responsible for the formation of β-1,3-D-glucan.
- **Cell Wall Interruptions**: Echinocandins interfere with glucan synthase's ability to produce glucan from glucose connected to the cell membrane, resulting in interruptions in the physical structure of the fungal cell wall due to inadequate production of β-1,3-D-glucan
- **Cell Death and Lysis**: When the cell wall gets damaged, and becomes less resistant to outside stressors, which eventually leads to lysis of the cells, and when interfering with the fungal life cycle, this technique kills the fungus.

5.1.3 Membrane Interruption

Amphotericin B and other polyene antibiotics attach to the ergosterol in the fungal membrane, enabling it to become unstable by creating pores that allow cell contents to pass through (Pemán et al., 2009). The outcome of this disturbance is cell death.

5.1.4 Mechanism of Action

- **Ergosterol binding**: Ergosterol is highly preferred by amphotericin B. After being given, it attaches itself to ergosterol molecules and incorporates into the fungal cell membrane.
- **Pore Formation**: Once bound to ergosterol, amphotericin B begins to form multi-molecular aggregates that traverse the lipid bilayer of the fungi's cell membrane. The complexes represent physical barriers between the inside and outside of the fungal cell and create a series of pores/channels through which material can move freely between the two environments.
- **Loss of Cellular Contents**: As a result of the formation of these pores, potassium ions and other electrolyte ions, as well as a variety of small organic molecules, begin to leak out of the fungal cells uncontrollably. Consequently, the osmotic balance of the fungal cell ceases, resulting in severe metabolic dysfunction.
- **Cell Death**: The consequence of the fungal cell membrane disruption is to result in a loss of structural integrity of the fungal cell, leading to cell death via dehydration (water loss), inability to maintain homeostasis, and depletion of energy reserves.

5.1.5 Inhibition of Nucleic Acid Synthesis

Griseofulvin and other secondary metabolites affect the ability of fungi to divide and grow through inhibition of cell division through microtubule disruption. An example of this would be how griseofulvin disrupts fungal cell division by affecting the formation of microtubules. The application of griseofulvin as a form of antifungal therapy is one way to show how targeting specific structural/functional elements that exist only in fungi can lead to further development of herbal-based therapies that might be used to treat infections caused by mushroom-based organisms without damaging or harming non-fungal life forms (Cowen et al., 2015).

5.1.6 Mechanism of Action

- **Binding to Microtubules**: Griseofulvin binds to tubulin, a protein that polymerizes to form microtubules. This binding disrupts the assembly of the mitotic spindle, which is crucial for chromosome segregation during mitosis.
- **Suppression of Mitosis**: Grizeofulvin stops fungal cells from undergoing mitosis by disrupting cytoskeleton activity, which results in the stoppage of cell division and also fungal growth.
- **Targeted Toxicity**: Griseofulvin reduces its negative impact on human cells by selectively attacking fungal microtubules since fungi and mammals have different tubulin structures.

5.1.7 Restriction of Enzymes

Many of the secondary metabolic products of antifungals block enzymes that regulate fungal life. For example, azoles block lanosterol 14α-demethylase, which prevents the synthesis of ergosterol necessary to maintain the stability of fungal membranes. In some cases, azoles limit ergosterol development necessary for the integrity of the membrane by blocking lanosterol 14α-demethylase (Pemán et al., 2009).

5.1.8 Mechanism of Action

- **Addressing 14α-Demethylase of Lanosterol**: Azoles attack lanosterol 14α-demethylase by adhering to it with their azole ring, and then it forms a complex with the heme group of the enzyme, and this prevents the enzyme from demethylating lanosterol, considered an essential step in the production of ergosterol, as given in the Fig. 2.
- **Ergosterol Synthesis Inhibited**: Once lanosterol 14α-demethylase is blocked, ergosterol, which is necessary for preserving membrane fluidity and integrity, is lost. Toxic sterol intermediates accumulate in and cause the fungal membrane to become unstable.
- **Membrane Instability and Fungal Death**: Without ergosterol, the fungal membrane becomes compromised, resulting in increased permeability, loss of essential ions and molecules, cellular dysfunction, and eventual cell death.

6 Clinical Applications of Antifungal Compounds

Antifungal secondary metabolites are routinely used in clinical practice to treat a variety of fungal infections. Antifungal compounds are crucial for treating fungal infections, which range from superficial conditions like athlete's foot to life-threatening systemic infections in immunocompromised individuals. The choice of antifungal agents depends on the infection's severity, the fungal species, and patient-specific factors. Some of the most well-known compounds include.

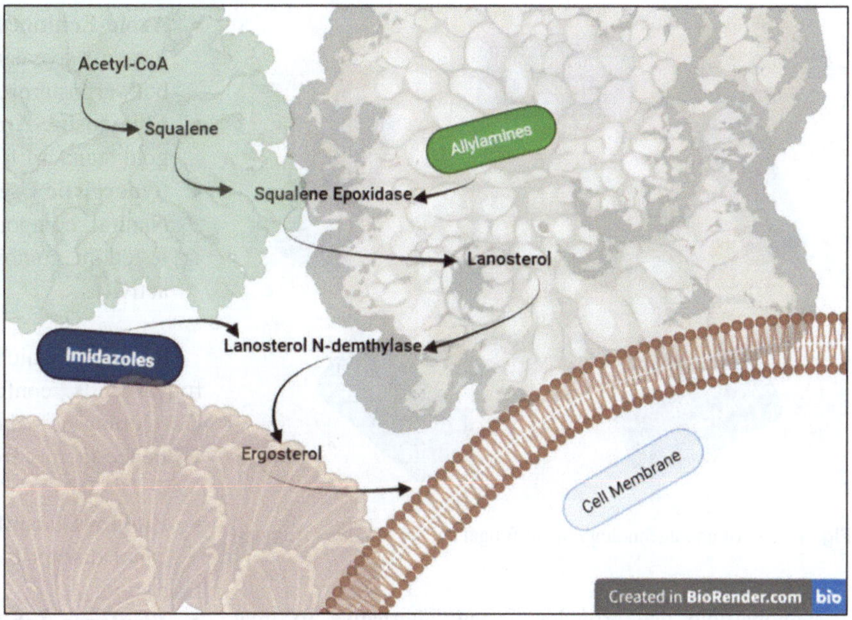

Fig. 2 Mechanism by which azoles inhibit lanosterol 14α-demethylase, preventing ergosterol synthesis (*Source* Authors)

6.1 Polyene Antibiotics

Amphotericin B is a potent antifungal agent used for severe systemic infections, including those caused by *Cryptococcus neoformans* and *Aspergillus* spp. Despite its efficacy, its nephrotoxicity limits its use. Polyene antibiotics are also effective against Candida species, especially azole-resistant strains (*Candida glabrata* and *C. krusei*) (Maertens, 2016), Used in invasive candidiasis and esophageal candidiasis, and also an alternative or adjunct in aspergillosis therapy.

6.2 *Azoles (E.G., Fluconazole, Voriconazole, Posaconazole)*:

- **Fluconazole**: Widely used for mucosal infections (e.g., oropharyngeal candidiasis, vulvovaginal candidiasis). Effective against *Cryptococcus neoformans* in meningitis.
- **Voriconazole**: First-line treatment for invasive aspergillosis.
- **Posaconazole**: Prophylaxis in high-risk neutropenic patients or those undergoing hematopoietic stem cell transplants (Perfect, 2017).

Fluconazole and *Itraconazole* are widely prescribed for treating superficial and systemic fungal infections. They act by inhibiting the synthesis of ergosterol, disrupting the integrity of the fungal cell membrane.

6.3 Echinocandins

These compounds, such as *Caspofungin*, are used to treat infections caused by *Candida* spp., particularly in immunocompromised patients. They target the synthesis of β-glucan, essential for the fungal cell wall. Other examples are, e.g., Micafungin, Anidulafungin

6.4 Griseofulvin

Used mainly for dermatophyte infections, *Griseofulvin* interferes with fungal cell division and is effective in treating conditions like ringworm (Del Rosso & Gupta, 2013). However, its use is limited by side effects such as gastrointestinal disturbances and the need for prolonged treatment.

7 Emerging Trends in Antifungal Therapies

Fungal infection management is being advanced through the implementation of innovative approaches that improve the effectiveness of antifungals while simultaneously minimizing the potential adverse consequences of antifungal drugs and addressing antifungal resistance. The following are three of the most significant advances in the management of fungal infections, with additional examples and literature.

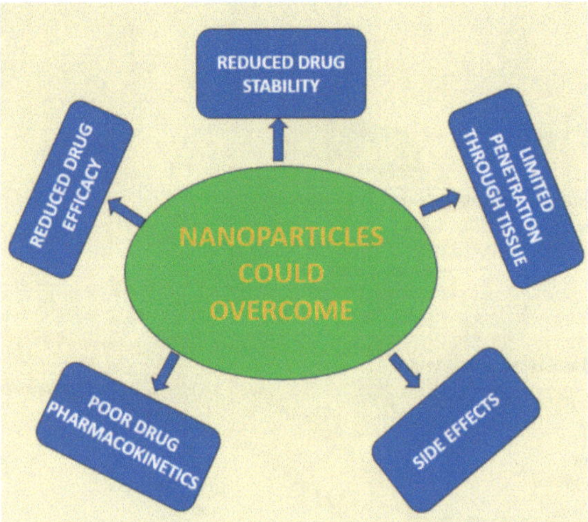

Fig. 3 Use of nanotechnology in antifungal delivery (*Source* Authors)

Nanoparticle use provides a new alternative to treat fungal infections. Because of the unique characteristics of nanoparticles, they provide an increased inhibition capability compared to traditional antifungal drug treatments, while providing the same effective dose when administered at a lower concentration than standard antifungal drugs, resulting in an improved level of performance. Nanoparticles may be used to deliver higher concentrations of antifungals than possible with standard methods and can penetrate deeper into fungal biofilms and, consequently, provide higher levels of bioavailability. A report shows that silver nanoparticles increased the activity of fluconazole against Candida albicans and that, because of their superior efficacy, the dose needed to achieve effective results was lower than that with conventional methods. The result of this will be reduced side effects of antifungal therapy (Hussain et al., 2019) (Fig. 3).

- **Liposomal Amphotericin B**: A liposomal formulation of amphotericin B reduces nephrotoxicity and is used for systemic fungal infections.
- **Chitosan Nanoparticles**: Zinc-chitosan nanocomposites enhance antifungal activity against plant pathogens, offering potential applications in human therapeutics.
- **Silver Nanoparticles**: These are being explored for their antifungal properties and ability to synergize with traditional antifungal agents.

To increase the efficacy of treating fungal infections and to prevent or delay antifungal resistance, combination therapy using antifungal agents with different modes of action is recommended. The benefits of this approach include reducing the amount of drug(s) needed and thus minimizing the potential for toxicity. Examples of combination therapies include:

- **Azole-Echinocandin Combinations**: Used for refractory invasive candidiasis and aspergillosis, as they target both ergosterol synthesis and cell wall biosynthesis.
- **Flucytosine-Amphotericin B**: This combination is the gold standard for treating cryptococcal meningitis.
- **Synergistic Use of Natural Products and Antifungals**: Natural compounds, such as essential oils, are being tested in combination with azoles to boost antifungal activity.

Biofilms, which are protective structures formed by fungal cells, confer resistance to traditional antifungals. Developing new therapies based on disrupting biofilms and virulence factors from fungi (Flemming, 2016).

- **Biofilm Disruptors**: Compounds like farnesol and dodecanol disrupt biofilms, enhancing antifungal drug penetration.
- **Virulence Inhibitors**: Targeting fungal enzymes (e.g., proteases, hydrolases) and signaling pathways (e.g., quorum sensing) reduces pathogenicity without affecting fungal viability, limiting resistance.
- **CRISPR-Cas9 Technology**: CRISPR-Cas9 provides a method to investigate fungal virulence and biofilm development.

8 Resistance Mechanisms in Pathogens

Infections caused by fungi can develop resistance to antifungal medications, like the way bacteria develop antibiotic resistance, which will help reduce effective treatment options available today and thus lead to unsuccessful attempts to treat the infected patient and require alternative forms of treatment (Czajka et al., 2023).

The antifungal medications currently available are categorized into four main categories based on their mode of action. Antifungal drugs may block the synthesis of important biological molecules (for example, flucytosine), disrupt the membrane barrier (for instance, polyenes), prevent ergosterol production through their action on ergosterol biosynthesis pathways (for example, allylamines, thiocarbamates, azole derivatives, and morpholines), or disrupt the microtubule function (griseofulvin) (Araújo et al., 2024).

8.1 Evolution of Antibiotic Resistance

Several steps are taken when developing the antifungal resistance method via fungal mutation methodologies.

- **Targeted Mutation**: The mutations in a fungus' genome (in an enzyme), such as the enzymes that generate ergosterol, would decrease the effective treatment effect of antifungal medications (ie, decrease drug binding to the enzyme). A specific example of this is the mutation of the 14-alpha demethylase enzyme caused by the use of azole medications. The 14-alpha demethylase enzyme is essential in creating fungal ergosterols. It is made from the ERG11 gene that encodes the 14-alpha demethylase enzyme (Jahanshiri et al., 2022). By forming different conformational shapes (modifying their basic structure) due to mutations that occur within the ERG11 gene, the 14-alpha demethylase enzyme reduces azole medications' ability to bind to the enzyme, thus creating azole resistance. These changes are often followed by further processes to increase resistance that include upregulation of efflux pumps and modifications in membrane composition (Li et al., 2021).

- **Efflux Pumps**: The efflux mechanism is one of the most common ways that many fungi can use to get rid of the antifungal drugs that are found within the cell. As antifungal drugs are pumped out by efflux pumps, these proteins are found within the cell membrane of many fungal cells. As antifungal drugs are pumped out of fungal cells, the amount of antifungal drugs found in the cells is reduced, and hence, the antifungal drugs become less effective against those cells (Fig. 4). Resistance to antifungal drugs in fungi is primarily due to the function of efflux pumps. By pumping antifungal drugs out of fungal cells, these proteins reduce the concentration of antifungal drugs that are present in the cells, which means that the drug is not as effective when treating the infection. Efflux pumps can be classified into two main categories:

1. **ATP-Binding Cassette (ABC) Transporters**: Such pumps strongly move medicines and other materials from the cell through ATP hydrolysis. Cdr1p and Cdr2p from Candida albicans represent important examples.
2. **Major Facilitator Superfamily (MFS) Transporters:** All of these propel drug efflux through a proton gradient. Mdr1p from the yeast *Candida albicans* serves as an example (Czajka et al., 2023).

Antifungal therapy often triggers efflux pumps to become excessively expressed, leading to the emergence of multidrug resistance (MDR). Research on antifungal resistance focuses heavily on comprehending the management and functioning of these pumps in order to produce regulators that would reverse their impact and re-establish antifungal efficacy (Abbes et al., 2021). The technique through which efflux pumps progressively decrease the effectiveness of antifungal medications:

- **Biofilm Formation**: Many types of fungi have developed biofilm capabilities, which consist of a single layer of fungal cells surrounded by a gelatinous substance called an extracellular matrix. This biofilm provides a layer of protection for many fungal cells against antifungal therapies. These biofilms are commonly associated with numerous healthcare-associated infections that involve the utilization of implanted medical devices and prosthetics (Czajka et al., 2023) (Fig. 5).
- **Altered Membrane Composition**: Changing the amount of sites that can bind antifungals decreases a fungi's susceptibility to antifungal drugs (including polyene & azole antifungals). This is how fungi develop resistance to many different types of antifungal drugs. Changes in the lipid composition (including sterols) and protein content of the fungal cell membrane are thought to be the primary ways through which antifungal resistance occurs. Different types of antifungal medications will not work as effectively if they target aspects of the fungal cell membrane (Araújo et al., 2024).

9 Strategies to Overcome Resistance Using Fungal Compounds

It is critical to develop effective strategies to address antifungal resistance in order to maintain the efficacy of antifungal medications. Fungal resistance to antifungal medications is an important problem in the clinical setting. In addition, by utilizing fungal substances (either directly or as a source of inspiration for new medications), there is an opportunity to develop novel methods to address antifungal resistance. The research community is currently researching a variety of different approaches to combat antifungal resistance, some of which include:

9.1 Combination Therapy with Fungal-Derived Synergistic Agents

- **Mechanism**: Pairing existing antifungal drugs with fungal-derived compounds that enhance their efficacy or inhibit resistance mechanisms.
- **Examples**:
  - Fungi produce secondary metabolites called fumagillin that can act on the growth and metabolism of fungi (Choudhary et al., 2022).
  - Gliotoxins are an example of the way that fungi can enhance the azole class of antifungals by modifying the effects of oxidative stress on the fungus (Wang et al., 2023).

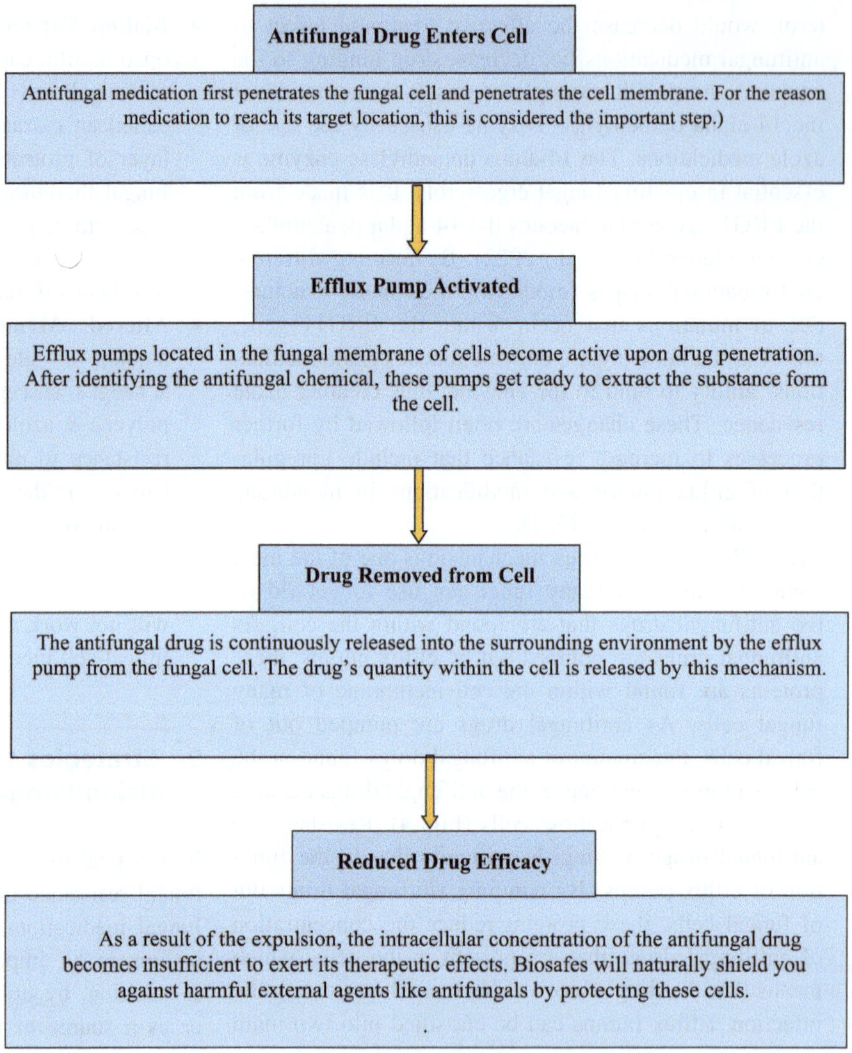

Fig. 4 Antifungal agents may be actively exported by fungal cells via the use of efflux pumps. This greatly impairs the efficiency of antifungals (*Source* Authors)

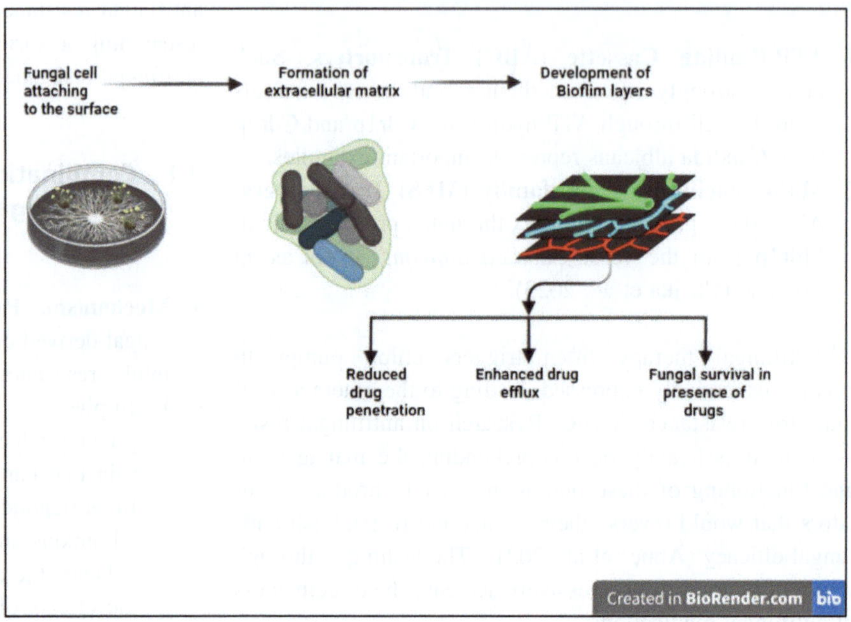

Fig. 5 Biofilms help protect fungi from being treated by antifungal medications (*Source* Authors)

- **Benefits**:
  - Overcomes biofilm-associated resistance.
  - Minimizes dosage of medications to lessen associated toxicity (side effects).

9.2 Inhibitors of Efflux Pumps

- **Mechanism**: Fungal compounds can be used to inhibit efflux pumps, preventing drug expulsion and restoring intracellular drug concentrations.
- **Examples**:
  - **Cladosporin**: Inhibits fungal ATP synthase, potentially impacting energy-dependent efflux mechanisms (Choudhary et al., 2022).
- **Benefits**:
  - Combats multidrug-resistant (MDR) fungal strains.

9.3 Targeting Cell Wall or Membrane Biosynthesis

- **Mechanism**: Utilize fungal compounds that inhibit essential pathways for cell wall or membrane synthesis.
- **Examples**:
  - Echinocandins are a type of lipopeptide derived from fungi that are able to stop or lower the production of β-glucan synthase, which makes up part of the cell wall of fungi (Bodade & Lonkar, 2024).
- **Benefits**:
  - More effective than conventional antifungal agents against both resistant and biofilm-forming fungal strains.

9.4 Disruption of Stress Response Pathways

- **Mechanism**: Many resistant fungi upregulate stress response pathways to survive antifungal treatment. Fungal compounds can target these adaptive responses.
- **Examples**:
  - **Radicicol**: A fungal metabolite that inhibits heat shock protein 90 (Hsp90), a critical player in fungal stress response and resistance (Raffa & Keller, 2019).
- **Benefits**:
  - Enhances susceptibility to existing antifungals.

9.5 Modulation of Fungal Signaling Pathways

- **Mechanism**: Fungal compounds can interfere with quorum sensing or signaling pathways that regulate resistance mechanisms.
- **Examples**:
  - **Farnesol**, a quorum-sensing molecule produced by *Candida albicans*, inhibits biofilm formation and enhances drug sensitivity (Kong et al., 2017).
- **Benefits**:
  - Reduces virulence and resistance in fungal communities.

9.6 Development of Novel Antifungal Agents

- **Mechanism**: By isolating novel fungal metabolites that contain antifungal characteristics, a completely new class of antifungal drugs will be created.
- **Examples**:
  - **Myriocin**, produced by the fungus Myriococcum albomyces, targets sphingolipid biosynthesis in fungi (Choudhary et al., 2022).
- **Benefits**:
  - The development of new types of antifungal drugs using the method above fills existing gaps in the existing antifungal arsenal.

9.7 Enhancing Host Immune Response

- **Mechanism**: Fungal compounds can act as immunomodulators, enhancing the host's ability to fight fungal infections.
- **Examples**:
  - **Pneumocandins**, precursors of echinocandins, boost host immune responses (Bodade & Lonkar, 2024).
- **Benefits**:
  - Supports antifungal drug action by leveraging the immune system.

10 Potential in Combating Multidrug-Resistant Infections

Emerging multidrug-resistant (MDR) fungal diseases due primarily to infections with Candida and Aspergillus species are becoming an increasing public health issue. Multiples of both the above fungi have shown resistance to many classes of antifungal medications, resulting in a very limited number of therapeutic options; this creates a serious and increasing threat to global public health (Perlin et al., 2017). Combating these diseases will require an extensive strategy that combines novel therapeutics, preventive measures, and policy intervention strategies.

10.1 Exploration of Novel Metabolites for MDR Pathogens

The discovery of novel fungal secondary metabolites is crucial for addressing MDR infections. There are several promising compounds have been isolated from fungi, such as Fungal Peptides (Antifungal peptides from fungi, such as those from *Trichoderma* spp., show broad-spectrum activity against MDR fungi by disrupting cell wall integrity) (Goyal et al., 2021), Terpenoids and Alkaloids (such are produced by *Penicillium* spp., exhibit antifungal activity through multiple mechanisms, including membrane disruption and inhibition of ergosterol biosynthesis) (Ding et al., 2020) and Endophytes (Fungi from plant tissues, known as endophytes, are being investigated for their potential to produce novel bioactive metabolites. Some endophytes produce compounds that are effective against MDR fungi (Venugopalan & Srivastava, 2015).

10.1.1 Examples of Promising Novel Metabolites

- **Teixobactin**: A naturally occurring antibiotic with activity against Gram-positive MDR pathogens by binding to highly conserved regions of bacterial peptidoglycan precursors (Ling et al., 2015).
- **Lugdunin**: A peptide antibiotic produced by *Staphylococcus lugdunensis* that shows activity against *Staphylococcus aureus* (Zipperer et al., 2016).
- **Malacidins**: A family of cyclic peptides discovered through metagenomic analysis, effective against Gram-positive bacteria, including MDR strains (Hover et al., 2018).

11 Combination Therapies with Fungal Antibiotics

Combination therapies have gained attention as a strategy to treat MDR infections. By targeting multiple pathways within the pathogen, combination therapies can enhance efficacy and reduce the risk of resistance development (Perfect, 2017). When utilized together with other antifungal medications, fungal secondary metabolic products can exhibit synergistic effects that enhance the therapeutic potency and reduce resistance risk. By employing medications that disrupt many fungal mechanisms, such as DNA replication, cell wall formation, and membrane integrity, combination therapy can more successfully overcome resistance by focusing on several routes (Roemer & Krysan, 2014). Although preserving the therapeutic benefits, minimizing the prescribed amount of each medication in the combination therapy may reduce the risk of overdose and adverse effects. For instance, the combination of echinocandins and azoles has shown synergistic effects against *Candida* species, improving treatment outcomes (Kontoyiannis & Lewis, 2017).

11.1 Advantages of Combination Therapies

1. Range of Action: Works with microorganisms having various resistance mechanisms or infections that are in combination (Whaley et al., 2017).
2. Reduction of Resistance Development: Addressing multiple systems in series minimizes the potential of resistance forming (Perfect, 2017).
3. Enhanced Efficacy: Synergistic combinations could be far more effective in overcoming infections than monotherapy (Roemer & Krysan, 2014).
4. Decreased toxic effects: Using smaller dosages of certain therapies to achieve the planned effect, negative effects can be reduced (Lewis & Kontoyiannis, 2017).
5. Management of Biofilm-Associated Diseases: Biofilms being formed are naturally immune to specific drugs and can be broken down with coupled therapy (Sherry et al., 2017).

Although promising, combination therapies have several types of issues that need to be properly addressed to ensure both safety and efficacy. Drug interactions can lead to antagonism, compromising treatment because the combined effect of the medications is less beneficial than their individual actions. Another risk associated is toxic substances, particularly with heavy medications like amphotericin B, which can have substantial side effects (Odds et al., 2003). These threats can rise if other medications are consumed at the same time. The price as well as availability of the combination medicines sometimes restricts their usage in resource-constrained instances in which advanced treatments may be both economically and logistically unaffordable. In addition, to prevent undesirable outcomes or negative consequences, dosage and timing optimization is a challenging process that calls for an in-depth knowledge of pharmacokinetics and pharmacodynamics. The majority of combination medications have the objective to prevent resistance; misuse or insufficient dosage could lead to bacteria with resistance developing, making potential therapies harder to attain. Addressing the challenges to combination therapies and increasing the value of these types of drugs, and also decreasing the risks associated with combination drugs/therapy, requires extensive research, clinical validation, and strategic use. The measurement of the effectiveness of the duration of oral intake can be a complex process that requires knowledge of pharmacokinetics and pharmacodynamics in order to avoid negative side effects and severe complications. While the goal of many combination drugs is to prevent the development of resistance, misuse or overuse of drugs may result in the formation of resistant pathogens, which would complicate future treatment of infection. Addressing these barriers to combination drug use and maximizing the potential benefits of combination drugs requires extensive research, clinical validation, and strategic use.

12 Conclusion

Microbial secondary metabolites (metabolites produced by the use of antibiotics) represent an underutilized source of potential solutions to emerging issues such as AMR (Antimicrobial Resistance) due to their functionality as antibiotics. With the rapid growth of technologies in genomics, informatics, and molecular identification, researchers are now positioned to discover new antibiotic compounds with various modes of action. Unique to many of these antibiotics is their absence of pre-existing resistance, since they target non-essential factors in a microbe's life cycle. Current methods of targeting emerging organisms through metagenomics and self-resistance analysis may allow researchers to identify and exploit beneficial compounds produced by fungi and bacteria. To mitigate the potential negative impact of AMR on global health, agricultural and medical legislation must support strict usage requirements for antibiotics, while also integrating traditional methods with new technology. Continued research and development on secondary metabolites will be essential in ensuring the maintenance of healthcare systems and the effectiveness of the antibiotics used within these systems. Future research should focus on identifying, utilizing, and developing new antibiotics through the combination of innovative technology, utilizing untapped and under-researched fungi and bacteria, and promoting integration of multiple disciplines in order to expedite the identification of new antimicrobial compounds.

References

Abbes, S., Sellami, H., Neji, S., Trabelsi, H., Makni, F., & Ayadi, A. (2021). Implication of efflux pumps and *ERG11* genes in resistance of clinical *Trichosporon asahii* isolates to fluconazole. *Journal of Medical Microbiology, 70*(3), Article 001236.

Abraham, E. P., & Chain, E. (1940). An enzyme from bacteria able to destroy penicillin. *Nature, 146*(3713), 837.

Abraham, E. P., & Newton, G. G. F. (1961). The structure of cephalosporin C. *Biochemical Journal, 79*(2), 377.

Al-Fakih, A. A., & Almaqtri, W. Q. A. (2019). Overview on antibacterial metabolites from terrestrial *Aspergillus* spp. *Mycology, 10*(4), 191–209.

Aminov, R. I. (2010). A brief history of the antibiotic era: Lessons learned and challenges for the future. *Frontiers in Microbiology, 1*, 134.

Araújo, T. C., de Araújo, G. F. L., Santos, K. da S., Veloso, T. R. M., Costa, L. R., Dourado, J. dos S., ... Serra, I. C. P. B. (2024). Mecanismos de resistência antifúngica: exploração dos aspectos moleculares e genéticos. *Foco, 17*(9), e006.

Assaf, C. E. H., Zetina-Serrano, C., Tahtah, N., Khoury, A. E., Atoui, A., Oswald, I. P., Puel, O., & Lorber, S. (2020). Regulation of secondary metabolism in the penicillium genus. *International Journal of Molecular Sciences, 21*(24), 9462.

Barrenechea, V., Vargas-Reyes, M., Quiliano, M., & Milón, P. (2021). A complementary mechanism of bacterial mRNA translation inhibition by tetracyclines. *Frontiers in Microbiology, 12*.

Bennett, J. W., & Chung, K. T. (2001). Alexander Fleming and the discovery of penicillin. *Advances in Applied Microbiology, 49*, 163–184.

Bentley, R. (2000). Mycophenolic acid: A one-hundred-year odyssey from antibiotic to immunosuppressant. *Chemical Reviews, 100*(10), 3801–3826.

Bérdy, J. (2005). Bioactive microbial metabolites. *The Journal of Antibiotics, 58*(1), 1–26.

Berger, R. S., & Paparella, A. S. (2013). Topical antifungal agents for dermatophytosis. *Journal of the American Academy of Dermatology, 29*(1), 99–102.

Bhattacharya, S., Sae-Tia, S., & Fries, B. C. (2020). Candidiasis and mechanisms of antifungal resistance. *Antibiotics, 9*(6), 312.

Bills, G. F., & Gloer, J. B. (2016). Biologically active secondary metabolites from the fungi. *Microbiology Spectrum, 4*(6).

Blin, K., Shaw, S., Kloosterman, A. M., Charlop-Powers, Z., Van Wezel, G. P., Medema, M. H., & Weber, T. (2021). antiSMASH 6.0: Improving cluster detection and comparison capabilities. *Nucleic Acids Research, 49*(W1), W29–W35.

Bodade, R. G., & Lonkar, K. (2024). Antifungal activity of microbial secondary metabolites. In *Microbial Biotechnology in Healthcare and Sustainable Agriculture* (pp. 105–128). Elsevier.

Brakhage, A. A. (1998). Molecular regulation of B-Lactam biosynthesis in filamentous fungi. *Microbiology and Molecular Biology Reviews, 62*(3), 547–585.

Brakhage, A. A. (2013). Regulation of fungal secondary metabolism. *Nature Reviews Microbiology, 11*(1), 21–32.

Chain, E., Florey, H. W., Gardner, A. D., Heatley, N. G., Jennings, M. A., Orr-Ewing, J., & Sanders, A. G. (1945). Penicillin as a chemotherapeutic agent. *The Lancet, 246*(6320), 226–228.

Choudhary, M., Kumar, V., Naik, B. S., Verma, A., Saris, P. E. J., Kumar, V., & Gupta, S. (2022). Antifungal metabolites, their novel sources, and targets to combat drug resistance. *Frontiers in Microbiology, 13*, 1061603.

Clardy, J., Fischbach, M. A., & Currie, C. R. (2009). The natural history of antibiotics. *Current Biology, 19*(11), R437–R441.

Cowen, L. E., Sanglard, D., Howard, S. J., Rogers, P. D., & Perlin, D. S. (2015). Mechanisms of antifungal drug resistance. *Cold Spring Harbor Perspectives in Medicine, 5*(7), Article a019752.

Czajka, K. M., Venkataraman, K., Brabant-Kirwan, D., Santi, S. A., Verschoor, C., Appanna, V. D., Singh, R., Saunders, D. P., & Tharmalingam, S. (2023). Molecular mechanisms associated with antifungal resistance in pathogenic *Candida* species. *Cells, 12*(22), 2655.

Davies, J., & Davies, D. (2010). Origins and evolution of antibiotic resistance. *Microbiology and Molecular Biology Reviews, 74*(3), 417–433.

Del Rosso, J. Q., & Gupta, A. K. (2013). Oral griseofulvin therapy in dermatology. *Journal of Clinical and Aesthetic Dermatology, 6*(9), 10–16.

Demain, A. L., & Elander, R. P. (1999). The β-Lactam Antibiotics: Past, Present, and Future. *Anton Van Leeuwenhoek, 75*(1–2), 5–19.

Demain, A. L., & Fang, A. (2000). The natural functions of secondary metabolites. *Advances in Biochemical Engineering/biotechnology, 69*, 1–39.

Demain, A. L., & Sanchez, S. (2009). Microbial drug discovery: 80 years of progress. *The Journal of Antibiotics, 62*(1), 5–16.

Ding, Z., Tao, T., Wang, F., Zhao, X., Yuan, J., & Dai, Y. (2020). Antifungal activities of terpenoids and alkaloids from *Penicillium* species. *Frontiers in Microbiology, 11*, 2010.

Duggar, B. M. (1948). Aureomycin: A product of the continuing search for new antibiotics. *Annals of the New York Academy of Sciences, 51*(2), 177–181.

Elander, R. P. (2003). Industrial production of β-lactam antibiotics. *Applied Microbiology and Biotechnology, 61*(5–6), 385–392.

Fleming, A. (1929). On the antibacterial action of cultures of a Penicillium, with special reference to their use in the isolation of B. influenzae. British Journal of Experimental Pathology, 10(3), 226–236.

Fleming, A. (1945). The Penicillin Production Process. *British Journal of Medicine, 2*(436), 645–650.

Flemming, H. C. (2016). Biofilms: An emergent form of bacterial life. *Nature Reviews Microbiology, 14*(9), 563–575.

Goyal, S., Lambert, C., Clapp, M., & Timonen, S. (2021). Antifungal peptides from *Trichoderma* spp.: Mechanisms and applications. *Applied Microbiology and Biotechnology, 105*(15), 5905–5921.

Guzmán-Guzmán, P., Kumar, A., de los Santos-Villalobos, S., Sánchez-González, E., Cruz-Morales, P., & Villegas, J. (2023). *Trichoderma Species: Our Best Fungal Allies in the Biocontrol of Plant Diseases — A Review. Plants, 12*(3), Article 3032.

Hang, S., Lu, H., & Jiang, Y. (2024). Marine-derived metabolites act as promising antifungal agents. *Marine Drugs, 22*(4), 180.

Hawksworth, D. L. (2012). Global species numbers of fungi: Are tropical studies and molecular approaches contributing to a more robust estimate? *Biodiversity and Conservation, 21*(9), 2425–2433.

Hossain, M. S., Al Amin, M., Islam, S., Imam, H., Das, L. C., & Mahmud, S. (2025). Analyzing the challenges and opportunities associated with harnessing new antibiotics from the fungal microbiome. *Frontiers in Fungal Biology, 6*, 1476.

Hover, B. M., Kim, S.-H., Katz, M., Charlop-Powers, Z., Owen, J. G., Ternei, M. A., Maniko, J., Estrela, A. B., Molina, H., Park, S., & Brady, S. F. (2018). Culture-independent discovery of the malacidins as calcium-dependent antibiotics with activity against multidrug-resistant Gram-positive pathogens. *Nature Microbiology, 3*(4), 415–422.

Hussain, M. A., Ahmed, D., Anwar, A., Perveen, S., Ahmed, S., Anis, I., Shah, M. R., & Khan, N. A. (2019). Combination therapy of clinically approved antifungal drugs is enhanced by conjugation with silver nanoparticles. *International Microbiology, 22*(2), 277–285.

Ishak, A., Mazonakis, N., Spernovasilis, N., Akinosoglou, K., &Tsioutis, C. (2024). Bactericidal versus bacteriostatic antibacterials: Clinical significance, differences and synergistic potential in clinical practice. *Journal of Antimicrobial Chemotherapy, dkae380.*

Jahanshiri, Z., Manifar, S., Arastehnazar, F., Hatami, F., & Lotfali, E. (2022). Azole resistance in *Candida albicans* isolates from oropharyngeal candidiasis is associated with *ERG11* mutation and efflux overexpression. *Jundishapur Journal of Microbiology, 15*(11), Article e131046.

Katz, L., & Baltz, R. H. (2016). Natural product discovery: Past, present, and future. *Journal of Industrial Microbiology & Biotechnology, 43*(2–3), 155–176.

Katzung, B. G. (2017). *Basic and clinical pharmacology* (14th ed.). McGraw-Hill Education.

Keller, N. P., Turner, G., & Bennett, J. W. (2005). Fungal secondary metabolism—from biochemistry to genomics. *Nature Reviews Microbiology, 3*(12), 937–947.

Koch, A. L. (2019). *Bacterial growth and form.* Springer Science & Business Media.

Kong, E. F., Tsui, C., Kucharíková, S., Van Dijck, P., & Jabra-Rizk, M. A. (2017). Modulation of *Staphylococcus aureus* response to antimicrobials by the *Candida albicans* quorum sensing molecule farnesol. *Antimicrobial Agents and Chemotherapy, 61*, e01573-e1617.

Kontoyiannis, D. P., & Lewis, R. E. (2017). Combination chemotherapy for invasive fungal infections: What laboratory and clinical studies tell us so far. *Drug Resistance Updates, 32*, 1–17.

Krishna, C. (2005). Solid-state fermentation systems—An overview. *Critical Reviews in Biotechnology, 25*(1–2), 1–30.

Kusari, S., Hertweck, C., & Spiteller, M. (2012). Chemical ecology of endophytic fungi: Origins of secondary metabolites. *Chemistry & Biology, 19*(7), 792–798.

Leekha, S., Terrell, C. L., & Edson, R. S. (2011). General principles of antimicrobial therapy. *Mayo Clinic Proceedings, 86*(2), 156–167.

Lewis, R. E., & Kontoyiannis, D. P. (2017). Invasive aspergillosis in glucocorticoid-treated patients. *Medical Mycology, 55*(1), 82–89.

Li, J., Coste, A. T., Liechti, M., Bachmann, D., Sanglard, D., & Lamoth, F. (2021). Novel *ERG11* and *TAC1b* mutations associated with azole resistance in *Candida auris. Antimicrobial Agents and Chemotherapy, 65*(3), e02663-e2720.

Ligon, B. L. (2004). Penicillin: Its discovery and early development. *Seminars in Pediatric Infectious Diseases, 15*(1), 52–57.

Ling, L. L., Schneider, T., Peoples, A. J., Spoering, A. L., Engels, I., Conlon, B. P., Mueller, A., Schäberle, T. F., Hughes, D. E., Epstein, S., Jones, M., Lazarides, L., Steadman, V. A., Cohen, D. R., Felix, C. R., Fetterman, K. A., Millett, W. P., Nitti, A. G., Zullo, A. M., Lewis, K., et al. (2015). A new antibiotic kills pathogens without detectable resistance. *Nature, 517*(7535), 455–459.

Maertens, J. A., et al. (2016). Posaconazole versus voriconazole for primary treatment of invasive aspergillosis. *The Lancet, 387*(10020), 1611–1620.

Mahajan, G. B., & Balachandran, L. (2015). Antibacterial agents from actinomycetes - A review. *Frontiers in Bioscience (Elite Edition), 7*(2), 577–609.

Mantravadi, P. K., Kalesh, K. A., Dobson, R. C. J., Hudson, A. O., & Parthasarathy, A. (2019). The quest for novel antimicrobial compounds: Emerging trends in research, development, and technologies. *Antibiotics, 8*(1), 8.

Mc Guire, J. M. (1952). Erythromycin, a new antibiotic. *Antibiotics and Chemotherapy, 2*(9), 281–283.

Odds, F. C., Brown, A. J. P., & Gow, N. A. R. (2003). Antifungal agents: Mechanisms of action. *Trends in Microbiology, 11*(6), 272–279.

Olson, A. C., Chen, J., & Ruan, S. (2000). Fumagillin: An antibiotic with antiangiogenic activity. *Journal of Biological Chemistry, 275*(34), 25291–25295.

Pankey, G. A., & Sabath, L. D. (2004). Clinical relevance of bacteriostatic versus bactericidal mechanisms of action in the treatment of gram-positive bacterial infections. *Clinical Infectious Diseases, 38*(6), 864–870.

Pemán, J., Cantón, E., & Espinel-Ingroff, A. (2009). Antifungal drug resistance mechanisms. *Expert Review of Anti-Infective Therapy, 7*(5), 453–460.

Perfect, J. R. (2017). The antifungal pipeline: A reality check. *Nature Reviews Drug Discovery, 16*(9), 603–616.

Perlin, D. S., Rautemaa-Richardson, R., & Alastruey-Izquierdo, A. (2017). The global problem of antifungal resistance: Prevalence, mechanisms, and management. *The Lancet Infectious Diseases, 17*(12), e383–e392.

Raffa, N., & Keller, N. P. (2019). A call to arms: Mustering secondary metabolites for success and survival of an opportunistic pathogen. *PLoS Pathogens, 15*(4), Article e1007606.

Roemer, T., & Krysan, D. J. (2014). Antifungal drug development: Challenges, unmet clinical needs, and new approaches. *Cold Spring Harbor Perspectives in Medicine, 4*(5), Article a019703.

Sherry, L., Rajendran, R., Lappin, D. F., Borghi, E., Perdoni, F., Falleni, M., Tosi, D., Smith, K., Williams, C., Jones, B., & Ramage, G.

(2017). Biofilm-forming capability of highly virulent, multidrug-resistant *Candida auris*. *Emerging Infectious Diseases, 23*(2), 328–331.

Sutton, D. A., Fothergill, A. W., & Rinaldi, M. G. (1994). *Guide to clinically significant fungi*. Williams & Wilkins.

Uddin, T. M., Chakraborty, A. J., Khusro, A., Zidan, B. R. M., Mitra, S., Emran, T. B., Dhama, K., Ripon, Md. K. H., Gajdács, M., Sahibzada, M. U. K., Hossain, Md. J., & Koirala, N. (2021). Antibiotic resistance in microbes: History, mechanisms, therapeutic strategies and future prospects. *Journal of Infection and Public Health, 14*(12), 1750–1766.

Ventola, C. L. (2015). The antibiotic resistance crisis: Part 1. *Pharmacy and Therapeutics, 40*(4), 277–283.

Venugopalan, A., & Srivastava, S. (2015). Endophytic fungi: Novel sources of anticancer lead molecules. *Applied Microbiology and Biotechnology, 99*(2), 571–582.

Waksman, S. A. (1947). What is an antibiotic? *Mycologia, 39*(5), 565–569.

Waksman, S. A., & Woodruff, H. B. (1947). Actinomyces venezuelae, a new soil organism antagonistic to pathogenic and non-pathogenic microorganisms. *Journal of Bacteriology, 54*(5), 549–556.

Walsh, C. (2003). *Antibiotics: Actions, origins, resistance*. ASM Press.

Wang, J., et al. (2023). Gliotoxin and related metabolites as zinc chelators: Implications and exploitation to overcome antimicrobial resistance. *Essays in Biochemistry, 67*(2), 263–275.

Whaley, S. G., Berkow, E. L., Rybak, J. M., Nishimoto, A. T., Barker, K. S., & Rogers, P. D. (2017). Azole antifungal resistance in *Candida albicans* and emerging non-*albicans Candida* species. *Frontiers in Microbiology, 7*, 2173.

Wrighton, K. H. (2018). Discovering antibiotics through soil metagenomics. *Nature Reviews Drug Discovery, 17*(4), 241.

Ziemert, N., Alanjary, M., & Weber, T. (2016). The evolution of genome mining in microbes—A review. *Natural Product Reports, 33*(8), 988–1005.

Zipperer, A., Konnerth, M. C., Laux, C., Berscheid, A., Janek, D., Weidenmaier, C., Burian, M., Schilling, N. A., Slavetinsky, C. J., Marschal, M., Willmann, M., Kalbacher, H., Schittek, B., Brötz-Oesterhelt, H., Grond, S., & Peschel, A. (2016). Human commensals producing a novel antibiotic impair pathogen colonization. *Nature, 535*(7613), 511–516.

Fungal Secondary Metabolites in Cancer Therapy

Rupsa Roy, Sampurna Dutta, and Arun Kumar Dutta

Abstract

Nature-derived substances have significantly contributed to fighting against cancer for the past fifty years. Among these, the use of medicinal plants and microorganisms stands out as significant examples. Products derived from fungi constitute more than half of all clinical medications used worldwide. Over 60% of clinically approved cancer therapies originated from natural sources. Fungi represent the largest group within the eukaryotic kingdom that are capable of synthesizing numerous secondary metabolites. In recent years, numerous studies have been conducted by researchers to investigate the anti-tumor properties of secondary metabolites derived from fungus. These compounds include terpenoids, alkaloids, phenyl quinines, ligase, isocoumarin, steroids, polysaccharides, and propanoids. Notably, fungal polysaccharides have gained more attention due to their distinctive biological activities. The enzyme asparaginase extracted from strain such as *Aspergillus terreus, A. niger, A. tubingensis, Trichoderma viride,* and *Tamaromyces pinophilis* showed highest anticancer efficacy. Various genera of *Penicillium* also possess the significant quantity of secondary metabolites exhibiting anticancerous properties. This chapter explores the screening of various fungal secondary metabolites exhibiting cytotoxic properties, elucidating the mechanism of action in cancer therapy, and discussing the current status of fungal metabolites in clinical studies.

R. Roy
Department of Botany, West Bengal State University, Barasat, Kolkata, India

S. Dutta · A. K. Dutta (✉)
Molecular and Applied Mycology Laboratory, Department of Botany, Gauhati University, Guwahati, Assam, India
e-mail: arun.botany@gmail.com

Keywords

Anticancerous compound · Clinical trials · Fungi · Secondary metabolite · Treatment

1 Introduction

Cancer is a life-threatening condition resulting from the unregulated proliferation of cells and is one of the primary causes of mortality worldwide (Saravanakumar et al., 2019). It remains one of the most formidable threats to human life, with an estimated 7.6 million fatalities annually and 12.7 million patients being affected every year (Chen et al., 2016). Chemotherapy is commonly employed in cancer treatment; yet, its significant adverse effects are of considerable concern (Yan et al., 2017). Thus, naturally derived resources are explored as anticancer agents. Over the past fifty years, these bioresources substances have aided humanity in the fight against cancer. Approximately 60% of clinically approved oncological pharmaceuticals are of natural origins, with fungal-derived compounds accounting for about 50% of clinically utilized pharmaceuticals worldwide (Shamsaei et al., 2021). Fungi are regarded as a diversified source of secondary metabolites among microorganisms. Recent investigations have suggested various effects of these substances. They have demonstrated antioxidant, antimicrobial, anti-inflammatory, anti-aging, and immunomodulatory efficacy. Particularly, in recent years, extensive studies have been undertaken on the anti-tumor properties of substances produced from fungi. Among these, bioactive chemicals include terpenoids, alkaloids, polysaccharides, and isocoumarins (Salehi et al., 2018).

Fungal-derived bioactive compounds are categorized into two principal groups: low-molecular-weight substances such as phenolic compounds and terpenoids and large molecular weight substances including polysaccharides and various

© The Author(s), under exclusive license to Springer Nature Switzerland AG 2026

A. Prakash and M. K. Dubey (eds.), *Advances in Fungal Secondary Metabolites: Bioactive Attributes and Biotechnological Potential of Industrial Interest*, Sustainable Economy and Ecotechnology, https://doi.org/10.1007/978-3-032-23507-7_4

enzymes (Matuszewska et al., 2019). Among these, polysaccharide extracted from *Astragalus membranaceus* exhibited promising anti-tumor properties and immunomodulating effects by reduction of cell proliferation, cell cycle arrest, and modulation of apoptotic-related genes (Li et al., 2019). Moreover, various cytotoxic metabolites have been isolated from various marine fungal sources. For example, a compound known as simplicilliumtide isolated from the marine fungus *Simplicillium obclavatum* has demonstrated cytotoxic properties against human leukemia cell lines (Deshmukh et al., 2018). Similarly, a diterpenoid compound, asperolide E, derived from the deep-sea fungus *Aspergillus wentii,* exhibits cytotoxic properties against human breast cancer cell lines (Deshmukh et al., 2018). Another secondary metabolite, Ophiobolin A, obtained from the endophytic fungus *Bipolaris* sp., also exhibits anticancer properties (Maehara et al., 2020).

Thus, a plethora of anticancer compounds has been isolated from diverse fungal sources. This chapter primarily focuses on the identification of fungal compounds with cytotoxic properties. It delves into the details of fungal-derived secondary metabolites as anticancer agents, elucidates the mechanism of action of fungal secondary metabolites in cancer treatment, and discusses clinical trials associated with fungal-derived cancer therapies.

2 Identification of Fungal Compounds with Cytotoxic Properties

Identification of fungal compounds with cytotoxic properties entail the use of anti-proliferative assay using different cancer cell lines. High throughput screening aids in speeding up the assessment process of the compounds with possible anticancer activity. Marine fungi, particularly genus *Penicillium* and *Aspergillus,* are well known to produce a variety of natural products with cytotoxic activity (Gao et al., 2024). Endophytic fungi are also good sources of cytotoxic substances. They are also a rich reservoir of beneficial products that can be altered in order to create new options for chemotherapeutic agents (Uzma et al., 2018).

2.1 Screening Techniques for Anticancer Activity

Several methods are employed to assess the anticancer properties of the metabolites synthesized by fungi with the following being particularly noteworthy (Fig. 1).

2.1.1 Evaluation of Cytotoxicity of the Fungal Metabolites

The 3-(4, 5-dimethylthiazol-2-yl)-2, 5-diphenyltetrazolium bromide test (MTT) is extensively utilized in cytotoxicity research owing to its great precision, speed, and relative ease of use. The MTT assay is a colorimetric test that utilizes the conversion of the MTT molecule into formazan in the presence of enzymes produced by viable cells. The extent of colour change of the MTT molecule is assessed against a control set to estimate the proportion of viable cells present after treatment with different concentrations of the test compounds (Ganot et al., 2013). The data is represented in terms of IC_{50} values that determine the anticancerous efficacy of the compound.

2.1.2 Nuclear Integrity Assessment Through DAPI Staining

Apoptosis is a genetically regulated cellular process that induces structural as well as metabolic changes in the cells. Diverse cellular events modify the DNA bringing about changes in the nucleus and consequently affecting the entire cell, resulting in dysfunction of the organism (Atale et al., 2014). Studies have shown that the treatment of numerous cancer cell lines with different fungal extracts, followed by DAPI staining, leads to significant morphological alterations in the cells. This encompasses cytoplasmic shrinkage and chromatin condensation, as evidenced by treatment with the ethanolic extract of *Phellinus ignarius* (Wang et al., 2018).

2.1.3 Apoptotic Morphological Detection by Dual Staining Assay

The dual staining assay is a rapid and cost-effective technique to evaluate apoptotic damage in cancer cells after exposure to fungal extract. Acridine orange permeates both non-viable and viable cells, emitting green fluorescence upon intercalation with DNA, and red fluorescence upon associating with RNA. Ethidium-bromide intercalates with DNA and emits red fluorescence when cellular membranes are altered. Four distinct cell types can be identified based on their fluorescence and the structural changes in the nucleus:

1. Vibrant green nuclei arranged in orderly manner.
2. The cells approaching early apoptosis exhibit irregularly shaped green nucleus, with condensed chromatin manifesting as luminous green apoptotic bodies.
3. The cells approaching late apoptosis display orange to red colored nucleus with significantly dispersed chromatin.
4. Nucleus displaying a consistent orange to red coloration with a structured arrangement, resulting from necrotic cells (Ciniglia et al., 2010).

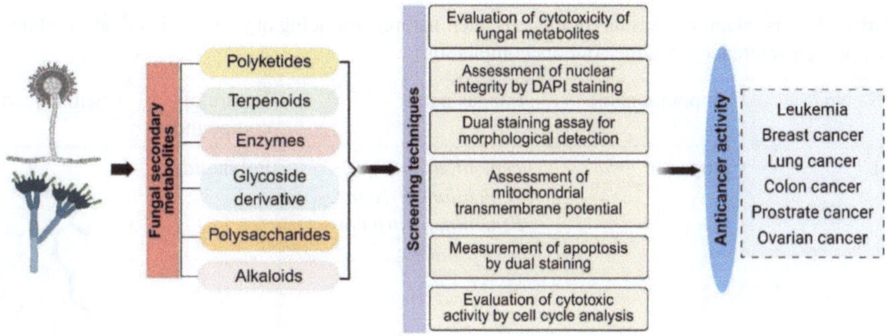

Fig. 1 Overview of fungal secondary metabolites and screening techniques used to evaluate their anticancer activity against various cancer types (Source *Authors*)

2.1.4 Assessment of Mitochondrial Transmembrane Potential

A comprehensive understanding of apoptotic signaling pathways and the causes of apoptosis resistance is important for developing new therapeutic targets. The integration of potential-dependent probes with fluorescence microscopy facilitates the investigation of transmembrane potentiality within the mitochondrial membrane of both single living cells and large cell populations. Permanent cationic fluorescent probes are extensively utilized to assess mitochondrial transmembrane potential and its variations (Zorova et al., 2018). These methodologies for analyzing membrane potential are highly valuable for gaining insights into both fundamental energy metabolism and its malfunction in cancer cells.

2.1.5 Measurement of Apoptosis by Dual Staining

The apoptotic impact of the fungal extract is evaluated using Fluorescein isothiocyanate (FITC) and propidium iodide (PI) staining. To determine the proportion of live cells post-treatment with extract, PI, a membrane-impermeable dye typically rejected from live cells, is employed. The cells are grown in a medium with varying doses of the fraction for different timepoints (Fang et al., 2006b). Further, analysis of the obtained data is represented in the form of dot plots that depicts the tumor cells undergoing apoptosis. The overall cells undergoing apoptosis include both early and late apoptotic phase cells, which are studied using a flow cytometer.

2.1.6 Evaluation of Cytotoxic Activity of Fungal Metabolites Through Cell Cycle Analysis

Unregulated cellular proliferation, a hallmark of cancer and cancerous cells, generally exhibits gene mutations that directly govern the cell cycle (Sherr, 1996). Apoptosis is a meticulously regulated self-immolative process, while the cell cycle comprises several highly organized phases that culminate in cell division. The commencement of the cell cycle, enabling cellular proliferation, is contingent upon the G1 checkpoint, which assesses whether cells have reached the S phase. This process is governed by cell cycle linked proteins, namely CDKs and cyclins. Cyclin D1 is a crucial regulator that facilitates the G1 to S phase cell transition. Enhanced levels of CDK4, CDK 6, and cyclin D1 signal prime chromosomes for replication (Marzec et al., 2006). The overexpression of cyclin D1 correlates with the onset and advancement of certain malignancies, and several anticancer drugs have demonstrated effectiveness in facilitating the degradation of the cyclin (Alao, 2007). Mushroom extracts cause arrest of the cell cycle by the inhibition of CDKs and cyclin D1, as observed in the human colorectal adenocarcinoma cell line when treated with ethanolic extract of *Innotus obliquus*, resulting in reduced expression of cdk4, cdk2, and the cyclin D1 levels and thereby inducing cell cycle arrest at the G1 phase (Lee et al., 2015).

2.2 Database of Known Fungal Secondary Metabolites as Anticancer Agents

Purified bioactive compounds extracted from medicinal mushrooms represent a significant category of novel anticancer drugs. In fungi, secondary metabolite produced through specific biosynthetic pathways can alter cell shape and induce DNA fragmentation, ultimately resulting in the death of the cancer cells (Noman et al., 2021). Among the fungal secondary metabolites, high molecular weight polysaccharides isolated from members of Basidiomycetes exhibit anticancerous properties by inhibition of metastasis following immunomodulatory pathways, whereas the low molecular weight secondary metabolite derived from fungi exhibits potential anticancer effects by targeting numerous processes, cell cycle regulation, metastasis, and signal transduction pathways (Balde et al., 2010). A comprehensive tabulation of some of the anticancerous compounds produced by fungi, along with their nature, structure (PubChem) and the types of cancer cells they affect is presented (Table 1):

Table 1 List of anticancerous compounds, their nature, producing organism, chemical structure, and activity against cancer types along with their supporting references (*Source* Authors compilation)

S. No.	Compound name	Fungus name	Nature of compound	Chemical structure	Types of cancer	Referencew
1	Compactin	*Penicillium solitum, Penicillium hirsutum, Penicillium citrinum*	Polyketides	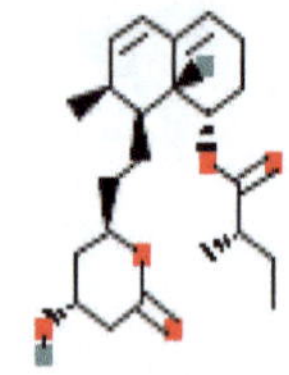	Acute myeloid leukemia	Frisvad and Filtenborg (1989), Larsen et al. (2007)
2	Lovastatin	*Aspergillus terreus*	Statins	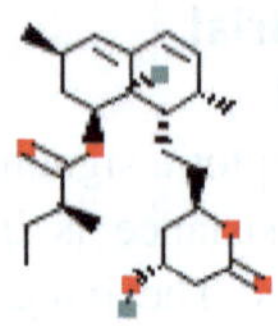	Cancer in regions of breast, lung, liver, and colon	Xie et al. (2021)
3	Simvastatin	*Aspergillus terreus*	Statins	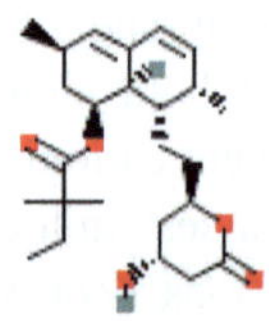	Mantle cell lymphoma	Santos et al. (2022)
4	Terrein	*Aspergillus terreus*	Polyketides	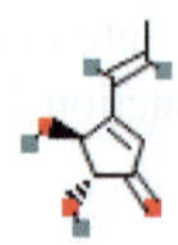	Human lung cancer	Srivastava et al. (2024)
5	Brefeldin A	*Penicillium janthinellum*	Substituted cycloalkane	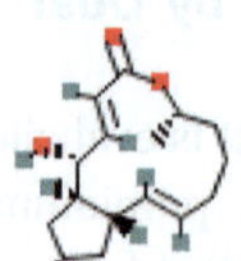	Cancer in regions of the colon	Wang et al. (2024)
6	Asperlin	*Aspergillus nidulans*	Polyketide	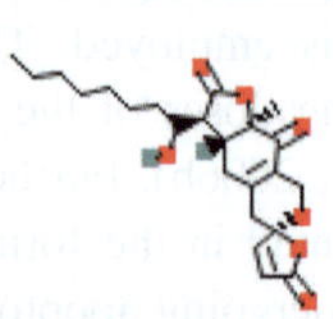	Human cervical carcinoma	He et al. (2011)
7	Griseofulvin	*Penicillium griseofulvum*	Polyketide	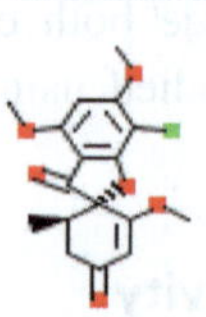	Human cervical cancer	Yu et al. (2024)
8	Sequoiamonascin A	*Aspergillus parasiticus*	Polyketide	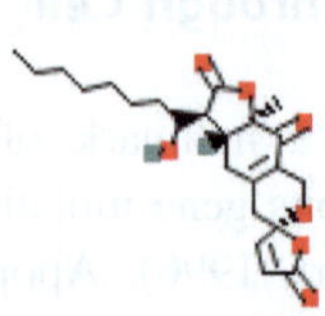	Breast cancer, CNS cancer, lung cancer	Stierle et al. (2003)

(continued)

Table 1 (continued)

S. No.	Compound name	Fungus name	Nature of compound	Chemical structure	Types of cancer	Referencew
9	Penisimplicissin	*Penicillium pinophlium*	Polyketide	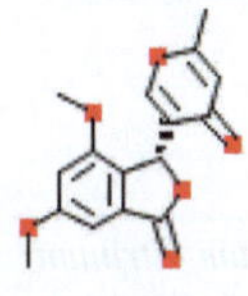	Leukemia	Buommino et al. (2004)
10	Norsolorinic acid	*Aspergillus parasiticus, Aspergillus nidulans*	Polyketide	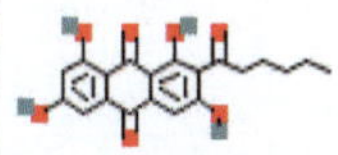	Human bladder cancer, human breast cancer	Wang et al. (2008)
11	Austocystin D	*Aspergilluspseudoustus*	Polyketide	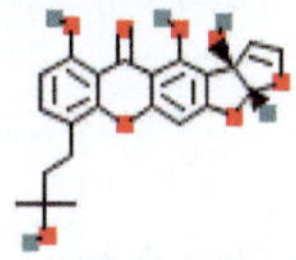	Human colon carcinoma	Gohar et al. (2020)
12	Paclitaxel	*Taxus brevifolia, Taxomyces andreanea*	Terpene	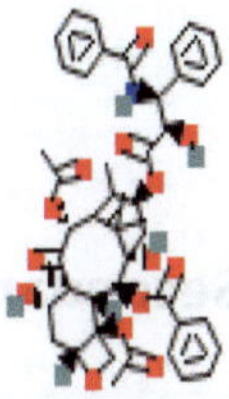	Ovarian cancer, lung cancer, breast cancer	Niedens et al. (1999)
13	Etoposide	*Fusarium solani*	Glycoside derivative	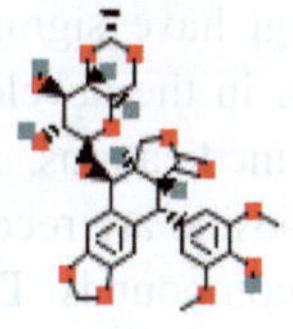	Lung cancer, lymphoma, leukemia	Meresse et al. (2004)
14	Chloctanspirone	*Penicillium chrysogenum*	Polyketide	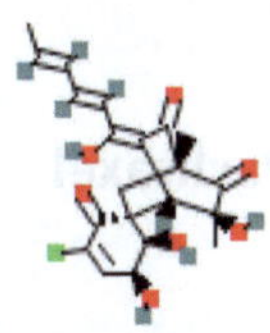	Human leukemia, lung cancer	Bladt et al. (2013)
15	Prenylterphenyllins	*Aspergillus candidus*	Polyketide	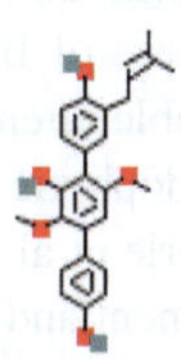	Lung cancer, human leukemia	Bladt et al. (2013)
16	Xanthocillin	*Dichotomomyces albu, Penicillium chrysogenum*	Alkaloid		Human cervical cancer, liver cancer, breast cancer	Fang et al. (2006a)

(continued)

Table 1 (continued)

S. No.	Compound name	Fungus name	Nature of compound	Chemical structure	Types of cancer	Referencew
17	Wentilactones	*Aspergillus wentii*	Terpenes	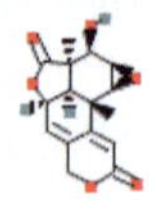	Human hepatoma cancer	Sun et al. (2012)
18	Tricitrinol	*Penicillium citrinum*	Polyketide	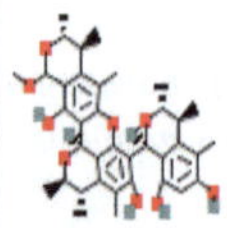	Human leukemia, colon cancer, cervical cancer	Li et al. (2011)
19	Spirotryprostatin	*Aspergillus fumigatus*	Alkaloid	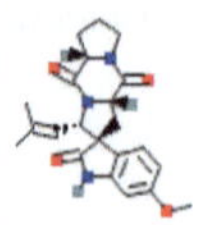	Hepatocellular carcinoma, prostate cancer	Cui et al. (1995)
20	Stephacidin B	*Aspergillusochraceus, Aspergillus westerdijkiae*	Alkaloid	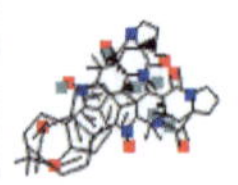	Colon cancer, lung cancer, ovarian cancer	Yamazaki & Miyaki (1971), Abraham and Arfmann (1990)

3 Notable Anticancer Fungal Secondary Metabolites

Natural compounds derived from fungi have significantly contributed to cancer therapy, resulting in the development of numerous therapeutically effective medications. Among these, paclitaxel, griseofulvin, and etoposide are recognized as particularly notable anticancerous compounds. Detailed descriptions of these compounds are provided below.

3.1 History and Discovery of Paclitaxel from Fungal Endophytes

As research on the identification of natural compounds expands, use of endophytic fungi as a source of bioactive secondary metabolites has gained considerable attention. In 1993, *Taxomyces andreanae,* the first endophytic fungus that produces paclitaxel was identified (Stierle et al., 1995). This study emphasized the capability to ferment and modify fungi genetically to produce paclitaxel. Despite extensive efforts to enhance fungal taxol production through methods such as mutagenesis, gene overexpression, cocultures, and genome shuffling, the molecular mechanism underlying taxol biosynthesis and its regulation mechanisms largely remain uncharacterized (Yin et al., 2024). Endophytic fungi offer several advantages, including rapid growth in basic culture conditions, the scalability in bioreactors, and direct taxol production. Fungi associated with both non-*Taxus* and *Taxus* species synthesized taxol. Although the biosynthetic genes for taxol have been characterized at the mRNA level, comprehensive genomic sequencing data is still lacking. The mechanism by which the biosynthetic pathway for taxol production in fungi evolved remains unclear (Subban & Kempken, 2023). However, a widely accepted hypothesis suggests that horizontal gene transfer was prevalent among the Taxus species and their endophytic fungi, enabling the latter to produce taxol (Kusari et al., 2014). It is noteworthy to observe the ubiquity of orthologs of taxol biosynthesis genes among various fungi (Subban & Kempken, 2023). Thus, synthesis of taxol from endophytic fungi paved a way in research on pharmacological metabolites derived from endophytic sources. Additionally, studies have demonstrated that endophytes exhibit remarkable metabolic mimicry enabling them to synthesize, inducing, or modulate the metabolic chemical entities within their hosts (Shankar, 2019).

3.2 Other Noteworthy Compounds like Griseofulvin and Etoposide

Griseofulvin, derived from *Penicillium griseofulvum,* has recently gained attention as a potent anticancer agent due to its efficacy in restraining the proliferation of various cancer cell types while exhibiting low toxicity (Zhang et al., 2017). The therapeutic potential of griseofulvin has been substantiated by several preclinical studies, emphasizing its function in modulating diverse cancer effects, including the suppression of cellular proliferation, interruption of the cell cycle, and facilitation of apoptosis (Ho et al., 2001). Research indicates that griseofulvin impedes tumor progression by disrupting spindle microtubules dynamics during mitosis, with effect reported in liver, breast, lung, and colorectal cancers (Yu et al., 2024). Etoposide, obtained from *Fusarium solani,* is a crucial component of combination chemotherapies regimes for several malignancies, including leukemia, lung cancer, and lymphoma (Zhang et al., 2021). This compound also has been applied in breast cancer treatment. The clinical trial involving untreated patients demonstrated approximately 15% response rate. Oral etoposide has been the subject of numerous clinical trials investigating its efficacy in treatment of ovarian cancer. The rate of response reported across these studies ranges from 20.4% to approximately 30%. The effectiveness of etoposide is significantly influenced by the extent of preceding treatment (Ozols, 1999). The mode of functioning of etoposide was seen as analogous to that of podophyllotoxin. All these chemicals inhibit microtubule assembly by blocking tubulin polymerization and subsequently leading to the degradation of spindle fibers (Zhang et al., 2021).

4 Mechanism of Action in Cancer Treatment

4.1 Apoptosis Induction by Fungal Secondary Metabolites

Studies have shown that dicitrinone B, a secondary metabolite, extracted from *Penicillium citrinum* triggered apoptosis, at least partially, through two apoptotic mechanisms. The initial pathway involved the intrinsic mitochondrial mechanism associated with ROS, while the subsequent pathway engaged the extrinsic apoptotic route (Chen et al., 2014). Pretreatment with dictrinone B significantly increased ROS production in a dose – dependent manner. Increased oxidative stress results in the opening of the mitochondria permeability transition pore (MPTP) leading to permeability of mitochondrial membrane. Mitochondria are essential intracellular organelles responsible to produce ATP, apoptosis regulation, and synthesis of various metabolites. Subsequent experiments demonstrated that pretreatment of dicitrinone B dramatically enhanced the rapid production of ROS in a way dependent on dosage (Chen et al., 2014). The release of cytochrome c from mitochondria initiates caspase activation, which may subsequently facilitate either apoptosis or essential processes like differentiation and proliferation. The activation of various caspases activates intrinsic as well as extrinsic pathways of apoptosis by mediating the cleavage of poly (ADP-ribose) polymerase (PARP) protein. The extrinsic apoptosis pathway is regulated by a signaling complex composed of procaspase-8 and a Fas-associated death domain (FADD). This death-inducing complex switches on caspase-8 and subsequent caspases (Sang et al., 2012). Consequently, the regulation of caspase-8 suggests that the extrinsic apoptosis pathways may be pivotal in the induction of apoptosis by dicitrinone B.

Greensporone A, a secondary metabolite obtained from the *Halenospora* sp., has demonstrated anticancer activity against multiple leukemia cell lines (Prabhu et al., 2019). It reduces cell proliferation and initiates apoptosis mediated cell death in organisms. Fungal metabolite downregulates antiapoptotic genes and its subsequent target proteins such as, GSK3, FOXO1 and others and thereby mediates the therapy of leukemia cells (Farhan et al., 2017). Further, the cell lines of leukemia when treated by greensporone A exhibit an enhanced expression of some of the pro-apoptotic member proteins such as Bax, that belong to Bcl-2 family. Greensporone A increases the Bax/Bcl-2 ratio, promoting mitochondrial membrane permeabilization and accumulation of cytochrome c. The accumulation of cytochrome c generated by greensporone A activates the cascade of caspases that results in cleavage of the effector, PARP, leading to apoptosis. Furthermore, three anthraquinone-type secondary metabolites were extracted from *Penicillium pinophilum.* These compounds elicit apoptosis in pancreatic cells via the mitochondrial route (Koul et al., 2016).

4.2 Anti-proliferative Mechanisms by Fungal Secondary Metabolites

Extracts from *Cordyceps* were assessed on HeLa cell lines, resulting in the identification of active extracts, namely gliocladicillins A and B. These compounds exhibited antitumor activity e by arresting the cell cycle at G2/M phase. Additionally, they induced apoptosis via activating *p53*, *p21*, and *cyclin B*, as well as the caspases-8, -9, and -3 pathways. This comprehensive strategy triggered tumor cell death via extrinsic and intrinsic pathways (Chen et al., 2009). Further, several endophytic fungi, including *Taxomyces andreanae, Alternaria alternata, Seimatoantlerium nepalense,* and *Chaetomella raphigera* have been reported to produce paclitaxel as potent anticancer agent (Chen et al., 2014). Paclitaxel exerts its effects by binding to tubulin, thereby inhibiting microtubule depolarization during mitosis. The interaction disrupts microtubular networks and induces elevated levels Bcl-2 and *Bax*, leading to cell cycle arrest at the G2 and M phase transition. These processes also caused apoptosis during the G1 phase of mitosis. It induced cell cycle arrest during the mitotic phase, resulting in cell death (Kousar et al., 2022). Endophytic fungi such as *Sinopodophyllum hexandrum* and *Dysosma veitchii* can produce adequate amounts of anticancerous medication podophyllotoxin, another potent anticancer compound (Xianzhi et al., 2003). Podophyllotoxin inhibits DNA replication by attaching to topoisomerase II and inhibits its activity. Etoposide, a precursor of podophyllotoxin, is highly toxic to topoisomerase II, and its strong binding causes damage to the DNA double strand and consequently elevating the DNA damage risk. The significant accumulation of DNA damage causes cell death due to alterations in the DNA generated by insertions, deletions, and genetic recombination (Kumar et al., 2021a). *Thielavia subthermophila*, an endophytic fungus, synthesized hypericin within its host plant *Hypericum perforatum* (Kusari et al., 2008). Hypericin is a potent anticancerous agent, and its effectiveness against T-cell lymphoma has been well investigated. Hypericin functions by inhibiting multiple genes and activating the caspase cascade. It interacts with reactive oxygen species (ROS), resulting in DNA damage, and elevates the level of p*21*, *p27*, and cyclin B/D1/E. Cytochrome c is liberated from the inner mitochondrial membrane, initiating caspases 3 and 8. These molecular mechanisms enhanced the expression of PARP. The functioning of the cell cycle interrupted at the G2 phase and caused cellular demise (Agostinis et al., 2002). Fusarubin and anhydrofusarubin, cytotoxic agents is isolated from *Cladosporium* sp., (endophytic fungi), influence molecular events during the cell cycle. These compounds possessed anticancer properties against the leukemia cell line OCI-AML3 by inhibiting proliferation through cell cycle arrest. Fusarubin demonstrated anti-proliferative effects on OCI-AML3 cells, mostly through its interaction with *p53* and *p21*. It also triggered apoptosis in metastatic cells through Fas ligand binding and subsequent *caspase-8* activation (de Wit et al., 2016). Cladosporols another group of secondary metabolites from *Cladosporium cladosporioides* have shown significant efficacy against several cancer cell lines. These compounds have distinct mechanisms of action and target cells of HT-29 colon cancer by switching on the halted G1 phase of these cells and hence exhibiting anti-tumor properties (Li et al., 2017).

5 Clinical Trials and Fungal-Derived Cancer Therapies

The secondary metabolites, along with their other derivatives produced by fungi, have both beneficial and detrimental effects on human health. This duality was revealed by the aflatoxin poisoning incident, commonly known as the Turkey X illness, which occurred during the 1960s. This phenomenon was also evident in the widespread use of the very first discovered, broad-spectrum antibiotic, penicillin, which was considered as a "wonder drug" at the times of World War II. Since ancient times, secondary metabolites of fungi, such as mycophenolic acid, penicillin, cyclosporin, and statins, have been utilized as potent drugs in the health industry. Consequently, these compounds have been produced in large quantities. Simultaneously, researchers are actively exploring novel methodologies and techniques enhancing the biosynthesis and characterization of these fungal metabolites (Keller, 2019). Due to their wide-ranging pharmaceutical properties, fungal metabolites are considered strong candidate for drug development. However, to transition from a bioactive compound to an approved therapeutic agent, these substances must undergo a multi-phase clinical trial process. Recently, as widely acknowledged, cancer has emerged as an eminent cause of mortality, significantly hindering the global pursuit of higher life expectancy. In 2019, World Health Organization (WHO) estimated that in the age group ranging from 70 to 112, cancer is the primary cause of death, across 183 countries (Sung et al., 2021). Thus, in the formulation of oncological therapeutics, natural substances undergo numerous stages of clinical trials.

5.1 Current Status of Fungal Metabolites in Clinical Trials

5.1.1 Overview of Various Clinical Cases (I, II, and III)

Case I

This case includes the trials conducted to ascertain the maximum dose that is safe for the body, commonly termed as Maximum Tolerated Dose (MTD). The estimation of MTD is conducted with the assistance of a small cohort of patients who are treated with progressively increasing doses of the drug in progress until the number of patients experiencing a specific level of toxicity reaches a predetermined threshold. Once this threshold is met, the preceding lower dose is considered as MTD. However, the statistical analysis of this data is often not done. There are very few methods which involve estimating MTD using statistical models (Robbins & Monro, 1951). In the phase I trials, several challenges and difficulties are not adequately addressed in the existing literature. These include a limited number of subjects in the trial, extended evaluation times for each subject, ethical concerns regarding the careful determination of the MTD, the subjective nature of assessing toxicity (toxic vs. non-toxic) compared to clear outcomes such as life or death, and differences among the subjects in the trial, as well as difficulties in assessing responses due to participants dropping out for reasons unrelated to the drug's toxicity. The above given traditional design for estimating MTD is reasonable, but it does not give any specific percentile; hence, there are several other simple alternative designs that are discussed below:

Design A (Traditional 3 + 3 Design): In this design, groups of three patients are treated at a time. If none of the three individuals encounters toxicity, the dosage is increased. Upon observing toxicity in one or more individuals, a new cohort of three patients is administered the same dosage. If merely one of the six patients exhibits toxicity, the dosage is yet escalated. If a single patient of the three exhibits toxicity, the dosage is nonetheless escalated. If two or more patients exhibit toxicity, the experiment is terminated.

Design B (Single-Patient Escalation): In this design, a single patient receives a specific dosage, and their condition is closely monitored. If any adverse effects are observed, the dosage is reduced for the next patient. Conversely, if no discomfort or toxicity is reported, the dosage is increased for the next patient. This process continues until safe and effective dose is reached. The approach emphasizes individual dose adjustment based on immediate toxicity observation.

Design C (Modified Single-Patient Design): This design bears a striking resemblance to design B. In this approach, the initial patient is administered a specific, predetermined dosage, and their response is closely monitored for any signs of toxicity. If toxicity is observed, the dosage is subsequently reduced for the next patient. Conversely, if no toxicity is detected, the dosage remains unchanged and is administered to the next patient. This process continues until two consecutive patients demonstrate no adverse effects at the same dosage. At that point, the dosage is gradually increased for the subsequent patient. The distinguishing feature of this design lies in the requirement for two non-toxic responses before the dosage is escalated, and de-escalation is promptly initiated upon the occurrence of toxicity.

Design D (Three Patient Cohort Rule-Based Design): In this case, three patients are treated with a particular dose, and the group is subsequently observed. If no toxicity is observed in any of the three patients, the dose is increased for the subsequent group. Conversely, if more than one patient in the group exhibits toxicity, the dose is decreased for the next group. If a single patient is classified as hazardous, the following cohort of three patients will get the identical dosage.

Case II

Trials are typically classified into three categories: design single-arm, randomized, and purpose-based. Some trials are used for screening new drugs and selecting tumor types, while others are used to make affirmative or negative decisions. Although the concept of dividing trials into "screening" and "decision-making" phases was considered, it was deemed inefficient for most drugs. This approach may be appropriate in specific cases, such as when the tumor type is unclear or when validating a biomarker. Ideally, these goals should be integrated within a single trial, with single-arm and randomized designs remaining the primary classifications (Seymour et al., 2010).

1. **Single-Arm Trial:** In this method, participants receive the experimental treatment without a control group. The purpose is to assess the treatment's efficacy, often measured by tumor response or progression-free survival. It is a simple design, and fewer participants are needed. However, this method presents several drawbacks, notably the absence of a control group. This absence renders it challenging to ascertain whether the observed effects are attributable to the treatment or other extraneous factors.
2. **Randomized Trials:** Participants are randomly allocated to various experimental treatments or a control group (e.g., placebo or other standard treatments). The primary objective of this study is to evaluate the efficacy of various treatments or establish a preliminary comparison with the established standard of care. One advantage of this approach is that it generates more robust data for comparison purposes. However, it also entails certain limitations, such as the requirement for larger sample sizes and more intricate logistics compared to single-arm trials.
3. **Adaptive Phase II Trials:** The trial design is flexible, allowing modifications (e.g., dose adjustments, dropping ineffective arms) based on interim results. Here the purpose is to optimize the study design and improve efficiency. The advantage of this method is it can save time and resources by focusing on promising treatments. However, this method has some limitations—it requires advanced planning and statistical expertise.
4. **Seamless Phase Trials:** This method integrates both phases into a single trial, allowing for a transition to the subsequent phase if the phase II portion demonstrates promising outcomes. The principal aim of this methodology is to expedite the drug development process. The advantages of this method include time savings and cost-effectiveness, as it eliminates the gap between phases. However, there are limitations to consider, as insufficient phase II data may pose risks in justifying the transition to phase III.
5. **Biomarker-Driven Phase II Trials:** It focuses on patients with specific biomarkers or genetic characteristics. The purpose of this test is to give targeted therapies on populations likely to help them. The advantages here are higher likelihood of success in targeted populations, but there is one drawback; that is, it requires validated biomarkers and sufficient eligible patients.

6. **Window-of-Opportunity Trials:** Short trials are conducted prior to standard treatment to assess the efficacy of a new therapy on biomarkers or the tumor. The primary objective is to gain early insights into treatment mechanisms. A notable advantage of short trials is their brevity and resource efficiency. However, a limitation is their limited scope and generalizability.

Case III trials

These studies are designed to rigorously evaluate both the safety and efficacy of any newly presented treatments relative to the existing standard of care. These are typically the final phase preceding regulatory approval and involve a larger and diverse patients' cohort to validate the findings of earlier phases. The primary aim of these clinical trials is to assess the therapeutic effectiveness treatment and validate the findings from earlier trials on a broader scale. This entails comparing the new treatment with the established standard of care or a placebo. Additionally, phase III trials seek to thoroughly analyze the safety profile of the medication, encompassing the evaluation of infrequent or prolonged side effects. The main goals of phase III trials are to establish the efficacy and safety of a treatment, with a focus on identifying benefits that are superior to those already available. Additionally, phase III trials serve as a platform for monitoring adverse effects, particularly those that may not be detected in earlier phases. Furthermore, dose optimization is another objective of phase III trials (Eisenhauer et al., 2009).

There are different types of phase III clinical trials:

1. **Superiority Trials:** These clinical trials demonstrate the superiority of the novel treatment over the existing ones.
2. **Non-inferiority Trials:** This research illustrates that innovative therapy is not substantially inferior to conventional treatment in terms of efficacy, while simultaneously providing additional advantages such as reduced adverse effects, cost-effectiveness, and enhanced patient adherence.
3. **Equivalence Trials:** These trials demonstrate that the efficacy of a novel treatment is comparable to that of an existing treatment with a predetermined margin of therapeutic equivalence.
4. **Adaptive Trials:** This framework facilitates modifications to the trial design, including sample size, treatment arms, and dosing based on interim results, without compromising the trial's effectiveness. It is employed to optimize trial efficiency and minimize resource utilization in trials with uncertain outcomes.
5. **Pragmatic Trials (Real-World Trials):** This method is utilized to evaluate the effectiveness of a treatment in clinical practice contexts beyond the controlled environment of conventional trials.
6. **Registration Trials:** These trials are used to generate the clinical data which is required for regulatory approval of the treatment for commercial use.
7. **Dose–Response Trials**: The aim of these phase III trials is to assess and identify the dosage that exhibits the greatest efficacy. These trials are typically combined to yield fruitful results based on diverse research objectives (Umscheid et al., 2011).

5.2 Approved Fungal-Derived Cancer Therapies

Despite its ongoing status as one of the most formidable and arduous diseases, cancer has witnessed significant advancements in recent times. Numerous anticancer agents have been developed, enabling patients to effectively combat cancer within certain limits and extend their life expectancy while simultaneously enhancing their quality of life. Advances in biotechnology have resulted in the creation of various novel pharmaceuticals, with fungi serving as a vital source of secondary metabolites. These metabolites function as precursor chemicals in the development of efficacious drugs. Notably, several fungal secondary metabolites have been approved as effective cancer therapies:

Paclitaxel (Taxol): Paclitaxel, also referred to as taxol, was first extracted from *Taxus brevifolia* and can furthermore be synthesized by endophytic fungi, including species of *Penicillium*. It is widely used in medication for the treatment of various malignancies, including ovarian, breast, and lung cancers.

Lentinan: This supplement is derived from the shiitake mushroom (*Lentinula edodes*). It has been approved as a supportive therapy for cancer patients in several countries, including Japan. It is used to enhance the immune response of cancer patients.

Polysaccharide-K (PSK, Krestin): In countries like Europe and Japan, PSK is used as an immunotherapeutic agent in cancer treatment which is extracted from the fungus *Trametes versicolor*.

Etoposide: Etoposide, a semisynthetic derivative of podophyllotoxin, is extracted from the mayapple plant. However, certain fungi, namely *Fusarium solani*, *F. oxysporum*, *F. proliferatum*, *Phialocephala fortinii*, and *Trametes hirsuta*, possess the ability to produce podophyllotoxin, which serves as a precursor in etoposide synthesis. Etoposide is extensively employed in the management of lung and testicular malignancies.

6 Challenges in Drug Development for Cancer

The development of anticancer drug is often hindered by pharmacokinetics, bioavailability, and toxicity challenges. Consequently, effective drug discovery and improvement necessitate addressing these critical elements to enhance the therapeutic efficacy and safety of anticancer agents. These challenges are prevalent in both synthetic and fungal-derived anticancer therapies.

6.1 Overcoming Pharmacokinetic and Bioavailability Issues

Pharmacokinetics holds paramount significance in determining the therapeutic efficacy of anticancer drugs. Several factors, including absorption, distribution, metabolism, and excretion (ADME), substantially affect the efficacy of the drug and safety as well.

Absorption and Distribution

Prescott et al. (2023) highlight the complexities associated with reaching foremost drug absorption and systemic distribution. Factors like drug solubility, balance inside the gastrointestinal tract, and their interaction with transport proteins often restrict bioavailability. Moreover, ensuring adequate drug awareness at tumor web sites without off-target results becomes a major task. Venturella et al. (2021) primarily focus on the position of physicochemical residences in drug biodistribution. Lipophilicity and molecular size drastically affect the capability of drugs to traverse biological membranes and accumulate in tumor tissues.

Metabolism and Excretion.

Drug metabolism can lead to the formation of inactive or toxic metabolites, thereby decreasing therapeutic effectiveness and increasing adverse effects. Interpatient variability in enzyme function is a key limiting factor in achieving accurate dosage adjustments (Venturella et al., 2021). In addition, rapid renal or hepatic clearance may necessitate common dosing, increasing the chance of toxicity.

Strategies to Overcoming Pharmacokinetic Barriers.

In response to the demanding situations, modern solutions, including enzyme inhibitors, prodrug designs, and advanced formulation technology, are being explored. Enzyme inhibitors can modulate metabolic pathways to prolong drug half-existence, at the same time as prodrug strategies can decorate the stableness and solubility of lively compounds.

Bioavailability Issues

Enhancing bioavailability is very important for maximizing and improving the therapeutic impact of anticancer drugs. Poor solubility and stability, especially in fungal-derived therapies, are major challenges.

Fungal-Derived Therapies

Oliveira et al. (2023) explore bioavailability challenges unique to fungal-derived anticancer compounds. Among those compounds, many show poor aqueous solubility, which therefore hinders absorption. Therefore, strategies consisting of nanoparticle-based delivery systems and chemical changes are being developed to overcome these limitations.

Drug Delivery Systems

Advanced drug delivery systems can deal with bioavailability constraints by enhancing solubility, stability, and focused transport. Prescott et al. (2023) highlight the capacity of liposomes, polymer-based totally nanoparticles, and prodrug formulations to decorate the pharmacokinetic profiles of anticancer sellers.

Emerging Techniques

The latest emerging techniques which include lipid-primarily based providers and stable dispersions are gaining importance in enhancing drug bioavailability. These strategies not only enhance solubility but also offer controlled release, thus lowering the frequency of dosing and enhancing patient compliance.

6.2 Stability and Toxicity Consideration

Due to narrow therapeutic window of most anticancer drugs, toxicity remains a critical concern in drug development.

General Toxicity Issues

Tsoupras and Davi (2024) highlight the dual challenges of maximizing therapeutic efficacy while minimizing toxicity. Many anticancer agents show dose-dependent toxicity, affecting not only malignant cells but also rapidly dividing healthy.

Fungal-Derived Compounds

Fungal-derived anticancer agents often face unique balance and toxicity challenges. Li et al. (2022) evaluate the susceptibility of those compounds to degradation below physiological conditions, leading to reduced efficacy and potential poisonous byproducts. Strategies along with encapsulation in biocompatible carriers or structural modification can mitigate these problems.

Stability Considerations

The stability of anticancer pills without delay affects their safety and efficacy. According to several researchers, fungal-derived compounds degrade unexpectedly underneath physiological situations, necessitating stabilization techniques (Zhang et al., 2019). These consist of chemical derivatization and the use of stabilizing excipients in formulations.

Reducing Off-Target Effects

One promising method to minimize toxicity is the improvement of centered treatment plans, together with antibody–drug conjugates and ligand receptor targeting system. The purpose of this technique is to selectively deliver therapeutic agents to cancer cells while sparing healthy tissues. Additionally, the incorporation of biomarkers to identify patient selection enhances treatment specificity, improving therapeutic outcomes and minimizing side effects.

6.3 Integrative Strategies

An integrative approach that combines advanced drug design, formulation technology, and personalized remedy is important to overcoming those challenges. Li et al. (2022) endorse for the incorporation of predictive fashions and excessive-throughput screening strategies to optimize pharmacokinetics and decrease toxicity at some point of the early degrees of drug development.

Personalized Medicine

Personalized medication, guided through genetic and molecular profiling, gives a direction to tailoring treatments to individual patient needs. This approach not only enhances healing effects but also reduces the probability of unfavorable reactions by figuring out the best drug and dose for each affected person.

Multidisciplinary Collaboration

Effective drug development calls for collaboration throughout disciplines, inclusive of pharmacology, toxicology, bioinformatics, and clinical medicine. By leveraging the knowledge of numerous fields, researchers can cope with the multifactorial challenges of anticancer drug improvement more successfully.

For anticancer medicines to be progressed, pharmacokinetic, bioavailability, and harmfulness issues must be tended to. Contemporary research paradigms emphasize the critical role of innovative methodologies in developing fungal-derived pharmaceuticals. The industry can only move toward compelling and secure cancer treatments by utilizing progressions in medicate transport frameworks, stabilization procedures, and customized medication. Future considerations should focus on emerging technologies, such as artificial intelligence (AI) and machine learning (ML), to enhance pharmaceutical development and discovery processes. In the long term, the incorporation of AI and ML could enhance the prognosis for cancer patients globally by accelerating the identification of potential medications, assessing pharmacokinetic properties, and optimizing formulations.

7 Challenges and Future Opportunities

Although there were huge improvements in most cancer treatments over the years, troubles together with toxic fine, side consequences, and constrained specificity stay most important limitations. Among the wide range of natural substances being explored for their therapeutic potential, fungal secondary metabolites have emerged as promising candidates. These metabolites exhibit many bioactivities, including anticancer properties, and are translocated throughout organisms as a defensive mechanism. This section explores the prospects, challenges, and implications for managing fungal metabolites in cancer therapy, emphasizing strategies to overcome associated restrictions and highlighting successful therapeutic advancements. A type of auxiliary metabolites, together with polyketides, alkaloids, and terpenoids, is introduced via fungi, in particular endophytic species, and they have strong anticancer residences. These substances induce apoptosis, disrupt the boom of cancer cells, and balance resistance responses. One notable example is paclitaxel (Taxol) derived from yew trees, located in endophytic fungi, showcasing the capability of fungal resources in producing well-known anticancer capsules.

With the advancement of metabolomics and synthetic biology, scientists now have unraveled the biosynthetic pathways of those metabolites, which has allowed huge-scale manufacturing and amendment. The reactivation of silent biosynthetic gene clusters in fungi has enabled the discovery of novel compounds, thereby broadening the scope of cancer therapies.

7.1 Overcoming Toxicity and Side Effects

While fungal secondary metabolites hold therapeutic promise, their clinical utility is limited because of their common toxicity. It is important to control facet results and consist of organ damage, immunosuppression, and stale-target interactions. Patient-focused trials at the side of careful profiling of those medicines can help to decide their long-time period toxicity as observed upon examining patients.

Drug shipping techniques like liposomes and nanoparticles were used to lessen undesirable effects. Toxic metabolite encapsulation directs the drugs to tumor tissues while decreasing systemic publicity. Fungal metabolite liposomal compositions, as an example, have proven decreased toxicity without sacrificing effectiveness.

Strategies to Minimize Off-Target Effects

1. Precision-Engineered Therapies

Recent improvements in proteolysis-targeting chimeras (PROTACs) have enabled a unique focus on most cancer cells, thereby decreasing off-tissue effects. By conjugating fungal-derived active compounds focused on moieties, researchers have developed treatment plans with superior specificity.

2. CRISPR and Gene Editing

CRISPR technology has significantly enhanced the specificity of most cancer therapies. Fungal secondary metabolites when combined with CRISPR engineered delivery systems, ensures targeted action against cancer cells while sparing the healthy tissues.

Case Studies of Successful Drug Modifications

1. Enhancing Efficiency Through Nanomedicine.

The application of nanotechnology along with fungal secondary metabolites has revolutionized drug transport nowadays. Recent case studies have shown that how encapsulating fungal-derived compounds in nanoparticles can improve solubility, stability, and tumor. For instance, a nanomedicine component of a fungal polyketide can provide considerably greater healing effects in preclinical trials (Kelany et al., 2023).

2. Importance of Fungal Metabolites in Combination Therapy

The incorporation of fungal metabolites into customized and personalized immunotherapy has also yielded promising results. Case studies of patients and their process treatment with modified fungal compounds have demonstrated improved survival prices and decreased aspect results, showcasing the synergy between natural products and advanced therapeutic methods (Ulrich-Merzenich et al., 2010).

3. Drug Repurposing Using Fungal Compounds

Repurposing fungal metabolites at first intended for non-oncological programs has furnished new avenues in most cancers remedy. One case examines the use of fungal-derived cyclosporine, formerly used as an immunosuppressant, is now employed as an adjunct therapy to mitigate tumor boom in strong cancers (Fojnica et al., 2024). The repurposed drug validated stepped forward consequences in clinical trials.

4. Advanced Delivery Systems

Incorporating advanced delivery structures consisting of hydrogels and micelles has in addition superior the efficacy of fungal metabolites. For example, the fungal metabolite lovastatin encapsulated inside a hydrogel device, confirmed sustained launch, and advanced tumor suppression in vivo as compared to unfastened drug administration (Ali Mujtaba et al., 2024).

7.2 Development of Targeted Drug Delivery Systems

Targeted drug delivery systems have recently developed pharmaceutical technologies that are designed to transfer therapeutic agents to specific cells or tissues, which reduce side effects and enhance treatment effectiveness.

1. Innovations in Delivery Technologies

Targeted drug delivery structures have revolutionized cancer therapy with the aid of enhancing the precision and efficacy of treatments. These systems employ providers which include nanoparticles, liposomes, and dendrimers to supply fungal secondary metabolites at once to tumor sites. By exploiting the microenvironment of tumors, consisting of acidic pH and overexpressed receptors, those systems ensure selective drug release, minimizing systemic side consequences.

2. Role of Bioconjugation

Bioconjugation techniques, such as antibody–drug conjugates (ADCs) and ligand-receptor interactions, were incorporated with fungal metabolites. This approach improves the specificity of drug shipping, as visible in recent studies wherein fungal-derived compounds conjugated with tumor-unique ligands proven more advantageous therapeutic results. Such improvements preserve promise for overcoming the constraints of traditional chemotherapy. Fungal secondary metabolites continue to offer significant potential in meeting the unmet needs of most cancers therapy. While demanding situations along with toxicity and off-goal consequences persist, improvements in drug transport, precision medicine, and bioengineering provide possible solutions. By leveraging the strength of fungi and modern-day science, the future of cancer therapy could be both effective and safe.

3. Use of Liposomes and Nanocarriers for Precision Therapy

Nanocarriers are tiny delivery systems (1–100 nm) designed to transport drugs, genetic material, or other therapeutic molecules to specific sites in the body, thus minimizing side effects and improving treatment outcomes (Smith et al., 2020). Nanoparticles that are developed to encapsulate drugs or therapeutic agents are called nanocarriers. The applications of the nanocarriers are vast in the field of medical science. One such nanoparticle is a liposome, which is a spherical vessel with one or more phospholipid bilayers. These liposomes are also used as carriers for drugs or genetic material.

There are various types of nanocarriers, each with distinct properties that render them suitable for specific applications:

- **Liposomes**: Phospholipids containing spherical vesicles possess the unique property of encapsulating both hydrophilic and hydrophobic drugs, making them highly promising in the pharmaceutical industries (Liu et al., 2022).
- **Polymeric Nanoparticles**: These are biodegradable polymer particles that provide controlled and prolonged drug release (Lee et al., 2020).
- **Dendrimers:** These are highly branched, tree-like structures that possess functional groups suitable for drug attachment (Garcia et al., 2015).
- **Micelles**: These are small amphiphilic structures that spontaneously assemble to encapsulate hydrophobic drugs (Kumar et al., 2021b).
- **Metallic Nanoparticles**: Metallic nanoparticles are frequently employed in theragnostics, a field that merges therapeutic and diagnostic capabilities, exemplified by gold nanoparticles (Singh et al., 2021).
- **Lipid Nanoparticles (LNPs)**: Widely used for RNA delivery, as seen in mRNA vaccines (Wang et al., 2021).

Some of the major functions of liposomes as nanocarriers:

- **Enhanced Solubility and Stability**: There are various drugs or therapeutic agents with poor solubility that may get destructed when introduced in aqueous environment. Nanocarriers enhance the solubility of hydrophobic pharmaceuticals and safeguard them against degradation (Masters et al., 2015).
- **Targeted Drug Delivery**: Some modification on the surface of the nanocarriers is used for the delivery of drugs to a particular targeted site (Clark et al., 2020).
- **Controlled Release**: Liposomes ensure slow and prolonged release of drugs that ensures consistent drug levels at the target site.
- **Precision in Cancer Therapy**: Liposomes facilitate the accumulation of chemotherapeutic drugs within tumors by enhancing permeability and retention (EPR) effects (Murciano-Goroff et al., 2020).
- **Gene and RNA Delivery**: Liposomes are also used in the delivery of siRNA, mRNA, or DNA (Zhang et al., 2011).
- **Reduced Side Effects:** By concentrating drugs at the target site, liposomes minimize exposure to healthy tissues, reducing unwanted side effects.

8 Conclusion

Fungal secondary metabolites hold significant promise in fulfilling the unmet therapeutic needs of various cancer treatments. The compounds produced during the secondary metabolic process can alter cell shape and induce DNA fragmentation, resulting in the apoptotic death of cancer cells. Fungi serve as a substantial source of bioactive compounds with immunostimulatory and anticancer properties, making them highly effective natural supplements in cancer treatment. This fact continues to drive molecular therapy, which employs low-molecular-weight drugs to disrupt intracellular signal transduction pathways, potentially enhancing the efficacy of cancer treatment. The potential of marine fungi as a source of anticancer medications is attributed to their ability to synthesize diverse chemicals and enzymes that effectively inhibit the proliferation of cancer cells. Further exploration and screening of fungal sources are necessary for the discovery of novel cancer drugs. By harnessing the unique biological properties of fungi and incorporating them with modern scientific advancement, the development of safe, effective, and targeted cancer therapies becomes increasingly feasible. Such integrative strategies hold the promise of transforming the landscape of oncology in the years to come.

References

Abraham, W. R., & Arfmann, H. A. (1990). 12, 13-Dihydroxy-fumitremorgin C from *Aspergillus fumigatus*. *Phytochemistry, 29*(3), 1025–1026.

Agostinis, P., Vantieghem, A., Merlevede, W., & de Witte, P. A. (2002). Hypericin in cancer treatment: More light on the way. *The International Journal of Biochemistry & Cell Biology, 34*(3), 221–241.

Alao, J. P. (2007). The regulation of cyclin D1 degradation: Roles in cancer development and the potential for therapeutic invention. *Molecular Cancer, 6*, 1–16.

Ali Mujtaba, M., Alam, S., & Alotaibi, N. M. (2024). Formulation Development and Characterization of Lovastatin Nanogel for the

Treatment of Hyperlipidemia. *Oriental Journal of Chemistry, 40*(4), 945–951.

Atale, N., Gupta, S., Yadav, U. C. S., & Rani, V. (2014). Cell-death assessment by fluorescent and nonfluorescent cytosolic and nuclear staining techniques. *Journal of Microscopy, 255*(1), 7–19.

Balde, E. S., Andolfi, A., Bruyere, C., Cimmino, A., Lamoral-Theys, D., Vurro, M., Evidente, A., et al. (2010). Investigations of fungal secondary metabolites with potential anticancer activity. *Journal of Natural Products, 73*(5), 969–971.

Bladt, T. T., Frisvad, J. C., Knudsen, P. B., & Larsen, T. O. (2013). Anticancer and antifungal compounds from *Aspergillus*, *Penicillium* and other filamentous fungi. *Molecules (Basel, Switzerland), 18*(9), 11338–11376.

Buommino, E., Nicoletti, R., Gaeta, G. M., Orlando, M., Ciavatta, M. L., Baroni, A., & Tufano, M. A. (2004). 3-O-methylfunicone, a secondary metabolite produced by *Penicillium pinophilum*, induces growth arrest and apoptosis in HeLa cells. *Cell Proliferation, 37*(6), 413–426.

Chen, L., Gong, M. W., Peng, Z. F., Zhou, T., Ying, M. G., Zheng, Q. H., Zhang, Q. Q., et al. (2014). The marine fungal metabolite, dicitrinone B, induces A375 cell apoptosis through the ROS-related caspase pathway. *Marine Drugs, 12*(4), 1939–1958.

Chen, L., Zhang, Q. Y., Jia, M., Ming, Q. L., Yue, W., Rahman, K., Han, T., et al. (2016). Endophytic fungi with antitumor activities: Their occurrence and anticancer compounds. *Critical Reviews in Microbiology, 42*(3), 454–473.

Chen, Y., Guo, H., Du, Z., Liu, X. Z., Che, Y., & Ye, X. (2009). Ecology-based screen identifies new metabolites from a *Cordyceps*-colonizing fungus as cancer cell proliferation inhibitors and apoptosis inducers. *Cell Proliferation, 42*(6), 838–847.

Ciniglia, C., Pinto, G., Sansone, C., & Pollio, A. (2010). Acridine orange/Ethidium bromide double staining test: A simple In-vitro assay to detect apoptosis induced by phenolic compounds in plant cells. *Allelopathy Journal, 26*(2), 301–308.

Clark, T., Martin, L., & Zhao, X. (2020). What We Need to Know about Liposomes as Drug Nanocarriers. *International Journal of Molecular Sciences, 21*(19), 7098.

Cui, C. B., Kakeya, H., Okada, G. E. N., Onose, R. I. E., Ubukata, M., Takahashi, I., Osada, H., et al. (1995). Tryprostatins A and B, novel mammalian cell cycle inhibitors produced by *Aspergillus fumigatus*. *The Journal of Antibiotics, 48*(11), 1382–1384.

de Wit, P. J. (2016). *Cladosporium fulvum* effectors: Weapons in the arms race with tomato. *Annual Review of Phytopathology, 54*(1), 1–23.

Deshmukh, S. K., Prakash, V., & Ranjan, N. (2018). Marine fungi: A source of potential anticancer compounds. *Frontiers in Microbiology, 8*, 2536.

Eisenhauer, E. A., Therasse, P., Bogaerts, J., Schwartz, L. H., Sargent, D., Ford, R.,... & Verweij, J. (2009). New response evaluation criteria in solid tumours: revised RECIST guideline (version 1.1). *European Journal of Cancer, 45*(2), 228–247.

Fang, L. Z., Qing, C., Shao, H. J., Yang, Y. D., Dong, Z. J., Wang, F., Liu, J. K., et al. (2006a). Hypocrellin D, a cytotoxic fungal pigment from fruiting bodies of the ascomycete *Shiraia bambusicola*. *The Journal of Antibiotics, 59*(6), 351–354.

Fang, N., Li, Q., Yu, S., Zhang, J., He, L., Ronis, M. J., & Badger, T. M. (2006b). Inhibition of growth and induction of apoptosis in human cancer cell lines by an ethyl acetate fraction from shiitake mushrooms. *Journal of Alternative & Complementary Medicine, 12*(2), 125–132.

Farhan, M., Wang, H., Gaur, U., Little, P. J., Xu, J., & Zheng, W. (2017). FOXO signaling pathways as therapeutic targets in cancer. *International Journal of Biological Sciences, 13*(7), 815.

Fojnica, A., Gromilic, Z., Mohamed, Y. A. A., Akhtar, S., & Vranic, S. (2024). The potential role of cyclosporine A in cancer treatment: A comprehensive literature review. *Contemporary Oncology/ współczesna Onkologia, 28*(4), 271–282.

Frisvad, J. C., & Filtenborg, O. (1989). *Terverticillate penicillia*: Chemotaxonomy and mycotoxin production. *Mycologia, 81*(6), 837–861.

Ganot, N., Meker, S., Reytman, L., Tzubery, A., & Tshuva, E. Y. (2013). Anticancer metal complexes: Synthesis and cytotoxicity evaluation by the MTT assay. *Journal of Visualized Experiments: Jove, 81*, 50767.

Gao, Y., Wang, J., Meesakul, P., Zhou, J., Liu, J., Liu, S., Cao, S., et al. (2024). Cytotoxic compounds from marine fungi: Sources, structures, and bioactivity. *Marine Drugs, 22*(2), 70.

Garcia, R., Taylor, M., & Patel, S. (2015). Dendrimers: Advanced Nanocarriers for Drug Delivery Applications. *Frontiers in Pharmacology, 6*, 286.

Gohar, U. F., Mukhtar, H., Mushtaq, A., Farooq, A., Saleem, F., Hussain, M. A., & Ghani, M. U. (2020). Fungi: A potential source of biopharmaceuticals. *Agrobiological Records, 2*, 49–62.

He, L., Nan, M. H., Oh, H. C., Kim, Y. H., Jang, J. H., Erikson, R. L., Kim, B. Y., et al. (2011). Asperlin induces G2/M arrest through ROS generation and ATM pathway in human cervical carcinoma cells. *Biochemical and Biophysical Research Communications, 409*(3), 489–493.

Ho, Y. S., Duh, J. S., Jeng, J. H., Wang, Y. J., Liang, Y. C., Lin, C. H., Lin, J. K., et al. (2001). Griseofulvin potentiates antitumorigenesis effects of nocodazole through induction of apoptosis and G2/M cell cycle arrest in human colorectal cancer cells. *International Journal of Cancer, 91*(3), 393–401.

Kelany, N. A., El-Sayed, A. S., & Ibrahim, M. A. (2023). *Aspergillus terreus* camptothecin-sodium alginate/titanium dioxide nanoparticles as a novel nanocomposite with enhanced compatibility and anticancer efficiency in vivo. *BMC Biotechnology, 23*(1), 9.

Keller, N. P. (2019). Fungal secondary metabolism: Regulation, function and drug discovery. *Nature Reviews Microbiology, 17*(3), 167–180.

Koul, M., Meena, S., Kumar, A., Sharma, P. R., Singamaneni, V., Riyaz-Ul-Hassan, S., Singh, S., et al. (2016). Secondary metabolites from endophytic fungus *Penicillium pinophilum* induce ROS-mediated apoptosis through mitochondrial pathway in pancreatic cancer cells. *Planta Medica, 82*(04), 344–355.

Kousar, R., Naeem, M., Jamaludin, M. I., Arshad, A., Shamsuri, A. N., Ansari, N., Al-Harrasi, A., et al. (2022). Exploring the anticancer activities of novel bioactive compounds derived from endophytic fungi: Mechanisms of action, current challenges and future perspectives. *American Journal of Cancer Research, 12*(7), 2897.

Kumar, S., Aharwal, R. P., Jain, R., & Sandhu, S. S. (2021a). Bioactive molecules of endophytic fungi and their potential in anticancer drug development. *Current Pharmacology Reports, 7*, 27–41.

Kumar, V., Singh, H., & Zhao, Y. (2021b). Amphiphilic Micelles in Drug Delivery: *Mechanisms and Applications. Chemical Reviews, 121*(9), 5379–5454.

Kusari, S., Lamshöft, M., Zühlke, S., & Spiteller, M. (2008). An endophytic fungus from *Hypericum perforatum* that produces hypericin. *Journal of Natural Products, 71*(2), 159–162.

Kusari, S., Singh, S., & Jayabaskaran, C. (2014). Rethinking production of Taxol®(paclitaxel) using endophyte biotechnology. *Trends in Biotechnology, 32*(6), 304–311.

Larsen, T. O., Lange, L., Schnorr, K., Stender, S., & Frisvad, J. C. (2007). Solistatinol, a novel phenolic compactin analogue from *Penicillium solitum*. *Tetrahedron Letters, 48*(7), 1261–1264.

Lee, H. S., Kim, E. J., & Kim, S. H. (2015). Ethanol extract of *Innotus obliquus* (Chaga mushroom) induces G1 cell cycle arrest in HT-29 human colon cancer cells. *Nutrition Research and Practice, 9*(2), 111–116.

Lee, J., Patel, S., & Brown, A. (2020). Biodegradable Polymeric Nanoparticles for Drug Delivery: A Review. *Materials Science and Engineering: C, 113*, Article 110996.

Li, D., Chen, L., Zhu, T., Kurtán, T., Mándi, A., Zhao, Z., Gu, Q., et al. (2011). Chloctanspirones A and B, novel chlorinated polyketides with an unprecedented skeleton, from marine sediment derived fungus *Penicillium terrestre. Tetrahedron, 67*(41), 7913–7918.

Li, H. L., Li, X. M., Mándi, A., Antus, S., Li, X., Zhang, P., Wang, B. G., et al. (2017). Characterization of cladosporols from the marine algal-derived endophytic fungus *Cladosporium cladosporioides* EN-399 and configurational revision of the previously reported cladosporol derivatives. *The Journal of Organic Chemistry, 82*(19), 9946–9954.

Li, H., Huang, H., Zhao, J., & Zhu, C. (2022). Overcoming stability and toxicity challenges in fungal-derived anticancer drugs. *Current Medicinal Chemistry, 29*(14), 2567–2586.

Li, W., Song, K., Wang, S., Zhang, C., Zhuang, M., Wang, Y., & Liu, T. (2019). Anti-tumor potential of astragalus polysaccharides on breast cancer cell line mediated by macrophage activation. *Materials Science and Engineering: C, 98*, 685–695.

Liu, P., Chen, G., & Zhang, J. (2022). A Review of Liposomes as a Drug Delivery System: Current Status of Approved Products, Regulatory Environments, and Future Perspectives. *Molecules (Basel, Switzerland), 27*(4), 1372.

Maehara, S., Yamane, C., Kitamura, C., Hinokuma, M., & Hata, T. (2020). High ophiobolin A production in endophytic fungus *Bipolaris* sp. associated with *Datura metel. Natural Product Research, 34*(20), 2990–2992.

Marzec, M., Kasprzycka, M., Lai, R., Gladden, A. B., Wlodarski, P., Tomczak, E., Wasik, M. A., et al. (2006). Mantle cell lymphoma cells express predominantly cyclin D1a isoform and are highly sensitive to selective inhibition of CDK4 kinase activity. *Blood, 108*(5), 1744–1750.

Masters, G. A., Krilov, L., Bailey, H. H., Brose, M. S., Burstein, H., Diller, L. R., Patel, J. D., et al. (2015). Clinical cancer advances 2015: Annual report on progress against cancer from the American Society of Clinical Oncology. *Journal of Clinical Oncology, 33*(7), 786–809.

Matuszewska, A., Stefaniuk, D., Jaszek, M., Pięt, M., Zając, A., Matuszewski, Ł, Bancerz, R., et al. (2019). Antitumor potential of new low molecular weight antioxidative preparations from the white rot fungus *Cerrena unicolor* against human colon cancer cells. *Scientific Reports, 9*(1), 1975.

Meresse, P., Dechaux, E., Monneret, C., & Bertounesque, E. (2004). Etoposide: Discovery and medicinal chemistry. *Current Medicinal Chemistry, 11*(18), 2443–2466.

Murciano-Goroff, Y. R., Taylor, B. S., Hyman, D. M., & Schram, A. M. (2020). Toward a more precise future for oncology. *Cancer Cell, 37*(4), 431–442.

Niedens, B. R., Parker, S. R., Stierle, D. B., & Stierle, A. A. (1999). First fungal aromatic L-amino acid decarboxylase from a paclitaxel-producing *Penicillium raistrickii. Mycologia, 91*(4), 619–626.

Noman, E., Al-Shaibani, M. M., Bakhrebah, M. A., Almoheer, R., Al-Sahari, M., Al-Gheethi, A., Abdulaal, W. H., et al. (2021). Potential of anti-cancer activity of secondary metabolic products from marine fungi. *Journal of Fungi, 7*(6), 436.

Oliveira, A. P., Carvalho, S. M., & Rodrigues, M. E. (2023). Addressing bioavailability issues in fungal-derived cancer therapies. *Frontiers in Pharmacology, 14*(125), Article 987654.

Ozols, R. F. (1999). Paclitaxel plus carboplatin in the treatment of ovarian cancer. *Seminars in Oncology, 26*(1 Suppl 2), 84–89.

Prabhu, K. S., Siveen, K. S., Kuttikrishnan, S., Jochebeth, A., Ali, T. A., Elareer, N. R., Uddin, S., et al. (2019). Greensporone A, a fungal secondary metabolite suppressed constitutively activated AKT via ROS generation and induced apoptosis in leukemic cell lines. *Biomolecules, 9*(4), 126.

Prescott, T. A., Hill, R., Mas-Claret, E., Gaya, E., & Burns, E. (2023). Fungal drug discovery for chronic disease: History, new discoveries and new approaches. *Biomolecules, 13*(6), 986.

Robbins, H., & Monro, S. (1951). A stochastic approximation method. *The Annals of Mathematical Statistics, 22*(3), 400–407.

Salehi, B., Bayat, M., Dezfulian, M., Sabokbar, A., & Tabaraie, B. (2018). The assessment of anti-tumoral activity of polysaccharide extracted from terrestrial filamentous fungus. *Saudi Journal of Biological Sciences, 25*(6), 1236–1241.

Sang, H., Zhang, L., & Li, J. (2012). Anti-benzopyrene-7, 8-diol-9, 10-epoxide induces apoptosis via mitochondrial pathway in human bronchiolar epithelium cells independent of the mitochondria permeability transition pore. *Food and Chemical Toxicology, 50*(7), 2417–2423.

Santos, J. C., Profitós-Pelejà, N., Ribeiro, M. L., & Roué, G. (2022). Antitumor activity of simvastatin in preclinical models of mantle cell lymphoma. *Cancers, 14*(22), 5601.

Saravanakumar, K., Shanmugam, S., Varukattu, N. B., MubarakAli, D., Kathiresan, K., & Wang, M. H. (2019). Biosynthesis and characterization of copper oxide nanoparticles from indigenous fungi and its effect of photothermolysis on human lung carcinoma. *Journal of Photochemistry and Photobiology B: Biology, 190*, 103–109.

Seymour, L., Ivy, S. P., Sargent, D., Spriggs, D., Baker, L., Rubinstein, L., Berry, D., et al. (2010). The design of phase II clinical trials testing cancer therapeutics: Consensus recommendations from the clinical trial design task force of the national cancer institute investigational drug steering committee. *Clinical Cancer Research, 16*(6), 1764–1769.

Shamsaei, S., Getso, M., Ahmadikia, K., Yarahmadi, M., Farahani, H. E., Aslani, R., Raiesi, O., et al. (2021). Recent findings on the role of fungal products in the treatment of cancer. *Clinical and Translational Oncology, 23*, 197–204.

Sherr, C. J. (1996). Cancer cell cycles. *Science, 274*(5293), 1672–1677.

Singh, G., Sharma, P. K., & Malviya, R. (2021). Biomedical applications and patents on metallic nanoparticles. *Nanoscience and Nanotechnology-Asia, 11*(2), 153–162.

Smith, J., Brown, A., & Johnson, K. (2020). Nanocarriers as Drug Delivery Systems: Enhancing Therapeutic Efficacy and Reducing Side Effects. *Journal of Drug Delivery Science and Technology, 55*, Article 101390.

Srivastava, R., Ahmad, F., Mishra, B. N., Mathkor, D. M., Singh, V., & Haque, S. (2024). Terrein: Isolation, chemical synthesis, bioactivity and future prospects of a potential therapeutic fungal metabolite. *Natural Product Research*, 1–13.

Stierle, A., Strobel, G., Stierle, D., Grothaus, P., & Bignami, G. (1995). The search for a taxol-producing microorganism among the endophytic fungi of the Pacific yew, *Taxus brevifolia. Journal of Natural Products, 58*(9), 1315–1324.

Stierle, D. B., Stierle, A. A., & Bugni, T. (2003). Sequoiamonascins A– D: Novel anticancer metabolites isolated from a redwood endophyte. *The Journal of Organic Chemistry, 68*(12), 4966–4969.

Subban, K., & Kempken, F. (2023). Insights into Taxol® biosynthesis by endophytic fungi. *Applied Microbiology and Biotechnology, 107*(20), 6151–6162.

Sun, T., Zheng, W., Peng, H., Zhang, A., Chen, Y., Tan, R., & Shen, P. (2012). A small molecule IFB07188 inhibits proliferation of human cancer cells by inducing G2/M cell cycle arrest and apoptosis. *Biomedicine & Pharmacotherapy, 66*(7), 512–518.

Sung, H., Ferlay, J., Siegel, R. L., Laversanne, M., Soerjomataram, I., Jemal, A., & Bray, F. (2021). Global cancer statistics 2020: GLOBOCAN estimates of incidence and mortality worldwide for 36 cancers in 185 countries. *CA: Aa Cancer Journal for Clinicians, 71*(3), 209–249.

Tsoupras, A., & Davi, K. G. (2024). Bioactive metabolites from fungi with anti-inflammatory and antithrombotic properties: Current status and future perspectives for drug development. In S. K. Deshmukh, J. A. Takahashi, & S. Saxena (Eds.), *Fungi bioactive metabolites: Integration of pharmaceutical applications* (pp. 427–494). Springer.

Ulrich-Merzenich, G., Panek, D., Zeitler, H., Vetter, H., & Wagner, H. (2010). Drug development from natural products: Exploiting synergistic effects. *Indian Journal of Experimental Biology, 48*(3), 208–219.

Umscheid, C. A., Margolis, D. J., & Grossman, C. E. (2011). Key concepts of clinical trials: A narrative review. *Postgraduate Medicine, 123*(5), 194–204.

Uzma, F., Mohan, C. D., Hashem, A., Konappa, N. M., Rangappa, S., Kamath, P. V.,... & Abd_Allah, E. F. (2018). Endophytic fungi—Alternative sources of cytotoxic compounds: a review. *Frontiers in Pharmacology, 9*, 309.

Venturella, G., Ferraro, V., Cirlincione, F., & Gargano, M. L. (2021). Medicinal mushrooms: Bioactive compounds, use, and clinical trials. *International Journal of Molecular Sciences, 22*(2), 634.

Wang, C. C., Chiang, Y. M., Kuo, P. L., Chang, J. K., & Hsu, Y. L. (2008). Norsolorinic acid from *Aspergillus nidulans* inhibits the proliferation of human breast adenocarcinoma MCF-7 cells via Fas-mediated pathway. *Basic & Clinical Pharmacology & Toxicology, 102*(6), 491–497.

Wang, F. F., Shi, C., Yang, Y., Fang, Y., Sheng, L., & Li, N. (2018). Medicinal mushroom *Phellinus igniarius* induced cell apoptosis in gastric cancer SGC-7901 through a mitochondria-dependent pathway. *Biomedicine & Pharmacotherapy, 102*, 18–25.

Wang, H., Li, Y., Wang, Y., Shi, T., & Wang, B. (2024). *Penicillium janthinellum*: A potential producer of natural products. *Fermentation, 10*(3), 157.

Wang, Y., Zhao, X., & Liu, H. (2021). Lipid Nanoparticles for RNA Delivery: Applications and Challenges. *Nature Reviews Drug Discovery, 20*, 801–828.

Xianzhi, Y., Shiping, G., Lingqi, Z., & Hua, S. (2003). Select of producing podophyllotoxin endophytic fungi from podophyllin plant. *Natural Product Research and Development, 15*(5), 419–422.

Xie, L., Zhu, G., Shang, J., Chen, X., Zhang, C., Ji, X., Wei, Y., et al. (2021). An overview on the biological activity and anti-cancer mechanism of lovastatin. *Cellular Signalling, 87*, Article 110122.

Yamazaki, M., Suzuki, S., & Miyaki, K. (1971). Tremorgenic toxins from *Aspergillus fumigatus* Fres. *Chemical and Pharmaceutical Bulletin, 19*(8), 1739–1740.

Yan, J. K., Pei, J. J., Ma, H. L., Wang, Z. B., & Liu, Y. S. (2017). Advances in antitumor polysaccharides from *Phellinus sensu lato*: Production, isolation, structure, antitumor activity, and mechanisms. *Critical Reviews in Food Science and Nutrition, 57*(6), 1256–1269.

Yin, J. Y., Lai, M., Yu, X. Y., Su, D. D., Xiong, X. Y., & Li, Y. L. (2024). Comprehensive strategies for paclitaxel production: insights from plant cell culture, endophytic microorganisms, and synthetic biology. *Horticulture Research, 12*(3), uhae346.

Yu, N., Fu, Y., Fan, Q., Lin, L., Ning, Z., Leng, D., She, T., et al. (2024). Antitumor properties of griseofulvin and its toxicity. *Frontiers in Pharmacology, 15*, 1459539.

Zhang, D., Zhao, L., Wang, L., Fang, X., Zhao, J., Wang, X., Yu, L., et al. (2017). Griseofulvin derivative and indole alkaloids from *Penicillium griseofulvum* CPCC 400528. *Journal of Natural Products, 80*(2), 371–376.

Zhang, L., Lu, Z., Zhao, Q., Huang, J., Shen, H., & Zhang, Z. (2011). Enhanced chemotherapy efficacy by sequential delivery of siRNA and anticancer drugs using PEI-grafted graphene oxide. *Small (Weinheim an der Bergstrasse, Germany), 7*(4), 460–464.

Zhang, P. H., Yu, X. Y., Weng, L. X., Sun, L. L., Mao, Z. C., & Zhang, Y. L. (2019). Degradation of ferulic acid by the endophytic fungus *Colletotrichum gloeosporioides* TMTM-13 associated with *Ostrya rehderiana* Chun. *ACS Omega, 4*(25), 21000–21004.

Zhang, W., Gou, P., Dupret, J. M., Chomienne, C., & Rodrigues-Lima, F. (2021). Etoposide, an anticancer drug involved in therapy-related secondary leukemia: Enzymes at play. *Translational Oncology, 14*(10), Article 101169.

Zorova, L. D., Popkov, V. A., Plotnikov, E. Y., Silachev, D. N., Pevzner, I. B., Jankauskas, S. S., Zorov, D. B., et al. (2018). Mitochondrial membrane potential. *Analytical Biochemistry, 552*, 50–59.

Immunomodulatory and Antiviral Fungal Metabolites

V. S. Lekshmi, Abhila Asi, U. M. Arya, Kumari Asha, and Binod Kumar

Abstract

Viral outbreaks and pandemics have created an increasing interest in finding new antiviral compounds from natural sources that offer a broad spectrum of pharmacological properties. Fungal secondary metabolites play an extensive role in human health and disease. The efficiency of fungi to adjust to various fermentation conditions makes them potent novel drug candidates. Only 3–8% of fungal species have been explored for their pharmacological activities, representing a significant gap in our understanding of their full potential. Ascomycota and Basidiomycota are the primary reservoirs of diverse fungal secondary metabolites (FSMs) with immunomodulatory and antiviral activities. These effects are largely mediated by modulating the cytokine production in the immune system. Fungal metabolites such as indole alkaloids (IA), non-ribosomal peptides (NRPS), polyketides (PKS), and terpenoids have been widely discussed for their antiviral activities against a wide variety of viruses, including Herpes simplex virus (HSV), Influenza A Virus (IAV), and Severe Acute Respiratory Syndrome Coronavirus 2 (SARS-CoV-2). The chapter discusses recent advancements in the area of immunomodulatory and antiviral activities of fungal secondary metabolites. It further emphasizes their mechanisms of action, synergistic effects in combination therapies, and the associated challenges.

Keywords

Pandemic · Antiviral compounds · Immunomodulation · Viruses · Influenza virus · HIV · Liposomes · Hydrogel · Micelles · Exosomes

1 Introduction

Immunomodulation refers to the process of reprogramming the immune system to either enhance or suppress its function when it is not operating properly. Over the years, various biological and synthetic immunomodulatory agents have been utilized; however, the use of some has been limited due to their systemic toxicity and narrow therapeutic index. Natural products can act as immunomodulatory agents, offering considerable benefits with minimal or no toxicity by influencing both innate and adaptive immunity. Plant-derived compounds such as epigallocatechin-3-gallate, quercetin, curcumin, and colchicine (Zebeaman et al., 2023), algal sulphated polysaccharides (SPs), Polyunsaturated fatty acids (PUFAs), and fungal secondary metabolites are some of the most discussed powerful immunomodulating candidates.

In general, immunomodulators are grouped into two distinct classes: immunostimulants and immunosuppressants. Immunostimulators help the body fight against infections or diseases by boosting the immune system. They can also restore balance in case of immunosuppression caused by certain pathological conditions, typically working in a non-specific manner. Examples of immunostimulants include interferons (IFNs), interleukins (ILs), bacterial vaccines, colony-stimulating factors, and other vaccines. On the contrary, immunosuppressants help in reducing the immune response in overactive conditions, such as autoim-

V. S. Lekshmi (✉)
Laboratory of Viral Immunology and Applied Virology, Pathogen Biology Division, Rajiv Gandhi Centre for Biotechnology, Thiruvananthapuram, Kerala, India
e-mail: lvsvanaja@gmail.com

A. Asi · U. M. Arya
Centre for Bio-Separation Technology, Vellore Institute of Technology, Vellore, Tamil Nadu, India

K. Asha · B. Kumar (✉)
Department of Cell and Biological Systems, Penn State University College of Medicine, Hershey, PA, USA
e-mail: binod.biochem@gmail.com

© The Author(s), under exclusive license to Springer Nature Switzerland AG 2026

A. Prakash and M. K. Dubey (eds.), *Advances in Fungal Secondary Metabolites: Bioactive Attributes and Biotechnological Potential of Industrial Interest*, Sustainable Economy and Ecotechnology, https://doi.org/10.1007/978-3-032-23507-7_5

mune disorders, allergies, and organ transplantation surgeries, primarily by mitigating the inflammatory conditions. Examples of immunosuppressants include antimetabolites like mycophenolate mofetil and azathioprine, corticosteroids, calcineurin inhibitors like cyclosporine and tacrolimus, mTOR (mechanistic target of rapamycin) and JAK (Janus kinase) inhibitors, as well as basiliximab, a monoclonal antibody. Additionally, adalimumab and infliximab are examples of commonly used biologics that act as immunosuppressants (Hadden, 1996).

Fungi are a naturally occurring reservoir of biologically active compounds of great health importance that remain largely unexplored. The kingdom fungi represents a wide range of secondary metabolites having immunomodulatory properties. Fungal secondary metabolites (FSMs) exert immunomodulatory activities through diverse mechanisms (Lebish & Moraski, 1987):

- Direct interaction with immune cells
- Regulation of cytokine production
- Induction of regulatory immune responses
- Modulation of immune cell trafficking

The immunomodulatory functions of fungal metabolites are mediated by various signalling pathways and cellular interactions that can enhance or suppress immune functions. The FSMs, such as ochratoxin A and patulin, have been reported to exhibit immunosuppressive effects by potentially compromising the immune functions (de Sales-Neto & Rodrigues-Mascarenhas, 2024). FSMs with antiviral properties target different phases of the viral life cycle, such as attachment, pentration, replication, viral protein synthesis, and viral particle assembly. For example, the antiviral activity of triterpenoids from *Ganoderma lucidum* has been demonstrated against enterovirus 71 (Zhang et al., 2014). The compound cordycepin, obtained from *Cordyceps militaris*, has demonstrated antiviral effects against a broad spectrum of viruses, including influenza and dengue virus (Lee et al., 2014). In the present chapter, we discuss in-depth mechanistic details of the biological activities of a wide variety of FSMs by highlighting their immunomodulatory and antiviral activities that provide opportunities for better therapeutic interventions.

2 Mechanism of Immunomodulation by Fungal Secondary Metabolites (FSMs)

The immune system plays a vital role in fighting against pathogens and preventing infections. The immunomodulatory action of FSMs involves various immune cells, including natural killer cells (NKs), macrophages, T-cells, B-cells, and dendritic cells. Immunostimulatory compounds activate these cells to secrete cytokines, antibodies, or chemokines to activate the immune system. Conversely, immunosuppressants inhibit the activation of immune cells and thus reduce their proliferation, differentiation, and further activation. Some FSMs can exhibit both immunostimulatory and immunosuppressive activities. This dual activity allows FSMs to regulate the immune system by modulating the activity and expression of transcription factors and signalling molecules associated with immunomodulation. This is particularly beneficial for autoimmune diseases, organ transplantation, chronic inflammatory conditions, obesity, neurodegeneration, and atherosclerosis. The FSMs exhibit these actions either by directly binding to the immune cell receptors, which are involved in antigen presentation, such as Toll-Like Receptors (TLRs) and Major Histocompatibility Complex (MHC), or by influencing the regulatory factors (Pillai & Antony, 2024).

Innate immunity functions as the body's first-line, non-specific defence mechanism against infection, which provides protection through epithelial barriers, the complement system, natural killer (NK) cells, and macrophages that are inherently tied to the induction of inflammation (Fig. 1).

Macrophages, as effector cells, defend the body through phagocytosis, proinflammatory cytokine production, and programmed cell death. NK cells respond rapidly to pathogens without needing antigen activation and can induce apoptosis by releasing apoptotic factors during infection (Hirayama et al., 2017). On the other hand, signals or antigens will be detected as pathogen/microbe-associated molecular patterns (PAMPs/MAMPs) or host-derived damage-associated molecular patterns (DAMPs) by pattern recognition receptors (PRRs) on phagocyte surfaces. These are then internalized, digested, and presented to adaptive immune cells via major histocompatibility complex II (MHCII). Upon PRR activation, NF-kB (Nuclear Factor-Kappa B) in B-cells initiates the transcriptional activity of proinflammatory and inflammasome-produced cytokines, such as NLRP3 (nucleotide-binding domain leucine-rich repeat-containing family, pyrin domain containing 3). The NLRP3-mediated inflammasome gets activated in response to pathogenic stimuli, triggering the induction of proinflammatory cytokines (Hayden & Ghosh, 2008).

The complement system (CS), an essential part of the host innate immune system, consists of proteins that help combat pathogens and promote healing. It can be activated via three distinct ways, namely the alternative pathway, the classical pathway, and the lectin pathway. The classical route requires an antigen–antibody (Ag-Ab) complex, while the alternative pathway relies on molecules on the pathogen's surface. The lectin pathway is mainly triggered

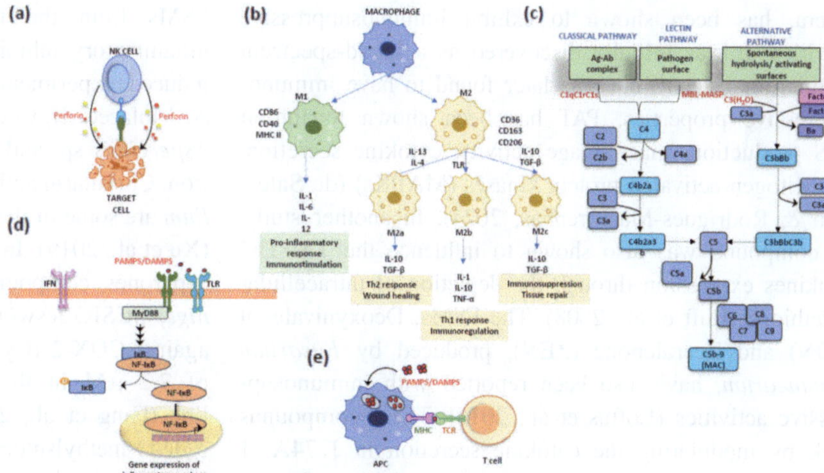

Fig. 1 Schematic diagrams showing the expected mechanisms involved in immunomodulatory activity of fungal secondary metabolites. **a–c** Innate immune response, **d, e** Adaptive immune response (*Source* Authors)

by mannan-binding lectins that bind to the microorganisms. However, some pathogens have evolved mechanisms to evade complement activation, using their own receptors to infect host cells (Rus et al., 2005).

Adaptive immunity is a highly specific and self-limiting immune response involving the T-cells (T-lymphocytes) and B-cells (B-lymphocytes). It recognizes a wide range of antigens, quickly eliminating them after infection while preserving immune memory. During cell-mediated immunity, Th1 cells (T helper cells) are activated, leading to antigen-presenting cells or APCs activation and cytotoxic T cell activation, while in humoral immunity, B-cells are stimulated to produce antibodies by the stimulation of Th2 cells. Th1 cells activate APCs by signalling through the CD40-CD40 (Cluster of differentiation 40) ligand upon receiving a pathogen signal via TCR (T cell receptor), and by releasing interferon-gamma (IFN-γ). This also activates the cytotoxic T-cells by secreting IL-2. Activated APCs present antigens to cytotoxic T-cells through MHC I, allowing the T-cells to identify infected cells. This triggers the release of granulysin, granzymes, and perforins, leading to cell apoptosis, DNA fragmentation, and cell death through death-inducing signalling complex (DISC) formation via Fas ligand interactions (Strzelec et al., 2023).

Several FSMs have already been shown to exert immunomodulatory functions in experimental models. Aflatoxin, a secondary metabolite of *A. parasiticus* and *Aspergillus flavus*, has been shown to have immunomodulatory actions by affecting the innate immunity and cell-mediated immunity, and to a larger degree than humoral immune responses in several experimental models such as poultry and pigs (Bondy & Pestka, 2000). Similarly, mycotoxin Ochratoxins A produced by *Aspergillus ochraceus* has been shown to inhibit humoral/antibody-mediated, cellular, and innate immune responses in poultry (Dwivedi & Burns, 1984). In summary, these diverse components of our immune system work together in harmony to ensure physiological homeostasis by fighting against pathogens. Fungal secondary metabolites are known to exert immunomodulatory effects by regulating one or more of the mechanisms outlined above.

2.1 Case Studies and Clinical Applications of FSMs

FSMs are high-value nutraceutical candidate drugs that exhibit both immunostimulatory as well as immunosuppressive activities. Cyclosporin A (CsA), a bioactive secondary metabolite produced by many fungi like *Tolypocladium inflatum, Aspergillus terreus, Fusarium oxysporum, Fusarium solani* and *Penicillium fellutanum* are excellent immunosuppressant widely employed in the treatment of autoimmune disorders and to minimize the risk of organ transplant rejection by the body, primarily inhibiting calcineurin or PP3 (Protein Phosphatase 3), an enzyme that is very important for cellular immunity (Kountouras et al., 2004). Antimetabolites such as methotrexate are used to treat rheumatoid arthritis by lowering immune activity. It is also used to treat cancer and psoriasis by slowing the cell growth. Cyclosporin A was also proven to be an effective alternative therapy in patients with rheumatoid arthritis during conditions in which the body was unable to adequately respond to methotrexate. The derivatives of mycophenolic acid, such as mycophenolate mofetil (MMF) derived from *Penicillium stoloniferum*is, are potent immunosuppressors commonly prescribed to organ transplant recipients (Hood & Zarembski, 1997). It acts as an immunosuppressant by inhibiting purine biosynthesis and preventing the proliferation of lymphocytes and T-cells, and also by restricting the antibody formation from B-cells. Similarly, Patulin (PAT), a FSM derived from numerous fungal species of *Penicillium* and *Aspergillus* genera, has been shown to exhibit immunosuppressive activities. It was initially discovered as a broad-spectrum antimicrobial agent, but was later found to have immunosuppressive properties. PAT has been shown to inhibit ROS production, macrophage activity, cytokine secretion, and Mitogen-activated protein kinases (MAPKs) (de Sales-Neto & Rodrigues-Mascarenhas, 2024). In another study, the compound was also shown to influence the Th1/Th2 cytokines expression through the depletion of intracellular glutathione (Luft et al., 2008). The FSMs, Deoxynivalenol (DON) and Zearalenone (ZEN), produced by *Fusarium graminearum*, have also been reported with immunosuppressive activities (Loftus et al., 2016). These compounds work by modulating the cytokine secretion in J774A. 1 murine macrophages; the cell lines were derived from a female BALB/c mouse.

Immunostimulant FSMs isolated from mushrooms are found to work by stimulating cytokine release through immune cell activation. These FSMs have therapeutic effects in the treatment of autoimmune diseases, allergy, and cancer. Levamisole, thalidomide, isoprinosine, immunocynin, and bestatin are some of the widely used immunostimulators. Recombinant cytokines and some complex polysaccharides from microbial sources are another category of immunostimulants that have been studied a lot (Jain et al., 2022). Among the fungal metabolites, polysaccharides, rather than secondary metabolites, have been studied more for their immunostimulatory activities. PG101, a water-soluble FSM isolated from the cultured mycelia of *Lentinus lepideus* was found to selectively activate transcription factors and cytokines such as TNF-α (Tumour necrosis factor-alpha), IL-β, IL-10, and IL-12 in Peripheral Blood Mononuclear (PBM) Cells, without affecting IFN-γ and IL-4 levels (Jin et al., 2003). In another study, polysaccharides from the fresh fruiting bodies of *Ganoderma lucidum* were shown to have antitumor activity by potentiating cytokine production by macrophages and T-lymphocytes (Wang et al., 1997). Likewise, the pyrrolidine alkaloid, preussins compounds obtained from the marine sponge-associated fungus *Aspergillus flocculosus* 16D-1 revealed anti-inflammatory activity by inhibiting IL-6 in THP1 cells with an half-maximal inhibitory concentration (IC_{50}) value of 0.11 μM when compared to the known inhibitor corylifol A with the calculated IC_{50} value of 0.67 μM (Gu et al., 2018). Similarly, Chrysamide C from *P. chrysogenum* SCSIO41001 is an FSM with anti-IL17 activity (Chen et al., 2016). The terpenoids Brasilanones and Dihydrobipolaroxins isolated from *Aspergillus* sp. showed anti-inflammatory activities in LPS (Lipopolysaccharide)-activated RAW264.7 cells by inhibiting the release of nitric oxide (NO) (Cheenpracha et al., 2010). Thomimarine, Graphostromane, Khusinol, Chondroterpenes, Hirsutanol A, and lovastatin are other FSMs from the terpenoid category that have shown inflammatory inhibitory activity by targeting NO in LPS-induced experimental models. VersicolactoneG, Territrem A, Violaceol II, Cordyol E, Cladosporin and Asperlin from *Aspergillus* sp. and Guaiadiol A, Citrinin H1, Penicillospiron, Curvularin and Pyrenocine from the species of *Penicillium* are some of the polyketides studied for anti-NO effects (Xu et al., 2019). In a study, Asperpyrone-type bis-naphtho-γ-pyrones compound from marine-associated fungus *A. niger* SCSIO Jcsw6F30 exhibited anti-inflammatory activity against COX-2 (Cyclooxygenase-2) with mean IC_{50} value of 7.2 μM in the RAW264.7 murine macrophage cell line (Fang et al., 2016) whereas the polyketides such as 3-de-O-methylsulochrin, Canthone A, Eurobenzophenone B, Dermolutein, Methylemodin and Yicathin B obtained from *Aspergillus* sp. have been found to inhibit NF-κB nuclear translocation to exhibit anti-inflammatory effect in LPS-activated SW480 cells (Xu et al., 2021). Also, Ko et al. (2016) studied the effects of alternaramide, a cyclic pentadepsipeptide from marine-associated fungi, on inflammation. From the experiments conducted in LPS-stimulated BV2 and RAW264.7 cells, they found that these FSMs inhibit MAPK and NF-κB signalling through the TLR4-MyD88 pathway.

Overall, between the period 2000–2018, more than 130 anti-inflammatory and hundreds of immunomodulatory FSMs belonging to diverse chemical groups have been reported, including terpenoids, peptides, alkaloids, and polyketides. It has been demonstrated that polyketides and terpenoid compounds derived mainly from *Penicillium* and *Aspergillus* spp. have potential as effective anti-inflammatory drugs. In summary, the research data suggested that the fungi-derived novel metabolites may be used as possible lead drug candidates for autoimmune diseases and other inflammatory conditions, which would be a low-cost and side-effect-free treatment strategy (Table 1).

3 Antiviral Properties of Fungal Metabolites

Since the discovery of Penicillin, the event that launched the antibiotic era, the global scientific community has started considering fungi as a promising reservoir of natural metabolites with significant biotechnological and health benefits. The interest reignited during the global COVID-19 (Coronavirus disease 2019) outbreak when the search for an effective antiviral drug became a global priority. Although fungal secondary metabolites (FSMs) play no direct role in the normal physiological functions of fungi, these compounds, however, provide several added benefits to mankind, especially in the pharmaceutical field, serving as antibiotics, immunomodulators, antivirals,

Table 1 Fungal secondary metabolites with Immunomodulatory activities (*Source* Authors' compilation)

Immunomodulatory FSMs	Source	Mechanism of action	References
Aflatoxin	*Aspergillus flavus* *A. parasiticus*	It primarily affects cell-mediated immunity and the function of phagocytic cells	Bondy and Pestka (2000)
Ochratoxin	*Aspergillus ochraceus*	Inhibits humoral, cellular, and innate immune responses	Dwivedi and Burns (1984)
Cyclosporin A	*Tolypocladium inflatum, Aspergillus terreus, Fusarium oxysporum, Fusarium solani Penicillium fellutanum*	It primarily inhibits calcineurin, also known as PP3 (Protein Phosphatase 3)	Kountouras et al. (2004)
Mycophenolate mofetil (MMF)	*Penicillium stoloniferumis*	It functions as an immunosuppressant by inhibiting purine biosynthesis, thereby preventing the proliferation of lymphocytes and T-cells, and by limiting the antibody formation from B-cells	Hood and Zarembski (1997)
Patulin (PAT)	*Penicillium expansum, Aspergillus clavatus, Byssochlamys fulva*	It inhibits phagocytosis by macrophages, as well as the release of cytokine, production ROS and also blocks the activation of NF-κB and MAPKs	de Sales-Neto and Rodrigues-Mascarenhas (2024); Luft et al. (2008)
Deoxynivalenol (DON) Zearalenone (ZEN)	*Fusarium graminearum*	Work by modulating the cytokine secretion	Loftus et al. (2016)
PG101	*Lentinus lepideus*	Selectively activate transcription factors and cytokines such as, IL-β, IL-10, IL-12 and TNFα	Jin et al. (2003)
Pyrrolidine alkaloid	*Aspergillus flocculosus 16D-1*	It shows anti-inflammatory activity by inhibiting IL-6 in THP1 cells	Gu et al. (2018)
Chrysamide C	*P. chrysogenum SCSIO41001*	Anti-IL17 activity	Chen et al. (2016)
Brasilanones and Dihydrobipolaroxins	*Aspergillus* sp.	Showed inflammatory inhibitory activities in LPS-activated RAW264.7 cells by inhibiting the release of nitric oxide	Cheenpracha et al. (2010)

anti-inflammatory, and antitumor agents. The secondary metabolites are small molecular weight compounds derived from the primary metabolites and are known for their broad diversity and inherent biocompatibility. A wide range of secondary metabolites has been uncovered from fungi, and the best known are classified as indole alkaloids (IA), polyketides (PKs), non-ribosomal peptides (NRPs), ribosomal peptides (RPs), terpenes, shikimate-derived compounds, and NRPS-PKS hybrids (Keller et al., 2005).

Following the SARS-CoV-2 outbreak, antiviral drug development became more prevalent, resulting in a desperate need for natural products for drug discovery as they are bioactive, safe, cost-effective, less toxic, and biodegradable. The continued emergence of viral diseases is a matter of serious public health concern and a never-ending battle, even after having a handful of therapeutic options, due to the continuous mutations and drug resistance. This section summarizes the current knowledge about the antiviral activities of FSMs against some major viral diseases. Furthermore, we highlighted the details of molecular mechanisms and targets involved in the antiviral activities of these compounds.

3.1 Overview of Antiviral Compounds from Fungi

Viruses are sub-microscopic infectious agents that are obligate intracellular parasites, with sizes ranging from about 20 to 400 nm (nm) in diameter. It can only replicate within a living system that includes prokaryotic organisms (bacteria and archaea) and eukaryotic organisms (algae, animals, fungi, plants, and protozoa) (Dimmock et al., 2015). Viruses have a simple structure composed of nucleic acid, either RNA or DNA, enclosed within a protein coat, with or without an envelope. The viral infection begins upon entry into a host cell, where the viral genetic material is uncoated, and the host cell machinery is hijacked to aid viral replication and produce viral proteins for the formation of new virions. Influenza, AIDS, COVID-19, dengue, chicken pox, and rabies are some of the prominent diseases caused by viruses that infect humans (Strauss & Strauss, 2008).

After decades of research on vaccines and antiviral drugs, viruses are still causing serious outbreaks and pose an economic burden all over the world. The major challenges behind the vaccine or antiviral drug development are the constant mutations and the advent of emerging of novel serotypes in the virus group, along with cases of drug resistance. To combat viral infections, vaccination strategies should be complemented with antiviral drugs, especially in immunocompromised populations and those with co-morbidities. The antiviral drugs primarily target either viral proteins or the host-derived cellular proteins utilized by the virus for productive infection (De Clercq, 2002). Any antiviral drug should meet some prerequisites, such as its non-toxicity and minimal or no side effects to the host cell. It should also completely inhibit viral infection, as partial inhibition frequently results in the emergence of resistant strains. The urgency of developing antivirals against emerging and re-emerging viruses and managing pandemics shifted the concept of 'one bug-one drug' to 'broad-spectrum antivirals', preferably from natural sources. The technological advances in microbial culturing and analytical techniques have addressed the challenges in using natural products and their structural analogues in pharmacotherapy. According to the reports, more than 40% of the modern drugs registered in the Food and Drug Administration (FDA) are based on natural sources or related compounds (Thomford et al., 2018).

Members of the kingdom fungi constitute an important repository vast array of secondary metabolites with a broad spectrum of bioactivities. However, the research on antiviral properties of fungal metabolites has a significant research gap that requires comprehensive investigation. Most of the antiviral studies have concentrated on FSM from the phyla Ascomycota and Basidiomycota (Fig. 2).

Bioactive secondary metabolites produced from the endophytic fungi, especially those from marine habitats, have been the focus of interest from the ascomycetes group, while the edible mushrooms were the prominent antiviral candidates from the basidiomycetes. The combined genome data with the traditional screening method suggest that ascomycetes are having larger number of secondary metabolites compared to basidiomycetes due to the differences in the number of biosynthetic gene clusters and their genome dominance (Bergmann et al., 2007; Clevenger et al., 2017).

Experimental results indicate that most of the DNA sequences encoding the production of diverse secondary metabolites exist as cryptic genes, the genes capable of coding for functions but remain silent under traditional fermentation conditions (Lim et al., 2012). Altering culture parameters like nutrient sources (carbon and nitrogen sources), pH, and temperature can act as an effective stimulus to activate these pathways, thereby resulting in better production of antiviral secondary metabolites (Brakhage, 2013).

Several bioactive fungal secondary metabolites have already been identified as potent antiviral agents against a varied range of viruses, especially against HIV and influenza viruses (Table 2).

Our focus in this section is on the virus-inhibiting activities of fungal metabolites with their underlying molecular mechanisms.

3.2 Mechanism of Antiviral Action

The mechanism of action of antiviral drugs differs for different viruses, depending on their structure, host invasion and replication cycle. Approaches to develop antiviral drugs for managing viral infections are mainly focused on two strategies: directly targeting the virus or regulating host factors involved in antiviral immune response (Fig. 3). The direct antiviral drugs target directly on viral particles, virus binding, entry, uncoating, or replication enzymes like polymerase, protease, and integrase (Ryu, 2016). Even after the development of inhibitors against key viral events-such as ritonavir (protease inhibitor), acyclovir (inhibitor of viral DNA polymerase), remdesivir (nucleoside analogue) and raltegravir (integrase inhibitor), all of which are listed among the top 200 drugs by sales in the recent years, there remains a lack of highly effective antiviral drugs or complete cure for many viral infections. This has accelerated the focus on host-directed therapies that target the host factors and pathways that the virus uses for its replication and propagation. The major areas are the host activities, such as interferon (IFN) response, as well as modulation of immune response during viral infection (Yu et al., 2016a, 2016b). So far, studies conducted to explore the mechanism of action of antiviral FSMs are scanty. Velutin, a ribosome-inactivating protein isolated from *Flammulina velutipes* fruiting bodies, commonly known as the winter mushroom, has been identified as an antiviral compound against HIV-1 through the inhibition of reverse transcriptase, β-glucuronidase, and β-glucosidase activities (Wang and Ng, 2001). Similarly, in 2000, a research group discovered that water and methanol soluble components of *Ganoderma lucidum* have antiviral activity towards Influenza A virus (IAV) along with vesicular stomatitis virus (VSV) (Eo et al., 2000).

In 2007, a novel FSM named GFAHP was extracted from the *Grifola frondose* fruiting bodies (Gu et al., 2007). It proved to be a potent antiviral agent against HSV-1

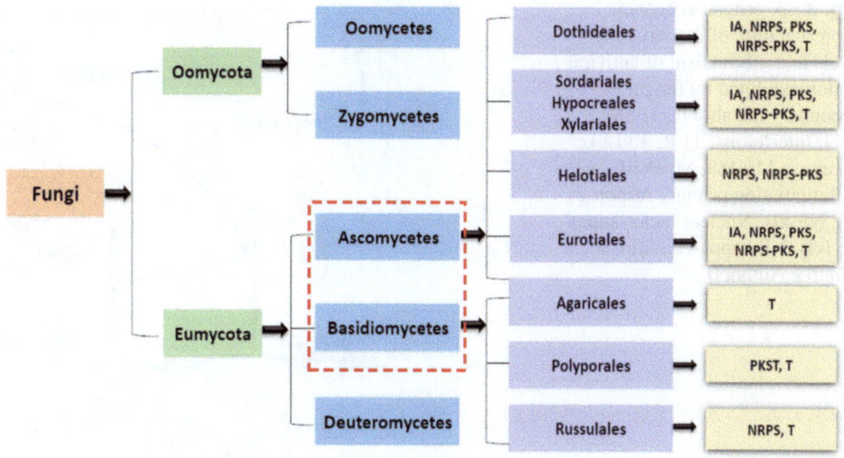

Fig. 2 General fungal classification and currently recognized fungal-derived antiviral molecules. T, terpenoids; NRPS, non-ribosomal peptides; PKS, polyketides; NRPS-PKS, hybrids; IA, indole alkaloids (*Source* Authors)

Table 2 Fungal secondary metabolites (FSMs) with anti-HIV activities (Roy, 2017)

Antiviral FSMs	Source	Mechanism of action
Isochromophilones I	*Penicillium nicillium*	Entry inhibitors
Variecolin Variecolol	*Emericella aurantiobrunnea*	
10-methoxydihydrofuscin Fuscinarin	*Oidiodendron griseum*	
Sch 210971 Sch 210972	*Chaetomium globosum*	
Phenylspirodrimanes Stachybotrins D	*Stachybotrys chartarum*	RT (reverse transcriptase) inhibitor
Altertoxins	*Alternaria tenuissima*	
Equisetin Phomasetin	*Fusarium heterosporum* *Phoma* sp.	HIV-1 integrase inhibitor
Cytosporic acid Australifungin	*Cytospora* sp. *Sporormiella australis*	
Funalenone Atrovenetinone methyl acetal	*Aspergillus niger FO-5904* *Penicillium* sp. *FKI-146387*	
Integracide A	*Fusarium* sp.	
Isochaetochromin B1, B2 and D1 Oxychaetochromin B	*Fusarium* sp.	
Doxorubicin	*Streptomyces peucetius*	
Exophillic acid Aquastatin A	*Exophiala pisciphila* *Fusarium aquaeductuum*	
Colossolactones	*Ganoderma colossum*	HIV-1 protease (HIV-1 PR) inhibitors
Ganoderic acid Ganomycin	*G. sinense* *G. lucidum*	
Isocochliodinol and didemethylasterriquinone D (asteriquinones)	*Chaetomium* sp. *Chrysosporium merdarium*	
Hinnuliquinone	*Nodulisporium hinnuleum*	
cytochalasins	*Hypoxylon fragiforme*	
Alachalasin A	*Stachybotrys charatum*	
Flazin and its synthetic amido analogue flazinamide	*Suillus granulatus*	HIV-1 inhibitory activity by potently reducing syncytium formation

by inhibiting its replication cycle in vitro with a therapeutic index > 29.3 and an IC_{50} value of 4.1 μg/ml. The research also found that this novel protein, at higher concentrations, inactivates HSV-1 particles directly while concurrently hindering HSV-1 from entering Vero cells. This suggests that this compound exhibits an antiviral mecha-

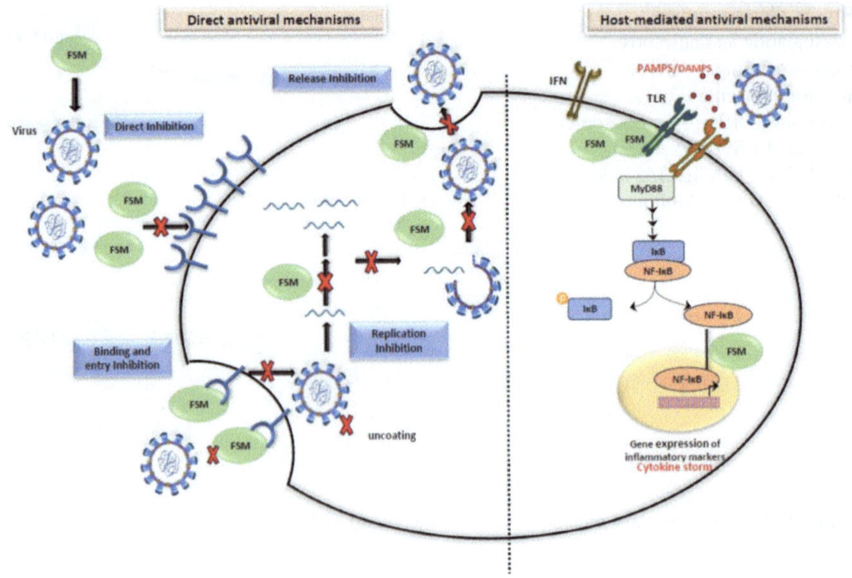

Fig. 3 A general schematic representation of the virus life cycle and mechanism of antiviral action exhibited by fungal secondary metabolites (FSMs). IFN, Interferons; TLR, Toll-like receptor; MyD88, Myeloid differentiation primary response 88; NF-κB, Nuclear factor-kappa B; IκB, NF-kappa-B inhibitor (*Source* Authors)

nism via three different mechanisms. If the virus particles escape the antiviral activity of FSM, they will be inhibited during the entry process; if they somehow enter the cells, the replication is prevented, ensuring the complete elimination of the virus particles from the system. In another study, researchers investigated the antiviral property of selected triterpenoids from isolated *G. lucidum* as entry inhibitors against enterovirus 71 (EV71) infection, which is the major cause of HFMD (Hand, Foot and Mouth Disease) (Zhang et al., 2014). In the very next year, Sacramento et al. (2015) discovered a fungi-derived tetrahydrofuran, Aureonitol, from *Chaetomium coarctatum* that could inhibit influenza virus replication by targeting its surface glycoprotein-Hemagglutinin (HA). In 2019, a computational-based drug discovery suggested the inhibitory potential of this compound against the dengue virus by inhibiting DENV-NS2B-NS3 protease activity (Bharadwaj et al., 2019). The protective inflammatory responses in COVID-19 patients administered with *G. lucidum* were reported by Al-Jumaili et al. (2020), and the outcome of this study revealed that it is through the modulation of haematological and immunological responses. During a virus infection, the components of the virus, including the genetic material, are recognized as PAMPS by the Pattern Recognition molecules. RNA viruses are primarily recognized by cytoplasmic RIG-I-like receptors (RLRs) and endosomal toll-like receptors (TLRs). DNA viruses are detected by various cytoplasmic DNA sensors, such as cyclic GMP-AMP synthase (cGAS) and by TLR9, the endosomal toll-like receptor. Upon viral entry, the innate immunity system of our body activates and functions as the first line of defence and triggers interferon (IFN) release. This triggers downstream signalling cascades that activate transcription factors such as NF-κB and interferon regulatory factors (IRFs) that lead to the induction of type-I IFNs and other proinflammatory cytokines (Sengupta & Chattopadhyay, 2024). The extensive data suggest that even the same or different secondary metabolites of the same fungal origin can have antiviral activity against multiple viruses by exhibiting diverse modes of inhibitory mechanism including virus as well as host-mediated factors.

3.3 Case Studies

3.3.1 Antiviral Fungal Secondary Metabolites Against Influenza

Influenza viruses, classified under Orthomyxoviridae family, have a negative-sense, single-stranded, segmented RNA genome and are one of the leading causes of respiratory viral infections worldwide. The major types encompass: Influenza A Virus (IAV), Influenza B Virus (IBV), Influenza C Virus (ICV), and Influenza D Virus (IDV). For IAV and IBV, there are eight viral RNA segments, while ICV has only seven segments and each codes for distinct viral proteins (Bouvier & Palese, 2008).

The flu, the disease caused by the influenza virus, is a communicable viral disease that represents a global burden on public health and affects the upper and lower respiratory tract. IAV, among all, represents the most common type that has a wide host range. It mutates frequently, driving the emergence of new influenza strains with unexpected pathogenicity. The classification of IAVs is based on their Hemagglutinin (HA), Neuraminidase (NA), and surface glycoproteins (Hutchinson, 2018). The 'Great Influenza' pandemic or Spanish flu pandemic of 1918, caused by the H1N1 strain of IAV, resulted in over 50 million deaths worldwide and is considered the most dangerous infectious disease outbreak in history. This pandemic was subsequently followed by more influenza pandemics caused by other strains. Similarly, the global spread of H1N1, the swine-origin influenza virus A, in humans attracted significant international focus in 2009. Likewise, the highly pathogenic avian influenza (HPAI) H5N1 has been circulating worldwide, causing mortality in birds. In 2021, HPAI H5N8 was reported in seals, wild foxes, and in a child from Laos, who was the first human case with H5N8. During the period 2003–2023, 878 cases of human H5N1 infection, including 52.16% fatalities, have been documented by WHO from over 23 countries, and it continues (Charostad et al., 2023). A few antiviral drugs against the NA protein, such as Relenza, Tamiflu, Zanamivir, and Oseltamivir, and against the matrix protein 2 (M2) protein, like adamantanes (amantadine and rimantadine), are available to combat influenza. However, the drug resistance, antigenic drift, and antigenic shift limit the current use of antiviral drugs and vaccination, and hence novel antiviral drugs that can overcome these issues are highly sought after and are being investigated.

The antiviral activity of several FSMs has been studied against influenza viruses. Two antiviral secondary metabolites, stachyflin and acetylstachyflin, were purified from *Stachybotrys* sp. RF-7260 through SSF (solid-state fermentation) and compared with the known drugs in MDCK cells. The study showed that stachyflin exhibited 77 times greater antiviral activity against H1N1 (A/WSN/33) virus than acetylstachflin, 1760 times greater than amantadine, and 250 times greater than zanamivir, due to the presence of a hydroxy group at the C-3 position. It was also observed that the antiviral activity of stachyflin was executed by inhibiting the conformational change in HA, which is essential for virus-cell membrane fusion (Minagawa et al., 2002; Yoshimoto et al., 1999).

In 2011, six isoindolone derivatives were extracted from *Emericella* sp. (HK-ZJ), a mangrove plant-derived endophytic fungus. Among all, Emerimidine A and B were reported to have moderate antiviral activity in MDCK cells (Zhang et al., 2011). In the same year, Wang et al. (2011) isolated some FSMs, among which only the polyketides exhibited anti-H1N1 activity with an average IC_{50} value of 67 μM (Wang et al., 2011). Similarly, the FSMs-oxoglyantripine, quinadoline, glyantrypine derivatives, tryptoquivaline, and deoxynortryptoquivaline were isolated from *Cladosporium* sp. PJX-41 worked against H1N1 with an average IC_{50} of 85 μM (Peng et al., 2013). In the year 2013, an antiviral cyclic tetrapeptide, asperterrestide A (1), obtained from the marine-associated fungus *Aspergillus terreus* SCSGAF0162, was reported to have inhibitory effects against H1N1 and H3N2 (He et al., 2013). In the same year, Gao et al. (2013) discovered that Aspulvinones from *Aspergillus terreus* Gwq-48, sourced from a mangrove rhizosphere soil, have anti-influenza A (H1N1) activity via neuraminidase (NA) inhibition. Computational docking analyses further showed that the antiviral activity was attributed to the trans-confirmation of the aryl group.

Wu et al. (2014) discovered hybrid polyketides Cladosins A-E from *Cladosporium sphaerospermum*, a deep-sea-associated fungus with moderate-level anti-influenza A H1N1 virus activity. In the same year, Lee et al. (2014) reported the anti-influenza activity of *Cordyceps militaris,* commonly known as caterpillar fungus. The investigation revealed that the fungi demonstrated antiviral activity against H1N1 in the DBA/2 mouse model through immunomodulation, mediated by NK cells and increased IL-12 expression. Subsequently, Peng et al. (2014) discovered FSMs Sorbicatechols A and B, obtained through cultivation of marine-derived fungal strain *Penicillium chrysogenum* PJX-17, for their activity against H1N1, with inhibitory concentrations (IC_{50}) values of 85 and 113 μM for each, respectively.

Similarly, in 2016, several novel antiviral compounds, including Neosartoryadins A and B, members of fumiquinazoline alkaloids, isolated from *Neosartorya udagawae* HDN13-313, showed their anti-H1N1 activity with IC_{50} values of 66 μM and 58 μM, respectively, in comparison to the reference drug ribavirin control of IC_{50} 94 μM (Yu et al., 2016a, 2016b). In the same year, mero-sesquiterpenoids from *Aspergillus ochraceopetaliformis* SCSIO 05702, an Antarctic fungal isolate, were assessed for activity against the H1N1 and H3N2 influenza viruses (Wang et al., 2016). A rare 14-nordrimane-type sesquiterpenoid, phomanolide, was isolated in 2019 from an endophytic fungus (*Phoma* sp.) inhibiting the *Aconitum vilmorinianum* roots. The phomanolides were found to have antiviral potency against influenza A virus strain A/Puerto Rico/8/34, H1N1, with corresponding IC_{50} values of 2.96 ± 0.64 μg/mL. The structurally diverse sesquiterpenoids, aspergillusene D and polyketides, also exhibited anti-influenza activity with corresponding IC_{50} values ranging from 2.17 ± 1.39 to 4.70 ± 1.11 μM (Liu et al., 2019).

3.3.2 Antiviral Fungal Secondary Metabolites Against HIV

Human Immunodeficiency Virus (HIV) belongs to the Retroviridae viral family. HIV particles consist of a lipid membrane with a capsid that holds a nucleoprotein complex containing nucleocapsid, integrase, reverse transcriptase, and two identical copies of RNA (Aiken & Rousso, 2021). Two primary forms of HIV have been recognized: HIV-1 and HIV-2, with divergent amino acid levels and similar genomic structures. HIV mainly targets our immune system, especially the CD4 + T-lymphocytes, which leads to continuous loss of cells and immune suppression. This eventually leads to AIDS, a condition in which the immune system loses its ability to fight against infection, leading to death due to opportunistic secondary infections (van Heuvel et al., 2022). The 2023 HIV data and statistics of the WHO estimate that 39.9 million people are affected with HIV, of which 1.4 million are children (0–14). It could lead to a global health crisis if left unchecked, but fortunately, the efficient use of highly active antiretroviral therapies (ARTs) has rendered the illness a manageable chronic disease. The introduction of combination antiretroviral regimens have significantly improved the life of HIV patients. This has been particularly achieved through the synergistic effects of combined therapy such as Highly Active Antiretroviral Therapy (HAART) targeting multiple viral mechanisms. However, the prevention of the disease remains a top priority due to the persistent underlying immunodeficiency and associated complications.

The HIV treatment involves several types of antiretroviral drugs such as the nucleoside/nucleotide reverse transcriptase inhibitors (NRTIs), non-nucleoside reverse transcriptase inhibitors (NNRTIs) (bind at a separate RTase site of HIV-1), INSTIs (inhibit the transcribed viral DNA from integrating into the host genome), and protease inhibitors (rendering assembled viral particles immature and non-infectious) (Swinkels et al., 2024).

Like influenza viruses, a number of fungal compounds with anti-HIV activity have also been discussed. The immunosuppressants, Cyclosporine A and mycophenolic acids, are two fungal metabolites approved as drug candidates by the United States Food and Drug Administration (FDA) against HIV-1, and Vesicular Stomatitis Indiana Virus (VSIV). Degré and Beck (1994) discovered that the 10-methoxydihydrofuscin, fuscinarin, and fuscin derived from the soil-inhibiting mitosporic fungus *Oidiodendron griseum* exhibited anti-HIV activity in combination with interferon (IFN), by competing with MIP-1R (Macrophage Inflammatory Protein-1 Receptor) for binding to human C–C chemokine receptor type 5 (CCR5). The first novel microbial binding inhibitors, isochromophilones I and II, were recovered from the broth culture of *Penicillium multicolor* FO-2338 against HIV with IC_{50} values of 6.6 and 3.9 μM, respectively. It strongly inhibits the interaction between the docking glycoprotein 120 (gp120) (on the HIV membrane) and CD4 (the receptor of HIV), thereby hindering its entry into the host cell (Matsuzaki et al., 1995).

Yoganathan et al. (2003) discovered three novel compounds from *Oidiodendron griseum;* 10-methoxydihydrofuscin, fuscinarin, and fuscin, that are antagonists of the CCR5 receptor in humans, a key anti-HIV-1 target involved in blocking HIV virus entry (Yoganathan et al., 2003). In the very next year, the same group discovered two fungal metabolites, variecolin and variecolol from *Emericella aurantiobrunnea* and found to be efficiently compete with MIP-1R for binding with CCR5, one of the co-receptors for gp 120-CD4, along with CXCR4, to exhibit anti-HIV activity by inhibiting its entry (Yoganathan et al., 2004).

Helotialean ascomycete-derived Helotialins A-C displayed an in vitro anti-HIV effect in C8166 cells through a replication-inhibitory mechanism with EC_{50} values of 8.01 and 27.9 nM, respectively, when compared to indinavir sulphate, the standard control drug with an EC_{50} value of 8.18 nM (Zou et al., 2009). On the other hand, Ma et al. (2013) purified stachybotrins D from the cultured broth of *Stachybotrys chartarum* MXH-X73 and tested for its activity against wild-type as well as NNRTI-resistant strains of HIV-1 and observed that the compound had an inhibitory activity on HIV-1 replication by suppressing the enzyme reverse transcriptase function (Ma et al., 2013). Another group of researchers screened a library of extracts from natural products derived from fungal endophyte *Alternaria tenuissima* QUE1Se residing in *Quercus emoryi,* and found that it completely inhibited HIV-1 replication (Bashyal et al., 2014). Inhibitors against HIV integrase, one of the three enzymes involved in HIV replication, are another class of antiviral compounds that can be used as an efficient tool with HAART to combat HIV infection. HIV DNA insertion into the host requires 3' processing, strand transfer, and gap filling, in which integrase catalyses the first two steps (Blanco et al., 2015). In a study, the enantiomeric homologous fungal metabolites from marine fungal sources *Fusarium heterosporum* and *Phoma* sp., phomasetin and equisetin, were shown to have anti-HIV-integrase inhibitory activity (Hazuda et al., 1999).

Of interest is the fact that Jayasuriya et al. (2003) discovered polyketide cytosporic acid from the filamentous fungus *Cytospora* sp. and Herath et al. (2004) discovered integrasone, a fungal polyketide metabolite featuring a dihydroxy bicyclic epoxy lactone structure that could inhibit the strand transfer reaction of HIV-1 integrase (Herath et al., 2004; Jayasuriya et al., 2003). Several other fungal secondary metabolites, like ophiobolin, aspernigrins, and funalenone from *Aspergillus niger* FO-5904 (Zhou et al., 2016), atrovenetinone methyl acetal from *Penicillium* sp. FKI-146387, phenalenones like erabulenon B from *Penicillium* sp. FO-5637 (Kitamura et al., 2011), integracide A, along with four tetracyclic triterpenoids obtained from the *Fusarium* sp. fermentation broth (Singh et al., 2003), integracins from *Cytonaema* sp. were also shown to have potent inhibitory activity against recombinant HIV-1 integrase in both coupled and strand transfer reactions (Singh et al., 2002). FDA-approved anticancer compound, doxorubicin, is now being tested to reduce HIV infection (Johansson et al., 2006). A six-membered unsaturated d-lactone group containing triterpenoids, isolated from *Ganoderma colossum,* has been discussed for its inhibitory activity against HIV-1 protease. Similarly, colossolactone, schisanlactone, isocochliodinol, triterpene-farnesyl hydroquinone conjugates, ganoderic acid, and ganoderiol are some of the highly discussed FSMs for their anti-HIV activity by inhibiting proteases (Roy, 2017).

3.3.3 Antiviral Fungal Secondary Metabolites Against Other Viruses

Herpes Simplex Virus (HSV) is a member of the alphaherpesviridae subfamily. It contains a linear dsDNA enclosed in an icosapentahedral capsid with a layer of proteins around it with an envelope. The two serotypes, HSV-1 and HSV-2, upon infection cause a wide range of symptoms, including painful ulcers or blisters, primarily in the orolabial and genital mucosa (Huber, 2003). The commonly used anti-HSV drug list includes acyclovir, penciclovir, and famciclovir; however, the requirement for novel drugs remains essential. From the group of naphthalene-derived fungal metabolites, balticols A-F, isolated from an Ascomycota fungal strain 222 from driftwood, baticol E showed strong anti-HSV-1 activity with IC_{50} values of 0.01 μg/ml (Shushni et al., 2009). Similarly, known metabolites ganoderone A, ergosta-7,22-dien-3beta-ol, and lucialdehyde B, isolated from *Ganoderma pfeifferi*, also exhibited anti-HSV activity. Ganoderone, lucidadiol, and applanoxidic are some other secondary metabolites isolated from these fungi with anti-HSV activity (Niedermeyer et al., 2005). Additionally, a novel polyketide–terpenoid mixed compound, hesseltin A, recovered from the filamentous fungal species *Penicillium hesseltinei* and pochonins from *Pochonia chlamydosporia* var. *catenulata* strain P 0297 have also been discovered for their anti-HSV activities through inhibition of the replication cycle of the virus (Hellwig et al., 2003; Phipps et al., 2004).

Infection caused by the Dengue virus (DENV), classified under the Flaviviridae family, represents another major global health concern. It has a viral genome and capsid enclosed in an envelope, giving it a roughly spherical shape, and contains non-structural proteins NS1, NS2A, NS2B, NS3, NS4A, NS4B, and NS5, which are key regulators of viral replication. The four serotypes of DENV exists as DENV-1, DENV-2, DENV-3, and DENV-4, with *Aedes aegypti* mosquitoes serving as the primary vector (Perera & Kuhn, 2008). There are no clinically approved antivirals or vaccines against DENV infection. Since the current treatment relies primarily on supportive care such as analgesics and fluid replacement, there is an urgent need for novel anti-DENV therapeutic candidates (Lee et al., 2023). In 2007, brefeldin A from *Penicillium* sp. FKI-7127 was discussed for its antiviral activity against DENV regardless of serotypes, and it also showed similar results for other related viruses, such as Japanese encephalitis virus and Zika virus, are inhibited by targeting the early stages of viral life cycle (Raekiansyah et al., 2017). Some pyranones, betulinic acid, asteroid named gilvsin E, two dehydro-aflavinine derivatives, and a series of 12 new trans-decalinecarneic acid analogues isolated from *Phomopsis* sp. (LAP1-7-32) and *Aspergillus* sp. (LAP1-7-57) are another list of fungal compounds assessed for their anti-DENV activity (Peyrat et al., 2016). Cordycepin, a natural adenosine derivative of *Cordyceps militaris,* known for its anti-influenza activity, also demonstrated its antiviral as well as anti-inflammatory activity against DENV. More recently, in 2024, a group of researchers found that cordycepin had direct as well as host-mediated anti-DENV activities. It could inhibit DENV replication by targeting the NS5 protein, virion formation, synthesis of viral protein, and DENV-induced cytokine and chemokine release by suppressing CCL5 expression and modulating the NF-κB pathway (Songprakhon et al., 2024).

The studies on the FSMs for their activity against SARS-CoV-2 during the COVID-19 pandemic were mainly based on computational methods. SARS-CoV-2 is a member of the family Coronaviridae, has a positive-sense ssRNA with four structural (envelope, spike, membrane, and nucleocapsid) and 16 non-structural proteins (NSP1-16). Even after the development of SARS-CoV-2 vaccines by world-leading manufacturers from various countries, the urgency of antiviral drugs against SARS-CoV-2 is still at its peak. FDA-approved fungal metabolites podophyllotoxin, lovastatin, and its derivative simvastatin, isolated from endophytic fungi, have shown antiviral activity against SARS-CoV-2 with a mechanism analogous to efavirenz, a non-nucleoside reverse transcriptase inhibitor (Puri & Verma, 2004). Using computational approaches such as docking and molecular dynamics, the fungal metabolites flaviolin and Pyranonigrin A were found to exhibit anti-SARS-CoV-2 activity by inhibiting $3CL^{pro}$ (Rao et al., 2020, 2022).

Likewise, jasmonic acid (*Stagonosporacirsii*), putaminoxin B (*Phomaputaminum*), notoamide I, emindole SB, colossolactone (*Ganoderma colosum*), ergosterol (*Auricularia polytricha, Flammulinavelutipes* and*Lentinula edodes*), heliantriol F (*Lignosusrhinoceros*), and velutin (*Flammulinavelutipes*) have been studied for their anti-SARS-CoV-2 activities by docking-based molecular interactions against the main protease (M^{pro}) and the receptor-binding domain (RBD) of spike (S1) glycoprotein. The FSMs exhibited a stronger binding affinity towards main protease than towards the S1 receptor-binding domain with good ADME properties (Padhi et al., 2021).

Also, in silico ADMET predictions, molecular docking analysis, and dynamics simulation suggested that the norquinadoline A, isochaetochromin D1, quinadoline B, scedapin C, and 11α-dehydroxyiso-terreulactone A may have antiviral activities against SARS-CoV-2 due to their binding affinity to PLpro, Mpro, RdRp, nsp15, and S binding domains. Of these, Quercetin and its derivatives are the attractive candidates for the therapeutic development targeting SARS-CoV-2 due to its direct as well as anti-inflammatory activity and are now in different phases of clinical trials for getting the label of an antiviral drug (Takahashi et al., 2021).

Overall, it has been observed that fungi and their bioactive compounds, especially the secondary metabolites, have long been known for their medicinal properties. Given the broad-spectrum antiviral activities demonstrated by these compounds, they hold potential as lead candidates for the development of novel antiviral drugs targeting emerging and re-emerging viruses. The lack of appropriate technological applications and pharmacological testing prevented most of these potent compounds from reaching clinical trials and thus the pharmaceutical market. Thorough research and investigations in this area are essential for the development of novel drugs derived from these fungal metabolites, enabling the efficient management and effective treatment of viral diseases for the benefit of humanity.

4 Synergy with Existing Antiviral Therapies

Synergistic effects refer to the enhanced impact observed when the combined activity of two or more drugs exceeds the effect of each agent alone. The combinations of existing antiviral drugs with antiviral microbial metabolites can potentially increase their therapeutic efficacy by effectively targeting infections. This is particularly important for viral infections and co-infections in immunocompromised patients. It also helps to improve drug potency and reduce toxicity by lowering the dose.

There are a variety of natural products that are excellent candidates for targeting specific biological factors through different mechanisms of action. Because of their structural diversity and complexity, they are a continuing valuable source of inspiration for the development of new antiviral medications. Studying the mechanistic synergy of drugs involves understanding how two or more drugs interact at the cellular or molecular level. It is particularly critical in the fields of medicine, such as infectious diseases, oncology, and chronic diseases, where drug combinations can make major impacts. According to pharmacology interactions of bioactive compounds, it is an important aspect that results in combined biological activity. A combination of drugs, however, can be synergetic and antagonistic at the same time.

4.1 Drug Synergy in Various Clinical Conditions

- In the treatment of cancer and severe infections, the combination of cytotoxic drugs helps to obtain better therapeutic effects with fewer side effects, even with the lower doses of each drug.
- Combined use of antibiotics can also result in better efficiency with less drug resistance and side effects.
- Many clinical cases require the use of combined drugs because of multiple clinical conditions, especially in the cases of trauma care. In such cases, the action of drugs cannot be completely called synergetic, but still, there can be interactions between these drugs that have to be studied before administration.
- In certain cases, the effect of one drug can be modified by another drug that possesses no independent bioactivity (Bürgi, 2012).

Different methods can be used to predict the synergic interactions between drugs. Clinical trials have traditionally been used to discover drug combinations, posing challenges in time and cost owing to the immense number of possible combinations. The use of high-throughput screening (HTS) is an alternative method for testing chemical and biological compounds against specific biological targets (Bajorath, 2002). The computational methods, such as machine learning models, have also become one of the most effective ways of analyzing the vast synergistic space of drug combinations (Preuer et al., 2018). Molecular interactions can be described from a comprehensive network perspective with system biology. Some system biology effects have been used to study the interaction of drugs to forecast the human body responses to drug combinations (Chen et al., 2015).

4.2 Synergistic Studies on Fungal Metabolites with Conventional Antiviral Drugs

Mycophenolic acid (MPA), is a secondary metabolite of fungal origin produced by members of the genus *Penicillium,* along with ribavirin, has been reported to show inhibitory activities against hepatitis B (HBV), Dengue, Japanese Encephalitis Virus (JEV), and in some orthopoxviruses like Monkeypox, Cowpox, Camelpox, and Vaccinia Virus (Chang et al., 2018a, 2018b; Smee et al., 2001; Takhampunya et al., 2006; Ying et al., 2007). The mechanism behind this was reported as the inhibition of Inosine monophosphate dehydrogenase (IMPDH) by MPA without requiring intracellular metabolism (Ransom, 1995). The active intracellular form of ribavirin, ribavirin monophosphate, also inhibited IMPDH by decreasing intracellular GTP or dGTP or by inhibiting viral and cellular DNA/ RNA and protein synthesis (Lowe et al., 1977; Streeter et al., 1973).

A combination of griseofulvin, an FSM from *Penicillium griseofulvum*, and IFNα has been shown to have a synergistic inhibitory effect on HCV replication in vitro. Griseofulvin has been an established therapeutic for decades in the management of some dermatophyte infections and ringworm in humans. Several studies indicate that the HCV replication complex is dependent on microtubule polymerization. Griseofulvin has been shown to arrest the G2/M phase of the human cell cycle by polymerizing microtubules and consequently influence HCV-RNA replication complex formation. In addition, the development of derivatives of this compound will be a potent therapeutic intervention for the management of chronic hepatitis (Jin et al., 2008). Molecular docking analysis further demonstrated that griseofulvin has a strong interaction with the main protease of SARS-CoV-2 with a docking score (− 6.8 kcal/mol), suggesting it could be a potential antiviral target (Aris et al., 2022).

The glucans are a group of immunomodulatory bioactive compounds that have been used successfully in clinical medicine (Vetvicka & Vetvickova, 2014). The β-glucans have high immunomodulatory activity, so they have been used in combination therapies with different antivirals to improve outcomes of antiviral therapies (Shi et al., 2022; Córdova-Martínez et al., 2021; van Steenwijk et al., 2021). Complement system receptor CR3, also known as CD11b/CD18, has been shown to be activated by some β-glucans (Chan et al., 2009; Hong et al., 2004). β-glucan has also been shown to provide protection against the influenza virus. In a study of swine influenza virus (SIV)-induced pneumonia, β-glucan treatment led to significantly higher levels of IFN-γ and nitric oxide (NO) in the bronchoalveolar lavage fluid of treated pigs compared to those untreated (Jung et al., 2004). In patients undergoing major surgeries, β-glucan administration enhanced monocyte and neutrophil functions and thereby reduced the risk of infectious complications. Therefore, pre-operative use of β-glucan has been recommended to lower the incidence of severe post-operative infections (Babineau et al., 1994; Dellinger et al., 1999). Additionally, recent in vitro studies suggest that β-glucan extracts exhibit pulmonary cytoprotective properties, supporting their further evaluation as a potential candidate for COVID-19 immunotherapy along with remdesivir (Murphy et al., 2020).

An adenosine analogue cordycepin, the fungal metabolite produced by the parasitic fungus *Cordyceps miltaris* (Cunningham et al., 1950), has been extensively studied for its antiviral activity against influenza virus, Dengue, HIV, Enterovirus, and SARS-CoV-2 (Lee et al., 2024; Müller et al., 1991; Panya et al., 2021; Rabie, 2022). The cordycepin has been shown to enhance the effect of doxorubicin in the treatment of Epstein-Barr virus-positive tumours (Du et al., 2016). It can be due to its immense role in regulating several biochemical reactions, including apoptosis, nucleic acid synthesis, inflammation, and cell cycle signalling (Tuli et al., 2013).

A cyclic undecapeptide, cyclosporin A, from *Trichoderma polysporum* and *Cylindrocarpon lucidum* is also a potent FSM that selectively acts on T-cells (Laupacis et al., 1982). This compound is reported as an effective immunosuppressant when synergistically used with tacrolimus (FK506) and rapamycin (Ho et al., 1996). The combined use of these compounds was also reported for HIV infection (Briggs et al., 1999). Therefore, the novel FSMs like cyclosporin A and cordycepin can be used as powerful, broad-spectrum antivirals that can address emerging and re-emerging viral infections through diverse mechanisms.

Overall, fungal metabolites, when combined with antiviral drugs, offer promising antiviral therapies. Nevertheless, further investigation is needed to completely understand the mechanisms of synergy, optimal combinations, and clinical applications of these natural products.

5 Challenges and Innovations

Fungal secondary metabolites are promising bioactive compounds with a diverse spectrum of drug-like activities. However, choosing an effective delivery system is important in addition to choosing the right therapeutic approach. The introduction of nanoparticles offers an excellent option for antiviral drug delivery. Various delivery systems, such as liposomes, micelles, and hydrogels, have been studied for their immunostimulatory properties and effectiveness in drug delivery by targeting specific cells and tissues (Fig. 4).

5.1 Delivery Systems for Immunomodulatory and Antiviral Fungal Drugs

5.1.1 Liposomes

Liposomal drug delivery systems have emerged as a revolutionary method for administering medications. These are nanoparticle carriers made of lipids that can encapsulate hydrophilic and lipophilic drugs. In the mid-1960s, the concept of liposomes was pioneered by British scientist Alec Bangham and his collaborators at Babraham, Cambridge, while the structure was published in 1964. Following their introduction, liposomes have been widely explored as delivery vehicles for proteins, nucleic acids, imaging agents, and especially for small molecular drugs.

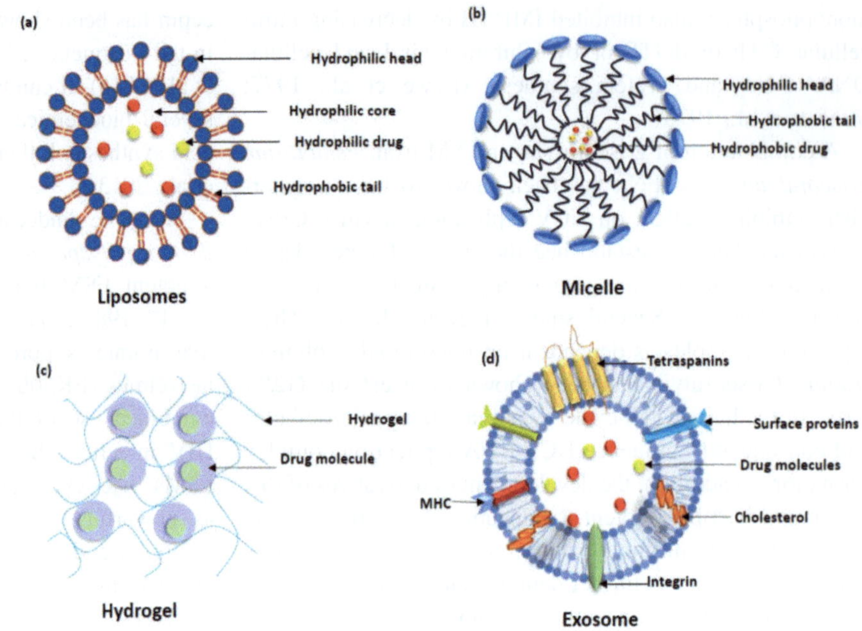

Fig. 4 Schematic representation of different drug delivery systems: **a** Liposomes, **b** Micelle, **c** Hydrogels, **d** Exosomes (*Source* Authors)

Liposomes are self-assembled vesicles formed of phospholipids that can form either a single bilayer (unilamellar) or multiple concentric bilayers (multilamellar), enclosing a central aqueous compartment. Their sizes range from 30 to 100 nm, with a phospholipid bilayer thickness of 4–5 nm (Liu et al., 2022).

They offer targeted drug delivery, improved pharmacokinetic profiles, protection from drug degradation, sustained drug release, and reduced clearance rates. This results in an optimized therapeutic index, characterized by maximal potency and reduced adverse events. These properties support the use of liposomes as an effective tool for delivering drugs. An established example is the amphotericin B-containing Ambisome, an approved liposome-based treatment for fungal infections (Swenson et al., 1995). Curcumin is a widely studied bioactive molecule with both anti-fungal and anti-bacterial properties. It has now been established for its antiviral activity also. Curcumin has the potential to inhibit replication across a broad range of viruses (Praditya et al., 2019). In 2018, a study showed that liposomes can solubilize curcumin within the phospholipid bilayer. This helps in its delivery into an aqueous medium that can greatly enhance its effectiveness (Chang et al., 2018a, 2018b). Numerous liposomal drug products have been successfully approved and implemented in clinical settings over the past few decades. The FDA and EMA (European Medicines Agency) have authorized around 14 different types of liposomal products. Doxil was the first FDA-approved liposomal drug for metastatic ovarian cancer patients who did not respond to platinum-based therapy or paclitaxel. The drug alone had some limitations, like poor stability and low water solubility, which can hamper its bioactivity. Various administration routes, including parenteral, pulmonary, oral, transdermal, ophthalmic, and nasal routes, have been designed to enhance the therapeutic efficacy and improve patient adherence. To maximize its therapeutic efficiency, it was encapsulated in a liposome that protected against enzymatic degradation and thereby enhanced the drug's half-life. The approval was based on three Phase II studies showing tumour regression in 20 of 145 women (13.8%). The therapeutic use of several known fungal metabolites is limited due to its low solubility and bioavailability. Similar to doxil, the limitations of these FSMs can also be overcome by incorporating them into liposomes that could enhance their potential as antiviral agents.

5.1.2 Micelles

Micelle is another drug delivery system that can deliver water-insoluble drugs. Micelles are self-organizing microarchitectures formed from surfactants within an aqueous environment. They typically have diameters less than 50 nm and can encapsulate a substantial amount of drugs. It offers excellent stability and the capability to solubilize substantial quantities of hydrophobic medications, making it a promising choice for administering drugs with poor solubility and bioavailability. Their compact size facilitates effective delivery by minimizing off-target effects and enhancing therapeutic efficiency (Lu et al., 2018).

In 1983, polymeric prodrug micelles were synthesized for the delivery of Acyclovir (ACV), a guanosine-based prodrug primarily used to treat infections caused by HSV, varicella zoster virus, cytomegalovirus, and Epstein-Barr virus (Richards et al., 1983). This method was found to be advantageous in enhancing drug-carrying capacity, eliminating drug-loading steps, and reducing production costs (Sawdon & Peng, 2014). Ahn et al. (2010) developed another polymeric micelle, PLLA-b-PEG-mRNA, that is encapsulated with amantadine (AMT). It targets influenza viruses effectively because of the presence of sialic acid derivative, a chemical modification at the "N-acetyl" position of mRNA that enhances antiviral activity and improves circulation time. Similar to liposomes, micelles have also been used to load curcumin to inhibit influenza A virus replication (Li et al., 2022a, 2022b). The finding showed that the micelles significantly increased the bioactivity/stability of curcumin, thereby reducing the influenza A virus infection.

5.1.3 Hydrogels

Hydrogels are a three-dimensional structure of a cross-linked network of water-soluble polymers. They can deliver pharmacological compounds effectively because they swell in aqueous environments and can release cargo in a controlled manner. Additionally, hydrogels can take many physical forms, including slabs, nanoparticles, coatings, and films (Hoare & Kohane, 2008). The unique property of hydrogel is the high level of water (typically 70–99%) that mimics the extracellular matrix (ECM). This similarity facilitates better integration with surrounding tissues, which enhances biocompatibility and reduces immune responses (Li & Mooney, 2016). Nucleoside Analogue Triphosphates (NATPs) are a class of nucleoside-based anticancer and antiviral agents that require intracellular activation through phosphorylation. Polyethylene Glycol-block-Polyethylenimine nanogels can deliver NATPs and facilitate their activation, inhibiting DNA replication and transcription (Kabanov & Vinogradov, 2009). The silver nanoparticles (AgNPs) derived from the aqueous extract of *Melaleuca alternifolia* leaves demonstrate potent antiviral activity against HSV-1, HSV-2, and various other microbes (Ramadan et al., 2020). Another study showed that a hydrogel composed of tannic acid and silver nanoparticles blocked HSV-1 infection to a great extent by reducing viral adherence, invasion, and post-infection (Szymańska et al., 2018). While direct studies connecting antiviral fungal metabolites with hydrogels are scarce, the success of hydrogels as antiviral delivery systems and the efficacy of fungal metabolites against HSV suggest a promising area for future research.

5.1.4 Exosomes

Exosomes are lipid bilayered vesicles, which were early considered "garbage of the cell", and later found to have a crucial role in cell communication, migration, angiogenesis, proliferation, and immunomodulation. Exosomes were discovered in 1983 during the study of the transferrin receptor in sheep erythrocytes. They are composed of cholesterol, sphingomyelin, and ceramide and have a size range of 30–100 nm. After exploiting their structure and functions, it was found that exosomes can carry lipids, proteins, mRNA, and other bioactive molecules. In comparison to synthetic drug delivery systems, exosomes are less toxic and more efficient because they exist naturally in the body. They ensure other beneficial traits such as targeted delivery through ligand-receptor interactions, reduced non-specific interactions, the ability to cross the blood–brain barrier, and a minimized adverse immune response (Li et al., 2022a, 2022b; Pan & Johnstone, 1983). Currently, there are no studies specifically addressing the use of exosomes as delivery vehicles for antiviral FSMs. However, research has explored exosomes as efficient carriers for drug molecules. Further pharmacological studies and clinical trials are needed to ensure their potency. The research focusing on exosome-mediated delivery of antiviral fungal metabolites could lead to novel therapeutic strategies against viral infections.

5.2 Development of Broad-Spectrum Antiviral Fungal Compounds

Antiviral fungal metabolites have become more popular in recent decades due to their low manufacturing cost compared with synthetic antiviral drugs. Marine fungi are gaining popularity due to their novel secondary metabolites, which show promise as antiviral drugs. Marine fungi like *Chaetomium cristatum* isolated from marine mudflat show inhibitory properties towards Vesicular stomatitis virus (Yun et al., 2016). A study done in 1960 showed that a polysaccharide statolon produced by *Pencillium stoloniferum* has been shown to induce murine cells to produce interferon against Friend Leukemia virus (Wheelock & Larke, 1967). Although researchers are more focused on developing broad-spectrum antiviral agents using fungal metabolites, some have succeeded.

Coumarins (benzo-α-pyrone) are one of the major secondary metabolites and have been proven to have antiviral activity against influenza viruses, dengue virus, enterovirus 71 (EV71), HIV, coxsackievirus A16 (CVA16), and Chikungunya virus. It takes several pathways and host enzymes to facilitate viral replication events. RNA viruses possess inosine 5′-monophosphate (IMP) dehydrogenase, Ebola has S-adenosyl homocysteine (SAH) hydrolase, orotidine 5′-monophosphate decarboxylase, CTP synthetase, and glycosylation pathways. Cumarine derivatives have shown antiviral effects by interrupting the above mechanisms (Mishra et al., 2020). Cumarine has been found to act on various targets related to viral infections, including HIV (Kudo et al., 2013). It inhibits HIV reverse transcriptase and herpes viral DNA polymerase, and functions as an adsorption inhibitor, potentially serving as a vaginal microbicide for AIDS prevention. Additionally, cumarine inhibits HIV-cell fusion by interacting with the viral protein gp41, and it affects the HIV nucleocapsid p7 through zinc finger-binding. It also shows inhibitory effects on HIV integrase, various viral proteases (including those from HIV, HSV, CMV, and HCV), and HIV-1 Vpr. Furthermore, cumarine acts as a binder to picornavirus capsids, inhibits the uncoating of influenza A viruses, and antagonizes HIV Tat and Rev proteins. Lastly, it inhibits the glycosylation of HIV and HBV and the neuraminidase activity in influenza A and B viruses. It has been shown that warfarin (4-Hydroxycoumarin (4-HC) Analogue), a coumarin derivative used in preventing thrombosis and thromboembolism, may be used in place of peptidomimetics, an HIV protease inhibitor (Hassan et al., 2016). Ansari et al. (year) conducted a computational analysis of 188 secondary metabolites obtained from basidiomycetes to evaluate their inhibitory capacity against the RdRp of the novel coronavirus. This study found that sterostrein B and sterostrein C exhibit pharmacokinetic properties that inhibit RNA-dependent RNA polymerase of SARS-CoV-2 (Ansari et al., 2023). Cordycepin, or 3′-deoxyadenosine, primarily derived from the fungus *Cordyceps militaris,* is a promising candidate for anticancer therapies (Yoon et al., 2018). Notably, it has shown the potential to inhibit the replication of the dengue virus, virion production, and viral protein synthesis. It also reduces the production of various cytokines and chemokines associated with the dengue virus.

Overall, FSMs have been proven as excellent compounds with broad-spectrum immunomodulatory as well as antiviral activities. It can beneficially modulate a wide variety of cellular factors and signalling pathways. The emergence of nanoparticles made the targeted delivery of fungal metabolites possible to specific cells. Nanoparticles such as liposomes, micelles, hydrogels, and exosomes demonstrate excellent drug delivery properties due to their biocompatibility, stability, and flexibility. Hydrogels, due to their structural similarity with the in vivo system, can mask the immune response, while the natural ability of exosomes to deliver drugs gives great promise for their drug delivery. The broad-spectrum antiviral properties of fungal metabolites, such as coumarins and cordycepin, should be studied to uncover the mechanistic pathways of these metabolites and to establish their drug-likeness. Using an improved metagenomic approach and recombinant DNA technology, the pharmaceutical benefits of FSMs can be explored as well as commercialized. This may allow us to explore an untapped pool of FSMs for the management of various diseases and pandemics.

6 Conclusion

The interaction of fungal secondary metabolites with the human immune system offers both therapeutic potential and practical challenges. The metabolic diversity and history of traditional medicines provided Ascomycetes and basidiomycetes a foundation for the scientific investigation into their pharmacological activities. The antiviral and immunomodulatory properties exhibited by the fungal secondary metabolites suggest potential applications, especially in synergistic combinations with current antiviral therapies. In the present chapter, we have reviewed the potential benefits of the FSMs as promising immunomodulators and antiviral drug candidates. The advancements in drug delivery, such as targeted release mechanisms and nanoscale carriers, made the administration and interaction of the drug more sophisticated with the body's immune system. These developments highlight that targeted drug delivery is also as important as the drug. These studies should be viewed as medical advancements rather than technical achievements. The therapeutic potential of FSMs can be unlocked to transform into novel drug candidates against viral threats and for pandemic preparedness. Achieving this would require a deeper understanding of virus-host interactions, host immune responses, and viral evasion strategies.

References

Ahn, Y. S., Jeong, Y. I., Choi, E., & Nah, J. W. (2010). Preparation of multifunctional polymeric micelles for antiviral treatment. *Macromolecular Research, 18*(8), 747–752.

Aiken, C., & Rousso, I. (2021). The HIV-1 capsid and reverse transcription. *Retrovirology, 18*(1), 29.

Al-Jumaili, M. M. O., Al-Dulaimi, F. K., & Ajeel, M. A. (2020). The role of *Ganoderma lucidum* uptake on some hematological and immunological response in patients with coronavirus (COVID-19). *Systematic Reviews in Pharmacy, 11*(8), 537–541.

Ansari, M., Alomary, M. N., Aldakheel, F. M., Almutairi, M. M., Alharthy, B. T., & Alamri, M. A. (2023). Computational investigation on the potential inhibitory effect of fungal metabolites against viral RdRp of SARS-CoV-2 and HCV. *Journal of Molecular Liquids, 390*, Article 123153.

Aris, P., Aydemir, M. N., Sayin, K., & Gul, H. I. (2022). In silico molecular dynamics of griseofulvin and its derivatives revealed potential therapeutic applications for COVID-19. *International Journal of Molecular Sciences, 23*(13), 7160.

Babineau, T. J., Hackford, A., Kenler, A., Bistrian, B., Forse, R. A., Fairchild, P. G., Heard, S., & Poole, M. (1994). Randomized phase I/II trial of a macrophage-specific immunomodulator (PGG-glucan) in high-risk surgical patients. *Annals of Surgery, 220*(5), 601–609.

Bajorath, J. (2002). Integration of virtual and high-throughput screening. *Nature Reviews Drug Discovery, 1*(11), 882–894.

Bashyal, B. P., Wellensiek, B. P., & Stierle, D. B. (2014). Altertoxins with potent anti-HIV activity from *Alternaria tenuissima* QUE1Se, a fungal endophyte of *Quercus emoryi. Bioorganic and Medicinal Chemistry, 22*(21), 6112–6116.

Bergmann, S., Schümann, J., Scherlach, K., Lange, C., Brakhage, A. A., & Hertweck, C. (2007). Genomics-driven discovery of PKS-NRPS hybrid metabolites from *Aspergillus nidulans*. *Nature Chemical Biology, 3*(4), 213–217.

Bharadwaj, S., Lee, K. E., Dwivedi, V. D., Yadava, U., Panwar, A., Lucas, S. J., & Kang, S. G. (2019). Discovery of *Ganoderma lucidum* triterpenoids as potential inhibitors against Dengue virus NS2B-NS3 protease. *Scientific Reports, 9*(1), 19059.

Blanco, J. L., Whitlock, G. G., & Milinkovic, A. (2015). HIV integrase inhibitors: A new era in the treatment of HIV. *Expert Opinion on Pharmacotherapy, 16*(9), 1313–1324.

Bondy, G. S., & Pestka, J. J. (2000). Immunomodulation by fungal toxins. *Journal of Toxicology and Environmental Health Part b: Critical Reviews, 3*(2), 109–143.

Bouvier, N. M., & Palese, P. (2008). The biology of influenza viruses. *Vaccine, 26*, D49–D53.

Brakhage, A. A. (2013). Regulation of fungal secondary metabolism. *Nature Reviews Microbiology, 11*(1), 21–32.

Briggs, C. J., Sharkey, M. A., Stevenson, M., & McConnell, D. J. (1999). Comparison of the effect of FK506 and cyclosporin A on virus production in H9 cells chronically and newly infected by HIV-1. *Archives of Virology, 144*(11), 2151–2160.

Bürgi, H. G. (2012). Drug synergy—Mechanisms and methods of analysis. In: Williams, A. (Ed.), *Toxicity and drug testing*, Ch. 7. IntechOpen.

Chan, G. C., Chan, W. K., & Sze, D. M. (2009). The effects of beta-glucan on human immune and cancer cells. *Journal of Hematology and Oncology, 2*, 25.

Chang, M., Wu, M., & Li, H. (2018a). Antitumor activities of novel glycyrrhetinic acid-modified curcumin-loaded cationic liposomes in vitro and in H22 tumor-bearing mice. *Drug Delivery, 25*(1), 1984–1995.

Chang, Q. Y., Liu, X. T., Bai, Y., Tao, X. H., Wang, X. Y., & Zhao, Y. (2018b). The IMPDH inhibitors, ribavirin and mycophenolic acid, inhibit peste des petits ruminants virus infection. *Veterinary Research Communications, 42*(4), 309–313.

Charostad, J., Rukerd, M. R. Z., Mahmoudvand, S., Bashash, D., Hashemi, S. M. A., Nakhaie, M., & Zandi, K. (2023). A comprehensive review of highly pathogenic avian influenza (HPAI) H5N1: An imminent threat at doorstep. *Travel Medicine and Infectious Disease, 55*, Article 102638.

Cheenpracha, S., Park, E. J., Rostama, B., Pezzuto, J. M., & Chang, L. C. (2010). Inhibition of nitric oxide (NO) production in lipopolysaccharide (LPS)-activated murine macrophage RAW 264.7 cells by the norsesterterpene peroxide, epimuqubilin A. *Marine drugs, 8*(3), 429–437.

Chen, D., Wang, M., Zhou, S., & Zhang, J. (2015). Systematic synergy modeling: Understanding drug synergy from a systems biology perspective. *BMC Systems Biology, 9*(1), 56.

Chen, S., Wang, J., Lin, X., Zhao, B., Wei, X., Li, G., & Liu, Y. (2016). Chrysamides A-C, three dimeric nitrophenyl trans-epoxyamides produced by the deep-sea-derived fungus *Penicillium chrysogenum* SCSIO41001. *Organic Letters, 18*(15), 3650–3653.

Clevenger, K. D., Bok, J. W., Ye, R., Miley, G. P., Verdan, M. H., Velk, T., & Kelleher, N. L. (2017). A scalable platform to identify fungal secondary metabolites and their gene clusters. *Nature Chemical Biology, 13*(8), 895–901.

Córdova-Martínez, A., Zavala-Sánchez, M. A., Martínez-Rodríguez, E. A., & Ramírez-Cabrera, M. A. (2021). β-Glucans could be adjuvants for SARS-CoV-2 virus vaccines (COVID-19). *International Journal of Environmental Research and Public Health, 18*(23), 12661.

Cunningham, K. G., Manson, W., Spring, F. S., & Hutchinson, S. A. (1950). Cordycepin, a metabolic product isolated from cultures of Cordyceps militaris (Linn.) Link. *Nature, 166*(4231), 949–949.

De Clercq, E. (2002). Strategies in the design of antiviral drugs. *Nature Reviews Drug Discovery, 1*(1), 13–25.

de Sales-Neto, J. M., & Rodrigues-Mascarenhas, S. (2024). Immunosuppressive effects of the mycotoxin patulin in macrophages. *Archives of Microbiology, 206*(4), 166.

Degré, M., & Beck, S. (1994). Anti-HIV activity of dideoxynucleosides, foscarnet and fusidic acid is potentiated by human leukocyte interferon in blood-derived macrophages. *Chemotherapy, 40*(3), 201–208.

Dellinger, E. P., Babineau, T. J., Bleicher, P., Kaiser, A. B., Seibert, G. B., Postier, R. G., Vogel, S. B., Norman, J., & Doyle, M. K. (1999). Effect of PGG-glucan on the rate of serious postoperative infection or death observed after high-risk gastrointestinal operations. *Archives of Surgery, 134*(9), 977–983.

Dimmock, N. J., Easton, A. J., & Leppard, K. N. (2015). *Introduction to modern virology* (7th ed.). Wiley-Blackwell.

Du, Y., Ma, X., Wang, Y., Wen, B., Zhang, Y., & Zeng, Q. (2016). Cordycepin enhances Epstein-Barr virus lytic infection and Epstein-Barr virus-positive tumor treatment efficacy by doxorubicin. *Cancer Letters, 376*(2), 240–248.

Dwivedi, P., & Burns, R. B. (1984). Effect of ochratoxin A on immunoglobulins in broiler chicks. *Research in Veterinary Science, 36*(1), 117–121.

Eo, S. K., Kim, Y. S., Lee, C. K., & Han, S. S. (2000). Possible mode of antiviral activity of acidic protein bound polysaccharide isolated from *Ganoderma lucidum* on herpes simplex viruses. *Journal of Ethnopharmacology, 72*(3), 475–481.

Fang, W., Lin, X., Wang, J., Liu, Y., Tao, H., & Zhou, X. (2016). Asperpyrone-type bis-naphtho-γ-pyrones with COX-2–inhibitory activities from marine-derived fungus *Aspergillus niger*. *Molecules, 21*(7), 941.

Gao, H., Liu, S., Wang, G., Xu, J., & Li, Q. (2013). Aspulvinones from a mangrove rhizosphere soil-derived fungus *Aspergillus terreus* Gwq-48 with anti-influenza A viral (H1N1) activity. *Bioorganic and Medicinal Chemistry Letters, 23*(6), 1776–1778.

Gu, B. B., Jiao, F. R., Wu, W., Jiao, W. H., Li, L., Sun, F., & Lin, H. W. (2018). Preussins with inhibition of IL-6 expression from *Aspergillus flocculosus* 16D–1, a fungus isolated from the marine sponge *Phakellia fusca*. *Journal of Natural Products, 81*(10), 2275–2281.

Gu, C. Q., Li, J. W., Chao, F., Jin, M., Wang, X. W., & Shen, Z. Q. (2007). Isolation, identification and function of a novel anti-HSV-1 protein from *Grifola frondosa*. *Antiviral Research, 75*(3), 250–257.

Hadden, J. W. (1996). *Immunomodulators*. Springer.

Hassan, M. Z., Osman, H., Ali, M. A., & Ahsan, M. J. (2016). Therapeutic potential of coumarins as antiviral agents. *European Journal of Medicinal Chemistry, 123*, 236–255.

Hayden, M. S., & Ghosh, S. (2008). Shared principles in NF-kappaB signaling. *Cell, 132*(3), 344–362.

Hazuda, D. J., Felock, P., & Wolfe, A. (1999). Isolation and characterization of novel human immunodeficiency virus integrase inhibitors from fungal metabolites. *Antiviral Chemistry and Chemotherapy, 10*(2), 63–70.

He, F., Bao, J., Zhang, X., Liu, X., Han, J., & Wang, C. (2013). Asperterrestide A, a cytotoxic cyclic tetrapeptide from the marine-derived fungus *Aspergillus terreus* SCSGAF0162. *Journal of Natural Products, 76*(6), 1182–1186.

Hellwig, V., Emmler, T., Mayer-Bartschmid, A., Schummer, D., & Müller, H. (2003). Pochonins A-F, new antiviral and antiparasitic resorcylic acid lactones from *Pochonia chlamydosporia* var. catenulata. *Journal of Natural Products, 66*(6), 829–837.

Herath, K. B., Jayasuriya, H., Guan, Z., & Zink, D. L. (2004). Isolation, structure, absolute stereochemistry, and HIV-1 integrase inhibitory activity of integrasone, a novel fungal polyketide. *Journal of Natural Products, 67*(5), 872–874.

Hirayama, D., Iida, T., & Nakase, H. (2017). The phagocytic function of macrophage-enforcing innate immunity and tissue homeostasis. *International Journal of Molecular Sciences, 19*(1).

Ho, S., Clipstone, N., & Crabtree, G. R. (1996). The mechanism of action of cyclosporin A and FK506. *Clinical Immunology and Immunopathology, 80*(3 Pt 2), S40–S45.

Hoare, T. R., & Kohane, D. S. (2008). Hydrogels in drug delivery: Progress and challenges. *Polymer, 49*(8), 1993–2007.

Hong, F., Yan, J., Baran, J. T., Allendorf, D. J., Hansen, R. D., Ostroff, G. R., Xing, P. X., Cheung, N. K., & Ross, G. D. (2004). Mechanism by which orally administered beta-1,3-glucans enhance the tumoricidal activity of antitumor monoclonal antibodies in murine tumor models. *Journal of Immunology, 173*(2), 797–806.

Hood, K. A., & Zarembski, D. G. (1997). Mycophenolate mofetil: A unique immunosuppressive agent. *American Journal of Health-System Pharmacy, 54*(3), 285–294.

Huber, M. A. (2003). Herpes simplex type-1 virus infection. *Quintessence International, 34*(6), 453–467.

Hutchinson, E. C. (2018). Influenza virus. *Trends in Microbiology, 26*(9), 809–810.

Jain, S., Purohit, A., Nema, P., Vishwakarma, H., & Jain, P. K. (2022). Asian Journal of Dental and Health Sciences. *Sciences, 2*(1), 33–39.

Jayasuriya, H., Zink, D. L., Basilio, A., & Vicente, F. (2003). Isolation, structure, and HIV-1 integrase inhibitory activity of cytosporic acid, a fungal metabolite produced by a *Cytospora* sp. *Journal of Natural Products, 66*(4), 551–553.

Jin, M., Jung, H. J., Choi, J. J., Jeon, H., Oh, J. H., Kim, B., & Kim, S. (2003). Activation of selective transcription factors and cytokines by water-soluble extract from *Lentinus lepideus.*

Jin, H., Zhao, Q., & Lin, J. (2008). Griseofulvin, an oral antifungal agent, suppresses hepatitis C virus replication in vitro. *Hepatology Research, 38*(9), 909–918.

Johansson, S., Golden, E., Matsuda, K., & O'Doherty, U. (2006). Elimination of HIV-1 infection by treatment with a doxorubicin-conjugated anti-envelope antibody. *AIDS, 20*(15), 1911–1915.

Jung, K., Ha, Y., Ha, S. K., Han, D. U., Kim, D. W., Moon, W. K., & Chae, C. (2004). Antiviral effect of *Saccharomyces cerevisiae β* -glucan to swine influenza virus by increased production of interferon-*γ* and nitric oxide. *Journal of Veterinary Medicine, Series B, 51*(2), 72–76.

Kabanov, A. V., & Vinogradov, S. V. (2009). Nanogels as pharmaceutical carriers: Finite networks of infinite capabilities. *Angewandte Chemie International Edition, 48*(30), 5418–5429.

Keller, N. P., Turner, G., & Bennett, J. W. (2005). Fungal secondary metabolism—From biochemistry to genomics. *Nature Reviews Microbiology, 3*(12), 937–947.

Kitamura, Y., Suzuki, M., Akao, S., Ogasawara, A., Watanabe, N., Nagai, K., & Nakao, Y. (2011). Fungal phenalenones inhibit HIV-1 integrase. *Molecular Biology of the Cell, 11*, 3765–3775.

Ko, W., Sohn, J. H., Jang, J. H., Ahn, J. S., Kang, D. G., Lee, H. S., & Oh, H. (2016). Inhibitory effects of alternaramide on inflammatory mediator expression through TLR4-MyD88-mediated inhibition of NF-κB and MAPK pathway signaling in lipopolysaccharide-stimulated RAW264. 7 and BV2 cells. *Chemico-Biological Interactions, 244*, 16–26.

Kountouras, J., Zavos, C., & Chatzopoulos, D. (2004). Immunomodulatory benefits of cyclosporine A in inflammatory bowel disease. *Journal of Cellular and Molecular Medicine, 8*(3), 317–328.

Kudo, E., Taura, M., Kariya, R., Goto, H., Okada, S., & Watanabe, Y. (2013). Inhibition of HIV-1 replication by a tricyclic coumarin GUT-70 in acutely and chronically infected cells. *Bioorganic and Medicinal Chemistry Letters, 23*(3), 606–609.

Laupacis, A., Keown, P. A., Ulan, R. A., McKenzie, N., & Stiller, C. R. (1982). Cyclosporin A: A powerful immunosuppressant. *Canadian Medical Association Journal, 126*(9), 1041–1046.

Lebish, I. J., & Moraski, R. M. (1987). Mechanisms of immunomodulation by drugs. *Toxicologic Pathology, 15*(3), 338–345.

Lee, H. H., Lee, S., Lee, J., Kim, Y., Park, C., & Kim, S. (2014). Anti-influenza effect of *Cordyceps militaris* through immunomodulation in a DBA/2 mouse model. *Journal of Microbiology, 52*(8), 696–701.

Lee, M. F., Wu, Y. S., & Poh, C. L. (2023). Molecular mechanisms of antiviral agents against dengue virus. *Viruses, 15*(3), 1–17.

Lee, Y. P., Hsu, Y. C., Chen, S. R., Chang, C. C., Lo, Y. C., & Lin, Y. L. (2024). Cordycepin inhibits enterovirus A71 replication and protects host cell from virus-induced cytotoxicity through adenosine action pathway. *Viruses, 16*(3).

Li, C. Z., Li, J., Yue, X. X., Zhou, Y., Wu, J. L., & Deng, H. B. (2022a). Curcumin-loaded oil-free self-assembled micelles inhibit the Influenza A virus activity and the solidification of curcumin-loaded micelles for pharmaceutical applications. *Pharmaceutics, 14*(11).

Li, J., & Mooney, D. J. (2016). Designing hydrogels for controlled drug delivery. *Nature Reviews Materials, 1*(12), 16071.

Li, T., Li, Y., Dai, S., He, J., & Mao, F. (2022b). The therapeutic potential and clinical significance of exosomes as carriers of drug delivery system. *Pharmaceutics*, 15(1).

Lim, F. Y., Sanchez, J. F., Wang, C. C., & Keller, N. P. (2012). Toward awakening cryptic secondary metabolite gene clusters in filamentous fungi. In *Methods in enzymology* (Vol. 517, pp. 303–324). Academic Press.

Liu, N., Zhang, X., Sun, W., Chen, M., & Zhang, W. (2019). Structurally diverse sesquiterpenoids and polyketides from a sponge-associated fungus *Aspergillus sydowii* SCSIO41301. *Fitoterapia, 135*, 27–32.

Liu, P., Chen, G., & Zhang, J. (2022). A review of liposomes as a drug delivery system: Current status of approved products, regulatory environments, and future perspectives. *Molecules, 27*(4).

Loftus, J. H., Kijanka, G. S., O'Kennedy, R., & Loscher, C. E. (2016). Patulin, deoxynivalenol, zearalenone and T-2 toxin affect viability and modulate cytokine secretion in J774A.1 murine macrophages. *International Journal of Chemistry*, *8*(2), 22–32.

Lowe, J. K., Brox, L., & Henderson, J. F. (1977). Consequences of inhibition of guanine nucleotide synthesis by mycophenolic acid and virazole. *Cancer Research, 37*(3), 736–743.

Lu, Y., Yang, J., Sega, E., & Gao, J. (2018). Strategies to improve micelle stability for drug delivery. *Nano Research, 11*(10), 4985–4998.

Luft, P., Oostingh, G. J., Gruijthuijsen, Y., Horejs-Hoeck, J., Lehmann, I., & Duschl, A. (2008). Patulin influences the expression of Th1/Th2 cytokines by activated peripheral blood mononuclear cells and T cells through depletion of intracellular glutathione. *Environmental Toxicology, 23*(1), 84–95.

Ma, X., Sun, W., Wang, J., Wang, B., & Zhang, W. (2013). Phenylspirodrimanes with anti-HIV activity from the sponge-derived fungus *Stachybotrys chartarum* MXH-X73. *Journal of Natural Products, 76*(12), 2298–2306.

Matsuzaki, K., Miyoshi, H., Murakami, K., & Yamamoto, S. (1995). Isochromophilones I and II, novel inhibitors against gp120-CD4 binding produced by *Penicillium multicolor* FO-2338. I. Screening, taxonomy, fermentation, isolation and biological activity. *Journal of Antibiotics* (*Tokyo*)*, 48*(7), 703–707.

Minagawa, K., Kouzuki, S., Yoshimoto, J., Kawamura, Y., Tani, H., Iwata, T., Terui, Y., Nakai, H., Yagi, S., Hattori, N., Fujiwara, T., & Kamigauchi, T. (2002). Stachyflin and acetylstachyflin, novel anti-influenza A virus substances, produced by *Stachybotrys* sp. RF-7260. I. Isolation, structure elucidation and biological activities. *The Journal of Antibiotics, 55*(2), 155–164.

Mishra, S., Pandey, A., & Manvati, S. (2020). Coumarin: An emerging antiviral agent. *Heliyon, 6*(1), Article e03217.

Müller, W. E., Weiler, B. E., & Bachmann, M. (1991). Cordycepin analogues of 2′,5′-oligoadenylate inhibit human immunodeficiency virus infection via inhibition of reverse transcriptase. *Biochemistry, 30*(8), 2027–2033.

Murphy, E. J., Masterson, C., Rezoagli, E., O'Toole, D., Major, I., Stack, G. D., Lynch, C., Laffey, J. G., & Rowan, N. J. (2020). β-Glucan extracts from the same edible shiitake mushroom *Lentinus edodes* produce differential in-vitro immunomodulatory and pulmonary cytoprotective effects—Implications for coronavirus disease (COVID-19) immunotherapies. *Science of the Total Environment, 732*, Article 139330.

Niedermeyer, T. H., Lindequist, U., & Mentel, R. (2005). Antiviral terpenoid constituents of *Ganoderma pfeifferi*. *Journal of Natural Products, 68*(12), 1728–1731.

Padhi, S., Masi, M., Chourasia, R., Rajashekar, Y., Rai, A. K., & Evidente, A. (2021). ADMET profile and virtual screening of plant and microbial natural metabolites as SARS-CoV-2 S1 glycoprotein receptor binding domain and main protease inhibitors. *European Journal of Pharmacology, 890*, Article 173648.

Pan, B. T., & Johnstone, R. M. (1983). Fate of the transferrin receptor during maturation of sheep reticulocytes in vitro: Selective externalization of the receptor. *Cell, 33*(3), 967–978.

Panya, A., Sawasdee, N., Boonyasuppayakorn, S., Choowongkomon, K., & Smith, D. R. (2021). Cordycepin inhibits virus replication in dengue virus-infected vero cells. *Molecules, 26*(11).

Peng, J., Zhang, X., Du, L., Wang, Y., Li, X., & Li, E. (2013). Antiviral alkaloids produced by the mangrove-derived fungus *Cladosporium* sp. PJX-41. *Journal of Natural Products, 76*(6),1133–1140.

Peng, J., Zhang, X., Du, L., Wang, Y., Li, X., & Li, E. (2014). Sorbicatechols A and B, antiviral sorbicillinoids from the marine-derived fungus *Penicillium chrysogenum* PJX-17. *Journal of Natural Products, 77*(2), 424–428.

Perera, R., & Kuhn, R. J. (2008). Structural proteomics of dengue virus. *Current Opinion in Microbiology, 11*(4), 369–377.

Peyrat, L. A., Coudouel, S., Meslin, C., Auvray, P., Le Pape, P., & Hutter, S. (2016). Search for anti-dengue secondary metabolites from two Diospyros endophytes, *Aspergillus* sp. and *Phomopsis* sp. *Planta Medica, 82*(S 01), P693.

Phipps, R. K., Barrow, R. A., & Glasby, K. (2004). Hesseltin A, a novel antiviral metabolite from *Penicillium hesseltinei*. *Organic Letters, 6*(20), 3441–3443.

Pillai, M. G., & Antony, H. (2024). Harnessing the immunomodulatory potential of natural products in precision medicine—A comprehensive review. *Exploration of Drug Science, 2*(3), 339–360.

Praditya, D., Kirchhoff, L., Brüning, J., Rachmawati, H., Steinmann, E., & Steinmann, J. (2019). Anti-infective properties of the golden spice curcumin. *Frontiers in Microbiology, 10*, 912.

Preuer, K., Lewis, R. P. I., Hochreiter, S., Bender, A., Bulusu, K. C., & Klambauer, G. (2018). DeepSynergy: Predicting anti-cancer drug synergy with deep learning. *Bioinformatics, 34*(9), 1538–1546.

Puri, J. L., & Verma, A. L. (2004). Endophytes for the production of podophyllotoxin. *U.P. Application.*

Rabie, A. M. (2022). Potent inhibitory activities of the adenosine analogue cordycepin on SARS-CoV-2 replication. *ACS Omega, 7*(3), 2960–2969.

Raekiansyah, M., Yamaji, R., Lim, C. K., Kondo, H., & Kobayashi, T. (2017). Identification of novel antiviral of fungus-derived brefeldin A against dengue viruses. *Tropical Medicine and Health, 45*, 1–8.

Ramadan, M. A., Abdel-Aziz, M. S., & Ali, A. E. (2020). Promising antimicrobial activities of oil and silver nanoparticles obtained from *Melaleuca alternifolia* leaves against selected skin-infecting pathogens. *Journal of Herbal Medicine, 20*, Article 100289.

Ransom, J. T. (1995). Mechanism of action of mycophenolate mofetil. *Therapeutic Drug Monitoring, 17*(6), 681–684.

Rao, P., Shukla, A., Parmar, P., Rawal, R. M., Patel, B. V., Saraf, M., & Goswami, D. (2022). Proposing a fungal metabolite-flaviolin as a potential inhibitor of 3CLpro of novel coronavirus SARS-CoV-2 identified using docking and molecular dynamics. *Journal of Biomolecular Structure and Dynamics, 40*(1), 348–360.

Rao, P., Shukla, A., Parmar, P., Rawal, R. M., Patel, B., Saraf, M., & Goswami, D. (2020). Reckoning a fungal metabolite, Pyranonigrin A as a potential Main protease (Mpro) inhibitor of novel SARS-CoV-2 virus identified using docking and molecular dynamics simulation. *Biophysical Chemistry, 264*, Article 106425.

Richards, D. M., Carmine, A. A., & Brogden, R. N. (1983). Acyclovir. A review of its pharmacodynamic properties and therapeutic efficacy. *Drugs, 26*(5), 378–438.

Roy, B. G. (2017). Potential of small-molecule fungal metabolites in antiviral chemotherapy. *Antiviral Chemistry and Chemotherapy, 25*(2), 20–52.

Rus, H., Cudrici, C., & Niculescu, F. (2005). The role of the complement system in innate immunity. *Immunology Research, 33*(2), 103–112.

Ryu, W. S. (2016). *Molecular virology of human pathogenic viruses*. Academic Press.

Sacramento, C. Q., Marttorelli, A., Fintelman-Rodrigues, N., de Freitas, C. S., de Melo, G. R., Rocha, M. E., & Souza, T. M. L. (2015). Aureonitol, a fungi-derived tetrahydrofuran, inhibits influenza replication by targeting its surface glycoprotein hemagglutinin. *PLoS ONE, 10*(10), Article e0139236.

Sawdon, A. J., & Peng, C.-A. (2014). Polymeric micelles for acyclovir drug delivery. *Colloids and Surfaces B: Biointerfaces, 122*, 738–745.

Sengupta, P., & Chattopadhyay, S. (2024). Interferons in viral infections. *Viruses, 16*(3), 451.

Shi, S., Liang, Y., Guo, Y., Cheng, L., & Yang, H. (2022). β-Glucans from *Trametes versicolor* (L.) Lloyd is effective for prevention of influenza virus infection. *Viruses*, 14 (2), 388.

Shushni, M. A. M., Singh, R., Mentel, R., & Laatsch, H. (2009). Balticols A-F, new naphthalenone derivatives with antiviral activity from an ascomycetous fungus. *Chemistry and Biodiversity, 6*(2), 127–137.

Singh, S. B., Zink, D. L., Liesch, J. M., Mosley, R. T., Trehan, A. K., & Greenlee, M. L. (2003). Integracides: Tetracyclic triterpenoid inhibitors of HIV-1 integrase produced by *Fusarium* sp. *Bioorganic and Medicinal Chemistry, 11*(7), 1577–1582.

Singh, S. B., Zink, D. L., Polishook, J. D., Dombrowski, A. W., Darkin-Rattray, S. J., Schmatz, D. M., & Goetz, M. A. (2002). Discovery, structure and HIV-1 integrase inhibitory activities of integracins, novel dimeric alkyl aromatics from *Cytonaema* sp. *Tetrahedron Letters, 43*(9), 1617–1620.

Smee, D. F., Bray, M., & Huggins, J. W. (2001). Antiviral activity and mode of action studies of ribavirin and mycophenolic acid against orthopoxviruses in vitro. *Antiviral Chemistry and Chemotherapy, 12*(6), 327–335.

Songprakhon, P., Phaonakrop, N., Pankla, R., Kiddeeh, A., Paemanee, A., & Roytrakul, S. (2024). Cordycepin exhibits both antiviral and anti-inflammatory effects against dengue virus infection. *iScience, 27*(9), 110711.

Strauss, J. H., & Strauss, E. G. (2008). *Overview of viruses and virus infection*. Elsevier.

Streeter, D. G., Witkowski, J. T., Khare, G. P., Sidwell, R. W., Bauer, R. J., Robins, R. K., & Simon, L. N. (1973). Mechanism of action of 1-β-D-ribofuranosyl-1,2,4-triazole-3-carboxamide (Virazole), a new broad-spectrum antiviral agent. *Proceedings of the National Academy of Sciences, 70*(4), 1174–1178.

Strzelec, M., Detka, J., Mieszczak, P., Sobocińska, M. K., & Majka, M. (2023). Immunomodulation—A general review of the current state-of-the-art and new therapeutic strategies for targeting the immune system. *Frontiers in Immunology, 14*, 1127704.

Swenson, C. E., Low, R. L., Perkins, W. R., & Lenk, R. P. (1995). Lipid-based formulations of amphotericin B. *Journal of Antimicrobial Chemotherapy, 35*(5), 709–711.

Swinkels, H. M., Vaillant, A. J., Nguyen, A., & Gulick, P. (2024). HIV and AIDS. In *StatPearls*. StatPearls Publishing LLC., Treasure Island (FL)

Szymańska, E., Winnicka, K., Karpinski, T. M., & Miltyk, W. (2018). Multifunctional tannic acid/silver nanoparticle-based mucoadhesive hydrogel for improved local treatment of HSV infection: In vitro and in vivo studies. *International Journal of Molecular Sciences, 19*(2), 387.

Takahashi, J. A., Barbosa, B. V., Fernandes, G. R., & Pimenta, R. S. (2021). Antiviral fungal metabolites and some insights into their contribution to the current COVID-19 pandemic. *Bioorganic and Medicinal Chemistry, 46*, Article 116366.

Takhampunya, R., Ubol, S., Houng, H. S., & Cameron, C. E. (2006). Inhibition of dengue virus replication by mycophenolic acid and ribavirin. *Journal of General Virology, 87*(7), 1947–1952.

Thomford, N. E., Senthebane, D. A., Rowe, A., Munro, D., Seele, P., Maroyi, A., & Dzobo, K. (2018). Natural products for drug discovery in the 21st century: Innovations for novel drug discovery. *International Journal of Molecular Sciences, 19*(6), 1578.

Tuli, H. S., Sandhu, S. S., & Sharma, A. K. (2013). Cordycepin: A bioactive metabolite with therapeutic potential. *Life Sciences, 93*(23), 863–869.

van Heuvel, Y., Wang, L., Xu, Y., & Patel, M. (2022). Infectious RNA: Human Immunodeficiency Virus (HIV) biology, therapeutic intervention, and the quest for a vaccine. *Toxins (Basel), 14*(2).

van Steenwijk, H. P., Bast, A., & de Boer, A. (2021). Immunomodulating effects of fungal beta-glucans: From traditional use to medicine. *Nutrients, 13*(4), 1333.

Vetvicka, V., & Vetvickova, J. (2014). Natural immunomodulators and their stimulation of immune reaction: True or false? *Anticancer Research, 34*(5), 2275–2282.

Wang, H., & Ng, T. B. (2001). Isolation and characterization of velutin, a novel low-molecular-weight ribosome-inactivating protein from winter mushroom (*Flammulina velutipes*) fruiting bodies. *Life Sciences, 68*(18), 2151–2158.

Wang, H., Zhang, X., Yu, H., Li, J., Lu, X., Tian, X., & Zhang, W. (2011). Anti-influenza virus polyketides from the acid-tolerant fungus *Penicillium purpurogenum* JS03-21. *Journal of Natural Products, 74*(9), 2014–2018.

Wang, J., Zhang, X., Liu, C., Li, G., Zhang, Y., & Tian, X. (2016). Antiviral merosesquiterpenoids produced by the Antarctic fungus *Aspergillus ochraceopetaliformis* SCSIO 05702. *Journal of Natural Products, 79*(1), 59–65.

Wang, S. Y., Hsu, M. L., Hsu, H. C., Lee, S. S., Shiao, M. S., & Ho, C. K. (1997). The anti-tumor effect of Ganoderma lucidum is mediated by cytokines released from activated macrophages and T lymphocytes. *International Journal of Cancer, 70*(6), 699–705.

Wheelock, E. F., & Larke, R. P. (1967). Prolongation of life in mice with established Friend virus leukemia by inoculation with Sendai virus, interferon, or statolon. *Science, 158*(3800), 537–538.

Wu, G., Xu, G., Wu, Z., Liu, C., & Guo, Y. (2014). Cladosins A-E, hybrid polyketides from a deep-sea-derived fungus, *Cladosporium sphaerospermum*. *Journal of Natural Products, 77*(2), 270–275.

Xu, J., Yi, M., Ding, L., & He, S. (2019). A review of anti-inflammatory compounds from marine fungi, 2000–2018.

Xu, Q., Qiao, Y., Zhang, Z., Deng, Y., Chen, T., Tao, L., et al. (2021). New polyketides with anti-inflammatory activity from the fungus *Aspergillus rugulosa*. *Frontiers in Pharmacology, 12*, Article 700573.

Ying, C., De Clercq, E., Neyts, J., & Zhang, H. (2007). Ribavirin and mycophenolic acid markedly potentiate the anti-hepatitis B virus activity of entecavir. *Antiviral Research, 73*(3), 192–196.

Yoganathan, K., Ross, S. A., Andersen, R. J., & Sweeney, M. (2003). 10-Methoxydihydrofuscin, fuscinarin, and fuscin, novel antagonists of the human CCR5 receptor from *Oidiodendron griseum*. *Journal of Natural Products, 66*(8), 1116–1117.

Yoganathan, K., Ross, S. A., Sweeney, M., & Andersen, R. J. (2004). Inhibition of the human chemokine receptor CCR5 by variecolin and variecolol and isolation of four new variecolin analogues, emericolins A-D, from *Emericella aurantiobrunnea*. *Journal of Natural Products, 67*(10), 1681–1684.

Yoon, S. Y., Park, S. J., & Park, Y. J. (2018). The anticancer properties of cordycepin and their underlying mechanisms. *International Journal of Molecular Sciences, 19*(10).

Yoshimoto, J., Kakui, M., Iwasaki, H., Fujiwara, T., Sugimoto, H., & Hattori, N. (1999). Identification of a novel HA conformational change inhibitor of human influenza virus. *Archives of Virology, 144*(5), 865–878.

Yu, D., Zhao, M., Dong, L., Zhao, L., Zou, M., Sun, H., & Zou, Z. (2016). Design and evaluation of novel interferon lambda analogs with enhanced antiviral activity and improved drug attributes. *Drug Design, Development and Therapy*, 163–182.

Yu, G., Li, J., Ma, L., Wang, X., & Zhang, W. (2016b). Neosartoryadins A and B, fumiquinazoline alkaloids from a mangrove-derived fungus *Neosartorya udagawae* HDN13-313. *Organic Letters, 18*(2), 244–247.

Yun, K., Kim, J., Lee, S. J., Lee, D., Kwon, H. C., & Yang, H. O. (2016). Cristazine, a new cytotoxic dioxopiperazine alkaloid from the mudflat-sediment-derived fungus *Chaetomium cristatum*. *Chemical and Pharmaceutical Bulletin (Tokyo), 64*(1), 59–62.

Zebeaman, M., Tadesse, M. G., Bachheti, R. K., Bachheti, A., Gebeyhu, R., & Chaubey, K. K. (2023). Plants and plant-derived molecules as natural immunomodulators. *BioMed Research International, 2023*(1), 7711297.

Zhang WenJing, Z. W., Tao JunYan, T. J., Yang XiaoPing, Y. X., Yang ZhuLiang, Y. Z., Zhang Li, Z. L., Liu HongSheng, L. H., & Wu JianGuo, W. J. (2014). Antiviral effects of two *Ganoderma lucidum* triterpenoids against enterovirus 71 infection.

Zhang, G., Sun, S., Zhu, T., Lin, Z., Gu, J., Li, D., & Gu, Q. (2011). Antiviral isoindolone derivatives from an endophytic fungus *Emericella* sp. associated with *Aegiceras corniculatum*. *Phytochemistry, 72*(11–12), 1436–1442.

Zhou, X., Fang, W., Chen, C., Chen, Q., Lin, X., Yang, B., Luo, X., & Liu, J. (2016). Aspernigrins with anti-HIV-1 activities from the marine-derived fungus *Aspergillus niger* SCSIO Jcsw6F30. *Bioorganic and Medicinal Chemistry Letters, 26*(2), 361–365.

Zou, X.-W., Li, S.-M., Zhang, L.-J., & Chen, X.-Y. (2009). Helotialins A-C, anti-HIV metabolites from a Helotialean ascomycete. *Chinese Journal of Natural Medicines, 7*(2), 140–144.

Mycotoxins: Fungal Metabolites and Human Health

Yashvardhan Raj, Swastick Bisht, Anuradha Singh, Chandan Mukherjee, and Karishma Tiwari

Abstract

Mycotoxins constitute a group of toxic secondary metabolites produced by fungi, mainly from the genera *Aspergillus, Penicillium,* and *Fusarium.* These compounds can contaminate a wide range of agricultural commodities both in laboratory settings and in industry, thereby degrading the quality of life. The metabolites produced by fungi are a major issue, mostly because they produce many extreme adverse health effects in animals and humans. Some of the well-known mycotoxins include aflatoxins, ochratoxins, fumonisins, and trichothecenes. Some of these were associated with specific pathological conditions like liver cancer and renal impairment, and some others with neurological syndromes. Normally, human exposure or contamination by mycotoxins primarily occurs by the ingestion of contaminated food; these exposures can also occur via inhalation and dermal exposure routes, especially at the workplace. The exposure to mycotoxins, therefore, usually causes long-term health impacts, and often there is a carcinogenic link that leads to death of the one exposed to it. There is also widespread evidence that mycotoxins contribute to the development of many diseases due to the depression of the immune system and altered metabolism. As a result, a great amount of scientific research focused on developing more sensitive methods for their detection, understanding of their toxicological mechanisms, and methods to prevent effectively, from both the agronomic and storage perspectives, has been going on to stop the threat of mycotoxins. This chapter offers a comprehensive overview of mycotoxin contaminants, detailing their toxic mechanisms, impacts on human health, and recent advances in detection and mitigation strategies aimed at protecting public health.

Keywords

Carcinogenicity · Detection techniques · Food safety · Fungal contamination · Hepatotoxicity · Neurotoxicity

1 Introduction

Fungal metabolites, in particular mycotoxins, remain very significant to public health and food safety issues today. Mycotoxins are toxic secondary metabolites produced mainly from fungi of the families *Aspergillus, Penicillium,* and *Fusarium* and often contaminate the food supply when crops and animal feed are sold and consumed (Bennett & Klich, 2003). Mycotoxins are common in warm, humid climates, further aggravated by storage and regulatory options. Contamination occurs with detrimental effects, i.e. health risks and substantial economic risk that are particularly severe for low to middle-income countries, not only to the integrity and safety of food supply (Atnkut et al., 2025a, 2025b). Health-related impacts from exposure to mycotoxins are manifold and serious in both humans and animals, including immune system suppression, cancer development, and subacute forms of poisoning (Al-Falih, 2024; Awuchi et al., 2022). Mycotoxins such as aflatoxins, ochratoxins, fumonisins, trichothecenes, zearalenone, and patulin have been extensively studied from a toxicological point of view (Hussein & Brasel, 2001). The first class of toxins, aflatoxins, is known as a carcinogen that is associated with the development of hepatocellular carci-

Y. Raj · S. Bisht · C. Mukherjee
Department of Life Sciences, School of Biosciences and Technology, Galgotias University, Greater Noida, Uttar Pradesh, India

A. Singh (✉)
Department of Biotechnology, Jaypee Institute of Information Technology, Greater Noida, Uttar Pradesh, India
e-mail: anuradha.singh@mail.jiit.ac.in; anuradhasingh.dr@gmail.com

K. Tiwari
University Centre for Research and Development (UCRD), Chandigarh University, Mohali, Punjab, India

© The Author(s), under exclusive license to Springer Nature Switzerland AG 2026

A. Prakash and M. K. Dubey (eds.), *Advances in Fungal Secondary Metabolites: Bioactive Attributes and Biotechnological Potential of Industrial Interest*, Sustainable Economy and Ecotechnology, https://doi.org/10.1007/978-3-032-23507-7_6

noma. The ochratoxins have been correlated with nephrotoxicity or damage to the kidneys and immunotoxicity in humans; while fumonisins cause damage by disrupting sphingolipid metabolism, leading to oesophageal cancer, and the trichothecenes have immunosuppressive and cytotoxic effects (Khan et al., 2024a, 2024b). Mycotoxins complicate the course of chronic diseases and immune responses, and can be exacerbated in chronic metabolic disorders, further exemplifying an important public health concern (Wang et al., 2025).

Despite the advancements made over the past several decades, the subject of mycotoxins presents many challenges that remain unsolved, primarily in the context of the population having no access to rapid detection methods for mycotoxins. Detecting mycotoxins with HPLC, GC–MS, or ELISA is the current gold standard methodology; however, these remarkable methods are costly, require specialized expertise, and laboratory spaces. These features limit their applicability in resource-poor settings, where detection methods must be hassle-free for monitoring to be accurate and timely, especially in food systems that regularly depend on subsistence farming, and communities consume contaminated products (Maphaisa et al., 2025a, 2025b). Moreover, while prevention measures with biological control agents should be guided by the principles of the Good Agricultural Practice (GAP), the Good Manufacturing Practice (GMP), and the Hazard Analysis and Critical Control Points (HACCP) systems (Jallow et al., 2021).

This chapter is about mycotoxins. It wants to help people understand what mycotoxins are, how they can harm people, and what effects they have on health. So we will look at ways to detect mycotoxins, like special tools called biosensors, lab-on-a-chip and next-generation sequencing. These new tools seem promising for quickly checking for mycotoxins. Mycotoxins are a problem, and we need to learn more about them. The new detection technologies for mycotoxins can really help us. We should think about how to prevent things from happening with the help of biological control agents, and how to handle food after it is picked and new ways to clean food. Mycotoxin contamination is a problem and we need to look at it in a new way. This part of the book wants to show where the old ways are not working and what new ideas might help. It wants to give people the information they need to do something about mycotoxin contamination. We need to look at this problem from different angles and work together. People who do research, people who make policies and people who work in industry. To solve the global problem of mycotoxins. Ultimately, the insights presented here contribute to the broader goal of safeguarding food security, protecting public health, and enhancing the resilience of agricultural systems in the face of an ever-evolving threat.

2 Classification of Major Mycotoxins

Mycotoxins are toxic secondary metabolites of different species of fungi. Because some of these metabolites always result in food and animal feed being contaminated, the mycotoxins bear a burden of risk to human and animal health (Singh et al., 2014).

There are different kinds of mycotoxins that we know about now. The main kinds of mycotoxins are aflatoxins, ochratoxins, fumonisins and trichothecenes. Each of these kinds of mycotoxins has smaller groups. Mycotoxins are a problem because they can get into food and animal feed. This can be bad for human health and animal health. Mycotoxins are a risk, to people and animals because of the way they can hurt us. Broadly classified, mycotoxins can be differentiated on the basis of three types: on their toxicity (Table 1), their stability (Table 2) and on their detection methods (Table 3).

Mycotoxins are made by fungi like *Aspergillus, Fusarium* and *Penicillium.* These mycotoxins include things like aflatoxins, fumonisins and ochratoxins. Mycotoxins are very different from each other when it comes to how toxic they're how they work. This is a deal for food safety and public health all around the world, as seen in Table 4. This table shows the organisms that produce things and the different subtypes. It helps us understand how many different types of these organisms exist and why they are important for farming and our health. The organisms are really important in agriculture. When we are sick.

The toxicity of major mycotoxins has been extensively described, with aflatoxins showing strong hepatocarcinogenic potential (Eaton & Gallagher, 1994), ochratoxins exerting nephrotoxic and immunotoxic effects (Heussner & Bingle, 2015), fumonisins associated with neural tube defects and hepatotoxicity (Henry & Wyatt, 2001), and trichothecenes impairing protein synthesis and immune responses (Mostrom, 2021). Table 1 summarizes the toxic effects of these key compounds.

Stability profiles vary markedly among mycotoxins: aflatoxins can persist in dry storage and resist moderate heating, ochratoxins show stability during processing and storage, while trichothecenes and fumonisins exhibit varying degrees of thermal resistance depending on matrix and conditions (Janik et al., 2021a, 2021b; Kiseleva et al., 2020). These characteristics are critical for food safety and detoxification strategies. Table 2 presents a comparative view of stability.

Mycotoxins represent a wide range of secondary, naturally occurring and practically unavoidable fungal metabolites. They contaminate various agricultural commodities like cereals, maize, peanuts, fruits, and feed at any stage in pre- or post-harvest conditions (Dubey et al., 2018). Consumption of mycotoxin-contaminated food and feed can cause acute or chronic toxicity in humans and animals.

Table 1 Toxicity of major mycotoxins

Mycotoxin	Producing organisms	Major types	Toxicity	References
Aflatoxins	*Aspergillus flavus* and*A. parasiticus*	AFB1, AFB2, AFG1, AFG2, AFM1	• AFB1: Highly toxic, hepatotoxic, tumorigenic; causes DNA mutations via cytochrome P450	Eaton et al. (2017)
			• AFM1: Hydroxylated metabolite of AFB1, found in milk; Group 1 carcinogen but less toxic	Marchese et al. (2018)
			• AFG1: Toxic but less potent than AFB1	Mahmoud et al. (2025)
			• AFB2, AFG2: Least toxic	Mahmoud et al. (2025)
			• Acute exposure: Aflatoxicosis, liver failure, death	Mahmoud et al. (2025)
			• Chronic exposure: Hepatocellular carcinoma, immunosuppression, poor growth	Mahmoud et al. (2025)
Ochratoxins	*Aspergillus* and *Penicillium* species	OTA, OTB, OTC, and OTD	• OTA: Most toxic, nephrotoxic, carcinogenic (Group 2B), linked to renal carcinoma; causes immunotoxicity and teratogenic effects	Heussner and Bingle (2015)
			• OTB: Less toxic; at high concentrations, it causes oxidative stress and liver toxicity	Heussner and Bingle (2015)
			• OTC, OTD: Less studied; believed to be less toxic than OTA	Heussner and Bingle (2015)
Fumonisins	*Fusarium verticillioides, F. proliferatum*	FB1, FB2, FB3	• FB1: Most toxic; linked to hepatotoxicity, nephrotoxicity, and esophageal cancer; classified as Group 2B carcinogen	Henry and Wyatt (2001)
			• FB2: Less toxic but causes liver and kidney damage	Henry and Wyatt (2001)
			• FB3: Least toxic, usually found in low concentrations	Henry and Wyatt (2001)
			• Mechanism: Inhibits ceramide synthase, leading to disrupted sphingolipid metabolism and cellular dysfunction	Henry and Wyatt (2001)
Trichothecenes	*Fusarium, Myrothecium, Stachybotrys*	Type A, Type B, Type C, Type D	• Type A (T-2, HT-2): Highly toxic; causes dermatonecrosis, gastrointestinal issues, immunosuppression	Mostrom (2021)
			• Type B (Deoxynivalenol, Nivalenol): Moderate toxicity; causes vomiting and feed refusal	Mostrom (2021)
			• Type C: Unknown epoxide group, less characterized	Mostrom (2021)
			• Type D (Verrucarins): Stable structure, low bioavailability	Mostrom (2021)
			• Pigs are highly susceptible to Type B; Type A affects poultry significantly	Mostrom (2021)

Source Authors' compilation

Table 2 Stability of major mycotoxins

Mycotoxin	Thermal stability	Chemical stability	Environmental factors	References
Aflatoxins	Degrade slightly above 150 °C; resistant to cooking, pasteurization, and autoclaving	Stable across acidic and neutral pH; ammonization reduces levels but may face restrictions	Thrive in poor storage conditions; encourage *Aspergillus* mould growth in grains and nuts	Kutasi et al. (2021)
Ochratoxins (OTA)	Highly stable in acidic environments; degrades with prolonged exposure to high temperatures and UV radiation	Stable in storage and processing, OTB is less stable than OTA	Persistent in foodstuffs; resistant to regular storage and processing conditions	Ding et al. (2023)
Fumonisins	Stable across wide pH and temperature ranges; not easily degraded by cooking or baking	Maintain stability in processed foods; long-term contamination risk in stored maize	Found in stored maize and cereals, they remain in food even after conventional preparation methods	Bryła et al. (2017)
Trichothecenes	Type A: Most resistant to acidic and thermal conditions; Type B: Less stable, resistant to slight heating; Type D: Highly stable	Retains toxicity in grains; it breaks down with moisture and UV irradiation	Very stable in dry conditions; difficult to decontaminate; heat processing has little effect	Meneely et al. (2023)

Source Authors' compilation

Table 3 Detection methods for major mycotoxins

Mycotoxin	Detection methods	Advantages	Disadvantages	References
Aflatoxins	Thin Layer Chromatography (TLC), High-Performance Liquid Chromatography (HPLC, often coupled with fluorescence or Mass Spectrometry [MS]), Enzyme-Linked Immunosorbent Assay (ELISA), Lateral Flow Device (LFD), Liquid Chromatography-Tandem Mass Spectrometry (LC–MS/MS), Near-Infrared Spectroscopy (NIR), and Fourier Transform Infrared Spectroscopy (FTIR)	HPLC and LC–MS/MS are highly sensitive; ELISA and LFD are fast and user-friendly	ELISA/LFD may yield false positives/ negatives; TLC is less sensitive; non-invasive techniques are less specific	Iqbal et al. (2014)
Ochratoxins	HPLC-FLD, HPLC–MS/MS, ELISA, TLC, PCR, immunosensors, and biosensors	HPLC variants are highly sensitive/ selective; biosensors are rapid, portable, and low-cost	ELISA has a higher chance of false positives; TLC is less sensitive; advanced techniques are expensive	Kos et al. (2016)
Fumonisins	HPLC, Gas Chromatography, ELISA, LC–MS/MS, surface plasmon resonance, and electrochemical sensors	LC–MS/MS offers excellent sensitivity; biosensors are fast and field-usable	Sample preparation for HPLC is time-consuming; biosensors are still in developmental stages	Ong et al. (2021)
Trichothecenes	HPLC, GC–MS, ELISA, LC–MS/MS, solid-phase/ liquid–liquid extraction for sample preparation	LC–MS/MS supports multi-toxin detection; ELISA is highly specific for Type A and B toxins	Detection in complex matrices is challenging; Type D toxins require advanced MS; co-occurrence complicates quantification	Janik et al. (2021a, 2021b)

Source Authors' compilation

Detection of mycotoxins relies on chromatographic and spectrometric methods (HPLC, GC–MS, LC–MS/MS) as well as emerging immunoassay and biosensor platforms. Each method differs in sensitivity, specificity, and cost, which guides application in regulatory or field testing (Janik et al., 2021a, 2021b; Singh & Mehta, 2020; Turner, 2024). Table 3 provides a consolidated summary of these methods.

Mycotoxins are potentially hazardous secondary metabolites produced by filamentous fungi (moulds). These small molecules, typically under 1000 Da are widespread in nature and virtually unavoidable. They can enter the food chain

Table 4 Overview of 20 mycotoxins: their producing organisms and subtypes

Mycotoxin	Producing organism	Subtypes	References
Aflatoxin	*Aspergillus flavus, A. parasiticus*	AFB1, AFB2, AFG1, AFG2, AFM1	Eaton et al. (2017)
Ochratoxin	*Aspergillus ochraceus, Penicillium verrucosum*	OTA, OTB, OTC	Heussner and Bingle (2015)
Fumonisin	*Fusarium verticillioides, F. proliferatum*	FB1, FB2, FB3	Henry and Wyatt (2001)
Deoxynivalenol (DON)	*Fusarium graminearum*	Type B Trichothecene	Mostrom (2021)
T-2 Toxin	*Fusarium sporotrichioides*	Type A Trichothecene	Mostrom (2021)
Zearalenone	*Fusarium graminearum, F. culmorum*	–	Ropejko and Twarużek, (2021)
Patulin	*Penicillium expansum*	–	Llobregat et al. (2025)
Citrinin	*Penicillium citrinum, Monascus purpureus*	–	Kamle et al. (2022)
Ergot Alkaloids	*Claviceps purpurea*	Ergometrine, ergotamine	Smakosz et al. (2021)
Sterigmatocystin	*Aspergillus versicolor*	–	Bullerman (2003)
Alternariol	*Alternaria alternata*	–	Wenderoth et al. (2019)
Fusaric Acid	*Fusarium oxysporum*	–	López-Díaz et al. (2018)
Enniatins	*Fusarium avenaceum*	Enniatin A, Enniatin B	Ivanova et al. (2006)
Beauvericin	*Fusarium proliferatum*	–	Plattner and Nelson (1994)
Moniliformin	*Fusarium verticillioides*	–	Hussein and Brasel (2001)
Verrucarin	*Myrothecium verrucaria*	–	Siciliano et al. (2017)
Roridin	*Myrothecium roridum*	Roridin A, Roridin E	Piao et al. (2013)
Gliotoxin	*Aspergillus fumigatus*	–	Knowles et al. (2020)
Cyclopiazonic Acid	*Penicillium commune, Aspergillus flavus*	–	Chang et al. (2009)
Penitrem	*Penicillium crustosum*	Penitrem A, Penitrem B	Kalinina et al. (2017)

Source Authors' compilation

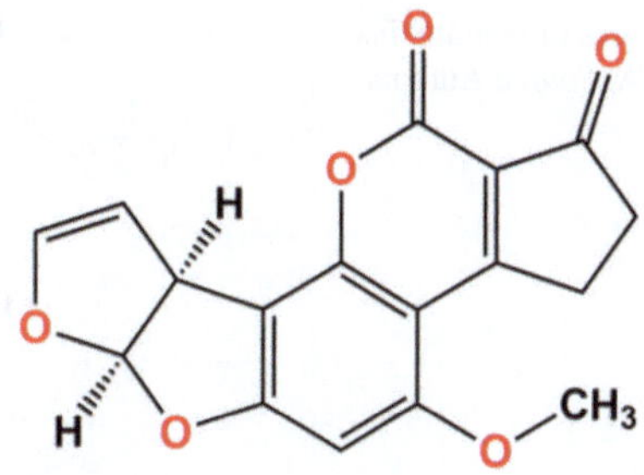

Fig. 1 Structure of aflatoxin. *Source* Authors

directly via contaminated plant-derived ingredients or indirectly through fungal growth on food products (El-Sayed et al., 2022). A detailed overview of major mycotoxins is provided in Table 4.

Specific fungal genera and species are known producers of the principal mycotoxin classes, with *Aspergillus* species yielding aflatoxins and ochratoxins, *Fusarium* producing fumonisins and trichothecenes, and *Penicillium* contributing additional ochratoxins and citrinin (Awuchi et al., 2022; Bennett & Klich, 2003; Hussein & Brasel, 2001). Table 4 outlines representative organisms and their toxin subtypes.

Aflatoxin, ochratoxins. fumonisins, trichothecenes, patulin, citrinin and zearalenone are commercially significant mycotoxins that have been identified in various research studies. Details of some of these toxins are provided below.

2.1 Aflatoxins

These are a family of very potent mycotoxins mainly produced by *Aspergillus flavus* and *Aspergillus parasiticus,* but a few among them show far more dramatic impact as potent carcinogenic, immunosuppressant, and toxic agents. Aflatoxin B1 is one of the highly synthesized aflatoxins. Aflatoxins are produced by fungi that infect maize, peanuts, and tree nuts, mainly those crops that grow in warm and humid climatic conditions. Aflatoxins cause DNA deletions or act as a mutagen, which leads to mutations and causes hepatitis carcinoma not only in humans but also in animals. Laws governing the quantity of aflatoxin present in dietary substances like cereals are very strict, and each nation has different parameters. They would also give it fewer chances by developing biocontrol strategies using non-toxigenic strains of *Aspergillus.* Developing resistant crops with good detoxification properties and preventive technologies would also help to resolve the overall threat of the mycotoxins (Chourabi et al., 2012). Figure 1 showcases the structure of aflatoxin.

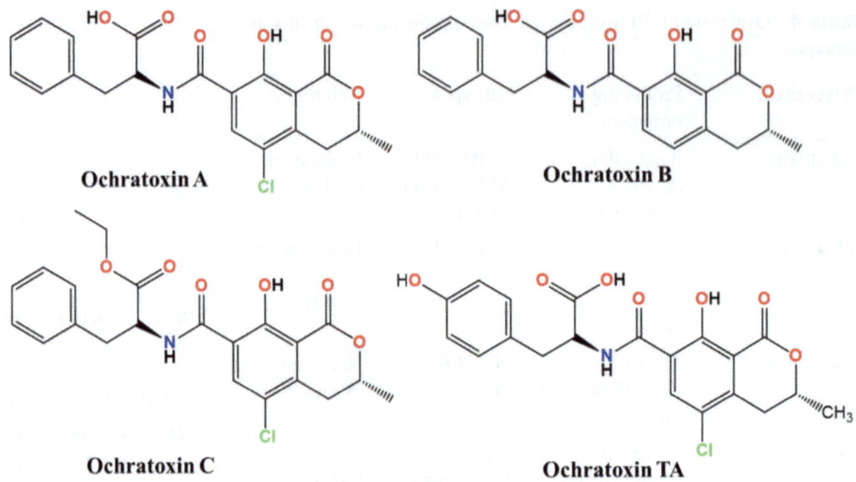

Fig. 2 Structures of ochratoxins A, B, C and TA. *Source* Authors

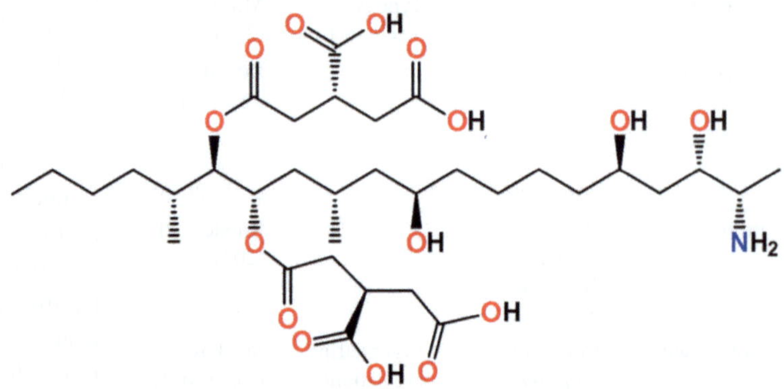

Fig. 3 Structure of fumonisin B1. *Source* Authors

2.2 Ochratoxins

Ochratoxins (OTA) are mycotoxins formed as secondary metabolic compounds by various species of *Aspergillus*, among them *A. ochraceus,* and some species of *Penicillium*, in particular, *P. verrucosum.* The fungi infect cereals, coffee, dried fruits, and wine, with high diffusion in temperate climates. OTA is naturally a potent nephrotoxic, hepatotoxic, and teratogenic, and it is also carcinogenic to humans. It mainly acts by interfering with protein synthesis, generation of oxidative stress, and disruption of mitochondrial activity. Its strong binding affinity to renal tissues supports the pathogenesis of chronic kidney disease. Proper agricultural practices during harvesting and storage, along with LC–MS detection, can avert or mitigate the condition. (Ding et al., 2023). The given Fig. 2, showcases the structure of various ochratoxins.

2.3 Fumonisins

Fumonisins are a type of mycotoxin mainly produced by *Fusarium verticillioides* (formerly *F. moniliforme*) and *Fusarium proliferatum.* These two toxigenic species are among all other toxigenic species of Fusarium, and they exacerbate the worst contamination of maize and products processed from it. The most frequently isolated and most toxic one is FB1. It has been well proven in various research studies that FB1 is one of the main inhibitors of ceramide synthase, which causes distortions in the sphingolipid metabolism to result in the dysfunction of cellular membranes. Health effects from fumonisins in humans are oesophagal cancer and equine leukoencephalomalacia. It has also been implicated in the production of pulmonary oedema in pigs. As fumonisin levels increase with humidity, the cycle of dry spells broken by rain seems to favour their increase. Prevention is possible through resistant crop varieties, careful drying, and storage practices. New biological control agents include certain strains of *Trichoderma* that could help control the problem (Alberts et al., 2016). The given Fig. 3 showcases the structure of fumonisin B1.

2.4 Trichothecenes

Trichothecenes refer to a broad group of mycotoxins whose main origin lies in species of the genus *Fusarium.* Some of them are *Fusarium graminearum, Fusarium culmorum, Fusarium sporotrichioides* and *Fusarium poae.* Less commonly known producers are *Myrothecium,*

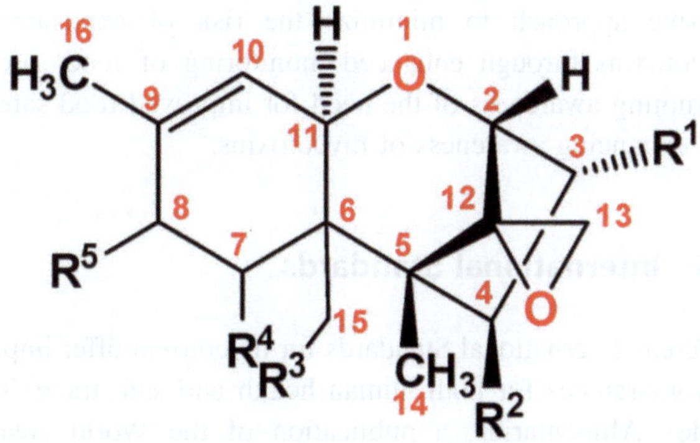

Fig. 4 Structure of trichothecene. *Source* Authors

Stachybotrys, and *Trichothecium* species. The best-known trichothecenes are Deoxynivalenol (DON), T-2 toxin, and HT-2 toxin. Generally, they are found in grains, that is, in wheat, barley, and oats. They are inhibitors of protein synthesis that can activate apoptosis and oxidative stress. These toxins also show immunosuppressive effects. Expressions of diseases in humans and animals are exhibited in clinical signs such as vomiting, diarrhea, and immunosuppression. Extraction of Trichothecene mycotoxins from agricultural products has been greatly complicated by the stability of these toxins under conditions of processing and thermal treatment. Preventive measures include the application of fungicides, crop rotation, and drying immediately after harvest. Having further implications for health in animals and trade consequences, mycotoxins have now become a serious challenge in food safety. This sets the stage for an argument in favour of a collective intervention through integrated approaches for monitoring and mitigation (Meneely et al., 2023). The given Fig. 4 showcases the structure of Trichothecene.

3 Global Impact of Mycotoxins on Food and Agriculture

Mycotoxins are toxic secondary metabolites produced naturally by different fungi, chiefly the genera *Aspergillus*, *Fusarium*, and *Penicillium*. These compounds pose a persistent global challenge, with recent studies highlighting their increasing prevalence due to climate change, which creates favourable conditions for fungal growth and mycotoxin production. It has a large impact on the environment and on food production as well as on food safety, food security, and the economy of the world. Additionally, with all of the methods we now have available to detect mycotoxins and the regulations that have been enacted to monitor them, it is now a priority to try to minimize the negative effects of mycotoxins on our health and to ensure that everyone has access to safe food.

3.1 Food Security

There is a large concentration of mycotoxins in staple food items, such as maize, wheat, and rice, as well as in some nuts, which ultimately reduces the amount of food we have to consume and decreases the overall nutritional value of the food we have available. In many countries, inadequate storage facilities are a major contributor to the loss of corn, wheat, and rice through spoilage/fermentation. The loss of this valuable resource through mycotoxins is a common topic of concern (Leslie et al., 2021).

3.2 Barriers to Trade

Mycotoxins represent an impediment to international trade because mycotoxins are classified as non-tariff barriers. The introduction of contaminated food products into a developed economy usually requires adherence to severe regulatory standards to prevent contamination; as a result of these strict regulations, the economies of the exporting nations and their farmers are adversely affected (Logrieco et al., 2018). The approximate dollar value of crop losses caused by mycotoxin contamination is estimated to be in the billions of dollars each year.

3.3 Livestock Industry

The consumption of contaminated livestock feed causes numerous health-related issues for livestock, including decreased productivity, reduced growth rate, immunity suppression, and death in extreme cases. Aflatoxins, ochratoxins and fumonisins are mycotoxins that have the potential to negatively impact metabolic and reproductive processes and create chronic economic losses for animal producers. Secondary contamination occurs when these toxins are transmitted through the food web (i.e. through contaminated feed that has been consumed by animals) and pose significant risks to human health via the consumption of meat, milk, and eggs of these livestock (Santos Pereira et al., 2019). The potential for bioaccumulation of mycotoxins in food animals demonstrates the importance of regulating and monitoring animal feed, in addition to developing detoxification methods to reduce the negative impact of mycotoxins on both human and animal health.

3.4 Economic and Health Impacts of Mycotoxin Contamination

3.4.1 Economic Impacts

Mycotoxin contamination has large financial and social ramifications for the agricultural sector. The most significant impact is crop loss because of reduced yield and quality of agriculturally-based products due to mycotoxin contamination; consequently, mycotoxin contamination affects both farmer income and food availability in agricultural systems impacted by mycotoxins (Zhang et al., 2024). The requirement to meet food safety standards means producer costs include additional expenses related to frequent testing, detoxification methods, and improved infrastructure to meet regulatory requirements. For smallholder farmers, these costs exert significant financial pressure relative to the higher minimum cost expected in large-scale operations. Trade losses are another major financial contribution to the economy. Countries with minimal mycotoxin surveillance and regulation are unable to export their crop products to areas where strict mycotoxin regulations, such as those in the United States and the European Union, are enforced. Export restrictions reduce agricultural exports as well as economically impact countries with low or medium incomes, when much of their GDP is generated through agriculture.

3.4.2 Health Impacts

Eating food that contains certain biological toxins (or mycotoxins) can cause harmful health effects due to both acute (short-term) and chronic (long-term) exposure to these substances. Some examples of mycotoxins are aflatoxin (a potent carcinogen, especially among people infected with the hepatitis B virus) and ochratoxin (nephrotoxic, immunosuppressive, and related to skeletal malformations). Additionally, fumonisin (another type of mycotoxin) has been identified as a cause of both developmental defects relating to the neural tube during pregnancy as well as immune system suppression. Chronic exposure to mycotoxins can result in irreversible changes to the body.

Some groups of individuals, including the malnourished, young children, old adults, and those with compromised immune systems, may experience a greater risk from exposure to mycotoxins because their bodies may not effectively eliminate these toxins.

Frequent consumption of food contaminated with mycotoxins is likely compounding the burden of foodborne diseases and chronic illnesses among low- and middle-income countries. Additionally, the negative effects of mycotoxins and foodborne illnesses are posing additional challenges on public health systems. Accordingly, comprehensive strategies are required to address the public health implications of mycotoxins and the need for a comprehensive approach to minimize the risk of exposure to mycotoxins through enhanced monitoring of food safety, promoting awareness of the need for improved food safety, and enhancing awareness of mycotoxins.

3.5 International Standards

Whereas International Standards for mycotoxin offer important guarantees for both human health and safe trade. The Codex Alimentarius, a publication of the World Health Organization (WHO) and Food and Agriculture Organization (FAO), has established guidelines regarding the maximum levels of mycotoxins that can legally be present in animal feeds and food products. These guidelines are the basis for harmonizing food safety regulations on an international scale. EU regulations are viewed as the strictest in the world and include very stringent threshold levels for example the maximum level allowable for Aflatoxin B1 in food is two Parts per Billion (ppb); In contrast, FDA regulations in the United States allow for a maximum of 20 ppb of aflatoxins in most food and feed products and a threshold level of 0.5 ppb in milk. In developing countries like India, on the other hand, oftentimes when there is no enforcement, no infrastructure, or no economic incentive to enforce stringent limits on aflatoxins; the maximum permissible levels may be much higher than those set by developed countries due to their limitations. Additionally, developing countries are frequently lacking the technological means of detecting mycotoxin residues at sufficiently low levels, and therefore are unable to effectively monitor and enforce regulations related to aflatoxins. Further compounding these difficulties, many of the effects of climate change are making mycotoxin contamination more complex, necessitating a continuous evaluation of the adaptation of International Standards to effectively respond to changing global conditions.

4 Mechanisms of Toxicity of Mycotoxins

Mycotoxins (secondary metabolites) are toxic by-products that are created using waste from the fungi that infest various food and feed sources. The toxicity of mycotoxins works through cellular events; they can cause extensive damage to the DNA and proteins of the cell, as well as to the phospholipid membranes (i.e. phospholipids). The end result of these cellular injuries may potentially result in immunosuppression, carcinogenicity, and other negative health impacts to both animals and humans. Figure 5 illustrates the interconnected pathway of initial exposure to mycotoxins and their influence on intracellular signalling pathways that ultimately determine the final product of the

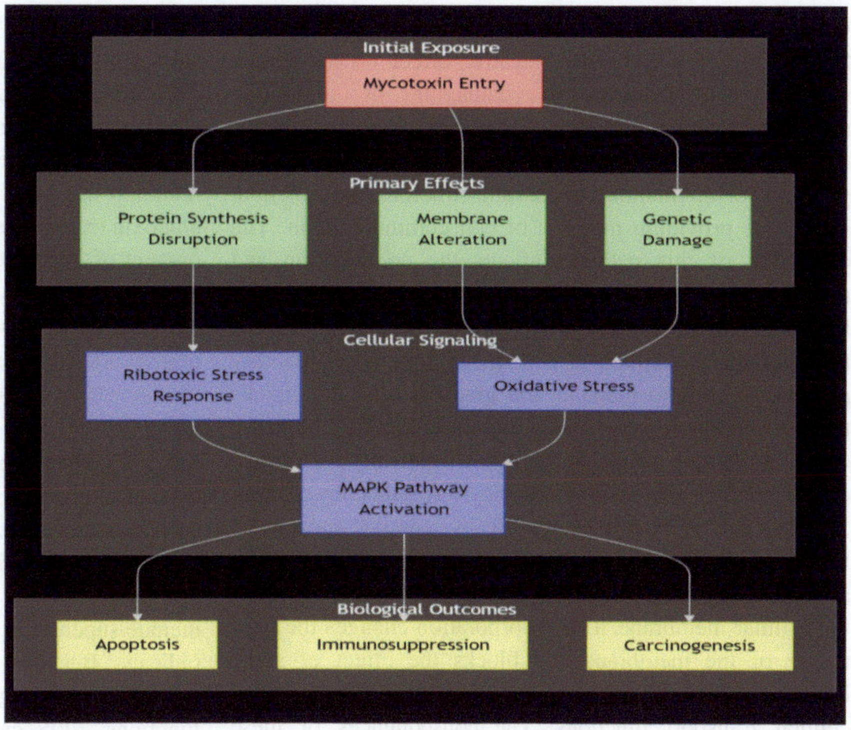

Fig. 5 The interconnective flow of mycotoxin initial exposure and its effects on cellular signalling governing its outcomes. *Source* Authors

toxic properties of mycotoxins through a bird's-eye view approach.

4.1 Pathways of Cellular Damage: Protein, DNA, and Membrane Targets

4.1.1 DNA Damage

Fungal mycotoxins such as aflatoxins and ochratoxins are potent genotoxic agents that exert damaging effects on cellular DNA. Among these, aflatoxin B1 (AFB1) produced predominantly by *Aspergillus flavus* and *Aspergillus parasiticus* is well recognized as a classic hepatocellular toxin. Inside the liver, AFB1 undergoes metabolic activation by cytochrome P450 enzymes, forming a reactive epoxide intermediate. This metabolite creates covalent bonds with guanine bases in DNA, resulting in the formation of DNA adducts that compromise genomic integrity and increase the risk of hepatocellular carcinoma (Hamid et al., 2013). Beyond adduct formation, these lesions interfere with DNA replication and transcription, frequently leading to point mutations such as guanine-to-thymine transversions. Such mutations are notably prevalent in populations exposed to high dietary levels of aflatoxins and are often observed in the TP53 gene, a key tumour suppressor involved in liver cancer pathogenesis (Alvarez et al., 2020). In addition, ochratoxin A (OTA), a mycotoxin synthesized by *Aspergillus* and *Penicillium* species, contributes to genotoxicity through oxidative mechanisms. OTA stimulates the generation of reactive oxygen species (ROS), which can lead to DNA strand breaks, base modifications, and chromosomal instability (Sorrenti et al., 2013). These modes of action highlight the critical role of mycotoxins in DNA damage and the associated risk of malignancies.

4.1.2 Protein Dynamics

The cellular protein functions that maintain cellular integrity and survival are severely impaired by the mechanism of mycotoxins. By binding to ribosomal peptidyl transferase centres, Trichothecenes, e.g. Deoxynivalenol (DON), made by the various Fusarium species, inhibit protein synthesis through halting translation, leading to ribotoxic stress that impacts cellular protein dynamics and activates stress response pathways (Pan et al., 2014). Fumonisin (a group of mycotoxins that includes fumonisin B1) disrupts sphingolipid metabolism by inhibiting ceramide synthase, which causes disruptions to membrane-bound protein function and to many of the key signalling cascades/signalling pathways. Disruption of the signalling pathways will lead to impaired cellular communication and disruption of homeostasis (Riley & Merrill, 2019). The regulation of apoptosis/pathways leading to apoptosis and the mitochondrial function are both regulated in some capacity through changes made by DON (Cao et al., 2021). These and other mechanisms highlight the numerous ways in which mycotoxins affect protein biosynthesis, signalling, and cellular survival.

4.1.3 Membrane Disruption

Several mycotoxins have a negative effect on membrane function and integrity by damaging or disrupting membranes. Fumonisin B1 (FB1) and other fumonisins disrupt sphingolipid metabolism, which is an important pathway for maintaining membrane structure and function, as well as providing methods of cell communication. The disruption of this pathway affects the composition and permeability of cellular membranes and can diminish their functions of ion transport and cellular communication. The result of this disruption in cell membranes is an inability of affected cells to maintain ion gradients and the signals they send to one another (including those received by nerve cells). This may result in necrosis, especially in critical organs such as the liver and kidneys. Another mycotoxin, zearalenone (ZEA), is produced by certain species of Fusarium and has estrogen-like properties, and also interacts with and incorporates itself into the lipid bilayer of the cellular membrane. This incorporation changes the fluid nature of the membrane and blocks the effect of some hormones on cellular receptors, thereby blocking some cellular transport functions. The consequences of these effects are an increased likelihood of cellular dysfunction and toxicity, as well as potential cellular death (Bulgaru et al., 2021).

5 Carcinogenic and Immunosuppressive Effects of Mycotoxins

5.1 Carcinogenic Effects

Some cancers have an association with both aflatoxins and ochratoxins due to their development from mycotoxins. The direct or indirect actions of the mycotoxin, depending on how the test was performed, determine whether or not it was found to be a carcinogen in any study. A summary of these actions is shown in Table 5.

5.1.1 DNA Adduct Formation and Mutagenesis

Aflatoxin B1 (AFB1), one of the most potent mycotoxins, exerts its carcinogenic effects primarily through the formation of DNA adducts, which play a critical role in mutagenesis. Upon entering the liver, AFB1 undergoes metabolic activation by cytochrome P450 enzymes, particularly CYP1A2 and CYP3A4, which convert it into a highly reactive AFB1-8,9-epoxide intermediate. This electrophilic compound forms covalent bonds with the N7 position of guanine bases in DNA, resulting in the formation of AFB1-DNA adducts. These adducts distort the DNA helix and interfere with replication and transcription, thereby increasing the likelihood of mutations. A notable mutational hotspot is codon 249 of the TP53 tumour suppressor gene, where the adducts frequently induce a G to T transversion. This mutation leads to an arginine-to-serine substitution, which compromises the normal function of p53 in DNA repair, cell cycle regulation, and apoptosis. The codon 249 mutation has been identified as a molecular fingerprint of aflatoxin exposure and is strongly associated with the development of hepatocellular carcinoma (HCC), particularly in regions with high dietary aflatoxin intake (Aguilar et al., 1993). The high specificity and prevalence of this mutation underscore the role of DNA adduct formation as a central mechanism in aflatoxin-induced carcinogenesis.

5.1.2 Oxidative Stress

OTA is a mycotoxin produced from certain fungal genera such as Aspergillus and Penicillium. OTA is strongly associated with oxidative stress-induced cell toxicity through the generation of reactive oxygen species (ROS), which include superoxide anions, hydrogen peroxide and hydroxyl radicals. Because ROS interacts with essential cellular machinery, causing damage, they adversely affect cellular functions. ROS can cause oxidative alterations to DNA, resulting in DNA breaks, oxidation of DNA bases and creation of mutated sites. In addition to the genotoxicity associated with ROS generated by OTA, oxidative stress-induced damage may peroxidize the membrane lipids, diminish the integrity of cellular membranes and reduce membrane fluidity. Disruption of protein ultrastructure/function also may occur as a result of oxidative stress caused by ROS generated by OTA, which may cause cellular pathways related to inflammation, apoptosis and ultimately cancer. Chronic oxidative stress is caused by a lack of balance between reactive oxygen species (ROS) production and a cell's protective mechanisms, in this case, when a cell is exposed to OTA. Chronic oxidative stress is one of the major factors responsible for the development of cancers and degenerative diseases. Recent studies indicate that oxidative stress induced by OTA is a central mechanism leading to the development of nephrotoxicity and carcinogenicity due to OTA, especially when conducted on the kidney, since it is one of the major sites of accumulation for OTA (Więckowska et al., 2024). The mechanism of oxidative stress induced by OTA calls into question the seriousness of OTA with respect to public health and serves as an indicator of concern for people in areas where food products are likely to have been ingesting products containing OTA.

5.1.3 Epigenetic Modifications

Mycotoxins like aflatoxin and fumonisins can disrupt gene expression through epigenetic means. They affect both the promoter and enhancer regions of a gene and are therefore able to bypass the need for base pairs to change (also known

Table 5 The carcinogenic and immunosuppressive effects of mycotoxins, toxic fungal compounds linked to various cancers

Mycotoxin	Fungal source	Affected organs	Mechanisms of action	Epidemiological evidence	Examples	References
Aflatoxins	*Aspergillus flavus, Aspergillus parasiticus*	Liver (Hepatocellular Carcinoma)	• Forms DNA adducts (e.g., AFB_1-N7-guanine) that cause TP53 mutations	Strong correlation with HCC in regions with high dietary exposure (e.g., maize and peanuts)	Study in China: Individuals with aflatoxin-albumin adducts were 3–6 times more likely to develop HCC	Pan et al. (2014)
			• Generates reactive oxygen species (ROS), leading to oxidative DNA and lipid damage	Risk amplified by Hepatitis B Virus (HBV) infection		Pan et al. (2014)
Ochratoxin A (OTA)	*Aspergillus, Penicillium*	Kidneys (Renal Cancer)	• Induces oxidative stress and lipid peroxidation	Linked to Balkan endemic nephropathy (BEN) and increased urothelial carcinomas	Studies in Serbia and Bulgaria: Higher prevalence of renal cancer associated with OTA exposure	Kőszegi and Poór (2016)
			• Disrupts mitochondrial function, impairing energy metabolism and triggering apoptosis	Chronic exposure is documented in populations consuming contaminated cereals and water		Kőszegi and Poór (2016)
Fumonisins	*Fusarium verticillioides*	Esophagus, Liver	• Inhibits ceramide synthase, disrupting sphingolipid metabolism	High oesophagal cancer rates in rural South Africa and China linked to fumonisin-contaminated maize consumption	Animal models: Dose-dependent development of liver and kidney cancers upon fumonisin exposure	Voss et al. (2001)
			• Leads to apoptosis dysregulation and increased tumorigenesis			Voss et al. (2001)
Zearalenone (ZEA)	*Fusarium*	Breast, Uterus (Hormone-Dependent Cancers)	• Mimics estradiol, binding to estrogen receptors	Limited human evidence	Chronic ZEA exposure in animals: Increased risk of breast and ovarian cancers	Voss et al. (2001)
			• Stimulates proliferation in hormone-sensitive tissues, altering hormonal balance	Animal studies link ZEA exposure to increased mammary and uterine tumours		Voss et al. (2001)

Source Authors' compilation

as DNA) to induce epigenetic alteration, in that they do not need to change the actual DNA sequence of a particular gene to produce an epigenetic effect on the way a gene is expressed. Mycotoxin contact and critical epigenetic pathways or mechanisms will result in DNA methylation changes which will in turn regulate histones; thereby, they change how genes are controlled and encoded, so that the cell's identity (i.e. cell type) and stability of its genome are maintained, as well as regulated transcription of the gene. Aflatoxins have been shown to convert hypermethylated tumour suppressor gene promoter regions to an inactive state. This alters expression of the TSGs. The maintenance of the correct function of these critical guard genes prevents uncontrolled cellular replication, as well as prevention of inappropriate repair of damaged DNA. Conversely, the hypomethylation effect of mycotoxins on oncogenes (OG) promotes the activation of oncogenes, further promoting the development of cancerous cells. Similarly, graphite-related fumonisins provide evidence for disruption of the normal histone acetylation and methylation profiles of chromatin, thereby modulating the structure and accessibility of chromatin to the transcriptional apparatus. Histone mark changes lead either to gene repression or gene activation, depending on the combination of histone modifications and the genomic background of the cell(s). The accumulation of epigenetic changes will reprogram the transcriptional profile of the cell toward a favourably aggressive tumorigenic and tumor development state. This means that epigenetic modifications provide potential for both a mechanism for toxicity and an important treatment target because they are potentially reversible. Therefore, the insight gained from the study of mycotoxin-induced epigenetic dysregulation provides significant knowledge about the long-term health impacts of mycotoxins and opens new options for prevention of and development of treatment options for mycotoxin-induced disease.

5.1.4 Evidence Relating to Human Cancers

Studies conducted on human cancer have consistently found an association between myctotoxins, such as aflatoxin B1 and the hepatitis B virus (i.e., HBV), and the development of various forms of cancer in both sub-Sahara Africa as well as Asia. Results indicate that people who consume aflatoxins B1 over a long period of time are at an increased risk for developing HCC if they have HBV because of the ability of HBV to impair the DNA's ability to repair itself and enhance the effects of aflatoxins on mutations at position 249 of the TP53 gene (Liu & Wu, 2010). Furthermore, ochratoxin A is believed to play a role in the genesis of renal cell tumours in various parts of the Balkan Peninsula, particularly those regions suffering from endemic nephropathy. Studies show that long-term consumption of OTA is likely responsible for this type of tumour development that is specific to a geographic location (Castegnaro et al., 2006).

Furthermore, fumonisins commonly found in maize contaminated with *Fusarium* species, have been associated esophageal cancer with increased incidence in populations with high maize consumption. The key pathogenic mechanisms include the disruption of sphingolipid metabolism, leading to deregulated cell growth, altered signalling pathways, and increased cellular proliferation. Collectively, these findings underscore the carcinogenic potential of mycotoxins in humans and emphasize the need for continued surveillance and intervention strategies in high-risk populations.

Aflatoxin

Thus, the association of a number of dietary carcinogens has been established as a major contributor to the etiology of liver cancer in areas of the world where dietary modifications have increased. In addition to the direct contribution of the carcinogenic effects of chronic exposure to aflatoxin, there is also a substantial number of mutations in TP53 due to chronic exposure to aflatoxins, specifically the transversion at codon 249 (R249S), and these mutations are enhanced by a synergistic effect of the hepatitis B virus (HBV) on the aflatoxin-caused oxidative injury to the DNA repair systems of cells (Han et al., 2020).

Ochratoxins

Ochratoxin A (OTA) is considered a nephrotoxic and carcinogenic agent mainly involved in causing renal cell carcinoma in humans. Prolonged exposure gives way to protein-mediated oxidative stress resulting in DNA damage with possible mutations involving either the oncogenes or tumour suppressor genes (Marin-Kuan et al., 2011).

Fumonisins

Although infrequent, a significant association with high fumonisin B1 (FB1) exposure has been positively correlated to oesophageal cancer in China and South Africa (Chen et al., 2021). FB1 causes sphingolipid disruption, resulting in the accumulation of toxic intermediates that induce cell proliferation and DNA damage.

Zearalenone

ZEA acts as an estrogen mimic and disrupts the endocrine signalling pathway. Chronic exposure has been related to the development of hormonal cancers, like *Carcinoma mammaria, Endometrial carcinoma* (Marhaba et al., 2024).

5.2 Immunosuppressive Effects

The presence of mycotoxin in the human body can decrease an individual's ability to fight off an infection and produce an effective vaccination. Figure 6 shows how various mycotoxins can negatively affect the human body in several ways, indicating how these mycotoxin exposures can be complicated by the amount and nature of the different types of mycotoxin and the timing of their exposure, all of which can create significant health risks.

5.2.1 Effects on Immune Cells

The immunosuppressive effects of mycotoxins are well established. Mould metabolites can act directly on components of the immune system, particularly those cell types involved in the host's adaptive immune response to disease-causing microbes. Trichothecene mycotoxins such as Deoxynivalenol (DON) act through direct inhibition of early proliferation and thus functional activity of lymphocytes. This increases, inhibiting any adaptive immune responses by decreasing the capacity of T and B lymphocytes to respond effectively to infection by inhibiting DNA and protein synthesis in rapidly dividing cells. Additionally, DON interferes with cytokine signalling pathways essential

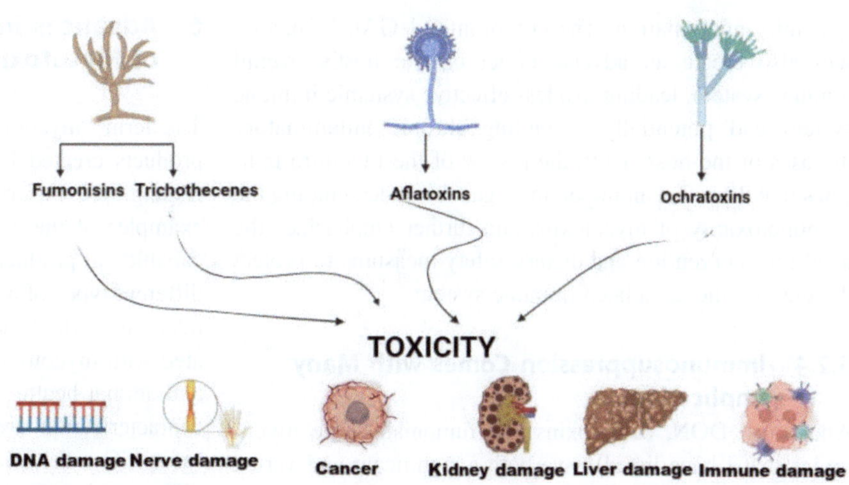

Fig. 6 Different mycotoxins integrated toxicity over different aspects of human health. *Source* Authors

to T-cell activation and differentiation, thus disrupting the effective orchestration of cell-mediated immunity (Pestka, 2010). Macrophages have their immune function compromised by mycotoxins like ochratoxin A and fumonisins. The production of cytokines, the ability to engulf foreign bodies, and the presentation of antigens are all needed to start and keep a successful immune response, which are diminished due to these toxins. When both the adaptive and innate immunity are suppressed together, it raises the risk of developing an infection, slows down healing time and may cause the advancement of chronic illnesses. The results of the studies indicate that mycotoxins can be harmful to the immune system and thus should be reduced as much as possible to maintain proper immune function in at-risk groups (children, elderly and immunocompromised people).

5.2.2 Cytokine Dysregulation

The cytokines are an important target for mycotoxins, particularly Deoxynivalenol (DON) since the body's normal balance (homeostasis) of cytokines is disrupted when DON is present at low to moderate (but non-lethal) amounts. DON's action is primarily through inducing a high level of cytokine activity (hyper-inflammatory), which results in an inflammatory (cytokine storm) that consists of excessive levels of pro-inflammatory cytokines, including but not limited to IL-6, TNF-α and IL-1 beta. These three pro-inflammatory cytokines act together to promote a systemic inflammatory reaction through their action on various cell types and tissues. When the immune system has been over-activated by cytokine production, the result can be cellular and/or tissue injury, cellular metabolic derangements and increased susceptibility to the development of secondary infections (Pestka, 2010). DON has an unusual feature of inhibiting cytokine production and killing off immune cell populations when large doses are given orally. DON has a dual effect, stimulating the immune system if consumed in small quantities and suppressing the immune system if taken in high doses. In addition, cytokine dysregulation results in changes to the way the immune system functions and ongoing inflammation, resulting in the formation of chronic inflammatory conditions or immune exhaustion, which could potentially lead to chronic inflammatory diseases (autoimmunity), increase the risk of developing infection, and accelerate the progression of cancer. Therefore, the importance of cytokine modulation as a component of the overall immunotoxicological effects of dietary mycotoxins is of utmost importance in order to understand the potential health risks of ingesting these substances.

5.2.3 Gut-Associated Lymphoid Tissue (GALT) Disruption

Mycotoxins are now considered to have major negative effects on GALT. GALT is an important component of the mucosal immune system that protects the intestinal tract from infection by pathogenic organisms by providing regulatory mechanisms. The mycotoxin's exposure has been linked to the breakdown of the integrity of the gut epithelial barrier, increased permeability, and susceptibility of the gut epithelial barrier to pathogen invasion. When the gut epithelial barrier is compromised due to mycotoxins, this allows for the movement of both bacteria and antigens through the intestinal barrier into the blood, causing disturbances to the balance in the gut and resulting in an imbalance in bacteria within the gut (dysbiosis). As a result of forming dysbiosis in the gut microbiome, there is also a risk of creating a dysfunctional GALT, which will present with decreased amounts of protective immunoglobulin A (IgA) and decreased antigen presentation to T-cells; T-cell responses will also become unbalanced. The reduced GALT functionality leads to significant impairment of the gut's ability to defend against enteric pathogens and increased susceptibility of the host to gastrointestinal infections and systemic inflammation. The compromised GALT function will also create an adverse effect on the host's overall immune system, leading to a less effective systemic immune system and potentially promoting chronic inflammatory diseases of the host. Thus, the review of the literature indicates that the gut is an important organ for understanding the immunotoxicity of mycotoxins and further emphasizes the need for intervention and dietary safety measures to protect the gut and the associated immune system.

5.2.4 Immunosuppression Comes with Many Implications

Aflatoxins, DON, ochratoxins and fumonisins (the mycotoxins that I have listed) have both a significant and varied impact on the immune system through their effects on both the innate (non-specific) and adaptive (specific) systems by causing immunosuppression. The mechanism of action for each of mycotoxins is different; for example, the toxin DON will inhibit B-cells and T-cells from growing and thereby reduce antibodies produced and result in weakened adaptive immunity; and ochratoxin A (OTA) will inhibit macrophages from phagocytizing and producing cytokines, and therefore weaken innate immunity. Other mycotoxins produce different effects on the immune system, including disruption of the cytokine network via several complex and contradictory mechanisms. For example, DON at low doses can induce hyperinflation and damage to host tissue; whereas at high doses, DON inhibits cytokine production and poses an even greater risk for damage to host tissue. Ultimately, the two separate mechanisms for immune dysregulation (one from excessive production of cytokines and the other from inhibited cytokine production) increase the likelihood of injury to host tissue and increase susceptibility to infections. Moreover, DON is likely to increase the risk for infectious diseases from polyvalent and respiratory pathogens. In this regard, fumonisins reduce macrophage and neutrophil function and therefore reduce the body's defence mechanisms against pathogens. Mycotoxins adversely affect both animals and public health because, in livestock, exposure to trichothecenes makes vaccines less effective due to the inability to create memory immune cells, resulting in lower levels of protection (antibody titers) after vaccination. There is also an increasing understanding of how immunosuppression is a significant factor in the development of cancer. Immunosuppression allows for cancerous or precancerous cells to grow, spread and develop due to the lack of natural killer cells and cytotoxic T lymphocyte activity because of compromised immune surveillance. Furthermore, the cumulative impacts of mycotoxin-induced immunosuppression on host defense, increased vulnerability to disease, and long-term health outcomes should not be underestimated.

6 Advances in Detection and Analysis of Mycotoxins

The term "mycotoxins" refers to the toxic metabolic by-products created by a number of fungal origins. In turn, Aspergillus, Fusarium, and Penicillium are simply three examples of these diverse numbers of fungi, and they are capable of producing mycotoxin contamination of many different types of agricultural products (for example, grains, nuts, etc.) when they are matured. The main risks associated with mycotoxins include the potential to affect human and animal health. Therefore, mycotoxin identification and characterization are essential components to ensure food safety. The detection of mycotoxins has evolved through different phases as technology has advanced and there has been increased use of automation.

6.1 Enzyme-Linked Immunosorbent Assay (ELISA)

The ELISA is an immunoassay-based detection method that is one of the preferred methods for detecting mycotoxin contamination in food and animal feed. The ELISA method detects both the specific presence and the amount of mycotoxins using the IgG antibody reaction to specifically bind to these toxins (mycotoxins). In a standard ELISA, mycotoxin is bound to the IgG antibody through a solid-phase immobilization (immobilization) before the addition of the second enzyme-linked antibody. This allows for the intermediary or capture/anchoring of mycotoxin while the other enzyme, horseradish peroxidase (HRP) or alkaline phosphatase (ALP), is being reacted with the target antigen. Once this sandwich structure has formed, the ELISA result becomes available based on the amount of generated colour during the reaction of the enzyme (HRP or ALP) with the substrate provided during the colorimetric detection. In the detection phase, chromogenic substrates are added, and an enzyme will cause a colour reaction that can be detected visually based on how dark the colour becomes (the higher the concentration of mycotoxin in the sample, the more actual colour in that particular test). To quantify, you compare your absorbencies of the test against a standard curve built with different concentrations of the same mycotoxin. The advantages of using ELISA are that ELISA is relatively inexpensive and easy to test with, which makes it an ideal option for screening and monitoring the occurrence of mycotoxins in agricultural products routinely (Janik-Karpińska et al., 2021).

6.2 High-Performance Liquid Chromatography (HPLC)

High-Performance Liquid Chromatography (HPLC) is an extremely sensitive analytical technique that can detect and quantify mycotoxins present in food, feed, and biological samples. It separates the metabolites of mycotoxins based on their physical and chemical characteristics, using a mobile liquid phase and a stationary phase contained within a chromatographic column. The sample is injected into the instrument, which exerts pressure to move the sample through the column while allowing for interaction with the stationary phase to separate each component based on retention time. Through this process, each respective substance is removed at a unique rate/ time; this will depend on how well each individual substance interacts with the stationary phase. The separated substances are usually detected via ultraviolet (UV) absorption or fluorescence, when the compounds leave the column; thus providing optical properties of all analytes (Singh & Mehta, 2020). Quantitative data for mycotoxins will be acquired by comparing the detector response (usually peak area or height) to a calibration curve prepared from standard solutions of known concentrations. The method presents a high degree of accuracy in terms of results obtained and repeatability of those results over time, as well as providing simultaneous detection of numerous mycotoxins in samples containing multiple matrices. When combined with more advanced detection technologies (e.g., mass spectrometry), HPLC has become the preferred choice for confirmation of mycotoxin residues within the laboratory setting (Singh & Mehta, 2020).

6.3 Gas Chromatography-Mass Spectrometry (GC–MS)

Gas Chromatography and Mass Spectrometry (GC–MS) is a useful means of analyzing and separating mycotoxin levels on a finer scale than conventional gas chromatography methods. Although typically employed for mycotoxin detection, GC–MS also provides the ability to quantify mycotoxins on an individual chemical basis using mass spectrometry. In this example of GC–MS, mycotoxin samples containing volatile, non-volatile and altered structures, i.e. derivatized mycotoxins, are vaporized and introduced into a gas chromatograph to perform the separation process in the gas chromatograph. As the sample travels through the gas chromatograph's column, each type of mycotoxin can be separated based on volatility, solubility in the stationary phase, etc. After exit, each individual compound will be evaporated, then analyzed by a mass spectrometer (Turner, 2024). Once analyzed, the mass spectrometer generates a unique spectral fingerprint for each compound based on the mass-to-charge ratio (m/z) of the respective protonated ions, thereby enabling the researcher to determine the structure of the individual compound and provide information regarding its potential toxic properties. By comparing peaks of the ion signals with standard or known concentrations of derivatization compounds, the quantification of these levels can be determined. Food safety and toxicology researchers rely heavily on Gas Chromatography-Mass Spectrometry (GC–MS) for the determination of trace levels of volatile mycotoxins present in complex food matrices such as grains, cereals, and many types of processed food. By providing very precise and sensitive simultaneous separation, identification, and quantification of multiple analytes, GC–MS provides food safety and toxicology researchers with an invaluable analytical technique. Additional details and performance parameters of this method can be found in (Turner, 2024).

6.4 Innovations in High-Throughput Screening (HTS)

Recent advances in mycotoxin detection are focused more and more on high-throughput systems for concurrent analysis of multiple samples in a single run. The impetus is there for quickly screening food and agricultural products and real-time monitoring of mycotoxin contamination.

6.4.1 Microarray-Based Detection

Microarrays consist of small spots arranged to form an array to hold various antibodies/receptors embedded within a small chip. Each spot captures a unique mycotoxin, and thus, is revealed if it has bound to an antibody/ receptor using fluorescent or chemiluminescent markers (Zhang et al., 2012). Microarrays also have advantages and disadvantages associated with them. Microarray-based detection of mycotoxins has many advantages, such as being able to screen multiple mycotoxins with more sensitivity and specificity than using methods that rely upon one at a time. Furthermore, microarrays use the same immobilized probes on a solid surface, permitting fast, simultaneous and low-cost analyses of complicated mixtures of samples, using very few reagents. Microarrays are small and hence very useful for large-scale screenings. Although microarrays have excellent technical attributes, there are significant limitations to using them for mycotoxin detection, which include the initial investment costs that are rather substantial, requiring a high-quality lab, and the possibility of cross-reactivity of the probes, which may negatively impact assay reliability. In addition, microarray assays have to be interpreted by highly skilled individuals, limiting their usage in developing countries or other resource-poor situations.

6.4.2 Biosensors and Lab-On-a-Chip Technology

A biosensor is a technology that detects mycotoxins via interactions between biosensors and biomaterials like antibodies or aptamers. These devices can be included in lab-on-chip systems (LOCs), which consolidate sample-processing, sample-detecting and sample-analyzing capabilities that fit within a smaller footprint (Khan & Song, 2020). LOC systems have some advantages and limitations related to their use as a mechanism for mycotoxin detection. As biosensors and LOC technologies provide faster detection of multiple mycotoxins and require very little preparation time, they are ideal for use both in field settings and in low-resource settings. LOCs have the advantage of portability, ease of use, and the ability to produce real-time results. The small footprint of LOCs allows designers to create a system that combines multiple detection steps into one device rather than using individual systems to perform detection, which also increases speed and reduces reagent usage. However, LOCs have some challenges, including limited multiplexing functionality, potential environmental effects on sensitivity, and the need for optimization to provide accurate results from a wide variety of sample types. In addition, the mass production of low-cost, durable biosensor platforms for commercial use is an area where further development is needed.

6.4.3 Next-Generation Sequencing (NGS)

Next-generation sequencing (NGS) allows for increased depth of sequencing of microbial DNA to identify fungal species responsible for producing mycotoxin and the genes associated with the synthesis of mycotoxin. Consequently, combining NGS data from fungi isolates collected from agricultural products can allow for predictions on which mycotoxins might be present in a crop/field (Cacho et al., 2015). NGS has inherent strengths and limitations. NGS provides a powerful tool for indirect assessments of mycotoxin risk through extensive profiling of mycotoxin-producing fungal communities using genetic data. The high-throughput nature of NGS allows for the simultaneous analysis of large numbers of DNA sequences and offers a very detailed view of fungal diversity, mycotoxin biosynthesis pathways, and potential sources of contamination. Thus, NGS has a unique advantage in the early detection of and predictive modelling in the realm of food safety. Unfortunately, there are some substantial limitations associated with NGS, including high costs, the need for highly sophisticated bioinformatics analysis tools, and trained staff to evaluate data. Further, NGS can only identify mycotoxin-producing species but cannot provide any quantitative information on the toxin; therefore, NGS needs to be integrated with chemical analysis methods to complete the risk determination process.

7 Emerging Technologies for Rapid Screening of Mycotoxins

Mycotoxins (the toxic metabolites from moulds) pose a serious threat to food safety and human health. There have recently been important advances in how mycotoxin detection is done, largely due to the development of new rapid screening technologies that allow for on-site testing with quicker results and less complex instrumentation. These new technologies will consist of portable devices utilizing biosensors as well as advancements in the point-of-care detection system. In this section, I will describe each of these approaches in more detail.

7.1 Biosensors and Portable Devices for Field Testing

Biosensors are small analytical devices that use biological materials to identify and measure specific chemicals (e.g., mycotoxins). Because biosensors are portable, easy to use, and provide quick results without a need for large laboratory setups (Khan & Song, 2020), they can be extremely helpful for field testing, as shown in Table 6. The biosensor's primary component, the biorecognition element, selectively attaches itself to the target mycotoxin with great strength. The biorecognition element interacts with the other biosensor components, such as the transducer, which turns the biochemical interaction into a quantifiable physical signal (e.g., an electrical impulse, an optical signal, or a physical force). After the creation of the physical signal, the amount of analyte can be quantified. Thus, the final result is either presented as a visible indication of something (e.g., a change in colour), or it can be retrieved digitally from a portable reading device or smartphone application. Therefore, biosensors are well-suited for on-site testing, especially in rural and agricultural areas where access to traditional laboratory methodologies may be limited.

Table 6 The examples of mycotoxins biosensors used in the detection

Mycotoxin detection	Description	References
Aflatoxin detection	Aflatoxin B1 immunosensor detects aflatoxin B1 in cereals, peanuts, and milk using specific antibodies	Yin et al. (2022)
Ochratoxin A	Biosensors with antibodies or DNA aptamers detect ochratoxin A in cereals, wine, and coffee	Xu et al. (2016)
Fumonisin detection	The graphene-based biosensor detects fumonisin in maize by producing electrical signals upon binding	Qin et al. (2022)

Source Authors' compilation

7.2 Advances in Point-Of-Care (POC) Mycotoxin Detection

Tests done at home (POCT) can prove to be effective in providing you with the results of your test within a very short period of time, which eliminates the need for you to drive to a laboratory to have a test done. In recent years, many developments have occurred with respect to the ability to detect mycotoxins by using POCT tools and systems, which have allowed for the rapid and reliable screening of agricultural products from anywhere on the globe (Suo et al., 2023). Table 7 shows some examples of POC. The testing steps include first obtaining a small amount of agricultural product from either one or multiple locations (e.g., field, storage, or processing facility) and sending it to a POC testing facility. Depending on the type of device being used at the POC facility, only limited sample preparation may be needed prior to beginning the detection process. For instance, many of these devices require only a few basic sample preparation processes, such as diluting the sample with water, filtering out any solids, or extracting the mycotoxin from the sample. The detection of mycotoxins, in general, occurs using a device that contains an integrated sensor (e.g., a mycotoxin detection device) or a small strip created for testing (e.g., an immunoassay strip or a lateral flow strip). When the mycotoxin interacts with the sensor or the strip, it provides the user with a measurable signal. There are several different types and formats of results that can be provided to users by POC devices to facilitate decision making in mycotoxin management. Hence, POCT tools are becoming widely used in industries where rapid decision making and intervention (e.g., processing, exporting) concerning mycotoxins are necessary. Point-of-care (POC) technologies offer rapid, user-friendly mycotoxin detection directly at the site of sample collection, making them ideal for use in remote or resource-limited areas. Their portability, minimal need for training, and quick readouts enable timely decision making and contamination control. However, POC devices often provide semi-quantitative results, may lack the sensitivity and specificity of laboratory-based methods, and are generally limited to detecting a narrow range of mycotoxins. Additionally, variability in sample matrices can affect accuracy and consistency.

Table 7 The point-of-care (POC) mycotoxin detection

Examples of POC devices for mycotoxin detection	Description	Example
Lateral flow immunoassay (LFIA)	A simple "dipstick" test using antibodies to detect mycotoxins; colour change indicates contamination	Aflatoxin test kits for detecting aflatoxin B1 in peanuts, corn, and other crops
Portable electrochemical sensors	Handheld sensors use electrochemical reactions to detect mycotoxins by measuring electrical signals	Potentiometric sensor works to detect ochratoxin. It is detected in wine and coffee
Microfluidic devices	Lab-on-a-chip integrates sample prep, detection, and signal processing for on-site testing	MIPs integrated into microfluidic chips for detecting multiple mycotoxins like aflatoxins

Source Authors' compilation

8 Control and Remediation of Mycotoxins

Mycotoxins are toxic secondary metabolites produced by some fungal species that contaminate agricultural commodities like cereals, nuts, fruits, and dairy products. This group of compounds endangers the health of humans and animals if they are present in food and feed and therefore must be effectively controlled and remediated. Both strategies refer to post-harvest detoxification or removal of mycotoxins and those that need to be executed during agricultural production; specific key aspects of these are discussed below.

8.1 Prevention and Detoxification Strategies

Prevention and detoxification are the first lines of defence against mycotoxin contamination. These approaches can be applied during the cultivation, harvest, storage, and post-harvest handling of crops to achieve decreased exposure to the agents. Significant strategies are the implementation of biological control agents and best post-harvest practices.

8.1.1 Biological Control Agents for Mycotoxin Reduction

Biological control refers to the use of natural organisms to reduce or eliminate the production of fungi that produce mycotoxins. It can be an environmentally friendly and sustainable alternative to chemical treatments. These agents include:

Fungal Antagonists

These are beneficial fungi that compete with or inhibit the pathogenic mycotoxin-producing fungi. They either prevent fungal establishment or decrease mycotoxin production. *Trichoderma* spp. is a well-known fungal biocontrol agent. In crops like maize, peanuts, and cotton, Trichoderma has been shown to compete with aflatoxin-producing fungi such as *Aspergillus flavus* and *Aspergillus parasiticus*, thus reducing their aflatoxin and ochratoxin production. Trichoderma species may produce enzymes that degrade fungal cell walls and may also produce antimicrobial metabolites that inhibit fungal growth (Kumar et al., 2022).

Bacterial Antagonists

Certain bacteria inhibit fungal growth and mycotoxin production. Their main mechanisms of suppression include the production of antifungal compounds or the establishment of the ecological niche through competitive exclusion for space and nutrients. *Bacillus* spp., such as *Bacillus subtilis* and *Bacillus pumilus*, have been used in the biological control of *Fusarium* species that produce fumonisins, zearalenone, and trichothecenes. *Bacillus* strains can compete for nutrients and space, and some produce lipopeptides and other antimicrobial compounds that suppress fungal growth (Bertuzzi et al., 2022).

Yeast-Based Biocontrol

A few yeasts, including *Saccharomyces cerevisiae* and *Pichia pastoris*, have been studied for their ability to reduce aflatoxin and other mycotoxin production by taking up mycotoxins in their environment and impeding their ability to be absorbed by plants and animals. Another study showed that the ability of *Saccharomyces cerevisiae* to reduce aflatoxin contamination in maize was due to its competing very well against *Aspergillus flavus* for space and nutrients (Mohammad Nejad Khiavi et al., 2020).

8.1.2 Mechanisms of Biological Control

Mycotoxin-producing fungi can be controlled biologically by utilizing living organisms, usually beneficial microorganisms, that hinder the ability of the pathogen to grow, reproduce, and/or produce toxins. The effectiveness of biocontrol agents to suppress mycotoxin contamination in agricultural settings is due to a number of different mechanisms.

Competition for Space and Nutrients

Competition is a vital method used by biocontrol agents to reduce fungal mycotoxin contamination. Trichoderma, Bacillus and non-toxigenic fungae (Aspergillus) are examples of beneficial microorganisms that compete for space and access to nutrients on host surfaces with pathogenic fungi. Environmental conditions allow these biocontrol agents to occupy the spaces in advance of the toxigenic fungi colonizing those spaces. By rapidly occupying the surfaces of either plant materials or grains, biocontrol agents prevent toxigenic fungi from establishing themselves. As a result, competition reduces both the biomass of the fungi present and their capacity to produce mycotoxins. This form of competition is a stable method that is both ecologically viable and cost-effective compared to the use of chemical control options.

Production of Antagonistic Compounds

Biocontrol agents have shown their ability to inhibit the growth of mycotoxigenic fungi by enabling the growth of biocontrol agents themselves through the production of chemicals that limit mycotoxin production and growth. These include antibiotics, hydrolases (like chitinases and glucanases), and organic acids (like lactic and acetic acids). These compounds can weaken or disrupt fungal cell walls, affect their metabolic pathways or lower pH levels in their environment, making it less favourable for fungal growth to grow and thrive. Bacillus subtilis has been shown to produce antimicrobial lipopeptides that inhibit the fungal production of aflatoxin. This antibiotic ability also helps increase the biocontrol agent's ability to resist competing with pathogens.

Induction of Host Plant Resistance

Systemic resistance responses in plants are often triggered by biocontrol agents as a consequence of their direct antagonistic effects. Many of the compounds that plants produce for biotic and abiotic defense possess anti-cancerous and antimicrobial properties (Singh et al., 2023). By inducing plant defenses such as the production of pathogenesis-related proteins or phytoalexins and oxidative bursts, these organisms induce an ISR in plants to prepare them to resist any future infections. Therefore, ISR reduces the incidence of future fungal infections, as well as limiting the amount of colonization and subsequent production of mycotoxins. Microbial elicitors, from organisms such as Pseudomonas fluorescens and Trichoderma harzianum, activate pathways associated with ISR and provide an additional layer of protection against fungal infections.

Induction of Host Plant Resistance

Some biocontrol agents possess the ability to catabolize mycotoxins that may already be present in crops and convert them to a less toxic or inert form. This is called mycotoxin biodegradation and it can take place through the action of specific enzymes that change the molecular structure of the mycotoxin. Several bacterial species can detoxify aflatoxins by cleaving the lactone ring, making them biologically inactive. Other mycotoxins, such as ochratoxins and fumonisins, also have been reported to be degraded by various yeast and fungal isolates, respectively. This type of action is most valuable during post-harvest management when residual mycotoxins can remain in crops despite the absence of active fungal growth.

8.1.3 Best Practices in Post-harvest Handling

Post-harvest handling plays a critical role in minimizing the risk of fungal growth and subsequent mycotoxin contamination. Since mycotoxin-producing fungi thrive in warm and humid environments, it is essential to implement effective management strategies immediately after harvesting. These include drying, proper storage, chemical control, and detoxification approaches, all of which help preserve the safety and quality of food and feed.

Drying of Crops

Proper drying of crops immediately after harvest is one of the key ways to prevent the production of mycotoxins. The majority of mycotoxin-producing fungi thrive in a warm, moist environment. Therefore, by drying grains to less than 13% moisture shortly after harvest, there will be a significant reduction in the possibility of fungal colonization and subsequent toxin production. Maize is one of the most susceptible crops to mycotoxin contamination from Aspergillus flavus and Fusarium spp., which produce aflatoxins and fumonisins, respectively. To prevent fungal proliferation and guarantee the storage of maize safely, prompt drying of maize to a moisture content of 12% or less within 24 h of harvest is critical (Habschied et al., 2021).

Storage Conditions

In addition, proper storage practices are extremely important in preventing the contamination of crops with mycotoxins during the post-harvest process. Proper storage facilities should include low humidity and temperature so that fungi do not thrive. Grains should be stored in airtight, adequately ventilated containers or silos to minimize moisture ingress. Conditions that provide poor ventilation, condensation, or temperatures greater than 15 °C will allow fungal development and the formation of mycotoxins. Researchers have substantiated that grains stored in poorly ventilated silos are highly susceptible to mycotoxin contamination. By employing proper air circulation/temperature control methods, such as maintaining storage environments below 15 °C, the risk of fungal invasion and mycotoxin production is significantly decreased (Dorn & Jackson-Ziems, 2013).

Use of Chemical Fungicides

Although biological approaches to mycotoxin management are becoming increasingly popular, chemical fungicides continue to be one of the most common ways to manage post-harvest mycotoxin contamination. Since chemical fungicides, including propiconazole and tebuconazole, can be applied directly to harvested grain to protect against fungal invasion and subsequent mycotoxin production, they continue to be utilized as a first line of defense against the two main types of toxin-producing fungi: Fusarium and Aspergillus. Definitive research has demonstrated that the application of these specific chemical fungicides to wheat and corn reduces the incidence of fungal contamination and therefore decreases the potential for mycotoxins (Kleber et al., 2023). However, there should be strict standards governing their use so that residues of these chemicals are not present on crops and to prevent the development of resistance.

Use of Mycotoxin Binders

Mycotoxin absorbents are used in the feed sector as a strategy to neutralize the impact of consumed toxins on animals. These substances, such as activated charcoal, bentonite clay and yeast cell wall derivatives, work by absorbing mycotoxin toxins in the gastrointestinal tract and reducing their absorption into the bloodstream. As a result, these substances help create a lower toxic burden for the livestock and alleviate some of the detrimental impacts of contaminated feed. A good example is bentonite clay, which has shown very good performance in binding aflatoxins and preventing their bioavailability in animal systems, thus making it one of the most widely used additives in animal nutrition programmes (Kihal et al., 2022).

8.2 Bioremediation Using Fungal Strains

Bioremediation is a method of using living organisms, primarily microorganisms, to destroy or render harmless environmentally hazardous substances. Recently, fungi have become recognized as an effective bioremediation tool due to their inherent ability to detoxify complex organic materials. Mycotoxins are potent biologically active substrates; produced by the Aspergillus, Fusarium and Penicillium families, and can contaminate food crops and cause significant health risks for humans and animals (Shekhar et al., 2025). The use of fungal strains for bioremediation presents a potential avenue for reducing mycotoxin levels to safe levels in contaminated environments, agricultural output and feed, as well as certain soils from which these agricultural outputs originate (Ghosh et al., 2023).

8.2.1 Fungi for Degradation and Detoxification of Mycotoxins

Different enzymes, such as hydrolases, peroxidases, and oxidases, exist in fungi and allow them to degrade or biotransform mycotoxins into less toxic compounds. The capacity of fungal strains to biotransform a variety of mycotoxins is among the most important characteristics of fungi utilized for bioremediation. Although different types of mycotoxins and fungal strains have different ways of acting on mycotoxins; generally, these mechanisms can be categorized as follows:

Fungal Enzymatic Degradation

Fungi secrete enzymes that degrade the chemical structure of mycotoxins into non-toxic or less toxic products. These mechanisms of degradation may include the cleavage of the toxic ring structure by fungal enzymes. *Aspergillus niger* has been shown to degrade aflatoxins through the action of laccases and peroxidases. These enzymes mediate the oxidation of the aflatoxin structure to a less harmful compound. In some cases, these degraded products remain biologically active but are far less toxic than the parent mycotoxin (Alberts et al., 2009).

Absorption and Biotransformation

Some fungi may also absorb mycotoxins present in the environment (for example, contaminated food or soil) and deactivate the toxins by transforming them into less harmful metabolites through metabolic pathways that typically involve oxidation, reduction, or hydrolysis reactions. *Trichoderma harzianum* has been highlighted as a potential mycotoxin-reducing biocontrol agent capable of absorbing ochratoxin A from the environment and biotransforming it into non-toxic metabolites. It is one of the possible candidates for ochratoxin A contamination reduction in wheat and barley and a good option for mycotoxin remediation (Wang et al., 2022).

Mycotoxin Degradation by Fungal Co-cultures

In some cases, a combination of fungal strains or co-cultures may be more effective at degrading mycotoxins than single strains. Co-cultures may provide synergism where each fungus degrades different parts of the mycotoxin or enhances the degradation process. For example, it was revealed that the combination of *Fusarium* and *Trichoderma* species demonstrated higher mycotoxin degradation potential compared to their individual strains. The synergy observed in these fungi enhanced detoxification (Lutz et al., 2003).

8.2.2 Success Stories in Field Applications

The use of various fungal strains in bioremediation has been demonstrated successfully in actual agricultural applications for degrading or detoxifying mycotoxins. The success stories, which illustrate this process, are listed in Table 8.

9 Conclusion

Mycotoxins are also one of the most enduring and complicated issues in food safety, human health, or agricultural sustainability. Human and animal exposures to these toxic secondary metabolites are highly impossible to avoid as a wide variety of staple foods and animal feeds are contaminated by these toxic secondary metabolites produced by fungi, including *Aspergillus, Fusarium,* and *Penicillium,* as discussed throughout this chapter. The chemical stability and different forms of toxicity of the mycotoxins, aflatoxins, ochratoxins, fumonisin, trichothecenes, among others, explain why the threat posed by mycotoxins is so severe around the world.

Medically, the chapter puts emphasis on the toxicological multidimensional nature of mycotoxins, including acute poisoning and long-term effects, including carcinogenicity, immunosuppression, endocrine disruption and developmental defects. The molecular pathways of these effects, such as the formation of DNA adducts, oxidative stress, protein synthesis disruption, membrane damage and the alteration of epigenetic patterns, prove the extent to which mycotoxins can disrupt basic cellular functions. Of particular concern is that there is a growing epidemiological and mechanistic roots of chronic exposure to some mycotoxins, including aflatoxin B1 and fumonisins, in cancers of the liver and the esophagus in high-risk populations in low- and middle-income nations.

In addition to health effects, the economic and social costs of mycotoxin contamination are high. Billions of dollars every year are employed in losses caused by low quality of crop, rejection in international markets, and low productivity of livestock. Climate change further adds to this issue as it provides more favourable environmental conditions to the growth of fungi and the production of toxins, therefore, raising the rate and geographic range of contamination. These facts underscore the fact that mycotoxins are not only a scientific or regulatory issue, but a general development challenge in the world, in terms of food security, trade equity and systems of population health.

Positively, there is also tremendous improvement in the front of detection, surveillance and mitigation strategies as shown in this chapter. The development of new methods of analysis, such as standard chromatography methods up to novel biosensors, lab-on-a-chip systems, and next-generation sequencing, is changing the way we can quickly and precisely identify mycotoxins. Meanwhile, the combination of control measures, especially the employment of biological control agents and enhanced pre- and post-harvest methods, provides long-term and eco-friendly ways of cutting contamination at its origin.

Finally, to solve the worldwide mycotoxin issue, a multidisciplinary method where scientific studies, technology development, government regulations, and cooperation among the stakeholders are combined is necessary. Further investment in inexpensive detection systems, resilient agricultural methods to climatic changes, and

Table 8 The success stories for using fungi in agriculture to degrade mycotoxins

Mycotoxin	Bioremediator	Crop	Field study example	Mechanism of action	References
Aflatoxins	*Aspergillus niger*	Peanuts	In Nigeria, *Aspergillus niger* applied to peanut crops reduced aflatoxin levels significantly. The strain demonstrated potential for bioremediation of contaminated crops by enzymatic degradation of aflatoxins	Enzymatic activities degrade aflatoxins	Fang et al. (2020)
Fumonisins (FB1)	*Trichoderma harzianum*	Corn	In the U.S., *Trichoderma harzianum* applied to cornfields reduced fumonisin B1 contamination. It inhibited *Fusarium* growth and detoxified fumonisin by transforming it into less harmful metabolites	Growth inhibition of *Fusarium* and enzymatic transformation into safe forms	Alberts et al. (2016)
Ochratoxin A (OTA)	*Penicillium roqueforti*	Barley	By enzymatically breaking down OTA post-harvest, *Penicillium roqueforti* has been shown to effectively detoxify it into non-toxic metabolites in barley crops. The significance of this bioremediation is critical for the safety of producing malt and beer	Enzymatic breakdown of OTA into non-toxic metabolites	Rodríguez et a. (2024)
Soil Mycotoxins	*Trichoderma viride, Aspergillus niger*	Soil	In Italy, fungal species were able to degrade aflatoxin residues from agricultural waste. These fungi had a significant impact on the reduction of soil concentrations of aflatoxin and demonstrated potential for preventing future crop contamination via mycotoxins	Degradation of aflatoxins in soil reduces environmental contamination	Jallow et al. (2021)
Zearalenone	*Alternaria alternata*	Wheat	In wheat fields, the fungal species *Alternaria alternata* was capable of reducing the levels of zearalenone by greater than 50%. The fungi were able to degrade zearalenone into less toxic by-products and also decreased the dry weight of *Fusarium* spp. fungi	Degradation of zearalenone into less toxic metabolites; reduction of Fusarium	Leslie et al., (2021)

Source Authors' compilation

awareness should be maintained especially in resource-constrained environments. The risks of mycotoxins, as well as ensuring the safety of food systems and the protection of human and animal health in the ever-challenging global environment, can be minimized by coordinating the work of research, industry, and policy.

References

Aguilar, F., Hussain, S. P., & Cerutti, P. (1993). Aflatoxin B1 induces the transversion of G–>T in codon 249 of the p53 tumour suppressor gene in human hepatocytes. *Proceedings of the National Academy of Sciences of the United States of America, 90*(18), 8586–8590. https://doi.org/10.1073/pnas.90.18.8586

Alberts, J. F., van Zyl, W. H., & Gelderblom, W. C. (2016). Biologically based methods for control of fumonisin-producing *Fusarium* species and reduction of the fumonisins. *Frontiers in Microbiology, 7*, 548. https://doi.org/10.3389/fmicb.2016.00548

Alberts, J., Gelderblom, W., Botha, A., & Van Zyl, W. (2009). Degradation of aflatoxin B1 by fungal laccase enzymes. *International Journal of Food Microbiology, 135*(1), 47–52. https://doi.org/10.1016/j.ijfoodmicro.2009.07.022

Al-Falih, A. M. (2024). Overview on mycotoxins health hazard to humans and animals. *Environmental Analysis & Ecology Studies, 12*(4). https://doi.org/10.31031/eaes.2024.12.000794

Alvarez, C. S., Ortiz, J., Bendfeldt-Avila, G., Xie, Y., Wang, M., Wu, D., Higson, H., Lee, E., Teshome, K., Barnoya, J., Kleiner, D. E., Groopman, J. D., Orozco, R., McGlynn, K. A., Gharzouzi, E., & Dean, M. (2020). Analysis of *TP53* aflatoxin signature mutation in hepatocellular carcinomas from Guatemala: A cross-sectional study (2016–2017). *Health Science Reports, 3*(2), Article e155. https://doi.org/10.1002/hsr2.155

Atnkut, B., Nigussie, A., Berza, B., Mikru, A., Zhao, B., & Astatkie, T. (2025a). Mycotoxins contaminations in Ethiopian food: Impacts, challenges, and mitigation strategies. *Food Control, 171*, Article 111066. https://doi.org/10.1016/j.foodcont.2024.111066

Atnkut, B., Nigussie, A., Berza, B., Mikru, A., Zhao, B., & Astatkie, T. (2025b). Mycotoxins contaminations in ethiopian food: Impacts, challenges, and mitigation strategies. *Food Control, 171*(111066), Article 111066.

Awuchi, C. G., Ondari, E. N., Nwozo, S., Odongo, G. A., Eseoghene, I. J., Twinomuhwezi, H., Ogbonna, C. U., Upadhyay, A. K., Adeleye, A. O., & Okpala, C. O. R. (2022). Mycotoxins' toxicological mechanisms involving humans, livestock and their associated health concerns: A review. *Toxins, 14*(3), 167. https://doi.org/10.3390/toxins14030167

Bennett, J. W., & Klich, M. (2003). Mycotoxins. *Clinical Microbiology Reviews, 16*(3), 497–516. https://doi.org/10.1128/CMR.16.3.497-516.2003

Bertuzzi, T., Leni, G., Bulla, G., & Giorni, P. (2022). Reduction of mycotoxigenic fungi growth and their mycotoxin production by *Bacillus subtilis* QST 713. *Toxins, 14*(11), 797. https://doi.org/10.3390/toxins14110797

Bryła, M., Waśkiewicz, A., Szymczyk, K., & Jędrzejczak, R. (2017). Effects of pH and temperature on the stability of fumonisins in maize products. *Toxins, 9*(3), 88. https://doi.org/10.3390/toxins9030088

Bulgaru, C. V., Marin, D. E., Pistol, G. C., & Taranu, I. (2021). Zearalenone and the immune response. *Toxins, 13*(4), 248. https://doi.org/10.3390/toxins13040248

Bullerman, L. B. (2003). *MYCOTOXINS | Classifications*. Elsevier.

Chang, P.-K., Ehrlich, K. C., & Fujii, I. (2009). Cyclopiazonic acid biosynthesis of aspergillus flavus and aspergillus oryzae. *Toxins, 1*(2), 74–99.

Cacho, R. A., Tang, Y., & Chooi, Y. H. (2015). Next-generation sequencing approach for connecting secondary metabolites to biosynthetic gene clusters in fungi. *Frontiers in Microbiology, 5*, 774. https://doi.org/10.3389/fmicb.2014.00774

Cao, L., Jiang, Y., Zhu, L., Xu, W., Chu, X., Zhang, Y., Rahman, S. U., Feng, S., Li, Y., Wu, J., & Wang, X. (2021). Deoxynivalenol induces caspase-8-mediated apoptosis through the mitochondrial pathway in hippocampal nerve cells of piglet. *Toxins, 13*(2), 73. https://doi.org/10.3390/toxins13020073

Castegnaro, M., Canadas, D., Vrabcheva, T., Petkova-Bocharova, T., Chernozemsky, I. N., & Pfohl-Leszkowicz, A. (2006). Balkan endemic nephropathy: Role of ochratoxins A through biomarkers. *Molecular Nutrition & Food Research, 50*(6), 519–529. https://doi.org/10.1002/mnfr.200500182

Chen, J., Wen, J., Tang, Y., Shi, J., Mu, G., Yan, R., Cai, J., & Long, M. (2021). Research progress on fumonisin B1 contamination and toxicity: A review. *Molecules (Basel, Switzerland), 26*(17), 5238. https://doi.org/10.3390/molecules26175238

Ding, L., Han, M., Wang, X., & Guo, Y. (2023). Ochratoxin A: Overview of prevention, removal, and detoxification methods. *Toxins, 15*(9), 565. https://doi.org/10.3390/toxins15090565

Dorn, T. W., & Jackson-Ziems, T. A. (2013). Grain storage management to minimize mold and mycotoxins. *Papers in Plant Pathology, Plant Pathology Department*. University of Nebraska—Lincoln. Retrieved from https://digitalcommons.unl.edu

Dubey, M. K., Aamir, M., Kaushik, M. S., Khare, S., Meena, M., Singh, S., & Upadhyay, R. S. (2018). PR toxin–biosynthesis, genetic regulation, toxicological potential, prevention and control measures: Overview and challenges. *Frontiers in Pharmacology, 9*, 288. https://doi.org/10.3389/fphar.2018.00288

Eaton, D., Beima, K., Bammler, T., Riley, R., & Voss, K. (2017). Hepatotoxic mycotoxins. *Comprehensive Toxicology (Third Edition)*, 483–521. https://doi.org/10.1016/B978-0-12-801238-3.64337-4

EFSA Panel on Contaminants in the Food Chain (CONTAM), Schrenk, D., Bignami, M., Bodin, L., Chipman, J. K., Del Mazo, J., Grasl-Kraupp, B., Hogstrand, C., Hoogenboom, L. R., Leblanc, J. C., Nebbia, C. S., Nielsen, E., Ntzani, E., Petersen, A., Sand, S., Schwerdtle, T., Vleminckx, C., Marko, D., Oswald, I. P., Piersma, A., Wallace, H. (2020). Risk assessment of aflatoxins in food. *EFSA Journal. European Food Safety Authority, 18*(3), e06040. https://doi.org/10.2903/j.efsa.2020.6040

El-Sayed, R. A., Jebur, A. B., Kang, W., & El-Demerdash, F. M. (2022). An overview on the major mycotoxins in food products: Characteristics, toxicity, and analysis. *Journal of Future Foods, 2*(2), 91–102. https://doi.org/10.1016/j.jfutfo.2022.03.002

Fang, Q., Du, M., Chen, J., Liu, T., Zheng, Y., Liao, Z., Zhong, Q., Wang, L., Fang, X., & Wang, J. (2020). Degradation and detoxification of aflatoxin B1 by tea-derived *Aspergillus niger* RAF106. *Toxins, 12*(12), 777. https://doi.org/10.3390/toxins12120777

Ghosh, S., Rusyn, I., Dmytruk, O. V., Dmytruk, K. V., Onyeaka, H., Gryzenhout, M., & Gafforov, Y. (2023). Filamentous fungi for sustainable remediation of pharmaceutical compounds, heavy metal and oil hydrocarbons. *Frontiers in Bioengineering and Biotechnology, 11*, 1106973. https://doi.org/10.3389/fbioe.2023.1106973

Habschied, K., Krstanović, V., Zdunić, Z., Babić, J., Mastanjević, K., & Šarić, G. K. (2021). Mycotoxins biocontrol methods for healthier crops and stored products. *Journal of Fungi (Basel, Switzerland), 7*(5), 348. https://doi.org/10.3390/jof7050348

Hamid, A. S., Tesfamariam, I. G., Zhang, Y., & Zhang, Z. G. (2013). Aflatoxin B1-induced hepatocellular carcinoma in developing countries: Geographical distribution, mechanism of action and prevention. *Oncology Letters, 5*(4), 1087–1092. https://doi.org/10.3892/ol.2013.1169

Han, C., Yu, T., Qin, W., Liao, X., Huang, J., Liu, Z., Yu, L., Liu, X., Chen, Z., Yang, C., Wang, X., Mo, S., Zhu, G., Su, H., Li, J., Qin, X., Gui, Y., Mo, Z., Li, L., & Peng, T. (2020). Genome-wide association study of the TP53 R249S mutation in hepatocellular carcinoma with aflatoxin B1 exposure and infection with hepatitis B virus. *Journal of Gastrointestinal Oncology, 11*(6), 1333–1349. https://doi.org/10.21037/jgo-20-510

Henry, M. H., & Wyatt, R. D. (2001). The toxicity of fumonisin B1, B2, and B3, individually and in combination, in chicken embryos. *Poultry Science, 80*(4), 401–407. https://doi.org/10.1093/ps/80.4.401

Heussner, A. H., & Bingle, L. E. (2015). Comparative ochratoxin toxicity: A review of the available data. *Toxins, 7*(10), 4253–4282. https://doi.org/10.3390/toxins7104253

Hussein, H. S., & Brasel, J. M. (2001). Toxicity, metabolism, and impact of mycotoxins on humans and animals. *Toxicology, 167*(2), 101–134.

Ivanova, L., Skjerve, E., Eriksen, G. S., & Uhlig, S. (2006). Cytotoxicity of enniatins A, A1, B, B1, B2 and B3 from Fusarium avenaceum. *Toxicon: Official Journal of the International Society on Toxinology, 47*(8), 868–876.

Iqbal, J., Asghar, M. A., Ahmed, A., Khan, M. A., & Jamil, K. (2014). Aflatoxins contamination in Pakistani brown rice: A comparison of TLC, HPLC, LC-MS/MS and ELISA techniques. *Toxicology Mechanisms and Methods, 24*(8), 544–551. https://doi.org/10.3109/15376516.2014.948247

Jallow, A., Xie, H., Tang, X., Qi, Z., & Li, P. (2021). Worldwide aflatoxin contamination of agricultural products and foods: From occurrence to control. *Comprehensive Reviews in Food Science and Food Safety, 20*(3), 2332–2381. https://doi.org/10.1111/1541-4337.12734

Janik, E., Niemcewicz, M., Podogrocki, M., et al. (2021a). The existing methods and novel approaches in mycotoxins' detection. *Molecules (Basel, Switzerland), 26*(13), 3981.

Janik, E., Niemcewicz, M., Podogrocki, M., Ceremuga, M., Gorniak, L., Stela, M., & Bijak, M. (2021b). The existing methods and novel approaches in mycotoxins' detection. *Molecules (Basel, Switzerland), 26*(13), 3981. https://doi.org/10.3390/molecules26133981

Janik-Karpińska, E., Niemcewicz, M., Podogrocki, M., Ceremuga, M., Górniak, L., Stela, M., & Bijak, M. (2021). The existing methods and novel approaches in mycotoxins' detection. *Molecules, 26*(13), 3981. https://doi.org/10.3390/molecules26133981

Kalinina, S. A., Jagels, A., Cramer, B., Geisen, R., & Humpf, H. U. (2017). Influence of environmental factors on the production of penitrems A-F by Penicillium crustosum. *Toxins, 9*(7). https://doi.org/10.3390/toxins9070210

Kamle, M., Mahato, D. K., Gupta, A., et al. (2022). Citrinin mycotoxin contamination in food and feed: Impact on agriculture, human health, and detection and management strategies. *Toxins, 14*(2), 85.

Khan, R., Anwar, F., & Ghazali, F. M. (2024a). A comprehensive review of mycotoxins: Toxicology, detection, and effective mitigation approaches. *Heliyon, 10*(8), Article e28361.

Knowles, S. L., Mead, M. E., Silva, L. P., Raja, H. A., Steenwyk, J. L., Goldman, G. H., Oberlies, N. H., & Rokas, A. (2020). Gliotoxin, a known virulence factor in the major human pathogen aspergillus fumigatus, is also biosynthesized by its nonpathogenic relative Aspergillus Fischeri. *mBio, 11*(1). https://doi.org/10.1128/mBio.03361-19

Khan, N. I., & Song, E. (2020). Lab-on-a-chip systems for aptamer-based biosensing. *Micromachines, 11*(2), 220. https://doi.org/10.3390/mi11020220

Khan, R., Anwar, F., & Ghazali, F. M. (2024b). A comprehensive review of mycotoxins: Toxicology, detection, and effective mitigation approaches. *Heliyon, 10*(8), Article e28361. https://doi.org/10.1016/j.heliyon.2024.e28361

Kihal, A., Rodríguez-Prado, M., & Calsamiglia, S. (2022). The efficacy of mycotoxin binders to control mycotoxins in feeds and the potential risk of interactions with nutrients: a review. *Journal of Animal Science, 100*(11), skac328. https://doi.org/10.1093/jas/skac328

Kiseleva, M., Chalyy, Z., Sedova, I., & Aksenov, I. (2020). Stability of mycotoxins in individual stock and multi-analyte standard solutions. *Toxins, 12*(2), Article 2. https://doi.org/10.3390/toxins12020094

Kleber, A., Gruber-Dorninger, C., Platzer, A., Payet, C., & Novak, B. (2023). Effect of fungicide treatment on multi-mycotoxin occurrence in French wheat during a 4-year period. *Toxins, 15*(7), 443. https://doi.org/10.3390/toxins15070443

Kos, J., Janić Hajnal, E., Jajić, I., Krstović, S., Mastilović, J., Šarić, B., & Jovanov, P. (2016). Comparison of ELISA, HPLC-FLD and HPLC-MS/MS methods for determination of aflatoxin M1 in natural contaminated milk samples. *Acta Chimica Slovenica, 63*(4), 747–756. https://doi.org/10.17344/acsi.2016.2451

Kőszegi, T., & Poór, M. (2016). Ochratoxin A: Molecular interactions, mechanisms of toxicity and prevention at the molecular level. *Toxins, 8*(4), 111. https://doi.org/10.3390/toxins8040111

Kumar, A., et al. (2022). Trichoderma species: Our best fungal allies in the biocontrol of plant diseases A review. *Plants, 12*(3), 432. https://doi.org/10.3390/plants12030432

Kutasi, K., Recek, N., Zaplotnik, R., Mozetič, M., Krajnc, M., Gselman, P., & Primc, G. (2021). Approaches to inactivating aflatoxins-a review and challenges. *International Journal of Molecular Sciences, 22*(24), 13322. https://doi.org/10.3390/ijms222413322

Leslie, J. F., Moretti, A., Mesterházy, Á., Ameye, M., Audenaert, K., Singh, P. K., Richard-Forget, F., Chulze, S. N., Ponte, E. M. D., Chala, A., Battilani, P., & Logrieco, A. F. (2021). Key global actions for mycotoxin management in wheat and other small grains. *Toxins, 13*(10), 725. https://doi.org/10.3390/toxins13100725

Liu, Y., & Wu, F. (2010). Global burden of aflatoxin-induced hepatocellular carcinoma: A risk assessment. *Environmental Health Perspectives, 118*(6), 818–824. https://doi.org/10.1289/ehp.0901388

Llobregat, B., Abad-Fuentes, A., Mercader, J. V., González-Candelas, L., & Ballester, A.-R. (2025). The role of penicillium expansum histone deacetylases HosA and HosB in growth, development, and patulin production. *Microbiological Research, 297*(128181), Article 128181.

Logrieco, A. F., Miller, J. D., Eskola, M., Krska, R., Ayalew, A., Bandyopadhyay, R., Battilani, P., Bhatnagar, D., Chulze, S., De Saeger, S., Li, P., Perrone, G., Poapolathep, A., Rahayu, E. S., Shephard, G. S., Stepman, F., Zhang, H., & Leslie, J. F. (2018). The mycotox charter: Increasing awareness of, and concerted action for, minimizing mycotoxin exposure worldwide. *Toxins, 10*(4), 149. https://doi.org/10.3390/toxins10040149

López-Díaz, C., Rahjoo, V., Sulyok, M., et al. (2018). Fusaric acid contributes to virulence of fusarium oxysporum on plant and mammalian hosts. *Molecular Plant Pathology, 19*(2), 440–453.

Lutz, M. P., Feichtinger, G., Défago, G., & Duffy, B. (2003). Mycotoxigenic Fusarium and deoxynivalenol production repress chitinase gene expression in the biocontrol agent Trichoderma atroviride P1. *Applied and Environmental Microbiology, 69*(6), 3077–3084. https://doi.org/10.1128/AEM.69.6.3077-3084.2003

Mahmoud, Y. A., Elkaliny, N. E., Darwish, O. A., Ashraf, Y., Ebrahim, R. A., Das, S. P., & Yahya, G. (2025). Comprehensive review for aflatoxin detoxification with special attention to cold plasma treatment. *Mycotoxin Research, 41*(2), 277–300. https://doi.org/10.1007/s12550-025-00582-5

Maphaisa, T. C., Akinmoladun, O. F., Adelusi, O. A., et al. (2025a). Advances in mycotoxin detection techniques and the crucial role of reference material in ensuring food safety—A review. *Food and Chemical Toxicology: An International Journal Published for the British Industrial Biological Research Association, 200*(115387), Article 115387.

Maphaisa, T. C., Akinmoladun, O. F., Adelusi, O. A., Mwanza, M., Fon, F., Tangni, E., & Njobeh, P. B. (2025b). Advances in mycotoxin detection techniques and the crucial role of reference material in ensuring food safety. A Review. *Food and Chemical Toxicology, 200*, Article 115387. https://doi.org/10.1016/j.fct.2025.115387

Marchese, S., Polo, A., Ariano, A., Velotto, S., Costantini, S., & Severino, L. (2018). Aflatoxin B1 and M1: Biological properties and their involvement in cancer development. *Toxins, 10*(6), 214. https://doi.org/10.3390/toxins10060214

Marhaba, Anjum, S., Mandal, P., Agrawal, S., & Ansari, K. M. (2024). Zearalenone promotes endometrial cancer cell migration and invasion via activation of estrogen receptor-mediated Rho/ROCK/PMLC signalling pathway. *Food and Chemical Toxicology: An International Journal Published for the British Industrial Biological Research Association, 193*, 115017. https://doi.org/10.1016/j.fct.2024.115017

Marin-Kuan, M., Ehrlich, V., Delatour, T., Cavin, C., & Schilter, B. (2011). Evidence for a role of oxidative stress in the carcinogenicity of ochratoxin a. *Journal of Toxicology, 2011*, Article 645361. https://doi.org/10.1155/2011/645361

Meneely, J., Greer, B., Kolawole, O., & Elliott, C. (2023). T-2 and HT-2 Toxins: toxicity, occurrence and analysis: A review. *Toxins, 15*(8), 481. https://doi.org/10.3390/toxins15080481

Mohammad Nejad Khiavi, N., Sowti Khiabani, M., Rezaei Mokarram, R., & Samadi Kafil, H. (2020). Reduction of aflatoxin M1 using a mixture of Saccharomyces cerevisiae and Candida albicans cell walls immobilized on silica nanoparticles entrapped in alginate gel. *Journal of Environmental Chemical Engineering, 8*(1), 103635. https://doi.org/10.1016/j.jece.2019.103635

Mostrom, M. S. (2021). *Trichothecene toxicosis in animals*. Professional Version. Reviewed/Revised November 2021. https://www.merckvetmanual.com/toxicology/mycotoxicoses/trichothecene-toxicosis-in-animals

Ong, J. Y., Pike, A., & Tan, L. L. (2021). Recent advances in conventional methods and electrochemical aptasensors for mycotoxin detection. *Foods (Basel, Switzerland), 10*(7), 1437. https://doi.org/10.3390/foods10071437

Pan, X., Whitten, D. A., Wilkerson, C. G., & Pestka, J. J. (2014). Dynamic changes in ribosome-associated proteome and phosphoproteome during deoxynivalenol-induced translation inhibition and ribotoxic stress. *Toxicological Sciences: An Official Journal of the Society of Toxicology, 138*(1), 217–233. https://doi.org/10.1093/toxsci/kft270

Pestka, J. J. (2010). Deoxynivalenol-induced proinflammatory gene expression: Mechanisms and pathological sequelae. *Toxins, 2*(6), 1300–1317. https://doi.org/10.3390/toxins2061300

Piao, M.-Z., Shen, L., & Wang, F.-W. (2013). A new trichothecene from Myrothecium roridum QDFE005, a symbiotic fungus isolated from Mactra chinensis. *Journal of Asian Natural Products Research, 15*(12), 1284–1289.

Plattner, R. D., & Nelson, P. E. (1994). Production of beauvericin by a strain of fusarium proliferatum isolated from corn fodder for swine. *Applied and Environmental Microbiology, 60*(10), 3894–3896.

Qin, Y., et al. (2022). A highly sensitive fluorometric biosensor for fumonisin B1 detection based on upconversion nanoparticles-graphene oxide and catalytic hairpin assembly. *Analytica Chimica Acta, 1207*, Article 339811. https://doi.org/10.1016/j.aca.2022.339811

Riley, R. T., & Merrill, A. H., Jr. (2019). Ceramide synthase inhibition by fumonisins: A perfect storm of perturbed sphingolipid metabolism, signalling, and disease. *Journal of Lipid Research, 60*(7), 1183–1189. https://doi.org/10.1194/jlr.S093815

Rodríguez, A., Magan, N., & Delgado, J. (2024). Exploring a cheese ripening process that hinders ochratoxin a production by *Penicillium nordicum* and *Penicillium verrucosum*. *Biology, 13*(8), 582. https://doi.org/10.3390/biology13080582

Ropejko, K., & Twarużek, M. (2021). Zearalenone and its metabolites-general overview, occurrence, and toxicity. *Toxins, 13*(1), 35.

Santos Pereira, C., C Cunha, S., & Fernandes, J. O. (2019). Prevalent mycotoxins in animal feed: Occurrence and analytical methods. *Toxins, 11*(5), 290. https://doi.org/10.3390/toxins11050290

Singh, A., Prakash, A., & Choudhary, R. (2023). Bioactive components having antimicrobial and anticancerous properties: A review. In M. Thakur & T. Belwal (Eds.), *Bioactive components: A sustainable system for good health and well-being* (pp. 271–299). Springer Nature. https://doi.org/10.1007/978-981-19-2366-1_17

Singh, J., & Mehta, A. (2020). Rapid and sensitive detection of mycotoxins by advanced and emerging analytical methods: A review. *Food Science & Nutrition, 8*(5), 2183–2204. https://doi.org/10.1002/fsn3.1474

Singh, V. K., Meena, M., Zehra, A., Tiwari, A., Dubey, M. K., & Upadhyay, R. S. (2014). Fungal toxins and their impact on living systems. In *Microbial diversity and biotechnology in food security* (pp. 513–530).

Sorrenti, V., Di Giacomo, C., Acquaviva, R., Barbagallo, I., Bognanno, M., & Galvano, F. (2013). Toxicity of ochratoxin a and its modulation by antioxidants: A review. *Toxins, 5*(10), 1742–1766. https://doi.org/10.3390/toxins5101742

Suo, Z., et al. (2023). Latest strategies for rapid and point of care detection of mycotoxins in food: A review. *Analytica Chimica Acta, 1246*, Article 340888. https://doi.org/10.1016/j.aca.2023.340888

Shekhar, R., Raghavendra, V. B., & Rachitha, P. (2025). A comprehensive review of mycotoxins, their toxicity, and innovative detoxification methods. *Toxicology Reports, 14*(101952), Article 101952.

Siciliano, I., Bosio, P., Gilardi, G., Gullino, M. L., & Garibaldi, A. (2017). Verrucarin A and Roridin E produced on spinach by Myrothecium verrucaria under different temperatures and CO_2 levels. *Mycotoxin Research, 33*(2), 139–146.

Smakosz, A., Kurzyna, W., Rudko, M., & Dąsal, M. (2021). The usage of ergot (Claviceps purpurea (fr.) Tul.) in obstetrics and gynecology: A historical perspective. *Toxins, 13*(7), 492.

Turner, D. (2024). GC-MS principle, instrument, and analyses, and GC-MS/MS. *Technology Networks*. Retrieved from https://www.technologynetworks.com/analysis/articles/gc-ms-principle-instrument-and-analyses-and-gc-msms-362513

Voss, K. A., Riley, R. T., Norred, W. P., Bacon, C. W., Meredith, F. I., Howard, P. C., Plattner, R. D., Collins, T. F., Hansen, D. K., & Porter, J. K. (2001). An overview of rodent toxicities: liver and kidney effects of fumonisins and Fusarium moniliforme. *Environmental health perspectives, 109 Suppl, 2*(Suppl 2), 259–266. https://doi.org/10.1289/ehp.01109s2259

Wang, J., Tolosa, J., Wang, W., & Yang, X. (2025). Mycotoxins and fungal toxins: Current status and future perspectives. *Toxins, 17*(4), 176. https://doi.org/10.3390/toxins17040176

Wang, L., Wang, Q., Wang, S., Cai, R., Yuan, Y., Yue, T., & Wang, Z. (2022). Bio-control on the contamination of Ochratoxin A in food: Current research and prospects. *Current Research in Food Science, 5*, 1539–1549. https://doi.org/10.1016/j.crfs.2022.09.007

Wenderoth, M., Garganese, F., Schmidt-Heydt, M., et al. (2019). Alternariol as virulence and colonization factor of Alternaria alternata during plant infection. *Molecular Microbiology, 112*(1), 131–146.

Więckowska, M., Cichon, N., Szelenberger, R., Gorniak, L., & Bijak, M. (2024). Ochratoxin A and its role in cancer development: A comprehensive review. *Cancers, 16*(20), 3473. https://doi.org/10.3390/cancers16203473

Xu, L., Zhang, Z., Zhang, Q., & Li, P. (2016). Mycotoxin determination in foods using advanced sensors based on antibodies or aptamers. *Toxins, 8*(8), 239. https://doi.org/10.3390/toxins8080239

Yin, S., Niu, L., & Liu, Y. (2022). Recent progress on techniques in the detection of aflatoxin B1 in edible oil: A mini review. *Molecules (Basel, Switzerland), 27*(19), 6141. https://doi.org/10.3390/molecules27196141

Zhang, C., Qu, Z., Hou, J., & Yao, Y. (2024). Contamination and control of mycotoxins in grain and oil crops. *Microorganisms, 12*(3), 567. https://doi.org/10.3390/microorganisms12030567

Zhang, Z., Li, P., Hu, X., Zhang, Q., Ding, X., & Zhang, W. (2012). Microarray technology for major chemical contaminants analysis in food: Current status and prospects. *Sensors (Basel, Switzerland), 12*(7), 9234–9252. https://doi.org/10.3390/s120709234

Fungal Secondary Metabolites in Agriculture

Keerthana Ramesh, Dikshaa Murthy, Sanusha Stalin, and Manoj Kumar Narasimhan

Abstract

Increasing health and environmental issues caused by chemical pesticides and fertilizers have spurred the adoption of sustainable alternatives in agriculture. Fungal secondary metabolites are the most important alternatives when compared to conventional agrochemicals. They can be used as eco-friendly substitutes in place of several other sustainable biopesticides, biofertilizers, and biocontrol agents as a highly promising solution for a sustainable agriculture. The bioactive chemicals generated by filamentous fungi serve a variety of purposes, such as managing pests, diseases, and promoting plant development. The intricate uses of secondary metabolites produced by fungi in the agricultural industry are discussed in this chapter. Examining the case studies provides crucial insights into practical approaches against disease and pest prevention. Mechanisms related to antagonistic interactions, competition, and the initiation of host defense systems are also studied. Furthermore, the section about mycorrhizal fungi will emphasize on how they aid in nutrient absorption, soil health promotion, and plant development, particularly in relation to sustainable agricultural systems. Genetic engineering and fermentation technologies are two recent innovations that will have a major influence on improving the production of fungal metabolites and, in turn, on raising agricultural productivity while reducing environmental impact.

Keywords

Biocontrol agents (BCAs) · Fungal secondary metabolites · Fungicides · Herbicides · Insecticides · Mycorrhizal fungi

1 Introduction

Agriculture is the foundation of human survival, providing the food, fiber, and resources essential for life. However, despite its critical importance, the future of agriculture is under threat due to overreliance on chemical fertilizers and pesticides. These synthetic chemicals, while being effective in increasing the yields, have led to significant environmental and health challenges. Over time, their consistent use has contributed to a decrease in soil fertility, nutrient imbalance, damage to the soil ecosystem, affecting non-target organisms and impairing human health. Many alternative solutions have been considered to overcome these challenges, among those the most promising amendment is the use of biopesticides, biofertilizers, and biocontrol agents from living organisms. A key area of focus is the use of fungal secondary metabolites, which are bioactive compounds produced by the fermentation process.

In general, secondary metabolites (SMs) are organic molecules synthesized by a varied group of organisms, including bacteria, fungi, plants, and animals. They are produced at the end of the growth phase or in the stationary phase. SMs production is not always essentially important for the growth and survival of the organism, but they have some ecologically important functions like initiating a competitive mechanism of defense against other organisms, as a transporting system, in building up a symbiotic relationship, expressed as sexual hormones and as effec-

K. Ramesh · D. Murthy · S. Stalin · M. K. Narasimhan (✉)
Department of Genetic Engineering, School of Bioengineering, College of Engineering and Technology, SRM Institute of Science and Technology, Kattankulathur, Chennai, Tamil Nadu, India
e-mail: manojkun@srmist.edu.in

K. Ramesh
e-mail: kr9925@srmist.edu.in

D. Murthy
e-mail: dm8385@srmist.edu.in

S. Stalin
e-mail: ss9245@srmist.edu.in

© The Author(s), under exclusive license to Springer Nature Switzerland AG 2026

A. Prakash and M. K. Dubey (eds.), *Advances in Fungal Secondary Metabolites: Bioactive Attributes and Biotechnological Potential of Industrial Interest*, Sustainable Economy and Ecotechnology, https://doi.org/10.1007/978-3-032-23507-7_7

tors. Among the wide range of SMs, fungal secondary metabolites are lower molecular weight compounds which is of unique importance as they promote the survival of organism in different ecological niches. There are many fungal species involved in production of SMs, out of this, a higher production is observed in filamentous Ascomycetes and Basidiomycetes, whereas lesser production is observed in unicellular Ascomycetes, Basidiomycetes, and also in Zygomycetes, Blastocladiomycota, and Chytridiomycota (Bills & Gloer, 2016; Demain & Fang, 2000). Unlike primary metabolite (PMs), genes coding for SMs in fungi are in the form of biosynthetic gene clusters, rather being distributed throughout the genome of the fungi. Fungal SMs are generally classified into four main groups they are the polyketides, non-ribosomal peptides, terpenoids, and indole alkaloids. They can be obtained from central metabolic pathways and PM's pools, with acyl-CoAs as an initial intermediate to synthesize polyketides and terpenes and SMs in combination with amino acids are used in producing non-ribosomal peptides (Keller, 2019).

Both ancient and contemporary times have seen the widespread use of fungi in food and medicine. Fungi can be used in the production of alcohol, antibiotics, enzymes, chemicals, and drugs. Besides these applications, their role in inhibiting pathogenic organisms and protection against the extensive resistance mechanisms makes them a useful candidate in agriculture by acting as a biocontrol agent in protecting the crops against plant diseases. They can also be used as natural biofertilizers, and some of them mimic the plant hormones which regulate plant growth under various kinds of biotic stress. The rising need for food crops to meet consumer demand is a serious concern. One of the various methods to increase agricultural productivity is the use of fungal secondary metabolites, which are intended to improve the efficient growth of food crops (Elhamouly et al., 2022).

2 Fungal Metabolites as Biocontrol Agents (BCAs)

Food production will need to be increased twofold to meet the demands of the growing population, while also considerably reducing the availability of arable land. Furthermore, the increased use of synthetic fertilizers in agriculture has damaged soil quality, disrupting the natural balance of soil and microbes, as well as their environment. As a result, utilizing biofertilizers rather than chemical fertilizers has been a better approach to minimizing environmental impacts and maintaining soil fertility and reproducibility. Fungal biofertilizers are biofertilizers that have achieved the recent popularity due to their ability to control plant diseases and sustainably boost plant growth. They utilize the symbiotic relationship between fungi and plants to improve soil fertility and crop production. Some core examples of fungi biofertilizers are plant growth-stimulating fungi (non-pathogenic fungi that promote plant growth and health), compost-producing enzymatic fungi, phosphate-solubilizing fungi, and potassium-solubilizing fungi. Furthermore, in this chapter, we will be seeing how fungal metabolites act as a biocontrol agent, their mechanism of action, their role in crop improvement, soil productivity, and future prospects (Aktar et al., 2009)

2.1 Fungi in Biological Pest Control

Compared to chemical pesticides, natural ways of controlling pests, weeds, and diseases are thought to be less effective and require more work. BCAs do, however, provide several advantages. Establishing and carrying out new crop protection programs requires significant effort and expense; BCA, however, provides an alternative to this situation. Since fungi are highly malleable and may change their shape in reaction to favorable or unfavorable circumstances, they are exceptionally successful soil residents, thus working as efficient biocontrol agents even in harsh conditions. Their capacity to generate a diverse range of extracellular enzymes allows them to decompose various organic substances, thereby reducing soil constituents and controlling the carbon and nutrition balance (Butt & Copping, 2000). Dead organic matter is broken down by fungi to produce organic acids, carbon dioxide, and biomass, thus serving as a further addition to the humus layer of the soil. Numerous fungal species can accumulate harmful elements like cadmium, copper, mercury, lead, and zinc in their fruiting bodies, thereby serving as an efficient biosorbent. Fungi have been used for over 150 years, especially in the management of insect-pathogenic organisms. *Beauveria, Metarhizium, Isaria, Hirsutella,* and *Lecanicillium* are among the most widely used strains in the control of biological pests. The most common method of applying them is inundation, which entails exposing the fungal spores to unfavorable humidity, temperature, and solar radiation levels. These abiotic variables lessen the fungi's ability to persist and cause harm to insects. To gain further understanding, let us learn about their mechanism of action and the key species used as fungal BCA.

2.2 Action Mechanism of Fungal Biocontrol Agents: Antagonism and Competition

Using a variety of approaches, microorganisms found in or on plants, soil, and air are identified, and their capacity to biocontrol the target pathogen or pathogens is evaluated concurrently. To fully utilize the potential of biotic biocontrol agents for crop disease management, it is imperative to comprehend their mechanisms of action, constraints, and prerequisites. Numerous studies have demonstrated that fungal biocontrol agents (BCAs) biocontrol processes are numerous and diverse. Within a family, various species of BCAs may function through different methods (Palmieri et al., 2022). For example, *Beauveria* may inhibit phytopathogens by two or more processes.

Fungal BCAs inhibit the pathogenic activity by three methods of antagonism:

1. Direct antagonism
2. Indirect antagonism
3. Mixed-path antagonism (MBCAs).

1. **Direct Antagonism**

BCAs when they come into direct contact with the pathogen causes damage to the structures of the pathogen through direct antagonism. Some of the mechanisms utilized in this method are parasitism, and cell wall degradation (Fig. 1). One example of parasitism is the action of BCA *Trichoderma virens* which parasitize pathogens by piercing their cell walls and siphoning off nutrients. T. *virens* invades *Rhizoctonia solani* (a plant pathogenic fungi) and develops haustoria (root-like structures that grow into or around another structure to absorb nutrients and water), which removes the pathogen's available resources and eventually leads to a hindrance in its growth. This action of *T. virens* invading *Rhizoctonia solani*, i.e., one fungus invading another is called mycoparatisim. Another example of this parasitic mechanism is *Pythium oligandrum* which attacks the fungal pathogen *Botrytis cinerea. Pythium oligandrum* encircles their hyphae, which results in protoplasm coagulation and thus kills the cell. On the other hand, in cell wall degradation, the BCAs produce enzymes that degrade the structural elements of the pathogen thereby causing cell wall degradation. Cell wall degradation by *Trichoderma harzianum* is through the production of chitinases and glucanases which degrade fungal cell walls (Rodríguez-Romero et al., 2023).

2. **Indirect Antagonism**

Direct contact between the pathogen and the BCA is not necessary for indirect antagonism to occur. By changing host defense or vying for some of the resources, it produces an environment that is not conducive to the pathogen. Competition, such as that between *Coniothyrium minitans* and pathogens for available nutrients and space, is one form of indirect antagonism. Due to their same biological niche, the pathogen is unable to obtain these essential materials, hence starving to death. In contrast, during host defense induction, the BCAs can cause the plant to develop a systemic resistance to infections. *Trichoderma asperellum*, e.g., causes the plant to produce defensive chemicals that strengthen its defenses. Finally, altering the environment: BCAs have the capacity to alter the rhizosphere's pH or nutrient availability, which inhibits the pathogens. For example, *Coniothyrium minitans* increases soil pH, and that in turn decreases the viability of *Sclerotinia sclerotiorum* (Ayaz et al., 2023).

3. **Mixed-Path Antagonism**

A mix of direct and indirect methods is known as mixed-path antagonism. The enzymatic and biochemical mechanisms of BCAs change the environment of infections and prevent their proliferation like parasitism; *Coniothyrium minitans*, for instance, parasitises *Sclerotinia sclerotiorum*'s sclerotia, lowering its soil inoculum. Competition is another approach; for instance, *Pseudomonas fluorescens* rivals *Fusarium oxysporum* for iron by producing siderophores, which ultimately results in resistance. Tomatoes that are exposed to *Trichoderma asperellum* develop a systemic acquired resistance to *Phytophthora infestans.* Alongside these three methods, the following can also be employed to restrict the growth and development of pests:

a. **Antibiotic Production**: BCAs such as *Trichoderma* spp. produce antifungal and antibacterial compounds. For instance, *it* produces antifungal metabolites which inhibit *Phytophthora capsici.*
b. **Secretion of Enzymes**: The enzymes involved are proteases and cellulases that kill the pathogen cells. For example, *Trichoderma koningii* secretes chitinolytic enzymes that lyse the hyphae of *Sclerotium cepivorum.* Production of Toxins: Many BCAs produce toxins that inhibit pathogens. One yeast BCA, *Pichia membranifaciens*, produces toxins that are toxic to *Botrytis cinerea* on fruits.

Hence, we can infer that the mechanisms of antagonism by BCAs include direct actions like parasitism and enzymatic degradation, indirect strategies like resource competition and induction of host resistance, and mixed-path methods involving antibiotic production and environmental alterations. These multiple mechanisms collectively suppress pathogen development and enhance crop protection.

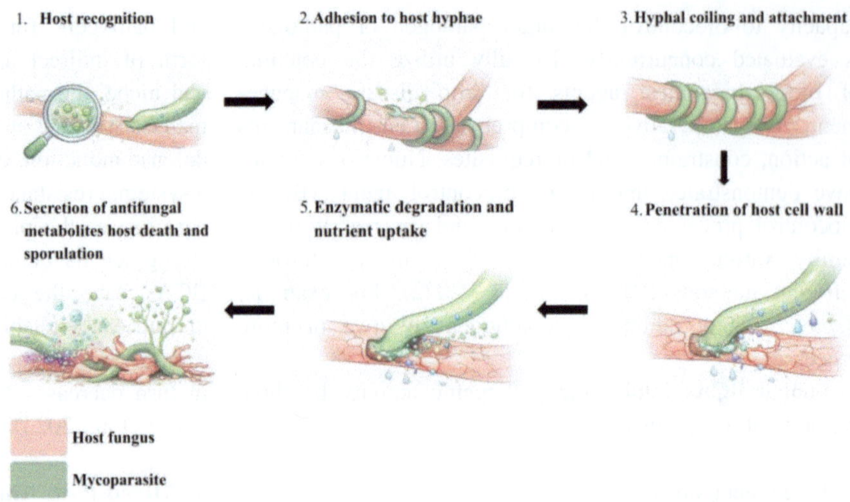

Fig. 1 Mycoparasitism (Direct antagonism). *Source* Authors

2.3 Key Species Used in Biocontrol

The important Fungal BCAs such as *Trichoderma* and *Beauveria bassiana* are very important for the sustainable management of pests as they offer environmentally benign alternatives to conventional chemical pesticides. The action mechanisms of these fungi are direct parasitism, synthesis of toxic metabolites, and stimulation of host plant defenses, as seen in the previous section regarding the mechanism of action of fungal metabolites. Among them, species of *Trichoderma* are highly valued as the leading biocontrol agents (Fig. 2). They have mycoparasitic properties because the given species penetrates the exoskeletons of pests. Penetration of *Trichoderma* into exoskeletons of the pest includes hydrolyzing a special structural component, the composition of exoskeletons made from chitin and glucan, using special enzymes. Thus, *Trichoderma* can not only regulate a population of pests but also positively influence plant growth, thus strengthening systemic mechanisms of defense. Specifically, *Trichoderma asperellum* has proven effective in controlling soil-borne pathogens and pests, specifically when applied as seed treatments to the crops or as soil inoculants (Rivera-Méndez et al., 2020). Similarly, in the case of *Beauveria bassiana*, it attacks a wide variety of insect pests, such as aphids, borers, and whiteflies, and attaches itself to the cuticle, germinating into the insect body. During this process, they degrade insect vital physiological functions through a toxic release. Among those pests, *B. bassiana* drastically reduced the degree of plant damage inflicted by the Asian corn borer, *Ostrinia furnacalis*, by intensifying the activity of antioxidants in the plant. Both *Trichoderma* and *B. bassiana* can be applied using various means: *Trichoderma* in seed application or drench application for soil, and *B. bassiana* applied through foliar application or as soil treatments. The cohesive use of *T. asperellum* and *B. bassiana* further adds a synergistic effect and increases their pest-suppressive power. Several studies confirm that the application of both at an equal level shows a mortality rate above 90%, thus successfully controlling these pests and reducing crop damage. Fungal BCAs also have multiple advantages, some of which include an increase in environmental safety as well as reduced pesticide resistance to pests, and other benefits such as promotion of plant growth and nutrient uptake. They are highly compatible with the principles of sustainable agriculture due to their varying mechanisms of action, efficacy, and ecological benefits. Fungal biological control agents such as *Trichoderma* and *Beauveria* are an important asset for the present agriculture sector, and Table 1 gives the key differences between *Trichoderma asperellum* and *Beauveria bassiana* (Thambugala et al., 2020).

2.4 Case Studies in Agricultural Applications

2.4.1 Successful Examples in Pest and Disease Management Using BCAs

Biological control agents (BCAs) are very effective in the sustainable management of pests and diseases. Fungal BCAs like *Trichoderma* and *Beauveria bassiana* have been widely used because of the difference in their mechanism of action and safety to the environment. The *Trichoderma* species with mycoparasitic activity represses the soil-borne pathogens like *Rhizoctonia solani* and *Fusarium oxysporum* (Abdel Ghany, 2015). This species infects the crop root system, breaks up the structure of the pathogen due to the use of special enzymes, and induces systemic resistance in crops. Similarly, the entomopathogenic fungus, *Beauveria bassiana* destroys various insect pests such as whiteflies and borers by adhering to the insect cuticles, producing toxic compounds, and multiplying in the insect. It has been used

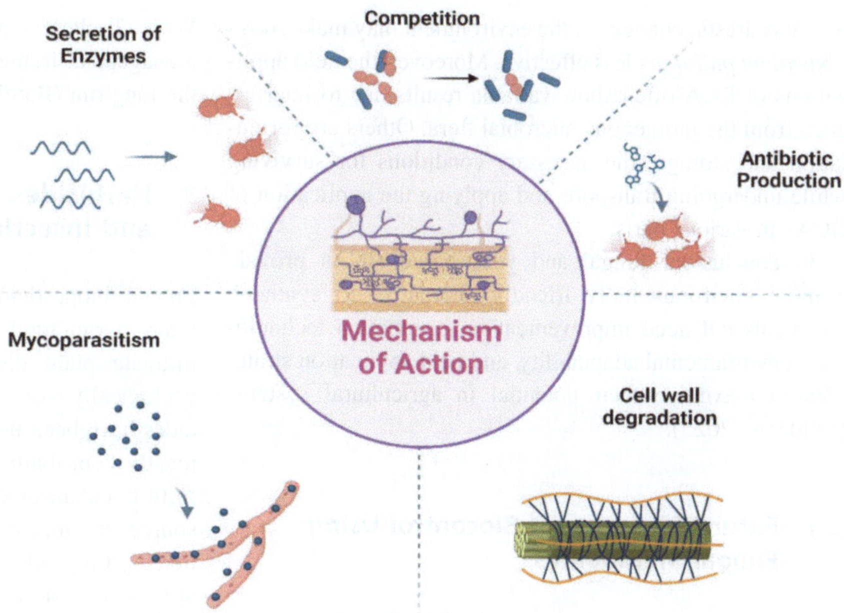

Fig. 2 Depicting mechanisms of action by Fungal BCA. *Source* Authors

Table 1 Successful BCAs in pest and disease management

BCA	Target pest/pathogen	Mode of action	Application method	References
Trichoderma spp.	*Fusarium, Rhizoctonia*	Mycoparasitism, resistance induction	Soil/seed treatment	Rivera-Méndez et al. (2020)
Beauveria bassiana	Whiteflies, borers	Cuticle penetration, toxin production	Foliar sprays, soil treatment	Anwar et al. (2018)
Metarhizium anisopliae	Termites, grasshoppers	Cuticle penetration, internal proliferation	Soil drenching, sprays	Dauda et al. (2018)
Clonostachys rosea	*Bemisia tabaci, Botrytis*	Toxin production, endophytism	Foliar sprays, seed treatment	Anwar et al. (2018)
Beauveria bassiana	Aphids, borers, whiteflies	Cuticle penetration, toxin production	Foliar sprays, soil treatment	Batool et al. (2020)

Source Authors compilation

widely in the management of pest populations in maize and cotton fields while maintaining beneficial organisms. Other BCAs include the *Metarhizium anisopliae,* which infects termites and grasshoppers through a similar infection process (Dauda et al., 2018). *Clonostachys rosea* can be bimodal and effective in antifungal metabolites as well as a plant endophyte. It is effective against the *Bemisia tabaci* (whitefly) and *Botrytis cinerea* (a necrotrophic fungus that affects many plant species, specifically wine grapes). These BCAs are sustainable alternatives to chemical pesticides and are in line with integrated pest management strategies (Anwar et al., 2018; Ayaz et al., 2023).

2.4.2 Benefits and Limitations of Biocontrol Agents

Fusarium oxysporum and *Rhizoctonia solani* are some of the effective soil-borne pathogens that can be managed using fungal BCAs like *Trichoderma* spp. These fungi directly parasitize the hyphae of the pathogens and produce hydrolytic enzymes, chitinases, and glucanases, to degrade the cell walls. They also stimulate plant defense through systemic resistance and, thus, offer both disease suppression and growth promotion. The postharvest pathogens, such as *Botrytis cinerea* and *Rhizopus stolonifer,* are well controlled by yeasts like *Aureobasidium pullulans.* Such BCAs work through quick colonization of fruit surfaces, production of extracellular polysaccharides, and nutrient competition, thus effectively stopping the establishment of the pathogen. These BCAs are nontoxic to human health and the environment, which makes them vital components of IPM systems (Rodríguez-Romero et al., 2023).

However, there are limitations of BCAs. The performance of fungi used is strictly dependent on their environment. For instance, the temperature, humidity, and pH of the soil determine its survival and activity. *Trichoderma* species need soil only damp enough to be heavily colonized, whereas drastic changes in the environment may make *Aureobasidium pullulans* less effective. Moreover, the field applications of BCA often show variable results due to interference from the indigenous microbial flora. Others are formulation and storage; the necessary conditions for surviving while undergoing transport, and applying the application of BCAs in various parts.

In conclusion, fungal and yeast-based BCAs provide a more environmentally friendly alternative to synthetic chemicals but need improvement in formulation technologies, environmental adaptability, and field application strategies to maximize their potential in agricultural systems (Evidente, 2023).

2.5 Future Prospects of Biocontrol Using Fungal Metabolites

2.5.1 Integration with Sustainable Farming Practices

Fungal metabolites have enormous potential in the improvement of biocontrol methods and integration into eco-friendly agricultural practices. The secondary metabolites of fungi, such as *Trichoderma* and *Penicillium*, have been shown to have excellent antifungal, antibacterial, and nematicidal properties, which are crucial for pest and disease control with the minimum use of chemical pesticides. For instance, *Trichoderma* spp. produces volatile organic compounds and hydrolytic enzymes that inhibit the growth of pathogens, strengthen plant immunity, and enhance root colonization. Such fungal metabolites are safe organic alternatives to synthetic pesticides-just what organic farming aimed for. Genetic and fermentation engineering shall come into the picture in the future in optimizing the manufacturing process of fungal metabolites with an increase in scalability and cost-effectiveness toward practical field application. Further advanced formulations, with encapsulated spores or seeds coated with metabolite, shall further improve these biologicals under diverse environmental conditions. Besides, compatibility with sustainable agriculture management practices like crop rotation, organic amendments, and reduced tillage helps in improving their performance through a balanced microbiome in soil. Through research collaboration in policy-making, the challenges of commercialization and the regulatory frameworks will be solved. Precision agriculture with the application of fungal metabolites by either sensors or drone technology is likely to increase crop yields while minimizing waste at the same time. Education and outreach will greatly aid in promoting fungal biocontrol in less-developed areas. In conclusion, fungal metabolites have the potential to transform sustainable agriculture by enhancing soil health, reducing pest and disease pressure, and improving agricultural productivity. With all these components integrated into a whole pest management framework, agriculture becomes sustainable in the long run (Bamisile et al., 2021).

3 Herbicides, Fungicides, and Insecticides from Fungi

Fungal biopesticides have lately gained prominence for many agricultural applications owing to their capacity to mitigate plant disease and enhance crop output in an ecologically sustainable manner. Numerous fungal biopesticides have been utilized for commercial production, significantly contributing to plant growth, productivity and soil fertility enhancement. In agriculture, fungi are an intriguing source of bioactive molecules that can be used as insecticides, fungicides, and herbicides. These natural products are valued for their low toxicity to non-target organisms and environmental friendliness. Their distinctive biochemical pathways enable the production of compounds that specifically target weeds, plant pathogens, and insect pests while minimizing environmental impact (Ferreyra-Suarez et al., 2024).

3.1 Herbicides

Weeds pose a serious risk to agricultural production. They compete with crops for nutrient sources and can become hosts to pests. Therefore, the administration of synthetic herbicides for pest control has been vital despite the undesirable side effects. Herbicides function to disrupt the metabolic process of plants, resulting in their demise or halting their growth. Chemical herbicides have detrimental effects on plants by disrupting the metabolic processes of the plant, soil health, polluting underground water, threatening aquatic life and causing oxidative damage and photobleaching (Harding & Raizada, 2015). Due to its toxicity, exposure to these harmful chemicals causes cancer, birth defects, DNA mutations, cell damage, and even death. As organic products are expanding market demand, farmers are shifting from hazardous synthetic chemicals to conventional bioherbicides. Biological weed control is a strategy to limit the germination and expansion of the weed population (Mohd Ghazi et al., 2023). Application of fungi, bacteria, and viruses as a bioherbicide has been notable over the decades. Mycoherbicides are biological herbicides that are fungal agents to control weeds in various settings. Several fungal species can emit biochemicals such as alcohol, fatty acids, phenolics, flavonoids, terpenoids, and steroids, which inhibit the production, growth, and development of nearby flora, weeds (Chakraborty & Ray, 2021).

The majority of mycoherbicides are a part of Hyphomycetes like *Aspergillus, Beauveria, Culicinomyces,* and *Hisrtulla.* Fungi from the *Coelomycetes* group - such as, *Colletotrichum, Septoria, Phoma, and Aschcocyta* also play an important role in biological weed control. Additionally several pathogenic strains of Ascomycetes such as *Sclerotinia, Balansia, Cryptomycina, Ceratocystis,* and *Nectaria* have been studied for their potential bioherbicidal agents (Singh et al., 2018).

3.1.1 Mechanism of Mycoherbicides Against Agricultural Pests

Mycoherbicide works through a process which involves a programmed attack on weeds by germination of spores, enzymatic infiltration, colonization, and synthesis of toxin leading to death of the target weeds.

1. Spore germination: In this process, mycoherbicide starts off as fungal spores drop onto a surface area of the selected weed, whether it is a leaf, stem, or soil next to its root. When they reach a proper environment, they start germinating.
2. Enzymatic invasion: Cutinase with pectinase as well as pressure of an appressorium infiltrates the fungal element, which enters the plant and causes infection.
3. Colonization: The fungus penetrates the plant and begins to grow and colonize the internal tissues. Fungal physiological stress reduces structural integrity and deprives the plant of its vital nutrients.
4. Toxin synthesis: Secondary metabolites are synthesized during colonization by the fungus. These metabolites cause direct damage to plant cells.
5. Death of the target weed: The interaction between nutrient deficiency, mechanical damage, and toxins mediates a metabolic disorder that causes the total collapse and death of the weed plant. It happens because this type of plant may not survive its physiological condition and may completely collapse through death.

This biological mechanism is host specific, eco-friendly, and sustainable, rendering mycoherbicides an alternative to synthetic herbicides in integrated weed management strategies.

3.1.2 Field Applications and Environmental Impact of Mycoherbicides

Mycoherbicides produce phytotoxins that harm plants directly during the infection process. The first commercially used mycoherbicides are DeVine and Collego.

DeVine is made of *Phytophthora palmivora* registered in 1981, marketed by Abott laboratories. It was used to control strangler vines species known as *Morinnia odorata* in Florida's citrus groves. De Vine is available in liquid suspension consisting of a large suspension of chlamydospores of the fungus. It is applied in the boom and nozzle sprayer system on the soil surface during wet conditions. It triggers the growth of strangler vine roots and encircles the stems of infected plants near the soil surface, resulting in the death of the seedlings. Young seedlings can be killed within 1 week after inoculation, and larger vines for about 4–6 weeks.

In 1982, Collego was formulated by Upjohn company, a dry powder consisting of 15% spores and 85% other ingredients that is rehydrated and resuspended in sugar solution within a spray tank. It consists of *Colletotrichum gloeosporioides* f. sp. *aeschynomene*, a causative organism of anthracnose disease in strawberry, blueberry, mandarin citrus, and chili. It is utilized to regulate the proliferation of joint vetch (species: *Aeschynomene virginica*) in rice and soybeans. Following inoculation, spore germination occurs, leading to the formation of appressoria, which penetrate the epidermis of the infected organism within 24 h. This application has the capability of killing plants up to 30–40 cm tall within 4–6 weeks (TeBeest & Templeton, 1985). Additional representatives that can be considered as a potential mycoherbicide are provided in Table 2.

Table 2 Potential mycoherbicide (Harding & Raizada, 2015)

Bioherbicide agent	Target weed	Field type
Alternaria destruens	Dodder species	Cranberries, carrots, peppers, tomatoes, and eggplant
Puccinia thlaspeos	*Isatis tinctoria*	For all types
Chondrostereum purpureum	Deciduous trees and shrubs	Forestry
Sclerotina minor	Dandelion, white clover, broadleaf plantain	Turf
Phoma macrostoma	Dicot plants	Turf
Colletotrichum gloeosporioides f. sp. *malvae*	*Malva pusilla*	Wheat, rye, flax, lentil, barley, canola, sunflower, oats, and mustard

3.2 Fungicides

Fungicides are natural chemicals that kill and prevent the growth of fungi and their spores. Antifungal chemicals produced by fungi to kill and control the growth of other fungi are called mycofungicides. Fungicides are classified into three categories based on their application: foliar, soil, and dressing fungicides.

3.2.1 Mechanism of Mycofungicide

Mycofungicide works in four major steps:

1. Interfering: Mycofungicide can interfere with the growth of other fungi.
2. Triggers plant defense mechanism: It triggers production of plant defense enzymes like peroxidases, chitinase, and glucanase, enhancing systemic resistance against pathogenic strain.
3. Production of antifungal compounds: Mycofungicide produces antifungal compounds/fungal secondary metabolites to inhibit pathogen growth.
4. Enhances plant growth: Mycofungicide increases plant growth and nutrient uptake (Boyetchko et al., 2004).

3.2.2 Field Application and Environmental Impact of Mycofungicide

Trichoderma harzianum a species which had the ability to biological control *Botrytis cinerea* (gray mold), a necrotrophic fungus, the most frequent cause of fungal rot in peas and beans. *T. harzianum* strain T-22 was created via protoplast fusion of strain of T-95 and T-12 of *T. harzianum.* Biworks, Geneva, NY, developed this strain in powder form called plant shield and granular form called root shield. *Trichoderma harzianum* recognizes *Botrytis cinerea* hyphae, and secretes its defense enzyme such as peroxidase, chitinase, and glucanase against Botrytis spp. This in turn activates the jasmonic acid and salicylic acid signaling, which are essential to break the cell wall of necrotropic pathogens like *Botrytis cinerea. T. harzianum* has been extremely effective mycofungicide due to their rapid growth, strong capacity for production, and wide range of inhibition of fungal disease, can alter rhizospheres, can endure adverse conditions, and encourage plant growth. It can also control *Fusarium* spp., *Phytophthora* spp., and *Rhizoctonia* spp. It is mostly used as bio-fungicides and biofertilizers (Kaewchai et al., 2009).

The *Fusarium* genus comprises both disease-causing and beneficial species. Among them, non-pathogenic strains, especially certain types of *Fusarium oxysporum*, have been shown to support plant health by suppressing soil-borne diseases like Fusarium wilt. These beneficial strains exert their effects through multiple mechanisms, including competition with pathogens and activation of the plant's defense systems. Non-pathogenic *F. oxysporum* produces enzymes such as chitinases and proteases, which help break down the cell walls of pathogenic strains. Additionally, it synthesizes both volatile organic compounds (VOCs) and non-volatile secondary metabolites that interfere with the growth and spore germination of harmful fungi. By forming a protective biofilm around plant roots, *F. oxysporum* also serves as a physical barrier, preventing the invasion of pathogenic strains. However, despite its potential, the commercial application of these strains remains limited due to insufficient understanding of their ecological behavior and genetic makeup.

One notable example is the strain Fo47, which is available as a liquid formulation under the trade name Fusaclean, developed by Natural Plant Products, France. Fo47 contributes to disease reduction by efficiently colonizing the root zone, competing with harmful *Fusarium* strains, and stimulating the plant's natural defense responses. This strain not only inhibits the growth of pathogens but also reinforces root protection. Moreover integrating *F. oxysporum* with other biocontrol organisms such as *Trichoderma* and *Pseudomonas* can further enhance disease suppression, offering a synergistic approach to sustainable crop protection (Rodrigo et al., 2022).

Gliocladium catenulatum infects *Sporidesmium sclerotiorum* and *Fusarium* spp. It eradicates the fungal host through direct hyphal contact and generates pseudo appressoria. It feeds on dead and decaying matter. The pesticide is made up of *G. catenulatum* strain JI446 into a wettable powder named Primastop by Kimra Agro Oy, Finland. *G. catenulatum* is also called *Clonostachys catenulata.* This solution can be utilized on all plant components, including leaves, soil, and roots to mitigate disease caused by *Pythium ultimum* and *Rhizoctonia solani*, primarily affecting greenhouse crops (Whipps, 1991).

3.3 Insecticides

Mycoinsecticides are generated from fungi, and it serves as microbial insecticides against agricultural pests, while preserving beneficial organisms. Upon, contact with an insect's body, the fungal spores in the plant would germinate and penetrate the cuticle. As the fungus develops within the host, it releases toxic compounds that ultimately lead to the insect's death. Additionally, some fungal strains emit compounds that can harm or even kill the insect. Fungi-based biopesticides that are available commercially are from *Beauveria, Metarhizium, Isaria,* and *Lecanicillium* species. Most widely used mycoinsecticides are *Beauveria bassiana* and *Metarhizium anisopliae* (de Faria & Wraight, 2007).

3.3.1 Mechanism of Action of Mycoinsecticides Against Agricultural Pests

The majority of Mycoinsecticides operate in five primary steps:

1. Spores are generated from the previously infected species. Alternatively, the spores can be purchased from retailers.

Table 3 Characteristic comparison between *Beauveria bassiana* and *Metarhizium anisopliae* (Biryol et al., 2022)

Characteristics	*Beauveria bassiana*	*Metarhizium anisopliae*
Target insect range	Soft bodied insects like beetles and flies	Hard exoskeleton insects like grasshoppers and weavers
Adaptability	Sensitive to heat and UV radiation, prefers cool environment	Heat tolerant
Disease caused	White muscardine disease	Fungal keratitis
Toxin produced	Beauverican	Destruxins
Application	Applied as foliar spray to control above-ground insects	Applied as soil insecticides to target dwelling insects
Life span	Short lived in soil	Long lived in soil
Other benefits	Endophyte	Promotes plant growth and nematode control

2. The spores meet the host organism through natural elements like wind, rain, or insect movement.
3. Spores enter the insect and germinate in the host body.
4. After the fungus has produced toxins and utilized the nutrients within the insect's cadaver, it transitions to a reproductive phase, emerging from the host's mummified body to release new spores into the environment.
5. Fungus emerges from the insect and thereby exposes it to the environment, and the cycle repeats.

This biocontrol agent targets pests like beetles, aphids, and caterpillars (see Table 3).

3.3.2 Field Application and Environmental Impact of Mycoinsecticides

In conclusion, *Beauveria bassiana* and *Metarhizium anisopliae* play vital roles in sustainable pest management by offering eco-friendly alternatives for crop management. Utilizing these entomopathogenic fungi, agriculture decreases chemical usage, mitigates environmental effects, and provides sustainable pest management for a healthy ecosystem.

3.4 Environmental Safety and Its Effect on Non-targeted Weeds

Pesticides derived from fungal metabolites, including herbicides, fungicides, and insecticides, reduce the need for synthetic pesticides. They have biodegradable properties, leaving no long-term residue. They are pathogen-specific and hence do not damage beneficial microbes. Fungal metabolites have a specific mechanism of action that promotes infection resistance. Using biopesticides can help agriculture, despite the growing need for vital minerals for plant development. These biocontrol agents can improve the ecological role of non-target plants, enhancing nutrient cycles and biodiversity by reducing invasive weed dominance, reducing insect pressure, and managing diseases. Each mycobiocontrol agent has individual benefits and limitations, but their combined potential emphasizes the need to incorporate them into crop pest and disease management systems.

While herbicides, fungicides, and insecticides serve as effective biocontrol agents, there are several environmental safety risks associated with their application:

- Persistence: Certain pesticides are recognized for their prolonged environmental presence, leading to lengthy exposure that may negatively impact non-target species.
- Bioaccumulation: A few pesticides are sequestered in the tissues of various species, such as weeds, resulting in an accumulation of concentration within food chains.
- Off-target movement: Pesticides can migrate to non-target locations and creatures by drift, runoff, or leaching (dela Cruz et al., 2022).

3.4.1 Effects on Non-target Weeds

- Direct poisoning: Pesticides directly kill or injure non-target weeds. This eliminates the weed species and alters biodiversity. This can alter the ecological balance and may affect ecosystems with associated services.
- Indirect Effects: Pesticides affect non-target weeds indirectly when the non-target weeds competitors, predators, or pollinators are affected. For instance, the application of insecticides can harm some pollinators that many non-target weeds rely upon for reproduction.
- Development of resistance: Overuse of pesticides leads to the development of resistance in the non-target weed species, which complicates the management and increases dependence on more dangerous chemical alternatives (Livingston, 2005).

3.4.2 Mitigating the Effects

- Integrated pest management (IPM) is a sustainable approach to pest control that focuses on the use of a variety of methods, including cultural, biological, and chemical controls, with the goal of minimizing pesticide use.
- Selective herbicides are the application of herbicides that target specific weed species to reduce the negative impact on non-target weeds.

- Buffer strips: The use of buffer strips around agricultural fields could efficiently reduce off-target drifts of pesticides.
- Supervision and evaluation: Continued monitoring of the pesticides applied and their effects on the weeds can determine possible problems, which can help inform the management decisions (Sánchez-Bayo, 2021).

The recent research indicates that the widespread use of herbicides in agricultural systems has significantly reduced the variety of non-target bee species in North America (Boutin et al., 2014). According to another research, fungicides may damage non-target bee species by upsetting their symbiotic connections with beneficial fungi (Schaeffer et al., 2016). Herbicides, fungicides, and insecticides can cause unintended impacts on non-target weeds in the form of reduced abundance, diversity, and ecosystem services.

3.5 Enhancing Crop Productivity Using Fungal Metabolites

The rising need for food crops to meet consumer demand is a serious concern. At present, production of agricultural materials suffers a heavy loss due to the attack of the crops by various plant pathogens, insect pests, and various abiotic stresses. And this situation necessitates a proper strategy that promotes sustainable agriculture. One of the various methods to increase agricultural productivity is the use of fungal secondary metabolites, which are intended to improve the efficient growth of food crops. Some fungal metabolites have shown potential for enhancing crop yield, particularly those from *Trichoderma* species. These metabolites enhance root development, stimulate plant hormone synthesis, and enhance nutrient uptake, thereby aiding in plant growth. Many fungal metabolites also have antibacterial properties that protect plants from pathogen-induced diseases (Segundo et al., 2022). For example, the metabolite harzianic acid produced by *T. harzianum* has been shown to induce seed germination and inhibit the growth of specific plant diseases (Vinale et al., 2013). Moreover, fungal metabolites may enhance plant resistance to abiotic stressors such as salinity and drought (Dufossé et al., 2021). The application of fungal metabolites for maximizing agricultural productivity and ensuring food security may be an environmental-friendly and sustainable approach (Tiwari & Park, 2024).

3.5.1 Role of Mycorrhizal Fungi and Secondary Metabolites

Plant root symbioses with fungi occur in several different forms and are referred to as mycorrhiza (from the Greek “mycos”, meaning fungus and “rhiza”, meaning root). In case of ectomycorrhiza, which is predominant on trees in temperate forests, the fungal partner remains outside of plant cells, whereas in endomycorrhiza, including orchid, ericoid, and arbuscular mycorrhiza (AM), the fungal hyphae penetrates inside the plant cells. AM is the most widespread terrestrial symbiosis and is formed by 70–90% of land and agricultural plant species with fungi that belong to a monophyletic phylum, the Glomeromycota (Parniske, 2008). In mycorrhizal interaction, the fungi receive carbon from the host plant and enhance the ability of the host plant to absorb nitrogen, phosphorus, water, and other nutrients. From current research, it was found that the mycorrhizal fungal species release lipo-chitooligosaccharide signal molecules, known as Myc-LCOs, which can be perceived by various receptors of plants and activate symbiosis (De Mita et al., 2014). Strigolactones are a family of plant hormones with a range of configurations. This hormone, secreted by the plant, enhances the effectiveness of symbiotic association by stimulating the branching of hyphae in mycorrhizae (Besserer et al., 2006). The formation of a unique cytoplasmic arrangement known as the pre-penetration apparatus facilitates the very first fungal penetration of the host root by directing mycorrhizal hyphae through the epidermis (Genre et al., 2005). Then hyphae would eventually expand to the root cortical cells and undergo profuse branching and emerge as arbuscules, and thus the fungal colonization takes place intercellularly. The arbuscules are separated from the cytoplasm by the peri-arbuscular membrane. The branching of the arbuscules creates a very large interface for efficient nutrient uptake (Wang et al., 2017).

Mycorrhizal Symbiosis in Nutrient Uptake

The nutrient uptake can take place through direct or the plant pathway, which involves uptake of nutrients through root hairs and epidermis. The other pathway, the mycorrhizal pathway, is an indirect mechanism of nutrient uptake from the external hyphae to the root cortical cells (as represented in Fig. 3). In the direct pathway, the direct absorption of nutrients like phosphorus (P) and nitrogen (N) takes place in the region where contact between the root and soil is established. Whereas in mycorrhizal pathway, the fungal extraradical mycelium (ERM) facilitates the uptake through the P, NO^{3-}, or NH^{4+} transporters. In the anabolic phase of the urea cycle, the conversion of “N” into arginine and in the ERM the conversion P into polyphosphate takes place. After that, the polyphosphate granules move from ERM to intraradical hyphae (IRM). In the IRM, arginine and P are released when the polyphosphate granules hydrolyze in the catabolic phase of the urea cycle conversion of arginine to NH^{4+}. This process is aided by P, NH^{4+}, and other amino acids efflux through the fungal plasma membrane into the interfacial apoplast. Using mycorrhiza-inducible P and NH^{4+} transporters, absorption of the nutri-

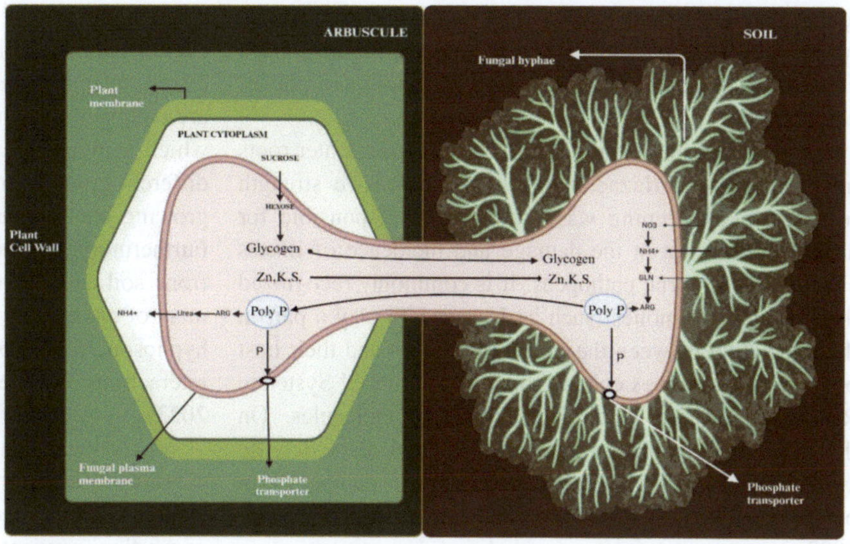

Fig. 3 Representation of nutrient uptake in plants during mycorrhizal symbiosis. *Source* Authors

ents by the plant from the mycorrhizal interface takes place. The enhanced nutrient supply promotes photosynthesis in the plant, which is made possible by the efflux of sucrose from the plant cell into the interfacial apoplast through the plant plasma membrane, the hydrolysis of sucrose in the interfacial apoplast mediated by apoplastic plant invertase, and the hydrolyzed hexoses are taken up by the mycorrhizae via various monosaccharide transporters (Huey et al., 2020).

Transport of Phosphorus

Mycorrhizal fungus increases the absorbing surface area and mobilizes sparingly accessible phosphorus, which helps plants to absorb phosphorus from the soil. Compared to the non-mycorrhizal controls, the mycorrhizal plants showed a greater phosphorus influx. About 70% of the total phosphorus received by mycorrhiza-colonized rice is supplied by the symbiotic fungus. The phosphorus uptake in plants associated with mycorrhizae is dependent on the distribution of hyphae in the soil. The phosphorus uptake in mycorrhiza-colonized plant takes place by suppressing or by down-regulating the originally available phosphate transporters in the plant by the mycorrhizae. Some of the phosphate transporter genes such as *OsPT2* and *OsPT6*, including the genes coding for phosphorus starvation response *IPS1* and *miR399* in the rice plant were found to be inhibited (non-functional) by the mycorrhizal P transporters. Mycorrhizal P transporters are located in the external hyphae of arbuscular mycorrhizal fungi, where they mediate phosphate absorption from the soil. Examples include GvPT from *Glomus versiforme* and GiPT from *Glomus intraradices* (now *Rhizophagus irregularis*) (Wang et al., 2017; Yang et al., 2012).

Transport of Nitrogen

Through the glutamine synthetase/gluamine oxoglutarate aminotransferase cycle, the inorganic nitrogen absorbed by the fungal ERM is firstly incorporated as a range amino acids and ultimately get transformed into arginine. The main amino acid in the ERM is arginine and it is transported by the fungal hyphae from the ERM to the IRM, where it is converted to urea and ornithine. Urea hydrolyzation produces ammonia, which is then released at the site of symbiotic contact and then it is absorbed by the plant (Govindarajulu et al., 2005). The ammonium transporters *GintAMT1, GintAMT2,* and *GintAMT3* of *Glomus intraradices* (also known as *Rhizophagus irregularis*), can be induced even at low concentration of NH^{4+}in the soil. The ammonium released from the mycorrhizal transporters at mycorrhizal-plant interface is taken up by the plants via various plant transporters. For instance, *GmAMT4* and *ATM2* are two genes coding for the transporters, in soybean they are usually expressed in cortical cells that contain arbuscules (Calabrese et al., 2016; Kobae et al., 2010).

Influence of Mycorrhizal Secondary Metabolites on the Host Plant

In general, mycorrhizal fungi and crop roots have a symbiotic connection in which the fungus exchange nutrients, water, and carbon for the crop's nutrition. Furthermore, the fungi generate a variety of secondary that have antibacterial and antifungal properties, which can help shield the plant against illnesses brought on by microorganisms. By creating a chemical defense, the addition of mycorrhizal fungi to *Ageratina adenophora* increased the invasiveness of the weed and increase its resistance to *Aphis gossypii* herbivory. The generalist herbivore *A. gosypii's* consumption was reduced by the colonization of the two major mycorrhizal species, *Septoglomus constrictum* and *Claroideoglomus etunicatum*, by increasing the plant's flavonoid and phenol content (Du et al., 2022).

There is minimal to no room for pathogens to enter roots infected by mycorrhizae. Interestingly, defensive strength from the plant priming was shown to be responsible for a general decrease in the damage and incidence of illness caused by soil-borne pathogens. It is commonly recognized that the phytohormones (such as JA and ET) play a part in the interaction between the symbiotic fungus and their host plant. Phytohormones enhance host plant Induced Systemic Resistance (ISR) by acting as signaling molecules. On the other hand, primed jasmonate-regulated plant defense systems limit the growth of necrotrophic pathogens in plant–fungal pathogen interaction signals (Elhamouly et al., 2022).

The mutualistic relationships between the legume plant soybean (*Glycine max*) and an arbuscular mycorrhizal fungus, which releases a range of metabolites into the soil, including saponins, have been studied. For soybean-arbuscular mycorrhizal fungal symbioses, flavanones (like strigolactones) are signal molecules that can enhance colonization and growth of ectomycorrhizal fungi. Furthermore, it has been demonstrated that flavanones promote the germination of ectomycorrhizal fungal spores in the genera *Pisolithus* and *Suillus* trigger the production of symbiotic effector protein in the Laccaria bicolor mushroom (Pei et al., 2020; Sugiyama, 2019).

Influence on Soil Microbiome and Plant Growth

In addition to supporting a range of biochemical reactions and essential ecological processes (including the solubilization of organic materials and the biogeochemical cycles of the elements), soil is thought as a biological reactor that is always in motion. The soil becomes more fertile due to microbial activity, which includes a variety of interactions like competition, parasitism, and cooperative microorganisms. The extraradical mycelium mycorrhizal fungi not only increases the nutrient absorptive surface area of host plant root systems for nutrient uptake, but it can also offer a direct route for the transport of carbon produced from photosynthetic processes to soil microenvironments. Important niches for bacterial colonization and development are created by the mycelium's enormous surface area interacting with the surrounding soil environment (hyphosphere) and the constant supply of energy-rich substances. Mycorrhizal fungi and other microbes interact more directly and reciprocally if live hyphae produce sugars (Toljander et al., 2007). It was observed that mycorrhizae and the bacterial (i.e., rhizobia) and fungal populations have a synergistic interaction that is dependent on the levels of soil's N and P (Li et al., 2023). Mycorrhizal association protected the colonized plants against various fungal and nematode pathogens (Veresoglou & Rillig, 2012). Mycorrhizal fungi (*Glomus intraradices*) also act against the fungi and bacteria which are pathogenic to the plants. It was found that the plants in which a mycorrhizal association was established hosted a different group of microorganisms as compared to normally growing plants (Hartman et al., 2023; Zhang et al., 2024). Furthermore, the recent research found that N_2O emissions from soil are decreased by mycorrhiza-mediated recruitment of completely denitrifying *Pseudomonas* bacteria. The hyphosphere microbiome is significantly shaped by the interactions between bacteria and mycorrhiza (Li et al., 2023).

Mycorrhizal fungi play an important role in ensuring sustainable agriculture. In the circumstances where some elements are scarce, mycorrhizae play a critical role in converting nutrients into forms that may be used. The extraradical mycelium of mycorrhizal fungi has improved soil quality and structural stability. It has been demonstrated that this promotes the development of plants. Since mycorrhizal fungal growth improves soil quality and structural stability, it is playing a bigger role in sustainable agriculture operations. By lowering the need for chemical inputs like fertilizers and pesticides, the use of these mycorrhizal connections would effectively increase production (Díaz-Urbano et al., 2023; Janowski & Leski, 2022). The plant's defense mechanism is strengthened by mycorrhizae, which makes the plant exceptionally capable of defending itself against diseases caused by microorganisms and pests. Apart from giving the plant nourishment, fungi also enhance its growth and development of the plant more quickly and efficiently. Fungi produce antibiotics, which aid in shielding the plant against particular diseases (Edlinger et al., 2022). Additionally, they make substances that attach to hazardous chemicals in the soil and lessen their toxicity, enhancing the fertility of the soil and thereby promoting the growth of the plant (Fall et al., 2022).

3.5.2 Applications in Organic Farming and Sustainable Agriculture

Modern agriculture must intensify crop output to meet the world's teeming population's high food demand, which entails the widespread use of synthetic fertilizers to increase crop yield. Despite their higher nutritional content and capacity to accelerate crop growth, chemical fertilizers have been found to be harmful to human health and the environment. They also pollute the atmosphere and groundwater in the future. Currently, biofertilizers have emerged as an effective alternative because of their advantages in terms of being economical, environmentally benign, and simple to use in agricultural settings. A mixture of several microorganisms known as biofertilizers can promote soil health and the development of the plant even in stressful conditions. Microbial inoculants containing live or dormant cells of efficacious N-fixing, P-solubilizing/mobilizing, and K-solubilizing strains are known as biofertilizers. This may include AMFs, rhizobacteria, and other beneficial microbial consortia that promote plant development (Fasusi et al., 2021).

Use of Biofertilizers and Soil Conditioners

AMFs biofertilizers are essential for the mobilization of nutrients in soil-plant systems. These biofertilizers greatly improvises the nitrogen phosphorus usage efficiency, and they are the most promising biofertilizers for increasing crop yields. With the goal of sustainable agriculture, it is becoming more and more important to employ AMF as a biofertilizer since proper management of these symbiotic fungi can reduce the need for agrochemicals. The primary method used to do this is the inoculation of a target soil with AMF propagules, or inoculum. Since AMFs are obligate symbionts, they cannot grow apart from their host plants in pure culture. Because of this limitation, producing AMF inoculum on a large scale is extremely difficult and complicated. AMF inoculum can be prepared through three primary strategies. First, as soil from a plant's root zone typically contains colonized root pieces, AMF spores, and hyphae, it may be employed as an inoculum. However, soil inoculum can be inaccurate and pose the danger of spreading diseases and weed seeds if exact information on the variety, quantity, and infectivity of the propagules is not available. Instead, soil-extracted spores can be utilized as precursors to produce crude inoculum. A host trap plant, which is a plant that can be extensively colonized by several AMF species, and a known isolate of AMF can be cultivated together in an inert medium that is tailored for AMF proliferation to yield a crude inoculum. This is the most often utilized inoculum type for large-scale crop inoculation because it typically contains a more concentrated collection of the same propagules as the one present in soil inoculum. Ultimately, a known AMF host's infected root fragments isolated from a trap plant culture can likewise be used as an inoculum source. Since large-scale inoculation is not technically and economically feasible all the time, alternative inoculation can be done on a small scale, which would establish a good mycorrhizal network and will also reduce the cost. In contrast to traditional fertilization, AMF inoculation can be financially advantageous, offering significant cost savings to producers and degraded land regeneration initiatives (Berruti et al., 2016; IJdo et al., 2011; Säle et al., 2015).

Fungal Metabolites as Natural Growth Promoters

Fungal metabolites are increasingly being recognized as natural plant growth stimulants, especially those generated by *Trichoderma* and *Penicillium* species. These helpful fungi produce a wide range of substances that significantly enhance the growth and well-being of plants. Much research has focused on the ability of *Trichoderma* species to produce phytohormones like auxins and gibberellins that promote cell division and elongation and improve root and shoot development. Besides this, it can solubilize inorganic phosphate. By doing so, its concentration in the soil is augmented and, thus, enhanced availability for the plants to absorb (Harman et al., 2004). *Penicillium* species are also used for their ability to manufacture penicillin. Their other metabolites can provoke plant growth. These organisms produce some plant hormones of the gibberellin type that induce seed germination and stem elongation. In addition, some metabolites of *Penicillium* can solubilize inorganic phosphate, making more nutrients available to the plants (Hedden & Phillips, 2000). It is environmentally friendly and renewable to boost agricultural output using metabolites of the fungi *Trichoderma* and *Penicillium* species. Better plant development and enhanced crop yields can be achieved with the inherent potential of these helpful fungi.

4 Conclusion

The use of fungal secondary metabolites is crucial in the pursuit of potentially sustainable agriculture. It provides an environment-friendly alternative method when compared to chemical pesticides to control pests on crops. These applications shall address the critical problems identified concerning soil degradation and decline in crop yields. For example, the metabolites in *Trichoderma*, *Beauveria bassiana*, *Metarhizium anisopliae*, and *Gliocladium catenulatum*, among others, have various useful means of action, which include parasitism, enzymatic degradation, production of toxins, and competition for key nutrients that inhibit the pathogens but promote growth and strengthen the resistance against biotic and abiotic stresses. In addition, mycorrhizal fungi significantly contribute to agricultural sustainability by maximizing nutrient uptake, improving soil structure, and promoting microbial diversity, thus setting up strong agricultural systems. Applications of fungal secondary metabolites range from biofertilizers to mycoherbicides and mycoinsecticides. These include enhancing growth in *Trichoderma* species that colonize the roots and produce enzymes for plant growth and induction of systemic resistance; *Beauveria bassiana* and *Metarhizium anisopliae* penetrate the insect cuticle to kill their hosts, thereby releasing toxins inside the host. *Clonostachys rosea* is both an endophyte and a biocontrol agent that provides several benefits for the host plant's health and suppression of disease. Mycorrhizal association enhances phosphorus and nitrogen uptake, boosts mechanisms of plant defense, and possesses resistance against environmental stresses like drought and salinity. With this integration of these fungal biocontrol agents into organic agriculture and other sustainable practices, it will be able to bring down the reliance of the agricultural sector on harmful chemicals, save biodiversity, and also ensure food security. Therefore, fungi's secondary metabolites are a very important tool in developing robust agroecosystems and achieving long-term sustainability economically and environmentally.

References

Abdel Ghany, T. M. (Ed.). (2015). *Entomopathogenic fungi and their role in biological control.* OMICS International. https://doi.org/10.4172/978-1-63278-065-2-66

Aktar, W., Sengupta, D., & Chowdhury, A. (2009). Impact of pesticides use in agriculture: Their benefits and hazards. *Interdisciplinary Toxicology, 2*(1), 1–12. https://doi.org/10.2478/v10102-009-0001-7

Anwar, W., Ali, S., Nawaz, K., Iftikhar, S., Javed, M. A., Hashem, A., Alqarawi, A. A., Abd Allah, E. F., & Akhter, A. (2018). Entomopathogenic fungus *Clonostachys rosea* as a biocontrol agent against whitefly (*Bemisia tabaci*). *Biocontrol Science and Technology, 28*(8), 750–760. https://doi.org/10.1080/09583157.2018.1487030

Ayaz, M., Li, C.-H., Ali, Q., Zhao, W., Chi, Y.-K., Shafiq, M., Ali, F., Yu, X.-Y., Yu, Q., Zhao, J.-T., Yu, J.-W., Qi, R.-D., & Huang, W.-K. (2023). Bacterial and fungal biocontrol agents for plant disease protection: Journey from lab to field, current status, challenges, and global perspectives. *Molecules, 28*(18), 6735. https://doi.org/10.3390/molecules28186735

Bamisile, B. S., Akutse, K. S., Siddiqui, J. A., & Xu, Y. (2021). Model application of entomopathogenic fungi as alternatives to chemical pesticides: Prospects, challenges, and insights for next-generation sustainable agriculture. *Frontiers in Plant Science, 12.* https://doi.org/10.3389/fpls.2021.741804

Batool, R., Umer, M. J., Wang, Y., He, K., Zhang, T., Bai, S., Zhi, Y., Chen, J., & Wang, Z. (2020). Synergistic effect of *Beauveria bassiana* and *Trichoderma asperellum* to induce maize (*Zea mays* L.) defense against the Asian corn borer, *Ostrinia furnacalis* (*Lepidoptera, Crambidae*) and larval immune response. *International Journal of Molecular Sciences, 21*(21), 8215.

Berruti, A., Lumini, E., Balestrini, R., & Bianciotto, V. (2016). Arbuscular mycorrhizal fungi as natural biofertilizers: Let's benefit from past successes. *Frontiers in Microbiology, 6*, 1559. https://doi.org/10.3389/FMICB.2015.01559

Besserer, A., Puech-Pagès, V., Kiefer, P., Gomez-Roldan, V., Jauneau, A., Roy, S., Portais, J. C., Roux, C., Bécard, G., & Séjalon-Delmas, N. (2006). Strigolactones stimulate Arbuscular mycorrhizal fungi by activating mitochondria. *PLOS Biology, 4*(7), e226. https://doi.org/10.1371/JOURNAL.PBIO.0040226

Bills, G. F., & Gloer, J. B. (2016). Biologically active secondary metabolites from the fungi. *Microbiology Spectrum, 4*(6). https://doi.org/10.1128/MICROBIOLSPEC.FUNK-0009-2016/ASSET/CC0BD2D9-5182-4DE6-9148-7DF9A9ECAB7E/ASSETS/GRAPHIC/FUNK-0009-2016-FIG6.GIF

Biryol, S., Demirbağ, Z., Erdoğan, P., & Demir, I. (2022). Development of Beauveria bassiana (Ascomycota: Hypocreales) as a mycoinsecticide to control green peach aphid, Myzus persicae (Homoptera: Aphididae) and investigation of its biocontrol potential. *Journal of Asia-Pacific Entomology, 25*(1). https://doi.org/10.1016/j.aspen.2022.101878

Boutin, C., Strandberg, B., Carpenter, D., Mathiassen, S. K., & Thomas, P. J. (2014). Herbicide impact on non-target plant reproduction: What are the toxicological and ecological implications? *Environmental Pollution, 185*, 295–306. https://doi.org/10.1016/j.envpol.2013.10.009

Boyetchko, S. M., Peng, G., Canada, A.-F., & Saskatchewan, C. (2004). Challenges and strategies for development of mycoherbicides agriculture and agri-food Canada challenges and strategies for development of mycoherbicides.

Butt, T. M., & Copping, L. G. (2000). Fungal biological control agents. *Pesticide Outlook, 11*(5), 186–191. https://doi.org/10.1039/b008009h

Calabrese, S., Pérez-Tienda, J., Ellerbeck, M., Arnould, C., Chatagnier, O., Boller, T., Schüßler, A., Brachmann, A., Wipf, D., Ferrol, N., & Courty, P. E. (2016). GintAMT3-a low-affinity ammonium transporter of the arbuscular mycorrhizal rhizophagus irregularis. *Frontiers in Plant Science, 7*, 182160. https://doi.org/10.3389/FPLS.2016.00679/BIBTEX

Chakraborty, A., & Ray, P. (2021). Mycoherbicides for the noxious meddlesome: Can colletotrichum be a budding candidate? In *Frontiers in microbiology* (Vol. 12). Frontiers Media S.A. https://doi.org/10.3389/fmicb.2021.754048

Dauda, Z., Maina, U. M., Um, M., Zakaria, D., Galadima, I. B., & Gambo, F. M. (2018). A review on the use of entomopathogenic fungi in the management of insect pests of field crops. *Article in Journal of Entomology and Zoology Studies, 6*(1). https://www.researchgate.net/publication/323445876

de Faria, M. R., & Wraight, S. P. (2007). Mycoinsecticides and mycoacaricides: A comprehensive list with worldwide coverage and international classification of formulation types. *Biological Control, 43*(3), 237–256. https://doi.org/10.1016/j.biocontrol.2007.08.001

De Mita, S., Streng, A., Bisseling, T., & Geurts, R. (2014). Evolution of a symbiotic receptor through gene duplications in the legume–rhizobium mutualism. *New Phytologist, 201*(3), 961–972. https://doi.org/10.1111/NPH.12549

dela Cruz, J. A., Camenzind, T., & Rillig, M. C. (Eds.). (2022). Sublethal fungicide concentrations both reduce and stimulate the growth rate of non-target soil fungi from a natural grassland. *Frontiers in Environmental Science, 10.* https://doi.org/10.3389/fenvs.2022.1020465

Demain, A. L., & Fang, A. (2000). The natural functions of secondary metabolites. *Advances in Biochemical Engineering/Biotechnology, 69*, 1–39. https://doi.org/10.1007/3-540-44964-7_1

Díaz-Urbano, M., Goicoechea, N., Velasco, P., & Poveda, J. (2023). Development of agricultural bio-inoculants based on mycorrhizal fungi and endophytic filamentous fungi: Co-inoculants for improve plant-physiological responses in sustainable agriculture. *Biological Control, 182*, 105223. https://doi.org/10.1016/J.BIOCONTROL.2023.105223

Du, E., Chen, Y., Li, Y., Zhang, F., Sun, Z., Hao, R., & Gui, F. (2022). Effect of arbuscular mycorrhizal fungi on the responses of *Ageratina adenophora* to *Aphis gossypii* herbivory. *Frontiers in Plant Science, 13*, 1015947. https://doi.org/10.3389/FPLS.2022.1015947/BIBTEX

Dufossé, L., Fouillaud, M., & Caro, Y. (2021). Fungi and fungal metabolites for the improvement of human and animal nutrition and health. *Journal of Fungi, 7*(4), 274. https://doi.org/10.3390/JOF7040274

Edlinger, A., Garland, G., Hartman, K., Banerjee, S., Degrune, F., García-Palacios, P., Hallin, S., Valzano-Held, A., Herzog, C., Jansa, J., Kost, E., Maestre, F. T., Pescador, D. S., Philippot, L., Rillig, M. C., Romdhane, S., Saghaï, A., Spor, A., Frossard, E., & van der Heijden, M. G. A. (2022). Agricultural management and pesticide use reduce the functioning of beneficial plant symbionts. *Nature Ecology & Evolution, 6*(8), 1145–1154. https://doi.org/10.1038/s41559-022-01799-8

Elhamouly, N. A., Hewedy, O. A., Zaitoon, A., Miraples, A., Elshorbagy, O. T., Hussien, S., El-Tahan, A., & Peng, D. (2022).

The hidden power of secondary metabolites in plant-fungi interactions and sustainable phytoremediation. *Frontiers in Plant Science, 13*, 1044896. https://doi.org/10.3389/FPLS.2022.1044896

Evidente, A. (2023). Specialized metabolites produced by phytopathogen fungi to control weeds and parasite plants. *Microorganisms, 11*(4), 843. https://doi.org/10.3390/microorganisms11040843

Fall, A. F., Nakabonge, G., Ssekandi, J., Founoune-Mboup, H., Apori, S. O., Ndiaye, A., Badji, A., & Ngom, K. (2022). Roles of Arbuscular mycorrhizal fungi on soil fertility: Contribution in the improvement of physical, chemical, and biological properties of the soil. *Frontiers in Fungal Biology, 3*, 723892. https://doi.org/10.3389/FFUNB.2022.723892/BIBTEX

Fasusi, O. A., Cruz, C., & Babalola, O. O. (2021). Agricultural sustainability: Microbial biofertilizers in rhizosphere management. *Agriculture, 11*(2), 163. https://doi.org/10.3390/AGRICULTURE11020163

Ferreyra-Suarez, D., García-Depraect, O., & Castro-Muñoz, R. (2024). A review on fungal-based biopesticides and biofertilizers production. In *Ecotoxicology and environmental safety* (Vol. 283). Academic Press. https://doi.org/10.1016/j.ecoenv.2024.116945

Genre, A., Chabaud, M., Timmers, T., Bonfante, P., & Barker, D. G. (2005). Arbuscular mycorrhizal fungi elicit a novel intracellular apparatus in medicago truncatula root epidermal cells before infection. *The Plant Cell, 17*(12), 3489–3499. https://doi.org/10.1105/TPC.105.035410

Govindarajulu, M., Pfeffer, P. E., Jin, H., Abubaker, J., Douds, D. D., Allen, J. W., Bücking, H., Lammers, P. J., & Shachar-Hill, Y. (2005). Nitrogen transfer in the arbuscular mycorrhizal symbiosis. *Nature, 435*(7043), 819–823. https://doi.org/10.1038/nature03610

Harding, D. P., & Raizada, M. N. (2015). Controlling weeds with fungi, bacteria and viruses: A review. In *Frontiers in plant science* (Vol. 6). Frontiers Research Foundation. https://doi.org/10.3389/fpls.2015.00659

Harman, G. E., Howell, C. R., Viterbo, A., Chet, I., & Lorito, M. (2004). Trichoderma species—Opportunistic, avirulent plant symbionts. *Nature Reviews Microbiology, 2*(1), 43–56. https://doi.org/10.1038/NRMICRO797

Hartman, K., Schmid, M. W., Bodenhausen, N., Bender, S. F., Valzano-Held, A. Y., Schlaeppi, K., & van der Heijden, M. G. A. (2023). A symbiotic footprint in the plant root microbiome. *Environmental Microbiome, 18*(1), 1–16. https://doi.org/10.1186/S40793-023-00521-W

Hedden, P., & Phillips, A. L. (2000). Gibberellin metabolism: New insights revealed by the genes. *Trends in Plant Science, 5*(12), 523–530. https://doi.org/10.1016/S1360-1385(00)01790-8

Huey, C. J., Gopinath, S. C. B., Uda, M. N. A., Zulhaimi, H. I., Jaafar, M. N., Kasim, F. H., & Yaakub, A. R. W. (2020). Mycorrhiza: A natural resource assists plant growth under varied soil conditions. *3 Biotech, 10*(5), 204. https://doi.org/10.1007/S13205-020-02188-3

IJdo, M., Cranenbrouck, S., & Declerck, S. (2011). Methods for large-scale production of AM fungi: Past, present, and future. *Mycorrhiza, 21*(1), 1–16. https://doi.org/10.1007/S00572-010-0337-Z/METRICS

Janowski, D., & Leski, T. (2022). Factors in the distribution of mycorrhizal and soil fungi. *Diversity, 14*(12), 1122. https://doi.org/10.3390/D14121122

Kaewchai, S., Soytong, K., Hyde, K. D., Rai, C., Kaewchai, T., Soytong, S., & Hyde, K. (2009). Mycofungicides and fungal biofertilizers.

Keller, N. P. (2019). Fungal secondary metabolism: Regulation, function and drug discovery. *Nature Reviews. Microbiology, 17*(3), 167. https://doi.org/10.1038/S41579-018-0121-1

Kobae, Y., Tamura, Y., Takai, S., Banba, M., & Hata, S. (2010). Localized expression of Arbuscular mycorrhiza-inducible ammonium transporters in soybean. *Plant and Cell Physiology, 51*(9), 1411–1415. https://doi.org/10.1093/PCP/PCQ099

Li, X., Zhao, R., Li, D., Wang, G., Bei, S., Ju, X., An, R., Li, L., Kuyper, T. W., Christie, P., Bender, F. S., Veen, C., van der Heijden, M. G. A., van der Putten, W. H., Zhang, F., Butterbach-Bahl, K., & Zhang, J. (2023). Mycorrhiza-mediated recruitment of complete denitrifying Pseudomonas reduces N_2O emissions from soil. *Microbiome, 11*(1), 1–18. https://doi.org/10.1186/S40168-023-01466-5/FIGURES/8

Livingston, J. V. (2005). *Agriculture and soil pollution: New research*. Nova Science Publishers.

Mohd Ghazi, R., Nik Yusoff, N. R., Abdul Halim, N. S., Wahab, I. R. A., Ab Latif, N., Hasmoni, S. H., Ahmad Zaini, M. A., & Zakaria, Z. A. (2023). Health effects of herbicides and its current removal strategies. In *Bioengineered* (Vol. 14). Taylor and Francis Ltd. https://doi.org/10.1080/21655979.2023.2259526

Palmieri, D., Ianiri, G., Del Grosso, C., Barone, G., De Curtis, F., Castoria, R., & Lima, G. (2022). Advances and perspectives in the use of biocontrol agents against fungal plant diseases. *Horticulturae, 8*(7), 577. https://doi.org/10.3390/horticulturae8070577

Parniske, M. (2008). Arbuscular mycorrhiza: The mother of plant root endosymbioses. *Nature Reviews Microbiology, 6*(10), 763–775. https://doi.org/10.1038/nrmicro1987

Pei, Y., Siemann, E., Tian, B., & Plants, J. D.-A. (2020). Root flavonoids are related to enhanced AMF colonization of an invasive tree. Retrieved December 17, 2024, from. https://academic.oup.com/aobpla/article-abstract/12/1/plaa002/5700701

Rivera-Méndez, W., Obregón, M., Morán-Diez, M. E., Hermosa, R., & Monte, E. (2020). Trichoderma asperellum biocontrol activity and induction of systemic defenses against Sclerotium cepivorum in onion plants under tropical climate conditions. *Biological Control, 141*, 104145. https://doi.org/10.1016/j.biocontrol.2019.104145

Rodrigo, S., García-Latorre, C., & Santamaria, O. (2022). Metabolites produced by fungi against fungal phytopathogens: Review, implementation and perspectives. In *Plants* (Vol. 11, No. 1). MDPI. https://doi.org/10.3390/plants11010081

Rodríguez-Romero, H., Rodríguez-Peláez, L., Reyes-Castro, A., Tlakaele Notario-Rendón, O., González-Peréz, M., Reza-Salgado, J., Antonio Rodríguez-Salazar, P., & Antonio Salazar-Magallón, J. (2023). Perspective chapter: Secondary metabolites of entomopathogens as biotechnological tools for the biological control of agricultural insect pests. In *Insecticides—Advances in insect control and sustainable pest management*. IntechOpen. https://doi.org/10.5772/intechopen.111465

Säle, V., Aguilera, P., Laczko, E., Mäder, P., Berner, A., Zihlmann, U., van der Heijden, M. G. A., & Oehl, F. (2015). Impact of conservation tillage and organic farming on the diversity ofarbuscular mycorrhizal fungi. *Soil Biology and Biochemistry, 84*, 38–52. https://doi.org/10.1016/J.SOILBIO.2015.02.005

Sánchez-Bayo, F. (2021). Indirect effect of pesticides on insects and other arthropods. *Toxics, 9*(8), 177.

Schaeffer, R. N., Vannette, R. L., Brittain, C., Williams, N. M., & Fukami, T. (2016). Non-target effects of fungicides on nectar-inhabiting fungi of almond flowers. *Environmental Microbiology Reports, 9*(2), 79–84.

Segundo, W. O. P. F., de Oliveira, R. S., Lima, R. M., Santiago, P. A. L., de Oliveira, L. A., Cortez, A. C. A., Lima, E. S., de Souza, É. S., Frickmann, H., & de Souza, J. V. B. (2022). Antimicrobial potential of metabolites in fungal strains isolated from a polluted stream: Annulohypoxylon stygium WL1B5 produces metabolites against extended-spectrum beta-lactamase-positive Escherichia coli. *Antibiotics, 12*(1), 27. https://doi.org/10.3390/ANTIBIOTICS12010027

Singh, M., Sehrawat, N., Kumar, V., Atri, N., Singh, R., Kumar Upadhyay, S., Kumar, S., & Yadav, M. (2018). Mycoherbicide control strategy: Concept, constraints, and advancements. https://connectjournals.com/02196.2021.17.29

Sugiyama, A. (2019). The soybean rhizosphere: Metabolites, microbes, and beyond—A review. *Journal of Advanced Research, 19*, 67–73. https://doi.org/10.1016/J.JARE.2019.03.005

TeBeest, D. O., & Templeton, G. E. (1985). Mycoherbicides: Progress in the biologic. *Plant Disease, 69*(1), 7.

Thambugala, K. M., Daranagama, D. A., Phillips, A. J. L., Kannangara, S. D., & Promputtha, I. (2020). Fungi vs. fungi in biocontrol: An overview of fungal antagonists applied against fungal plant pathogens. *Frontiers in Cellular and Infection Microbiology, 10*. https://doi.org/10.3389/fcimb.2020.604923

Tiwari, P., & Park, K.-I. (2024). Advanced fungal biotechnologies in accomplishing sustainable development goals (SDGs): What do we know and what comes next? *Journal of Fungi, 10*(7), 506. https://doi.org/10.3390/JOF10070506

Toljander, J. F., Lindahl, B. D., Paul, L. R., Elfstrand, M., & Finlay, R. D. (2007). Influence of Arbuscular mycorrhizal mycelial exudates on soil bacterial growth and community structure. *FEMS Microbiology Ecology, 61*(2), 295–304. https://doi.org/10.1111/J.1574-6941.2007.00337.X

Veresoglou, S. D., & Rillig, M. C. (2012). Suppression of fungal and nematode plant pathogens through Arbuscular mycorrhizal fungi. *Biology Letters, 8*(2), 214–217. https://doi.org/10.1098/RSBL.2011.0874

Vinale, F., Nigro, M., Sivasithamparam, K., Flematti, G., Ghisalberti, E. L., Ruocco, M., Varlese, R., Marra, R., Lanzuise, S., Eid, A., Woo, S. L., & Lorito, M. (2013). Harzianic acid: A novel siderophore from Trichoderma harzianum. *FEMS Microbiology Letters, 347*(2), 123–129. https://doi.org/10.1111/1574-6968.12231

Wang, W., Shi, J., Xie, Q., Jiang, Y., Yu, N., & Wang, E. (2017). Nutrient exchange and regulation in Arbuscular mycorrhizal symbiosis. *Molecular Plant, 10*(9), 1147–1158. https://doi.org/10.1016/J.MOLP.2017.07.012

Whipps, J. M. (1991). Effects of mycoparasites on sclerotia-forming fungi. In *Developments in agricultural and managed forest ecology* (Vol. 23, pp. 129–140). Elsevier.

Yang, S. Y., Grønlund, M., Jakobsen, I., Grotemeyer, M. S., Rentsch, D., Miyao, A., Hirochika, H., Kumar, C. S., Sundaresan, V., Salamin, N., Catausan, S., Mattes, N., Heuer, S., & Paszkowski, U. (2012). Nonredundant regulation of rice arbuscular mycorrhizal symbiosis by two members of the phosphate transporter1 gene family. *The Plant Cell, 24*(10), 4236–4251. https://doi.org/10.1105/TPC.112.104901

Zhang, C., van der Heijden, M. G. A., Dodds, B. K., Nguyen, T. B., Spooren, J., Valzano-Held, A., Cosme, M., & Berendsen, R. L. (2024). A tripartite bacterial-fungal-plant symbiosis in the mycorrhiza-shaped microbiome drives plant growth and mycorrhization. *Microbiome, 12*(1), 1–21. https://doi.org/10.1186/S40168-023-01726-4/FIGURES/8

Environmental Applications of Fungal Secondary Metabolites

Shivaji Misra, Anmol Adhikari, and Anupam Prakash

Abstract

Fungal secondary metabolites are low molecular weight (or bioactive) compounds synthesized by fungi not as a component of the more basic metabolic processes, but possibly with diverse bioactivities. They have diverse biological activity, including antioxidants, antimicrobial activity, anti-anti-inflammatory activity, and are used in many spheres, such as agriculture, cleaning the environment, medicine, human health, and so on. These metabolites are extremely crucial to the ecology of the fungi, like their self-defense mechanisms against other organisms, their food, and their association with plants. They are significant in environmental management such as the phytoremediation of the environment, plant growth, resistance to diseases, minimizing the utilization of synthetic pesticides. The fungus can be more widely used in improving the soil health, plant growth, and biocontrol agent in agriculture through the secondary metabolites produced by the useful fungi. In addition to this, fungal secondary metabolites play an important role in environmental control in the breakdown of toxic substances including organic xenobiotics and heavy metals. The chapter has provided an introduction on different environmental applications of fungal secondary metabolites that comprise bioremediation, biocontrol, generation of bio-based products, ecosystem restoration, better soil quality, plant growth, and sustainable waste management, etc.

Keywords

Bioremediation · Ecosystem recovery · Peroxidases · Laccases · Biodegradation · Phytoremediation · Nutrient cycling · Waste management · Electronic-waste · Green recycling · Arbuscular Mycorrhizal Fungi

S. Misra · A. Adhikari · A. Prakash (✉)
Department of Life Sciences, School of Biosciences and Technology, Galgotias University, Greater Noida, India
e-mail: prakash.anupam27@gmail.com

1 Introduction

The fungal secondary metabolites (SMs) are a mixture of compounds which has several biological functions and have significant environmental functions too. They are significant in the ecology of fungi because they help them defend against other organisms, including uptake of nutrients and interaction with plants, owing to their anti-inflammatory, anti-oxidant, and antimicrobial activity (Roy et al., 2021; Wadhwa et al., 2024). Following the discovery of penicillin, a metabolite generated by the fungus *Penicillium*, an event that signaled the dawn of antibiotic era, interest in the synthesis of metabolites by fungi gained particular significance (Bennett & Chung, 2001).

Currently, one of the most notable characteristics of fungi is their significant metabolic adaptability, which is demonstrated by the diversity of SMs found in different species (Keller et al., 2005). SMs are chemical compounds produced through biosynthetic pathways, and their synthesis is not essential for the normal growth and development of the fungus when grown in laboratory conditions. Nevertheless, these compounds are found in many species, indicating that their continued existence throughout evolution offers a competitive advantage in natural environments (Avalos & Limón, 2021).

SMs have numerous applications in promoting environmental sustainability. These substances, generated by fungi, contribute to Bioremediation, Ecological Restoration, and Waste Management, providing eco-friendly solutions to different environmental issues. There are different fungal species which are helpful in controlling different environmental pollutants, for example:

Penicillium chrysogenum can be helpful in degrading complex organic pollutants such as monocyclic aromatic

© The Author(s), under exclusive license to Springer Nature Switzerland AG 2026

A. Prakash and M. K. Dubey (eds.), *Advances in Fungal Secondary Metabolites: Bioactive Attributes and Biotechnological Potential of Industrial Interest*, Sustainable Economy and Ecotechnology, https://doi.org/10.1007/978-3-032-23507-7_8

hydrocarbons (e.g., benzene, phenol compounds, toluene, ethylbenzene, and xylene) (Abdulsalam et al., 2012; Pereira et al., 2014). *Penicillium ochrochloron* involves dyes bioremediation (Shedbalkar & Jadhav, 2011).

Aspergillus niger, Candida glabrata, Candida krusei, and Saccharomyces cerevisiae are effective in oil bioremediation (Burghal et al., 2016). *Saccharomyces cerevisiae* can bioacuinvolvesh as lead, mercury, and nickel (Chen & Wang, 2007).

Aspergillus versicolor, A. fumigatus, Paecilomyces sp., Paecilomyces sp., Terichoderma sp., Microsporum sp., Cladosporium sp. for utilizing heavy metal like cadmium (Mohammadian et al., 2015). Similarly, *A. niger*, *A. fumigatus*, *F. solani*, and *P. funiculosum* demonstrated a strong ability to break down and degrade hydrocarbon compounds (AI-Jawhari, 2014).

1.1 Bioremediation

Microorganisms like bacteria and fungi are used in the bioremediation process to eliminate or degrade environmental contaminants. Enzymatic activity of these microorganisms can transform hazardous substances into less toxic compounds. Bioremediation plays a significant role in the degradation, eradication, immobilization, or detoxification of various chemical wastes and physically harmful elements from the environment. The fundamental idea is to break down and change pollutants, including oil, heavy metals, hydrocarbons, dyes, pesticides, and more. Since it is done enzymatically through metabolism, it greatly aids in the resolution of several environmental issues (Abatenh et al., 2017). Bioremediation is one of the most cost-effective and ecologically beneficial biotechnological advancements. It can eliminate persistent organic contaminants, which are believed to be heterologous biological compounds that are difficult to decompose (Bala et al., 2022).

There are two main categories of bioremediation, as shown in Fig. 1.

The main principle of the bioremediation technique is biodegradation which involves complete transformation of harmful organic contaminants into non-toxic or naturally occurring substances, such as carbon dioxide, water, and inorganic substances that are safe for aquatic, plant, animal, and human life (Jain & Bajpai, 2012).

1.2 Ecological Restoration

Fungal SMs are essential for ecological restoration because they improve plant health, increase resistance to disease, and speed up bioremediation procedures. These metabolites contribute to sustainable agriculture methods and enhance soil quality by aiding in metal chelation and sequestration (ElHamouly et al., 2022).

1.3 Waste Management

Fungal SMs, such as enzymes, are essential in waste management as they facilitate biochemical reactions that break down stubborn pollutants and diminish their toxicity in polluted effluents, thereby improving the efficiency of bioremediation efforts (Unuofin & Kasonga, 2021). They can also assist in waste valorization, e.g., fungal secondary metabolites can be produced by *Trichoderma reesei,* which converts waste materials like fruit peels and barley straw into useful molecules. By transforming biomass into beneficial secondary metabolites, this technique is an innovative approach to f waste valorization (Shenouda et al., 2022).

2 Fungal Synergy with Bacteria

Synergistic interactions between bacteria and fungi are present in natural ecosystems and results in a greater functional outcome that cannot be achieved by both organisms individually. The ecological importance of this relationship is revealed by the fact that such mutual relationship has a significant influence on the health of a plant, human population, and restoration of environment. (Deveau et al., 2018). Together, the fungi and bacteria play a crucial role in the process of regeneration and nutrient cycle of soils and in degrading pollutants. The interactions result in enhanced wastewater treatment schemes, enhanced vegetation growth, and quick degradation of organic substances.

Certain important features of fungus bacteria synergy:

- Biodegradation: Complex organic pollutants can be broken down by both sequential and co-metabolic groups of fungi and bacteria. Indeed, it seems that lignin-rich conditions trigger fungus catabolic processes, which, in its turn, stimulate bacterial catabolism; this mutualism makes sure that the bacterial byproducts, in turn, catalyze fungal enzymatic activity that, in its turn, makes sure that the refractory compounds like lignin and hydrocarbons are more effectively degraded (Zhang et al., 2025). This mutual improvement of microbial taxa improves the activity of consortia designed to target specific environmental bioremediation.
- Soil Health and Nutrient Cycling: The connection between soil fungi and bacteria is essential to soil health particularly during the periods of soil stress that encompass drought. It is possible to use additives which include biochar and useful species of bacteria to enhance microbial diversity and create robust networks of microor-

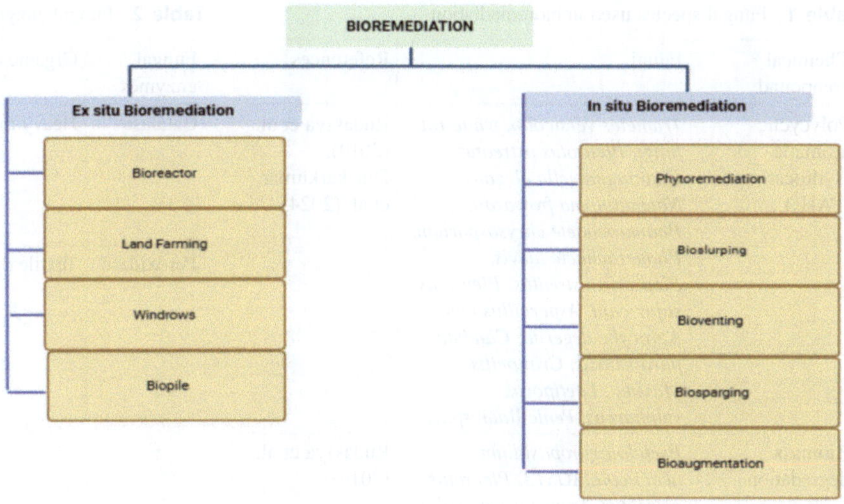

Fig. 1 Bioremediation techniques. *Source* Authors

ganisms that will then mobilize nutrients and plant resilience. The synergetic interactions are achieved with the assistance of these co-occurrence networks, and the application of this phenomenon is becoming the key to sustainable agriculture and environmental health (Ding et al., 2024).

- Wastewater treatment: Wastewater treatment when paired with a bacterium, fungi have been utilized in wastewater treatment where they were observed to improve dye decolorization, and the overall pollutant removal. Bacteria are useful in enhancing enzymatic degradation of the pollutants, whereas fungi are useful in enhancing sedimentation and flocculation. The simultaneous approach is more effective than the use of individual groups and encourages the implementation of new biotechnological methods of effluent treatment (Zhou et al., 2014).

Example of synergistic relationship: DDT Degradation: A combination of the bacteria *Ralstonia pickettii* and the brown-rot fungus *Fomitopsis pinicola* has demonstrated a synergistic impact in accelerating the breakdown of DDT. They illustrate how bacteria and fungi may cooperate to metabolize the persistent chemical DDT (Purnomo et al., 2020).

3 Fungal Metabolites in Bioremediation

Bioremediation requires fungal SMs since they provide new methods of detoxifying the environment. Depending on the fungal developmental growth responses, these compounds release bioactive chemicals that are biologically diverse, making them support the degradation and detoxification of hazardous environmental pollutants (Wadhwa et al., 2024). By participating in the degradation of organic pollutants, heavy metals, and xenobiotics, fungal metabolites contribute significantly to restoring ecological balance and remediating polluted environments.

Mode of action of fungal metabolites:

Enzyme Degradation: Enzymes constituting the fungal metabolites are capable of catalyzing metabolic processes; they mineralize stubborn pollutants, which results in mitigation of the toxic effects (Unuofin & Kasonga, 2021).

Biochemical Interaction: The metabolites produced by the fungi can interact with the pollutants and convert them into less toxic compounds; it can thus enhance the bioremediation process in entirety (Mohamed et al., 2024).

The various species of fungi have demonstrated efficacy in the bioremediation include *Aspergillus niger* and *Rhizopus microsporus*, which were found to effectively participate in the environmental detoxification and in wastewater treatment (Unuofin & Kasonga, 2021). Some of the fungal species such as *Phanerochaete chrysosporium* and *Trametes versicolour* are popularly known to degrade complex organic pollutants with the assistance of ligninolytic enzymes (Unuofin and Kasonga, 2021).

Some fungal species used in Bioremediation of different chemical compounds are given in Table 1.

3.1 Fungal Degradation of Environmental Pollutants

Mycoremediation is gaining acceptability as a feasible and sustainable tool for detoxification of various pollutants from the environment.

Mycoremediation is the use of fungi in environmental cleaning up is now a well-known desirable eco-friendly, and efficient strategy of cleaning up a variety of pollutants present in the environment.

Table 1 Fungal species used in bioremediation

Chemical compound	Fungi	References
Polycyclic aromatic Hydrocarbons (PAHs)	*Trametes versicolor, white-rot fungi, Pleurotus ostreatus, Cunninghamella elegans, Naematoloma frowardii, Phanerochaete chrysosporium, Phanerochaete laevis, Pleurotus ostreatus, Pleurotus sajor-caju, Aspergillus niger, Agrocybe aegerita, Candida parapsilosis, Crinipellis Maxima, Laetiporus sulphureus, Penicillium sp.,*	Rudakiya et al. (2019), Dinakarkumar et al. (2024)
Biomass degradation	*Pseudolagarobasidium acaciicola AGST3, Pleurotus ostreatus, Trametes versicolor, Tricholoma giganteum AGDR1*	Rudakiya et al. (2019)
Heavy metals	*Aspergillus, Pythium, Aspergillus flavus, Curvularia, Schizophyllum commune, Phanerochaete chrysosporium, Beauveria bassiana*	Dinakarkumar et al. (2024), Rudakiya et al. (2019)
Dyes	*Trametes hirsuta, White-rot fungi, Penicillium sp., Aspergillus niger, Pycnoporus cinnabarinus, Trichoderma sp., Ceriporiametamorphosa, Ganoderma sp., Bjerkanderaadusta, Aspergillus sojae, Aspergillus foetidus*	Dinakarkumar et al. (2024), Rudakiya et al. (2019)
Pesticide	*Aspergillus niger, Pleurotus ostreatus, Pleurotus sajor-caju, Trametes pubescens, Coriolus versicolor*	Rudakiya et al. (2019), Dinakarkumar et al. (2024)
Pharmaceutical active compounds	*Trametes versicolor*	Rudakiya et al. (2019)
Polychlorinated dioxins and dibenzofurans	*Phanerochaete velutina, Stropharia rugosoannulata*	Rudakiya et al. (2019)

Source Authors compilation

Table 2 Fungal enzymes mechanism of action

Fungal enzymes	Organic compounds	Mode of action	References
Catalase	Heavy metals	H_2O_2-based single-electron oxidation of aromatic compounds	Bosco and Mollea (2019)
Peroxidase	Textile dye	H_2O_2-dependent oxidation of aromatic compounds; organic molecules undergo further oxidation when Mn^{2+} undergoes single-electron oxidation driven by H_2O_2 and converts to Mn^{3+}	Jakovljević and Vrvić. (2018)
Manganese peroxidase (MnP)	Polyaromatic hydro-carbon (PAH), Di(2 ethylhexyl) phthalate, petroleum hydrocarbons, heavy metals	The oxidation of Mn^{2+} to Mn^{3+} is facilitated by H_2O_2 through a single-electron process, which subsequently leads to the oxidation of organic compounds	Larrondo et al. (2001)
Laccase	PCB, heavy metals, Pesticides, flexographic inks, total petroleum, hydrocarbons	O_2-based single-electron oxidation of organic molecules	Mouhamadou et al. (2013)
Lignin peroxidases	Bentazon, Pyrene, (herbicide), PAHs, aminodinitrotoluenes, phencyclidine (PCP)	Oxidation of single electrons with H_2O_2 on aromatic substances	Arora and Sharma et al. (2010)
Lipase	PAHs, dyes, pesticides, organic pollutants	H_2O_2-based single-electron oxidation of aromatic compounds	Solanki et al. (2019)

Source Authors compilation

3.1.1 Breakdown of Hydrocarbons, Pesticides, and Plastics

The metabolic roles of the fungal species are varied and hence can be used to degrade hydrocarbons, pesticides, plastics, heavy metals, and other recalcitrant pollutants.

The fungi possess certain enzyme structures that allow them to decompose structurally diverse and complex pollutants. In particular, it is successful in the environment in which xenobiotic compounds react.

Fungal Degradation Mechanism:

Enzymatic Action: Biodegradation of organic pollutants such as hydrocarbons, pesticides, and synthetic dyes is accompanied by fungal secondary metabolites and oxidative enzymes. These are laccases, peroxidases, and other enzymes of the ligninolytic type and they are involved in the breaking up of high-molecular-weight pollutants to the less dangerous and simplified forms (Gangola et al., 2022; Torres-Farradá et al., 2024; Tortella et al., 2005) (Fig. 2). Table 2 showing the different enzymes' mode of action.

Cometabolism: Fungi are capable of co-metabolism, wherein the breakdown of one chemical promotes the degradation of another. This mechanism is especially beneficial in settings where complex pollutant combinations are present.

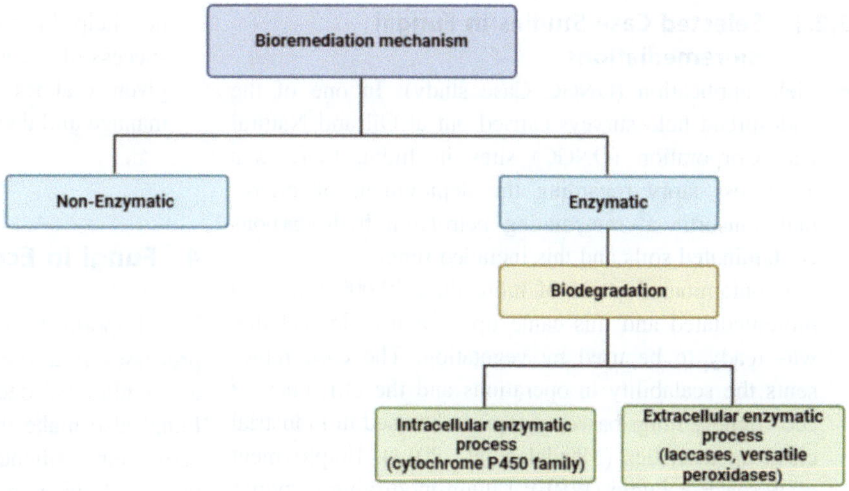

Fig. 2 Fungal bioremediation. *Source* Authors

Moreover, fungal synergy with other microorganisms, such as bacteria, can enhance the bioremediation process as a whole (Dinakarkumar et al., 2024).

Some benefits of mycoremediation are

- It is also shown that fungi can be used in the decomposition of several pollutants, including, medications and polycyclic aromatic hydrocarbons (PAHs). The fungi also play a significant role in reducing the level of pollutants through the release of the ligninolytic enzymes, breakage of lignin and PAHs through oxidation (Torres-Farrada et al., 2024; Singh & Roy, 2021). The synthetic polymers, including polyethylene, can be broken down by some fungi, which means that they can be used in combating plastic pollution (Zeghal et al., 2021).
- It has been discovered that many different types of fungi are capable of degrading hydrocarbons and pesticides, and this implies the biological diversity and flexibility of adaptation of fungi in different environments (Tortella et al., 2005; Vaksmaa et al., 2023).
- It has also been demonstrated that marine fungi can help in the decontamination of recalcitrant pollutants, and this implies that they can be utilized in aquatic and soil bioremediation processes (Zeghal et al., 2021).

3.1.2 Enzymes Involved in Biodegradation

Without fungi enzymes, it is not possible to carry out the biodegradation of various pollutants, polycyclic PAHs, plastics, and pharmaceutical residue. These enzymes play major role in the ecological detoxification processes hence facilitate the oxidation and hydrolytic degradation of complex organic materials to simpler and less toxic materials within the environment. Enzymes that are majorly biodegradative are:

Peroxidases: Manganese peroxidase (MnP) and lignin peroxidase (LiP) are the enzymes necessary to break down the PAHs and other organic pollutants. Such mechanism of action of these enzymes is related to the oxi-reduction mediated by the hydrogen peroxide that contributes to the decomposition of the stable aromatic rings (Kadri et al., 2017).

Laccases: These multicopper oxidases are really powerful in the degrading lignin, polyvinyl chlorine (PVC), poly ethylene (PE), and some xenobiotics. Their wide substrate preferences also render them useful in the degradation of pharmaceutical agents in the polluted settings (Temporiti et al., 2022; Nancy & Kumari, 2024).

Applications of these enzymes in bioremediations are:

Esterases and laccases are fungal enzymes that have properties of functionality best suited to deal with synthetic plastic, especially polyurethane (PUR) and polyethylene terephthalate (PET) (Temporiti et al., 2022). It is known that fungi can remove the pharmaceutical pollutants in wastewater via the biodegradation and biosorption process really well (Nancy & Kumari, 2024). As a result of their ligninolytic enzyme complex, white-rot fungi are especially effective at degrading diversity of organic pollutants, such as PAHs and heavy metals, in soils polluted by organic compounds (Chandra & Enespa, 2019).

3.2 Case Studies in Bioremediation

Mycoremediation is a subset of bioremediation. Fungal bioremediation and is one of the potential solutions that have been identified to be viable and environmentally effective in the removal of harmful contaminants in the environment, especially petroleum hydrocarbon and persistent organic pollutants.

The use in the real world also confirms its potential and reveals the effectiveness of the fungal systems in the case of remediation projects on a large scale.

3.2.1 Selected Case Studies in Fungal Bioremediations

- Field application (ONGC Case study): In one of the widespread field surveys carried out at Oil and Natural Gas Corporation (ONGC) sites in India, there was a success story regarding the deployment of microbial consortia at remediating petroleum hydrocarbons contaminated soils and this included fungi.
- Oil-contaminated waste of more than 30,000 tons was bioremediated and this came up as non-toxic soil that was ready to be used by vegetation. The case represents the scalability in operations and the efficiency of eco-friendly fungi-based systems to be used in industrial clean-up activities (Mandal et al., 2014). Employment of White-Rot Fungi (WRF): Lignin-modifying enzymes (LMEs), e.g., laccase and peroxidase, found among the white-rot fungi, facilitate their ability to break down a variety of heavy organic pollutants.
- The WRF enzyme can help oxidize lignin-like compounds in a range of xenobiotics because of their low substrate specificity, hence providing high flexibility regarding different bioremediation situations (Egwim et al., 2023).

3.2.2 Limitations and Challenges in Real-World Applications

Despite the potential in the fungal bioremediation, a number of problems such as operational, environmental and ecological barriers hinder the use of the technique in real-life applications.

- Ecological hindrance: Fungi may also be highly competitive with the local population of microorganisms in nature which may also inhibit the colonization and activity of the fungi. Moreover, the bioavailability of pollutants is not so easy because they are adsorbed to the particles of the soil or the complex chemical matrices, which causes the lower efficiency of the fungal degradation processes (Nandy et al., 2021).
- Environmental challenge: The abiotic factors, including the temperature, PH, moisture, and oxygen availability are highly susceptible to the fungus metabolism and the heterogeneous environmental conditions can barely be optimal and requires customizable interventions to address bioremediation (Singh et al., 2024).
- Lack of nutrient: Proliferation and enzyme activity of the fungi could also decrease significantly under the nutrient-deficient environments hence to maintain the fungal biomass and increase the rate of degradation of the pollutants, there may also be the requirement of the addition of carbon nitrogen or micronutrients.
- Gap in monitoring assessment: The traditional bioremediation monitoring procedures in most instances do not help in providing real time information on the success of the microbes to transform the pollutants. The given weakness makes field operations more difficult to manage and decide on (Singh et al., 2024).

4 Fungi in Ecological Restoration

The importance of fungi in the ecological restoration processes is a versatile and multifunctional role of fungi as symbionts, decomposers, and nutrient cycle agents. Fungi also make their contribution to the establishment of ecosystem resilience and high plant productivity, soil structure, and diversity. Fungi can make the degraded landscapes regenerate, prevent plant diseases, and accelerate the restoration of biogeochemical cycles. Having the power to release great nutrients, to destroy the organic residues, to enable the valuable combinations with the plants, they are indispensable forces in the restoration of the ruined ecosystems.

4.1 Role in Ecosystem Recovery

The fungus metabolites contribute to the decomposition, nutrients recycling, pollutant detoxification that is highly demanded in the polluted and damaged ecosystem recovery. Such bioactive compounds are helpful in:

- Growing plants at an increased rate and adding to the diversity of species,
- Enhancing the fertility and the stability of the soil,
- Repairing the degraded habitats and enhancing ecosystems to be more adaptable.

Fungal metabolites also directly protectorate the environment by decomposing harmful substances into non-hazardous substances. Besides, these secondary metabolites enhance fungal tolerance to abiotic stress, and it advances their symbiosis functions in ecosystems (Jha, 2024). They can also function as plant signaling molecules, resulting in their growth promotion and improved resistance to pathogens, which accelerates ecological recovery (ElHamouly et al., 2022).

4.1.1 Restoration of Degraded Lands Using Fungal Metabolite

The fungal secondary metabolites are considered the most efficient in land reclamation because of their ability to improve the soil quality and stimulate the vegetation growth. The fungi that are among the most ecologically useful

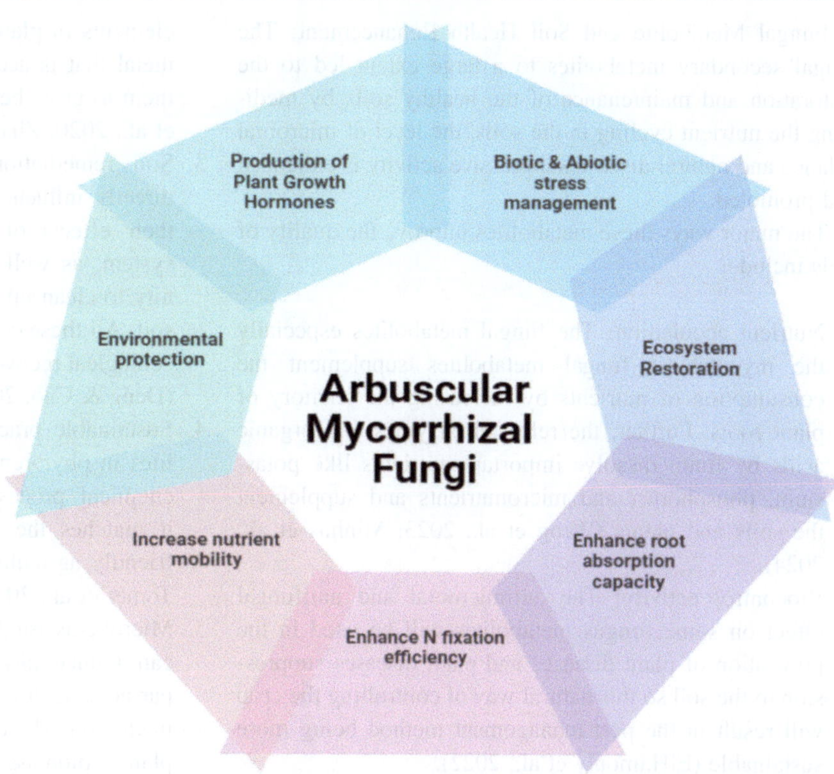

Fig. 3 Significance of AM fungi. *Source* Authors

fungi are Arbuscular Mycorrhizal Fungi (AMF); they have a mutualistic association with root tissue of the plants.

These associations:

- Improve the environment tolerance of plants,
- Promote nutritive absorption, and
- Enhance soil fabric and soil fertility (Shuab et al., 2017).

AMF application is also a green and sustainable alternative to synthetic fertilizer and pesticides as indicated in Fig. 3. This biological approach is rapidly being regarded as an effective technique to transform worn-out terrains and stimulate long-term ecological sustainability.

Role of Arbuscular Mycorrhizal Fungi

Arbuscular Mycorrhizal Fungi (AMF) has a centralized role in improving the soil condition, nurturing plants, and promoting the ecological stability of the landscape—especially that has been degraded. Their relationship with the plant roots maintains a mutual relationship between them, facilitates the exchange of nutrients, physical suitability of the soil, and offers general operations within the ecosystem.

1. In the nutrient-degraded or nutrient-deficient habitats, AMF causes the growth of soil fertility and crop production through the increased uptake and transport of nutrients to their host plants. (Asmelash et al., 2016; Shuab et al., 2017).
2. To enable the soil to aggregate, these fungi produce glycoprotein such as glomalin, which improves the sequestration of carbon by building stable soil organic matter (Pankaj et al., 2019).
3. Native AMF inoculation has the potential to both greatly enhance the productive capability of the plant communities in which excessively competitive foreign species may be held back and the potential of the latter have their native counterparts (Asmelash et al., 2016).
4. Besides, AMF hyphae help stabilize soil aggregates, enhance water infiltration or reduce erosion, and hydrate hydrological and structural stability of eroded soils (Asmelash et al., 2016).

4.1.2 Soil Health Improvement

The secondary metabolites of the fungi play an essential role in enhancing the soil health and phytoremediation which is the basis of sustainable farming and environment. Such substances are involved in rebuilding the soil under a variety of ways including being biocontrol agents against pathogens, being breakers of pollution or being those replenishing the nutrient cycle. The impact of the fungal metabolites is that they increase the richness of microbes, nutrient supply, and the functionality of the root microbes in ensuring that soils are healthy and resistant to the environmental disturbances and that in the long term they are able to support the vegetation in the ecosystems and the entire ecosystem.

Fungal Metabolite and Soil Health Enhancement: The fungal secondary metabolites to a large extent led to the restoration and maintenance of the healthy soils by mediating the nutrient cycling in the soils, the level of microbial balance and natural disease-suppressive activity is enhanced and promoted.

The major ways these metabolites improve the quality of soils include:

- Nutrient acquisition: The fungal metabolites especially the mycorrhizal fungal metabolites supplement the consumption of nutrients by increasing the territory of plant roots. Further, the released enzymes and organic acids by fungi dissolve important nutrients like potassium, phosphorus, and micronutrients and supplement the soils and plants (Weng et al., 2023; Minhas et al., 2024).
- Biocontrol activity: The antimicrobial and antifungal effect on some fungus metabolites will be used in the prevention of plant diseases and plant diseases suppression to the soil so this natural way of controlling the crop will result in the pest management method being more sustainable (ElHamouly et al., 2022).
- Increase in microbial diversity: Fungal products ensure functioning and diversified soil microbiome which is essential in nutrient cycling and plant–microbe symbiosis and long-term of soil productivity. Their presence and interrelations ensure the interactions and stability of the ecological balance of rhizosphere among the microbes (Weng et al., 2023; Minhas et al., 2024).

4.1.3 Phytoremediation

Phytoremediation is a more environmentally friendly form of bioremediation which involves the use of vegetation (usually with the assistance of soil microbes) to clean up contaminated lands, water and air. This kind of strategy offers an additional degradation, metal sequestration, and ecological restoration effects when combined with fungal metabolites.

Important role to phytoremediation by fungal known metabolites

1. Pollutant degradation: Fungal metabolites may also be useful in the degradation of organic forms of pollutants such as hydrocarbons, herbicides, and pesticides. The huge application of these compounds is the transformation of poisonous compounds to less poisonous or benign and thus aid in the purification of contaminated soils (Tomer et al., 2021).
2. Metal detoxification: The endophytic fungi produce some of the metabolites that are vital in the detoxification of heavy metal by immobilizing or chelating such elements in plant roots. This minimizes the quantity of metal that is accrued and is toxic in the plants helping them to grow healthily in polluted surroundings (Nandy et al., 2020; Zhao et al., 2024).
3. Soil remediation: The fungal secondary metabolites directly influence the improvement of the soil through their effects of greater plant growth, improved root system, as well as the ability of the microbe community to clean up the objects that were contaminating the soil. All these consequences introduce a remediation and ecological recovery of polluted soils on a long-term basis (Deng & Cao, 2017; Nandy et al., 2020).
4. Sustainable practices: Incorporation of fungal metabolites in phytoremediation systems minimizes the use of chemical processes of fertilizers and pesticides, thus it matches the norms of sustainable and ecologically friendly agricultural activities (ElHamouly et al., 2022; Tomer et al., 2021).
5. Microbe-assisted phytoremediation (MAP): Since MAP can induce phytoremediation in plants, the microbial partners, such as AMF, assist in the process of phytoremediation. The technique enhances better survival of the plants, biomass growth and facilitates carbon sequestration besides successfully eliminating the pesticides, metals and hydrocarbons in the contaminated area (Juwarkar, 2012).

4.2 Applications for Enhancing Soil Quality

Fungal metabolites are essential for improving soil quality through several processes, including increasing nutrient availability, promoting plant development, and aiding in detoxification of soil. These metabolites play a vital role in restoring soil health and promoting sustainable farming methods. Among the most important applications in the improvement of soil quality through fungal metabolites are: Soil health improvement (nutrient cycling, soil aggregation), promoting plant growth and diversity, bioremediation, and detoxification, etc.

4.2.1 Fungal Metabolites in Soil Aggregation and Nutrient Cycling

Fungal metabolites play a critical role in improving the health and productivity of ecosystems as they are involved in soil aggregation and cycling of nutrients. Fungi are also known to play a significant stabilizing role on the soil by formation of glycoproteins such as glomalin that attaches the aggregates of the soil and makes the soil more stable to retain water. Fungal communities are also involved in the absorption of nutrients and particularly phosphorus, and aid in formation of relationships with other communities in soil hence increases the process of nutrient cycling (Nichols, 2008).

Fungal Contribution in Soil Aggregation

Fungal metabolites play an important role in aggregation of soil such as:

- Glomalin production: AMF produces glomalin that coats soil aggregates to form a protective coating that makes them less susceptible to erosion (Nichols, 2008).
- Improving soil structure: Fungal hyphae develop webs that bind soil particles together thus improving permeability and aeration in the soil which play crucial roles in the growth of the plant (Nichols, 2008).

Role in Nutrient Cycling

The activities of the fungal metabolites in the nutrient cycles are as follows:

- Nutrient mobilization: Nitrogen and phosphorus cycle necessitates fungus especially ectomycorrhizal and saprotrophic fungi that augments the decrease of the organic tissue and liberation of nutrients. (Vietorisz et al., 2024).
- Decomposition: Fungi can decompose any type of organic remains through the creation of a great variety of extracellular enzymes, which disintegrates soil constituents and helps to restore the balance of carbon and nutrients (Zifcakova et al., 2016).
- Mycorrhizal fungi and nutrient transfer: Mycorrhizal fungi have the potential to enhance the uptake of nutrients by plants by dispersing malnutrienced hyphae into the soil, as well as by enhancing the ability of plants to access minerals, especially phosphorus (Alguacil et al., 2016; Nichols, 2008).

4.2.2 Role in Promoting Plant Diversity and Growth

The impact of fungal metabolites on plant ecosystem is because it enhances nutrient availability, causes systemic resistance and phenymimic growth regulation pathways. All these processes assist in augmented plant diversity and productivity especially in degenerated or marginal soils.

Plant Growth Promotion Mechanisms

Fungal metabolites promotes plant health and development by:

- Nutrient Solubilization: Fungi can also dissolve the required micronutrients, including potassium, zinc, and phosphorus and release them to the upper-most plants and enrich the soil (Devi et al., 2020).
- Production of Growth regulators: Fungal metabolites that resemble plant hormones such as auxins and gibberellins are produced (ElHamouly et al., 2022; Nigro, 2013).
- Biocontrol of pathogens: Many fungal lysates have antifungal activity that has been proven to suppress phytopathogens occurring in plants and reduce the prevalence of the disease (Mondal et al., 2000; Nigro, 2013).

Enhancing Plant Defense Mechanism

Fungal metabolites help in enhancing plant defenses by:

- Induction of plant resistance: By activating plant defense mechanism, fungal metabolites improve resilience against diseases and environmental stressors (ElHamouly et al., 2022; Nigro, 2013).
- Diversity of fungal endophytes: Through the production of numerous bioactive compounds that enhance growth and resistance, numerous fungal endophytes enhance the well-being of plants (Khalmuratova et al., 2021).

5 Fungal Metabolites in Waste Management

The recent years witnessed the use of fungal secondary metabolites as useful waste management solutions due to their multi-diverse enzymatic activities and biodegradation characteristics. These metabolites aid in the breakdown of complex pollution made of compounds such as plastics, heavy metals, and organic waste toward the sustainability of the earth and resource recovery. Its utilization includes plastic degradation, elimination of toxic chemicals, treatment of wastewater, and enhancement of agricultural remnants.

5.1 Degradation of Industrial and Agricultural Wastes

Fungal metabolites and enzymes work similarly during the breakdown of complex organic and inorganic waste materials. Fungi in bioremediation transform toxic or recalcitrant material into biodegradable but less risky substances that present an inexpensive, green solution to the traditional waste destruction technologies.

Agricultural waste: The fungal strains can be utilized to degrade lignocellulosic agricultural residuals and convert them to value added products like biofuels, compost and animal feed (Abdel-Azeem et al., 2020). It is this valorization that contributes to making agricultural activities circular and curbs the biomass overflow.

Industrial waste: Fungi have been known to remediate industrial effluents especially those of textile, pharmaceutical, and sugar-alcohol industries. They contain toxic compounds, such as phenols, dyes, and tannic acid, and, thus, decrease environmental poisoning and recover water quality (Singh et al., 2021).

5.1.1 Potential of Fungi in Processing Organic and Inorganic Waste

Fungi have a tremendous potential of processing, food or non-food, organic and chemical or non-chemical waste due to their peculiar biochemical and physiological functions. These features have left them as the main operators in most modern sustainable systems of waste control and reprocessing resources containing complex molecules that are difficult to break and require reshaping or formulation into other substances that are useful.

Organic Waste Processing

Bioremediation of organic contaminants: Fungal enzyme laccases and peroxidases fall critical in the process of remediating organic contaminants. It is through such oxidative enzymes that contaminated soils and waters are cleaned up due to the oxidation of hydrocarbons, pesticides, and phenolic compounds (Singh et al., 2021). Bioconversions and value added products: Fungi have the capability of transforming organic waste to commercially useful products including bio gas, fermentable sugars, organic acids, and fungal biomass rich in proteins that can be utilized as foods and feed (Bansfield et al., 2023; Chaurasia et al., 2025).Recycling of Organic waste: Recycling of organic waste by converting non-pathogenic indigenous fungi such as *Trichoderma* and *Penicillium* into nutrient-dense bio-organic fertilizer and global agricultural practices are associated with seed germination enhancement and the contributions made by each bio-fertilizer of organic waste are related to circular agricultural practices through this technique (Molla et al. (2022).

Decomposition: Fungi can break down any type of organic matter by producing a wide range of extracellular enzymes, which decomposes soil components and helps to regulate the carbon and nutrient balance (Žifčáková et al., 2016).

Inorganic Waste Remediation

Metal Removal and Detoxification: Fungi also have the capability to eliminate heavy metals in polluted sites by use of biosorption, intracellular reactions, and conversion to non-toxic or immobilized states. It is this competence which renders them useful in their use in cleaning up polluted water systems and soil.

Bioremediation of other inorganic compounds: The fungal species have also been found to be useful in reducing the effects of inorganic pollutants such as the radioactive elements and trace metals as they have extensive mycelia and versatility of their enzymes (Anand & Bharadvaja, 2021).

5.1.2 Breakdown of Xenobiotic Compounds

The numerous enzymes utilized by fungal metabolites in biotransformation processes are essential in the biodegradation of xenobiotic compounds. The fungi, especially the white-rot fungi, are essential due to their amazing properties to degrade toxic substances through their enzymatic reactions, thus they are significant in bioremediation efforts.

Enzymatic mechanism: Fungi possess a great diversity of enzymes like laccases, peroxidases, and cytochrome P450 that can decompose complex chemical structures.

Laccases and peroxidases: The two enzymes enhance oxidation of the xenobiotics and convert them to more palatable substances. Although laccases utilize molecular oxygen, but peroxidases rely on hydrogen peroxide that allows the effect of multiple degradation processes to be carried out (Baker et al., 2019). For example, *Pleurotus ostreatus* and *Trametes versicolor* are practically able to degrade xenobiotics, lignin, and other pollutant (Yang et al., 2017).

Cytochrome P450 monooxygenases: This group of enzymes is very important in the oxidative breakdown of complex substances into less toxic, simplified forms (Baker et al., 2019; Khan et al., 2017). For example, Cunninghamella genus (*C. elegans* and *C. echinulata*) contributes to the degradation of xenobiotics (Ibrahim et al., 2023).

Role of white-rot fungi: It is not unknown that lignin and numerous xenobiotics, including PAHs and chlorinated phenols, can be broken down using white-rot fungi (Dinakarkumar et al., 2024). For example, *Phanerochaete chrysosporium* has been examined in detail in terms of its ability to break down such contaminants like various pesticides and PAHs (Dinakarkumar et al., 2024). White-rot fungi were also used to degrade synthetic dyes and emerging contaminants and demonstrated their versatility in addressing different environmental pollutants (Torres-Farrada et al., 2024).

Biotransformation: Fungi can undergo xenobiotics metabolism and convert them into different forms (biotransformation) (Miglani et al., 2022). This mechanism is essential in detoxification since it has the ability to change lipophilic xenobiotics into hydrophilic metabolites, which are easily excreted. For example, many different fungi are also known to be able to bio transform xenobiotic compounds, especially those of the genera *Penicillium, Aspergillus, Fusarium*, and *Cunninghamella* (Khan et al., 2023).

5.2 Applications in Plastic and Chemical Waste Management

Plastic and chemical wastes are managed based on the enzyme property of the fungal metabolites which are applied in the breakdown of the man-made materials Fungus such as *Aspergillus* (*Aspergillus nomius, Aspergillus oryzae, Aspergillus niger and Aspergillus flavus*) and *Penicillium* (*Penicillium chrysogenum, Penicillium oxalicum and Penicillium simplicissimum*) generate the laccase, cutinase, peroxidase, and lipase enzymes, which are effective degraders of plastics (Mehta, 2024). Many of the enzymes capable of being produced by fungi are capable of degrading a wide range of chemical pollutants, such as pesticides, synthetic dyes, and PAHs.

Upcycling plastic: Innovative methods have been developed to convert polyethylene waste into beneficial fungal secondary metabolites such as asperbenzaldehyde and citreoviridin, using genetically modified strains of *Aspergillus nidulans* (Rabot et al., 2022). This procedure not only manages plastic waste but also increases the usefulness of waste plastics by producing pharmacologically active chemicals.

The potential of fungal metabolites, particularly its enzymes including laccases, peroxidases, and esterases, etc., is being studied in the management of plastic and chemical waste.

5.2.1 Biodegradation of Polymers or E-Waste

Fungal metabolites are essential to the biodegradation of electronic waste (e-waste) and polymer, using their enzymatic properties they can convert complex materials into simpler, non-toxic chemicals.

Biodegradation mechanism:

1. Fungi stick to the surface of polymers by using hydrophobic proteins.
2. After colonization, fungi release extracellular enzymes that attach to the polymer and help break it down.
3. These enzymes breakdown the polymer through oxidation and hydrolysis, causing many ruptures and erosion.
4. The smaller molecules produced can either be absorbed by the fungi or released for mineralization, eventually turning into CO_2 and water. (Bautista-Zamudio et al., 2023)

Fungal Enzymatic Action on Polymers

Fungi produce various enzymes that act on specific plastic polymers. For example, laccases and peroxidases are capable of breaking down polyethylene (PE) and polyvinyl chloride (PVC), while cutinases and lipases can decompose polyethylene terephthalate (PET) and polyurethane (PUR) (Temporiti et al., 2022).

E-Waste Degradation

Moreover, the fungi can break down e-waste, including the microplastics present in it. It is demonstrated that Penicillium brevicompactum can settle on and adsorb on PCB microplastics, depleting their mass, and causing fragmentation (Ferreira-Filipe et al., 2024). For example, *Aspergillus nomius, Cephalosporium* sp., *Colletotrichum fructicola, Diapportheitaliana, Stagonosporopsis citrulli, Trichoderma viride*, and others are also reported to have a high level of success working on the degradation of a large variety of plastic polymers (Varshney et al., 2023).

5.2.2 Use of Fungi for Green Recycling Processes

Recycling by applying fungi can offer new opportunities of handling waste materials and reuse resources in many industries. Fungi demonstrate outstanding bioaccumulation, biomass creation and the transformation of organic wastes into useful resources, which are beneficial to sustainable practices.

Some of the applications are:

- CO_2 recycling into biomass: Vesicular-AMF, like *Glomus fasciculatum*, improved photosynthesis and can convert CO_2 into biomass. In addition to helping sequester carbon, this method produced biomass that may be used as biofuel (Azmat & Moin, 2019).
- Fungal bioaccumulation from e-waste: Fungi can effectively recover vital or important raw materials from electronic waste (WEEE). Some strains exhibit high bioconcentration factors for elements such as palladium, copper, and yttrium. This highlights their potential use in environmentally sustainable recycling methodologies (Rosa et al., 2022).
- Production of flexible fungal material (FFM): FFMs are produced using fungi from organic waste, such as leftovers from agriculture. These materials support a circular economy and reduce environmental damage by serving as biodegradable substitutes in various sectors (Kapoor et al., 2024).
- Recycling of herbal residue: Endophytic fungi, such as *Aspergillus cristatus*, have the ability to decompose herbal remnants from traditional medicinal practices, thereby producing valuable bioactive compounds including anthraquinones. Such an effect enhances nutritional balance of the remaining materials besides diminishing organic waste accumulation. (Kong et al., 2019).
- Integrated agricultural system: With the help of adding fungi into the agricultural systems, cow dung and other organic wastes can be converted into fertilizer

and biogas. This symbiotic association increases crop production and enhances soil wellness which has proven to be a good way of managing waste. (Huang, 2015)

6 Conclusion

The potential of fungal secondary metabolites (SMs) in numerous environmental processes is particularly high in bioremediation and sustainable agricultural use. They serve as tools in the resolution of environmental issues since they are diverse in terms of their biological functions which include antibacterial, antifungal, and plant growth-promoting. There are also significant environmental applications of FMFs that are: bioremediation, waste management, interactions with microbes, sustainable agriculture, ecological functions, etc. Despite the promising future of the FSMs in the environment, there are concerns that still need to be addressed to make the best of its production and application and to accomplish this, additional research ought to be availed. To conclude, secondary metabolites of fungi offer a great source of natural products that can be used to solve various environmental problems and enhance ecological stability. The full potential of them concerning the creation of a more sustainable and resilient global ecosystem remains unexplored with the ongoing research on their biosynthetic pathways, the regulation process as well as their contributions to the ecology.

References

Abatenh, E., Gizaw, B., Tsegaye, Z., & Wassie, M. (2017). The role of microorganisms in bioremediation—A review. *Open Journal of Environmental Biology, 2*(1), 038–046.

Abdel-Azeem, A. M., Hasan, G. A., & Mohesien, M. T. (2020). *Biodegradation of agricultural wastes by chaetomium species* (pp. 301–341). Springer.

Abdulsalam, S., Adefila, S. S., Bugaje, I. M., & Ibrahim, S. (2012). Bioremediation of soil contaminated with used motor oil in a closed system. *Journal of Bioremediation and Biodegradation, 3*(12), 1–7.

Al-Jawhari, I. F. H. (2014). Ability of some soil fungi in biodegradation of petroleum hydrocarbon. *Journal of Applied & Environmental Microbiology, 2*(2), 46–52.

Alguacil, M. D. M., Torres, M. P., Montesinos-Navarro, A., & Roldan, A. (2016). Soil characteristics driving arbuscular mycorrhizal fungal communities in semiarid Mediterranean soils. *Applied and Environmental Microbiology, 82*(11), 3348–3356.

Anand, R., & Bharadvaja, N. (2021). *Fungi: The assorted bio-remediators, 1*(01).

Asmelash, F., Bekele, T., Birhane, E., & Birhane, E. (2016). The potential role of arbuscular mycorrhizal fungi in the restoration of degraded lands. *Frontiers in Microbiology, 7*, 1095.

Avalos, J., & Limón, M. C. (2021). Fungal secondary metabolism. *Encyclopedia, 2*(1), 1–13.

Azmat, R., & Moin, S. (2019). *An innovative method through fungal engineering for recycling of CO_2 into biomass* (pp. 239–269). Springer.

Baker, P., Tiroumalechetty, A., & Mohan, R. (2019). *Fungal enzymes for bioremediation of xenobiotic compounds* (pp. 463–489). Springer.

Bala, S., Garg, D., Thirumalesh, B. V., Sharma, M., Sridhar, K., Inbaraj, B. S., & Tripathi, M. (2022). Recent strategies for bioremediation of emerging pollutants: A review for a green and sustainable environment. *Toxics, 10*(8), 484.

Bansfield, D., Spilling, K., Mikola, A., et al. (2023). Growth of fungi and yeasts in food production waste streams: A feasibility study. *BMC Microbiology, 23*, 328.

Bautista-Zamudio, P. A., Flórez-Restrepo, M. A., López-Legarda, X., Monroy-Giraldo, L. C., & Segura-Sánchez, F. (2023). Biodegradation of plastics by white-rot fungi: A review. *Science of the Total Environment, 901*, Article 165950.

Bennett, J. W., & Chung, K. T. (2001). Alexander Fleming and the discovery of penicillin. *Advances in Applied Microbiology, 49*, 163–184.

Bosco, F., & Mollea, C. (2019). Mycoremediation in soil. In *Environmental chemistry and recent pollution control approaches*. IntechOpen.

Burghal, A. A., Abu-Mejdad, N. M. J. A., & Al-Tamimi, W. H. (2016). Mycodegradation of crude oil by fungal species isolated from petroleum contaminated soil. *International Journal of Innovative Research in Science, Engineering and Technology, 5*(2), 1517–1524.

Chandra, P., & Enespa. (2019). *Fungal enzymes for bioremediation of contaminated soil* (pp. 189–215). Springer.

Chaurasia, P. K., Bharati, S. L., Singh, S., Sivalingam, A. M., Shankar, S., & Mani, A. (2025). Fungal pretreatment methods for organic wastes: Advances and challenges in biomass valorization. *RSC Sustainability*.

Chen, C., & Wang, J. L. (2007). Characteristics of Zn2+ biosorption by Saccharomyces cerevisiae. *Biomedical and Environmental Sciences: BES, 20*(6), 478–482.

Deng, Z., & Cao, L. (2017). Fungal endophytes and their interactions with plants in phytoremediation: A review. *Chemosphere, 168*, 1100–1106.

Deveau, A., Bonito, G., Uehling, J., Paoletti, M., Becker, M., Bindschedler, S., Wick, L. Y., et al. (2018). Bacterial–fungal interactions: Ecology, mechanisms and challenges. *FEMS Microbiology Reviews, 42*(3), 335–352.

Devi, R., Kaur, T., Kour, D., Rana, K. L., Yadav, A. K., & Yadav, A. N. (2020). *Beneficial fungal communities from different habitats and their roles in plant growth promotion and soil health, 5*(1), 21–47.

Dinakarkumar, Y., Gnanasekaran, R., Reddy, G. K., Vasu, V., Balamurugan, P., & Murali, G. (2024). Fungal bioremediation: An overview of the mechanisms, applications and future perspectives. *Environmental Chemistry and Ecotoxicology*.

Ding, Y., Gao, X., Shu, D., Siddique, K. H. M., Song, X., Wu, P., & Zhao, X. (2024). Enhancing soil health and nutrient cycling through soil amendments: Improving the synergy of bacteria and fungi. *Science of The Total Environment*, 171332.

Egwim, E. C., Oyewole, O. A., & Yakubu, J. G. (2023). Fungal bioremediation of pollutants. In *Bioremediation for environmental pollutants* (pp. 181–237). Bentham Science Publishers.

ElHamouly, N. A., Hewedy, O. A., Zaitoon, A. M., Miraples, A., Elshorbagy, O. T., Hussien, S. M., El-Tahan, A. M., & Peng, D. (2022). The hidden power of secondary metabolites in plant-fungi interactions and sustainable phytoremediation. *Frontiers in Plant Science, 13*.

Ferreira-Filipe, D. A., Oliveira, L., Paço, A., Fernandes, A. J., Costa, F. M., Duarte, A. C., Silva, A. L. P., et al. (2024). Biodegradation of e-waste microplastics by *Penicillium brevicompactum. Science of the Total Environment, 935*, Article 173334.

Gangola, S., Bhatt, P., Joshi, S., Bhandari, N. S., Kumar, S., Prakash, O., Nainwal, D., et al. (2022). Isolation, enrichment, and characterization of Fungi for the degradation of organic contaminants. In *Mycoremediation protocols* (pp. 1–11). Springer US.

Huang, C. (2015). *Fungi, plant and animal recycling economy agricultural production system.*

Ibrahim, A. R. S., Mansour, M. K., Ahmed, M. M., Ulber, R., & Zayed, A. (2023). Metabolism of natural and synthetic bioactive compounds in Cunninghamella fungi and their applications in drug discovery. *Bioorganic Chemistry, 140*, Article 106801.

Jain, P. K., & Bajpai, V. (2012). Biotechnology of bioremediation—a review. *International Journal of Environmental Sciences, 3*(1), 535–549.

Jakovljević, V. D., & Vrvić, M. M. (2018). Potential of pure and mixed cultures of *Cladosporium cladosporioides* and *Geotrichum candidum* for application in bioremediation and detergent industry. *Saudi Journal of Biological Sciences, 25*(3), 529–536.

Jha, Y. (2024). *Differential fungal metabolite accumulation in response to abiotic and biotic stresses* (pp. 457–467). Elsevier BV.

Juwarkar, A. A. (2012). *Microbe-assisted phytoremediation for restoration of biodiversity of degraded lands: A sustainable solution, 82*(2), 313–318.

Kadri, T., Rouissi, T., Brar, S. K., Cledón, M., Sarma, S. J., & Verma, M. P. (2017). Biodegradation of polycyclic aromatic hydrocarbons (PAHs) by fungal enzymes: A review. *Journal of Environmental Sciences-China, 51*(1), 52–74.

Kapoor, S., Verma, N., Charak, H., Dutta, K., Gupta, A., & Gupta, V. B. (2024). *Flexible fungal materials* (pp. 154–175). Informa.

Keller, N. P., Turner, G., & Bennett, J. W. (2005). Fungal secondary metabolism—From biochemistry to genomics. *Nature Reviews Microbiology, 3*, 937–947.

Khalmuratova, I., Choi, D.-H., Yoon, H., Yoon, T.-M., & Kim, J.-G. (2021). Diversity and plant growth promotion of fungal endophytes in five halophytes from the Buan Salt Marsh. *Journal of Microbiology and Biotechnology, 31*(3), 408–418.

Khan, F., Farooqi, A., & Zafar, M. I. (2017). *Biodegradation of xenobiotics in soil by fungi* (pp. 235–242). Springer.

Khan, M. F., Hof, C., Niemcová, P., & Murphy, C. D. (2023). Recent advances in fungal xenobiotic metabolism: Enzymes and applications. *World Journal of Microbiology and Biotechnology, 39*(11), 296.

Kong, W., Huang, C., Shi, J., Li, Y., Jiang, X., Duan, Q., Huang, Y., Duan, Y., & Zhu, X. (2019). Recycling of Chinese herb residues by endophytic and probiotic fungus *Aspergillus cristatus* CB10002 for the production of medicinal valuable anthraquinones. *Microbial Cell Factories, 18*(1), 102.

Larrondo, L. F., Lobos, S., Stewart, P., Cullen, D., & Vicuña, R. (2001). Isoenzyme multiplicity and characterization of recombinant manganese peroxidases from *Ceriporiopsis subvermispora* and *Phanerochaete chrysosporium. Applied and Environmental Microbiology, 67*(5), 2070–2075.

Mandal, A. K., Sarma, P. M., Jeyaseelan, C. P., Channashettar, V. A., Singh, B., Agnihotri, A., Lal, B., & Datta, J. (2014). Large scale bioremediation of petroleum hydrocarbon contaminated waste at various installations of ONGC. India: Case studies. *Environmental Research, Engineering and Management, 68*(2), 41–52.

Mehta, P. (2024). From decay to decadence: Fungal alchemy in waste valorization. *Journal of Biomed Research, 5*(1), 130–135.

Miglani, R., Parveen, N., Kumar, A., Ansari, M. A., Khanna, S., Rawat, G., Panda, A. K., Bisht, S. S., Upadhyay, J., & Ansari, M. N. (2022). Degradation of xenobiotic pollutants: An environmentally sustainable approach. *Metabolites, 12*(9), 818.

Minhas, B., Katoch, P., Singha, A., Minhas, N., & Kaushik, N. (2024). *Microbial metabolites and their role to improve soil health* (pp. 67–95). Springer Nature.

Mohamed, H. I., Abdel Aal, M. H., & El-Mahdy, O. M. (2024). *Fungal metabolites and their role in remediation of environmental pollution* (pp. 283–315). Elsevier BV.

Mohammadian Fazli, M., Soleimani, N., Mehrasbi, M., Darabian, S., Mohammadi, J., & Ramazani, A. (2015). Highly cadmium tolerant fungi: their tolerance and removal potential. *Journal of Environmental Health Science and Engineering, 13*, 1–9.

Molla, A. H., Zahan, H., Oliver, M. M. H., & Mosharaf, M. K. (2022). Exploration of potential indigenous non-phytopathogenic fungi for bio-organic fertilizer recycling from organic waste. *Environment and Natural Resources Journal, 20*(6), 598–610.

Mondal, G., Dureja, P., & Sen, B. (2000). Fungal metabolites from Aspergillus niger AN27 related to plant growth promotion. *Indian Journal of Experimental Biology, 38*(1), 84–87.

Mouhamadou, B., Faure, M., Sage, L., Marçais, J., Souard, F., & Geremia, R. A. (2013). Potential of autochthonous fungal strains isolated from contaminated soils for degradation of polychlorinated biphenyls. *Fungal Biology, 117*(4), 268–274.

Nancy, & Kumari, U. (2024). Biodegradation of pharmaceutical pollutants using fungal enzyme. *International Journal for Research in Applied Science and Engineering Technology.*

Nandy, S., Andraskar, J., Lanjewar, K., & Kapley, A. (2021). *Challenges in bioremediation: From lab to land* (pp. 561–583). Elsevier.

Nandy, S., Das, T., Tudu, C. K., Pandey, D. K., Dey, A., & Ray, P. (2020). Fungal endophytes: Futuristic tool in recent research area of phytoremediation. *South African Journal of Botany, 134*, 285–295.

Nichols, K. A. (2008). *Indirect contributions of AM fungi and soil aggregation to plant growth and protection* (pp. 177–194). Springer.

Nigro, M. (2013). *Effect of metabolites produced by beneficial fungi on the plant metabolome, physiology and agronomic performance.*

Pankaj, U., Singh, G., & Verma, R. K. (2019). *Microbial approaches in management and restoration of marginal lands* (pp. 295–305). Elsevier.

Pereira, P., Enguita, F. J., Ferreira, J., & Leitão, A. L. (2014). DNA damage induced by hydroquinone can be prevented by fungal detoxification. *Toxicology Reports, 1*, 1096–1105.

Purnomo, A. S., Sariwati, A., & Kamei, I. (2020). Synergistic interaction of a consortium of the brown-rot fungus *Fomitopsis pinicola* and the bacterium *Ralstonia pickettii* for DDT biodegradation. *Heliyon, 6*(6).

Rabot, C., Chen, Y., Bijlani, S., Chiang, Y.-M., Oakley, C. E., Oakley, B. R., Williams, T. J., & Wang, C. C. C. (2022). Conversion of polyethylenes into fungal secondary metabolites. *Angewandte Chemie, 62*(4).

Rosa, E., Di Piazza, S., Cecchi, G., Mazzoccoli, M., Zerbini, M., Cardinale, A. M., & Zotti, M. (2022). Applied tests to select the most suitable fungal strain for the recovery of critical raw materials from electronic waste powder. *Recycling, 7*(5), 72.

Roy, A., Ahuja, S., & Garg, S. (2021). Fungal secondary metabolites: biological activity and potential applications. In *Recent trends in mycological research: Volume 1: Agricultural and medical perspective* (pp. 159–188).

Rudakiya, D. M., Tripathi, A., Gupte, S., & Gupte, A. (2019). Fungal bioremediation: A step towards cleaner environment. In *Advancing frontiers in mycology & mycotechnology: Basic and applied aspects of fungi* (pp. 229–249).

Shedbalkar, U., & Jadhav, J. P. (2011). Detoxification of malachite green and textile industrial effluent by *Penicillium ochrochloron. Biotechnology and Bioprocess Engineering, 16*, 196–204.

Shenouda, M. L., Ambilika, M., Skellam, E., & Cox, R. J. (2022). Heterologous expression of secondary metabolite genes in trichoderma reesei for waste valorization. *Journal of Fungi, 8*(4), 355.

Shuab, R., Lone, R., Ahmad, J., & Reshi, Z. A. (2017). *Retracted chapter: Arbuscular mycorrhizal fungi: A potential tool for restoration of degraded land* (pp. 415–434). Springer.

Singh Arora, D., & Kumar Sharma, R. (2010). Ligninolytic fungal laccases and their biotechnological applications. *Applied Biochemistry and Biotechnology, 160*, 1760–1788.

Singh, A., & Roy, A. (2021). *Fungal communities for the remediation of environmental pollutants* (pp. 127–165). Springer.

Singh, H., Kumar, S., & Arya, A. (2024). *Challenges in bioremediation* (pp. 143–162).

Solanki, M. K., Kashyap, B. K., Solanki, A. C., Malviya, M. K., & Surapathrudu, K. (2019). Helpful linkages of trichoderma s in the process of mycoremediation and mycorestoration. *Plant health under biotic stress: Volume 2: Microbial interactions* (pp. 51–64).

Temporiti, M. E. E., Nicola, L., Nielsen, E., & Tosi, S. (2022). Fungal enzymes involved in plastics biodegradation. *Microorganisms, 10*(6), 1180.

Tomer, A., Singh, R., Singh, S. K., Dwivedi, S. A., Reddy, C. U. kiran, Keloth, M. R. A., & Rachel, R. (2021). *Role of fungi in bioremediation and environmental sustainability* (pp. 187–200). Springer.

Torres-Farradá, G., Thijs, S., Rineau, F., Guerra, G., & Vangronsveld, J. (2024). White rot fungi as tools for the bioremediation of xenobiotics: A review. *Journal of Fungi, 10*(3), 167.

Tortella, G., Diez, M. C., & Durán, N. (2005). Fungal diversity and use in decomposition of environmental pollutants. *Critical Reviews in Microbiology, 31*(4), 197–212.

Unuofin, J. O., & Kasonga, T. K. (2021). *Fungi in the bioremediation of toxic effluents* (pp. 171–192). Academic Press.

Vaksmaa, A., Guerrero-Cruz, S., Ghosh, P., Zeghal, E., Hernando-Morales, V., & Niemann, H. (2023). Role of fungi in bioremediation of emerging pollutants. *Frontiers in Marine Science, 10*.

Varshney, S., Gupta, V., Yadav, A. N., Rahi, R. K., Devki, & Neelam, D. K. (2023). *An overview on role of fungi in systematic plastic degradation.*

Vietorisz, C., Policelli, N., Li, A., Adams, L. C., & Bhatnagar, J. (2024). *Microbial functional guilds and genes are key to explaining soil nutrient cycling alongside soil and plant variables.*

Wadhwa, K., Kapoor, N., Kaur, H., Abu-Seer, E. A., Tariq, M., Siddiqui, S., Yadav, V. K., Niazi, P., Kumar, P., & Alghamdi, S. (2024). A comprehensive review of the diversity of fungal secondary metabolites and their emerging applications in healthcare and environment. *Mycobiology, 52*(6), 335–387.

Weng, Z. (2023). *Fungal ministrations in soil detoxification, building, and health restoration* (pp. 61–95).

Yang, J., Li, W., Ng, T. B., Deng, X., Lin, J., & Ye, X. (2017). Laccases: Production, expression regulation, and applications in pharmaceutical biodegradation. *Frontiers in Microbiology, 8*, 832.

Zeghal, E., Vaksmaa, A., Vielfaure, H., Boekhout, T., & Niemann, H. (2021). The potential role of marine fungi in plastic degradation—A review. *Frontiers in Marine Science, 8*.

Zhang, W., Wen, Y., Wang, Z., Diao, C., & Liu, Z. (2025). The fungi-bacteria interaction mechanism of microbial consortium during efficient lignin degradation based on metabolomics analysis. *Molecules, 30*(3), 508.

Zhao, S., Yan, L., Kamran, M., Liu, S., & Riaz, M. (2024). Arbuscular mycorrhizal fungi-assisted phytoremediation: A promising strategy for cadmium-contaminated soils. *Plants, 13*(23), 3289.

Zhou, D., Zhang, X., Du, Y., Dong, S., Xu, Z., & Yan, L. (2014). Insights into the synergistic effect of fungi and bacteria for reactive red decolorization. *Spectroscopy, 2014*, 1–4.

Žifčáková, L., Vetrovský, T., Howe, A., & Baldrian, P. (2016). Microbial activity in forest soil reflects the changes in ecosystem properties between summer and winter. *Environmental Microbiology, 18*, 288–301.

Industrial Applications of Fungal Metabolites

Gulafsha Parveen, Priyanka Mishra, Waquar Akhter Ansari, Nishant Rai, Navin Kumar, Pankaj Gautam, and Durgesh Kumar Jaiswal

Abstract

Fungal metabolites are emerging as powerful tools in our efforts to tackle the world's most pressing environmental and sustainability challenges. With the tremendous rise in growth and population and increase in the consumption of resources, to fulfil the crisis for alternatives to fossil fuels and toxic chemicals becomes more urgent. Fungi, with their unique ability to naturally produce a wide range of useful molecules, including enzymes, pigments, and bioactive compounds, play a key role in this quest. Their diverse range of chemicals, including non-ribosomal peptides, polyketides, and terpenoids, enables them to adapt to their habitats and possess antioxidant, antibacterial, anti-inflammatory, and even anticancer properties. In addition to providing numerous opportunities for various sectors. These metabolites have the potential to improve the safety of food additives and medication, as well as produce eco-friendly colour for textiles/fabrics and cosmetics, all of which will make a significant contribution to environmental sustainability, but also hold the potential to make medicines and food additives safer and to create environmentally friendly colours for fabrics and cosmetics, thereby contributing significantly to environmental sustainability. Natural microbial pigments produced by fungi, such as melanins, carotenoids, are not only increasingly appealing alternative to synthetic dyes. They also show that the environmental benefits from fungus metabolite. These pigment not only contribute to the bright colour, they also function as antioxidant and provide UV protection fungal pigments that are highly appealing for sustainable manufacturing, as they can be easily scaled up and are not dependent on geography or season. This gives us confidence that they will last. Some fungi, such as *Trichoderma reesei* and *Aspergillus niger*, are currently powering big businesses by producing cellulase for biofuels and citric acid, respectively. Fungal metabolites are the focus to expanding research, which could lead to safer and more environmentally friendly substitutes to manufactured products. Utilising these natural resources will bring us closer to a more sustainable future and a bio-based economy. This highlights the significance of fungi and their incredible metabolites in maintaining the planet's health. This chapter provides in-depth details about the vital role of fungal metabolites, their application, and their importance to industry, while also highlighting their important function in maintaining the environment and promoting sustainable development. Understanding and harnessing the potential of fungal metabolites is a key first step towards a more sustainable future.

Keywords

Bioactive compound · Enzymes · Fungal metabolites · Pigments · Renewable energy · Sustainable solutions

G. Parveen · P. Mishra · N. Rai · N. Kumar
Department of Biotechnology, Graphic Era (Deemed to be) University, Dehradun, India
e-mail: navinkumar.bt@geu.ac.in

P. Gautam
Department of Microbiology, Graphic Era (Deemed to be) University, Dehradun, India

D. K. Jaiswal (✉)
Department of Biosciences, Graphic Era (Deemed to be) University, Dehradun, India
e-mail: durgeshjaiswal.bt@geu.ac.in

W. A. Ansari
Marwadi University Research Centre, Marwadi University, Rajkot, Gujarat, India

1 Introduction

The world's economy and population are expanding, and resource consumption and waste production also increase

© The Author(s), under exclusive license to Springer Nature Switzerland AG 2026

A. Prakash and M. K. Dubey (eds.), *Advances in Fungal Secondary Metabolites: Bioactive Attributes and Biotechnological Potential of Industrial Interest*, Sustainable Economy and Ecotechnology, https://doi.org/10.1007/978-3-032-23507-7_9

(Semaan et al., 2023). The food industry is one of the most significant sectors of the global economy (Bigliardi & Galati, 2013). Since the beginning of time, people have been consuming fungi as medicine, food and a source of psychoactive compounds. Over the past century, filamentous fungi have developed an increasing number of advantageous applications, making them one of the most important microbial groups in industries. (Saloheimo, 2025). These days, the ease of handling, processing and cultivating microorganisms such as bacteria, fungi and algae is a great alternative to natural colours. Microorganisms are more suitable for pigment production on a large scale. Numerous studies have been reported globally on the presence of on the occurrence of natural colours from microbes, especially from bacteria and fungi (Ramesh et al., 2019). Filamentous fungi have a significant impact on the structure and nutritional content of fungal mycelium due to their diverse enzymatic capacities and biochemical properties. This highlights their potential application in food, including the production of innovative products, plant-based proteins and functional foods. Protein, minerals, lipids, vitamins, antioxidants and essential amino acids are all abundant in fungal biomass (Elhalis, 2025). Aspergillus niger, a filamentous fungus, is an essential industrial cell factory and workhorse for the manufacture of citric acid. Microbial citric has a high economic value and is widely use in the beverage, food, detergent, cosmetic and pharmaceutical industries (Behera, 2020). To fulfil the demand of growing human population, sustainable oil production using novel microbes is essential for food fuels, feed and other lipid-based products (Qvirist et al., 2022). Since the industrial revolution, energy consumption, which utilises the environmental harmful fuels, has been a component of human advancement. Thus, it is the responsibility of science and engineering to find new, sustainable, useful and scalable energy source. Biofuels are an significant and fascinating alternative (Kamaraj & Rao, 2022). The second major group of eukaryotic, fungi are crucial for the biological system. These fungi function as cell factories that generate secondary metabolites that are advantageous to people, animals and the environment. Microorganisms are a great source of highly valuable naturally occurring substances (Sawant & Sajwan, 2022). Monascus pigments are high-quality, natural, and edible colourants produced by the filamentous fungus *Monascus* through fermentation. They serve as useful components in the food, cosmetic, pharmaceutical, textile, dye and chemical industries. The environmental benefits of using monascus pigment are significant, as they have numerous biological effects such as antioxidant, anti-inflammatory, antibacterial, cholesterol-regulating and anticancer properties (He et al., 2022). To combat global greenhouse gas emissions and prevent climate change, it is crucial for utilise alternative, low carbon transportation fuels especially biofuels (Froese, 2021). The filamentous Ascomycete species, Trichoderma reesei, that was initially discovered more than 70 years ago on the Solomon islands, has the unique ability to decompose and thrive on cellular textiles. This characteristic, which depends on hidden cellulose, especially in biorefineries that use T. reesei enzyme to saccharify lignocellulose from sustainable plant biomass in order to produce biobased chemicals and fuels (Bischof et al., 2016). It is a well-established fact that fungi have been reported to create a wide range of secondary metabolites, including colours, amino acids, vitamins, antibiotics and other chemicals. These metabolites exhibit numerous biological effects, including antifungal, anti-inflammatory, antioxidant, antibacterial, antioxidant and anticancer properties. Additionally, they hold enormous biotechnology potential that could revolutionise the agriculture, food, cosmetic and pharmaceutical sectors. Crucially, these metabolites, derived from the fungi, have the ability to replace synthetic chemicals, providing a more sustainable and eco-friendly alternative (Devi et al., 2020a, 2020b).

2 Diversity of Fungal Metabolites

The twentieth century has had a remarkably positive impact on human quality of life through the use of fungus metabolites. Over the past 50 years, the use of antifungals, antibiotics, cholesterol-lowering agents and immunosuppressants derived from fungal compounds has significantly improved human welfare and contributed to the remarkable increase in life expectancy in the latter half of the century. A comparatively small number of fundamental metabolic pathways, primarily non-ribosomal peptides, polyketides, terpenoids and combinations of these, are the source of the amazing range of chemical structures seen in fungal metabolites (Table 1). These pathways have become incredibly diversified throughout evolution (Pela, 2004). Fungal metabolites, with their diverse structures, have been a great source of novel medications. Several chemoinformatics analyses have shown the structural complexity and diversity of natural products. The diversity of authorised medications and commercial libraries for high throughput screening (HTS) that contain synthetic, semisynthetic, and natural chemicals was compared to the scaffold diversity of the 223 fungal metabolites (Gonzàlez-medina et al., 2017). HTS is a technology used in research on drug development (Hushpulian et al., 2024). These fungi have made significant contributions to the pharmaceutical industry by their extraction, modification and usage in research. The pharmaceutical, agriculture and medical sector characterise and use these chemicals. The pharmaceutical sector has benefited greatly from the contributions of these fungi, which are extracted, altered and used in research. These substances

Table 1 Bioactive fungal metabolites and their roles in plant-disease management

Class	Characteristics	Examples	References
Polyketides (PKs)	Synthesised by polyketide synthases (PKSs); exhibit diverse biological activities including antibiotic, antifungal and anticancer properties	Aflatoxin B1 (*Aspergillus flavus*), Lovastatin (*Pleurotus ostreatus*), Griseofulvin (*Penicillium* spp.)	Alves et al. (2025)
Nonribosomal peptides (NRPs)	Produced by nonribosomal peptide synthetases (NRPSs); often cyclic or modified peptides with significant bioactivities such as immunosuppressive and antibiotic effects	Cyclosporin A (*Tolypocladium inflatum*), Penicillin (*Penicillium* spp.), Echinocandin B (*Aspergillus* spp.)	Avalos et al. (2021)
Terpenoids	Derived from isoprene units; encompass a wide range of structures and functions, including roles as toxins, pigments and hormones	Ganoderic acid (*Ganoderma lucidum*), Trichothecenes (*Fusarium* spp.), Ergosterol (various fungi)	Shankar et al. (2022)
Alkaloids	Nitrogen-containing compounds; often exhibit potent biological activities including neurotoxicity and antimicrobial effects	Ergotamine (*Claviceps purpurea*), Fumitremorgin C (*Aspergillus fumigatus*), Paxilline (*Penicillium paxilli*)	Wadhwa et al. (2024)
Hybrid PKS-NRPS metabolites	Contain both polyketide and nonribosomal peptide moieties; biosynthesised by hybrid PKS-NRPS enzymes; often display complex structures and potent bioactivities	Fusarin C (*Fusarium* spp.), Sirodesmin PL (*Leptosphaeria maculans*)	Avalos et al. (2021)
Shikimic acid-derived compounds	Biosynthesised via the shikimate pathway; include aromatic compounds with various biological activities	Aflatoxin B1 (*Aspergillus flavus*), Patulin (*Penicillium* spp.)	Pusztahelyi et al. (2015)

Source Authors compilation

are described and used in the pharmaceutical, agricultural and medical sectors (Wadhwa et al., 2024). The primary producers of bioactive secondary metabolites are microorganisms that can be isolated from extreme environments, the marine environment or soil microbiota (Gakuubi et al., 2021). Secondary metabolites, sometimes also recognized as natural products, are low molecular-weight biomolecules that can increase a producer's survival rate by boosting their tolerance to various environmental stress and harsh conditions. Although they are not essential for their growth and development (Newman et al., 2020). According to recent research, nearly 20,000 compounds have already been discovered from endophytes, which are the largest producers of secondary metabolites among fungal species.

3 Fungal Pigments: Sustainable Alternatives

Numerous fungi produced different kinds of pigments and chemical compositions of organic dyes. These pigments serve several biological properties (Durán et al, 2002). The alternatives to synthetic pigments are becoming more popular due to concerns about the harmful effects of synthetic colourants on the environment and human health. Consequently, the global demand for natural colourant is increasing quickly in the food, cosmetic and textile industries. These natural colourants have many industrial applications, such as dyes for leather, paper, paints, coatings, food additives, cosmetics and textile and non-textile substrates (Kalra et al., 2020). Natural pigments, especially those derived from fungi, are gaining popularity and are in high demand worldwide. Fungal pigments are far more effective than other synthetic pigments. More alternatives have become available for a variety of application in several industries. Fungal pigments are used in more sectors due to their antibacterial, antioxidant, anticancer and cytotoxic properties (Lagashetti et al., 2019).

3.1 Types of Fungal Pigments

Over the last few decades, the usage of natural pigments and colourants has increased significantly due to their safe and environmentally friendly properties. Synthetic pigments are currently being replaced by natural pigments, as consumers desire more natural products. It has been demonstrated that filamentous fungi, especially ascomycetous fungi (*Monascus*, *Fusarium*, *Penicillium* and *Aspergillus*), produce secondary metabolites that contain a wide range of pigments, such as melanins and azaphilones (Mukherjee et al., 2017). The long history of allergies and toxicity associated with synthetic colourants makes these natural microbial pigments extremely intriguing (Lopes & Liagabue-Braun, 2021). Colourants are used in textiles, food, medicines, cosmetics and other industries for both social and commercial purposes. Chemically produced colourants are widely used across available options because of their low cost and versatile manufacturing methods; nevertheless, environmental and health concerns have prompted the value-adding of natural biopigment producers due to their ability to produce stable, inexpensive pigments with high yield titers throughout the year. One of the most important sources of natural colours in fungi (Meruvu & Dos Santos, 2021). The majority of fungal species generate biologically significant pigments. By protecting fungi from viruses, damaging UV rays, and environmental stressors, carotenoids, melanin, and lycopene can help them survive. Both bio-compounds and pigments have potential industrial uses. Agriculture and medicine can also be produced by endophytic fungi (Pagano et al., 2015). These various nutritional and functional qualities of these pigments make them important food components, utilised in the food industry. Some of the pigments include β-cyanin, flavonoids, anthocyanin, carotenoids, lutein, zeaxanthin, β-carotene, and chlorophylls (Nabi et al., 2019). Microbial pigments, used in food, medicine, colourants, and imaging, can be derived from a range of sources, including fungi, bacteria, yeast, basidiomycetes, parasites, and microalgae (Rather et al., 2023). Fungi are interesting among many microorganisms due to their rapid growth rates, which may also be genetically modified to encourage greater pigment synthesis (Meruvu & Dos Santos, 2021). Pigment-producing fungi inhabit diverse environments, including soil (dos Reis Celestino et al., 2014) and oceans (Fouillaud et al., 2017). Carotenoids are pigments that are terpenoids and lipophilic (Fig. 1). Filamentous fungi produce them and are widely distributed among photosynthetic organisms (Avalos et al., 2025). They are widely found in practically every taxonomic group. Fungal carotenoids have garnered particular interest due to their unique genetic makeup and biochemistry. The antioxidant capacity of carotenoids may extend the lifespan of fungi in their natural habitat (Naz et al., 2023). The chemical and nutritional properties of carotenoids can be used to classify them. They are divided into two major chemical classes: xanthophylls and carotene. Carotenoids are a well-known group of compounds with carbon (C) and hydrogen (H) in their chemical structure, such as α-carotene, β-carotene, γ-carotene, torulene and lycopene. The second class of compounds, known as xanthophylls, contains C, H and O in their structure, including lutein, zeaxanthin, canthaxanthin, torularhodin and astaxanthin (Cataldo et al., 2018). Due to their different structure, the molecules were separated into two groups, yellow xanthophylls and orange carotenes. The cosmetics industry extensively utilises them due to their unique antioxidant properties. As antioxidants, carotenoids are beneficial to human health. Nowadays, lycopene and carotene are the most commonly used chemicals in cosmetics. Following the advancement of biochemistry, chemistry, pharmacy and medicine, the field of cosmetology emerged (Baran et al., 2020). *Gibberella fujikuroi* (Ascomycetes), which produces the red polyketide pigment (Linnemannstonsen et al., 2002), offers potential as a food coloring. This applies to polyketide pigments from non-toxic, chemotaxonomically selected fungal strains, such as *Penicillium* and *Epicoccum* spp. The primary benefit of using fungi to produce colourants is that it allows manufacturers to be independent of seasonal raw material supply, thereby reducing batch-to-batch variability (Mapari et al., 2010). Melanins and anthraquinones (AQs) are pigments that function as virulence factors, protectants or antioxidants (Schmalhofer et al., 2024). All melanin forms occur in two stages: an uncontrolled polymerization of oxidised intermediates and an initial enzymatic phase, usually involving phenolase. Quinones produced by phenol oxidation are essential in that second stage. All melanins exhibit a protective role, a common characteristic, but they are not just pigments that defend against ultraviolet light (Solano, 2014).

3.2 Applications of Fungal Pigments in Food, Textiles, and Cosmetics Industries

The demand for colour is closely tied to food, textiles, medicines and cosmetics, driven by its rapid development. Natural pigments are being used by industry due to the health risks associated with manufacturing dyes; microbial pigments are the most significant and varied among them. The fungal pigments are highly sought after because they are primarily extracellular metabolites, are easily extractable, and have a wide range of biotechnological applications (Sajid & Akbar, 2018). Synthetic pigments and non-renewable colouring sources release toxic compounds into the environment. These synthetic pigments, utilised in the textile industry, pose serious ecological concerns. The use of natural pigments, such as microbial pigments, as a safer alternative to synthetic pigments has been investigated by industries and academia (Venil et al., 2020). The ability of microorganism are unique among natural biopigments producers because they can produced stable, low-cost pigments with high yield titers throughout the year. Natural pigments primarily found in fungi. These organisms are significant due to their diverse biological niche, biocatalytic processes and flexible metabolism. Important industrial bio-pigments produced by filamentous fungi, including melanins, azaphilones, quinones, carotenoids and riboflavins (Meruvu & Dos Santos, 2021). These have been consid-

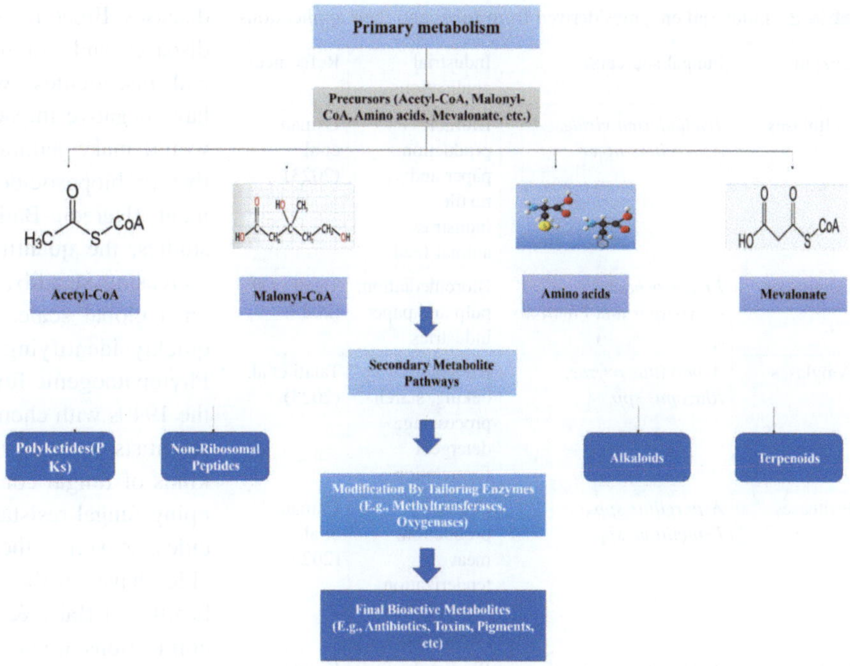

Fig. 1 Overview of fungal secondary metabolite pathways. *Source* Authors

ered as effective bio-compound from two different perspective: either due to their strong biological characteristics and chemical structure, or as colouring agents for products. Both characteristics are observed in their research and use in various field. These have been considered strong bio-compounds from two perspectives: either due to their strong biological properties and chemical structure, or as colouring agents for products. Both qualities are observed in their research and use in several fields (Muro Urista et al., 2016).

4 Fungal Bioactive Metabolites and Plant-Disease Management

One sector particularly vulnerable to environmental change is agriculture, which faces numerous challenges, including pollution, pathogenic attacks, salinity, drought, extreme heat and cold, among others. These issues impact yield (Yadav et al., 2020) and reduce overall agricultural output (Fuglie, 2018). The existence of the population and the stability of the economy depend on sufficient food production (Khan et al., 2025). By 2050 and beyond, developing a balanced strategy for disease prevention and sustainable food production will be a major challenge (Cai & Dimopoulos, 2025). Solutions are essential to resolving such problems. The application of fungi as biological fertilisers is one of the new fields that is gaining more attention as it demonstrates the value of improving plant growth and productivity through a variety of growth-promoting characteristics, such as the production of hydrolytic enzymes, siderophores and phytohormones, supplying various nutrients, and shielding plants from diseases (Kour et al., 2019). In this context, pesticides are crucial for minimising losses due to weeds, diseases and pest offering a reassuring solution to these problems. In India and globally, pesticide usage is focusing on the growing issues with food security caused by population growth and the rising need for crops that provide food and fibre (Sarica, 2024).

Furthermore, new agricultural techniques are developed in response to population and economic growth, aiming to boost production and meet demand. The large-scale use of chemical pesticides and fertilisers to boost productivity also has negative effects on the environment and various life ecosystems (Singh et al., 2020).

In nature, microbes are present everywhere, and one important tactic for colonising various ecosystems is the interaction between plants and microbes. Numerous bioactive substances and secondary metabolites produced by fungi are likely to substitute for traditional medications and antibiotics. Using various plant growth-promoting mechanisms, including the solubilization of phosphorus, zinc and potassium; the production of plant growth regulators (auxin, cytokinin, gibberellins, ethylene and abscisic acid); hydrolytic enzymes (laccase, xylanases, cellulases and pectinases); and siderophore, fungi are linked to crops and play a significant role in promoting plant growth and improving soil fertility. In addition to inducing a plant defence response against infections, PGP fungi promote plant growth and mitigate other abiotic stressors in severe conditions (Table 2) (Devi et al., 2020a, 2020b). Natural bioactive substances are crucial to agriculture because they can fight against a variety of biotic and abiotic

Table 2 Industrial enzymes derived from fungi and their applications

Enzyme	Fungal source(s)	Industrial applications	References
Cellulases	*Trichoderma viride, Aspergillus niger*	Biofuel production, paper and textile industries, animal feed	Usman et al. (2023)
Ligninases	*Phanerochaete chrysosporium, Pleurotus spp.*	Bioremediation, pulp and paper industries	Talati et al. (2023)
Amylases	*Aspergillus oryzae, Rhizopus spp.*	Brewing, baking, starch processing, detergent formulation	Talati et al. (2023)
Proteases	*Aspergillus spp., Penicillium spp.*	Cheese production, meat tenderization, brewing, leather processing	Usman et al. (2023)
Lipases	*Aspergillus niger, Candida rugosa*	Biodiesel production, food processing, cosmetics, detergents, pharmaceuticals	Kumar et al. (2023)
Pectinases	*Aspergillus niger, Aspergillus carneus*	Fruit juice clarification, wine production, textile processing	Usman et al. (2023)
Xylanases	*Aspergillus carneus, Trichoderma spp.*	Paper and pulp industries, animal feed, baking	Talati et al. (2023)
Laccases	*Trametes versicolor, Aspergillus spp.*	Textile dye decolorization, bioremediation, biosensors, pulp bleaching	Usman et al. (2023)
Lactase	*Aspergillus niger, Kluyveromyces lactis*	Dairy industry (lactose hydrolysis), production of lactose-free products	Ravindran et al. (2018)
β-Glucanases	*Trichoderma viride, Rhizomucor miehei*	Brewing industry (viscosity reduction), animal feed, wine clarification	Ravindran et al. (2018)

Source Authors compilation

diseases. Bioactive substances are useful in combating plant diseases and aid in the search for alternatives to chemical insecticides, which can be costly for growers and have negative impacts on health and the environment. This would make natural bioactive chemicals a viable alternative to biopesticides in integrated plant disease management (Heredia-Bàtiz et al., 2025). According to recent studies, the quantity of virulent phytopathogenic fungi is increasing steadily, resulting in substantial financial losses on a global scale. There are already several methods for quickly identifying and managing phytopathogenic fungi. Phytopathogenic fungi have been generally controlled since the 1940s with chemical and synthetic fungicides. However, scientists from all over the world are interested in various kinds of fungal control due to the significant rise in developing fungal resistance to these chemical controls or fungicides, as well as the detrimental effects that synthetic fungicides have on the environment, human health and animal health (El-Baky & Amara, 2021). Globally, inorganic salt applications are being used to reduce fungal disease in both the field and to protect crops as substitutes for synthetic fungicides (Deliopoulos et al., 2010). Plants are habitat to endophytic fungi, which are being increasingly researched for their potential to promote plant health and shield the host from diseases. An endophyte colonises the plant internally; it is sheltered from environmental stresses and variation, which may provide it a benefit over other biocontrol agents when used in disease control (Latz et al., 2018). To combat a variety of plant diseases, *Trichoderma* has been utilised extensively as a biological control agent. It promotes plant growth and suppresses plant disease. It can also easily activate other fungi and produce a variety of enzymes, including chitinase, protease and β-1,3-glucanase. It participates in a crucial detoxification process to reduce the toxicity released by plant pathogens (Thakur, 2021). *Aspergillus fumigatus, Penicillium* and Q strains of *Trichoderma virens* produce gliotoxin (non-ribosomal peptide), considered a useful component for inhibition of phytopathogenic organisms and as a biopesticide and antibiotic (Oviya et al., 2023).

5 Fungal Enzymes in Industrial Processes

Microorganisms have been used to produce industrial enzymes for over a century. Widely used fungal species in industry are ascomycetes, such as *Trichoderma reesei, Aspergillus niger* and *A. oryzae* (Arnau et al., 2020). Every kind of enzyme has a specific use, and includes the food, pulp and paper, textile, pharmaceutical, agricultural, biofuel and bioremediation industries. Enzymatic procedures offer more benefits than traditional chemical approaches, particularly in terms of waste material and by-products produced during the reaction, due to their superior qualities (de Souza et al., 2020). Obligate heterotrophs, including fungi, are recognised as the primary suppliers of enzymes. Due to environmental issues and rapid industrial growth, there is currently a quest for novel fungal enzymes. Fungal enzymes frequently used in industry include cellulase, amylase, β-glucosidase, glucose oxidase, protease, pectinase, laccase, lipase, invertase, xylanase, ligninase and chitinase (Dhevagi et al., 2021). Industrial enzymes are representative of all types of enzymes, except ligases. Hydrolases are used in all major sectors and are by far the most widely used class of enzymes. As a source of biological diversity and a host for the synthesis of industrial enzyme products, fungi have been and continue to be a significant source of enzymes (Østergaard & Olsen, 2010).

5.1 Applications of Fungal Enzymes in Biofuel, Detergent and Food Industries

The majority of the enzymes used commercially are synthesised from fungi. Hydrolytic depolymerases comprise the majority of enzymes currently used in the industry, and the addition of amylases and proteases to detergent formulations is the most important industrial application of enzymes. A variety of microbial genera, including fungal strains like Aspergillus, Rhizopus and penicillium can synthesise primary enzymes such as protease, xylanase and amylase that are utilised in commercial activities (Kelvey & Murphy, 2017). It has been widely recognised that enzymes play an essential role in a variety of activities, and that microorganism-derived enzymes are especially useful for baking, brewing, cheese synthesis and the production of alcohol, among many other uses. One of the most common and significant renewable biological resources, starch is a key component of the food chain for an extensive population (Patil et al., 2022). *Trichoderma* sp. and *Aspergillus* spp. produce cellulose and amylase respectively, which are used in the detergent, textile and bio-ethanol industries. Fungal proteases, such as keratinases, are utilised in waste management, the food industry, detergents, leather and pharmaceutical industries (Kango et al., 2019).

5.1.1 Biofuel Industries

The negative effects of non-renewable fossil fuels on the environment have led to the development of sustainable and renewable energy sources. The biofuel industries have made substantial use of cellulase in this context. The three primary enzyme components of the cellulose system efficiently transform cellulose substrates into sugars that can be fermented. Cellulase is commercially produced using mesophilic microbial strains under submerged fermentation (SmF) conditions, which are unsuitable or cost-ineffective. Meanwhile, the solid-state fermentation (SSF) method of producing fungal cellulases is more cost-effective and preferred for industrial applications (Srivastava et al., 2018). Anaerobic fungi, previously unexplored class of microorganism, hold exciting potential for use in biofuel production processes. Their remarkable ability to hydrolyse lignocellulose and convert this substance into H_2 and ethanol makes them a promising field for further studies and development (Fig. 3) (Saye et al., 2021).

5.1.2 Food Industries

Fungi are used in industrial fermentation to produce a variety of substances needed in food processing, including polyunsaturated fatty acids, flavours, enzymes, vitamins, acidulants and colourants. Higher yields of safe food-grade products were made possible by improvements in laboratory techniques for identifying non-toxic fungal strains, as well as by increasing understanding of the regulation of biosynthesis and the potential of genetic engineering (Copetti, 2019). Microbial fermentation (bacteria, fungi, yeast, actinomycetes and algae) also produces food enzymes, and fungal strains are frequently employed in this process. Using fungal enzymes in food production presents a viable way to enhance food preservation and shelf life without compromising the food's nutritional value and organoleptic properties (Da Silva et al., 2019). L-asparaginase is a common fungal enzyme used in food applications. It is regarded as generally recognised as safe (GRAS). It is used as a food additive to stop acrylamide from forming when food ingredients, such as amino acids and reducing sugars, react at high temperatures and low humidity. Desirable colour, taste and fragrance molecules are frequently produced by this reaction (Singh & Gaur, 2021). L-asparaginase aids in mitigation strategies to reduce acrylamide levels in food from two perspectives: either by intervening with the Maillard reaction or by converting L- L-asparagine into L-aspartic acid, which is non-toxic and prevents the formation of precursors without compromising the final products nutritional value, quality or flavour (Orabi et al., 2019). Glucoamylase, lipases, amylases, proteases, pectinases, tannases and many other enzymes are used in the food industry (Saxena et al., 2001). Enzymes are ideal tools for various industrial applications due to their affordability and ease of manufacture on a commercial scale. Food items from the dairy, baking and brewing industries are processed using microbial enzymes (Kumar et al., 2024).

5.1.3 Detergent Industries

The use of amylases, proteases, cellulases and lipases in the detergent industry has been widespread since the nineteenth century. Because α-amylases are alkalophilic, chelator-resistant and oxidant-insensitive, a detergent formulation incorporating them has shown promise for use in dish-washing (Samanta, 2020). Formulations of detergents based on alkaline protease enhance the breakdown capacity to extract proteinaceous material from stubborn stains (Savitha et al., 2011). Cellulases are utilised in the detergent industry, in addition to liquid laundry detergent, to keep clothes clean by removing debris from the interfibril gaps (Karmakar & Ray, 2011).

5.2 Mechanistic Insights and Commercial Enzymes

The robustness, substrate specificity and environmentally friendly features of fungal enzymes have made them essential catalysts in various industrial processes. Cellulases, amylases, laccases and proteases are fungal enzymes that function mechanistically by hydrolysing or reducing complex molecules into simpler ones. They are more effective in harsh pH and temperature environments because their catalytic activities are frequently mediated by highly conserved active sites and accessory domains (Sarrouh et al., 2012). The emphasis on their eco-friendly qualities reassures us of their sustainability. Industrial uses contain food fermentation, textile processing, paper bleaching and bioethanol production. Especially prized are enzymes from *Aspergillus niger*, *Trichoderma reesei* and Penicillium species. For instance, cellulases are crucial for saccharifying lignocellulosic biomass in the biofuel industries and aspergillus proteases are frequently used in detergent formulations. (El-Gendi et al., 2021). The efficient industrial application of microbial and fungal enzymes is demonstrated by commercial enzyme formulation including celluclast cellulase from *Trichoderma reesei* and Alcalase (protease from *Bacillus licheniformis*). The recombinant DNA technology and protein engineering's ability to increase enzyme yield and customising functionality provides an avenue towards more efficient and sustainable processes, offering hope for further developments. The effective industrial use of microbial and fungal enzymes is demonstrated by commercial enzyme formulations, such as Alcalase (protease from *Bacillus licheniformis*) and Celluclast (cellulase from *Trichoderma reesei*). The role of protein engineering and recombinant DNA technology in improving enzyme yields and customising functionality is a promising avenue for more sustainable and effective processes, instilling optimism for future advancements (Arnau et al., 2020).

6 Fungal Contributions to Renewable Energy

Fungi can help promote renewable energy in several ways, fungi may be added with soil carbon sequestration, produce biofuel and bioplastics, and produce electricity using fungal fuel cells. Additionally, they contribute to bioremediation by using their enzymatic systems to degrade pollutants and turn waste materials into bioenergy (Pearce et al., 2020). The ability of fungi to produce power, fuel and energy through metabolic processes makes them very useful for use in degradation and remediation. It is not well understood whether fungi can generate electricity during pollution remediation. According to reports, some species, including *Ganoderma lucidum*, *Trametes versicolor*, *Galactomyces reessii*, *Kluyveromyces smarxianus*, *Aspergillus* spp. and *Hansenula anomala*, may produce electricity at 207 mW/m^2, 1200 $mW/m^3/m^3$, 1163 mW/m^3, 850,000 mW/1h2, 438 mW/m^3, 2900 mW/m^3, respectively. Although fungi have the potential to be more environmentally friendly than traditional approaches, their significance remains largely unknown (Umar et al., 2024).

6.1 Use of Fungal Enzymes in Bioethanol Production

Fungal enzymes play a crucial role in producing bioethanol, particularly from lignocellulosic biomass. These enzymes facilitate the breakdown of complex plant polymers, such as cellulose and hemicellulose, into fermentable sugars. They are mostly cellulases and hemicellulases. Second-generation bioethanol production, which utilises agricultural byproducts and other non-food biomass, relies on this bioconversion. *Aspergillus niger*, *Trichoderma reesei* and *Penicillium janthinellum* are among the filamentous fungi that manufacture these enzymes in large quantities (Table 3). For example, *β-glucanase* produced by *Aspergillus niger* has shown great effectiveness in saccharification procedures, increasing the amount of fermentable sugar that can be obtained from lignocellulosic substrates (El-Shora et al., 2021). Solid-state fermentation is commonly used to improve the enzyme cocktails derived from these fungi, significantly enhances productivity and economic feasibility (Vasić et al., 2021). Furthermore, the lignin structure is modified by fungal pre-treatment techniques, which enhance enzymatic access and reduce the need for harsh chemical treatments (Robak & Balcerek, 2018). On-site enzyme manufacturing is another innovation that reduces expenses and streamlines logistics (Sørensen et al., 2011).

Table 3 Classification and examples of fungal metabolites

Metabolite	Fungal source	Bioactivity	Application in plant disease management	References
Griseofulvin	*Nigrospora* sp., *Penicillium brasilianum*	Antifungal; inhibits spore germination and hyphal growth	Effective against multiple phytopathogens including *Botrytis cinerea, Colletotrichum orbiculare,* and *Fusarium* spp.	Xu et al. (2021)
Trichothecenes	*Fusarium, Myrothecium, Trichoderma* spp.	Antifungal and phytotoxic; induces systemic resistance (ISR)	Enhances plant defense mechanisms and suppresses various plant pathogens	Xu et al. (2021)
Mycosubtilin	*Bacillus subtilis* BBG100	Lipopeptide with strong antifungal activity	Overproduction leads to enhanced biocontrol activity against *Pythium* infections in tomato seedlings	Leclère et al. (2020)
Sphaeropsidin A	*Diplodia corticola*	Antimicrobial, herbicidal and insecticidal properties	Potential use in controlling oak dieback and other plant diseases caused by *Diplodia* species	Salvatore et al. (2022)
Radicinin	*Alternaria radicina*	Phytotoxic; inhibits plant growth	Investigated as a target-specific bioherbicide for controlling invasive species like buffelgrass (*Cenchrus ciliaris*)	Masi et al. (2019)
Secondary metabolites from *Trichoderma* spp.	Indigenous *Trichoderma* strains	Diverse metabolites with antifungal and plant growth-promoting activities	Suppresses soil-borne pathogens and promotes growth in pulse crops; potential for sustainable agriculture	Mishra et al. (2022)
Alkaloid compounds	Various fungal endophytes	Antifungal; inhibits spore germination	Demonstrated efficacy against multiple phytopathogenic fungi; potential for development into natural fungicides	El-Baky et al. (2021)

Source Authors compilation

6.2 Advances in Lignocellulosic Biomass Conversion

The viability of producing sustainable biofuels and biochemicals has increased significantly due to recent developments in the conversion of lignocellulose, which is plentiful and renewable. However, its structural complexity makes it difficult to utilise effectively. Recent developments focus on the pre-treatment and hydrolysis phases to overcome this obstacle. Delignification and cellulose accessibility have been enhanced by pre-treatment methods, including ionic liquid treatment, microwave-assisted processing and organosolv (Haldar & Purkait, 2020). Modified enzymes and consolidated bioprocessing have significantly enhaced enzymatic hydrolysis, minimised cost and enhanced the productivity (Yadav, 2017). Additionally, to boost overall economic viability, integrated biorefineries are designed to extract a variety of value-added products and biofuels (Alonso et al., 2017). Genetically modified algae and bacteria are used in modern fourth-generation biofuels to more effectively turn sugar from biomass into fuel (Ashokkumar et al., 2024). All things considered these improvements are making the conversion of lignocellulosic biomass more commercially and environmentally viable.

7 Challenges in Industrial Applications

The historical importance of microbes in the production of food, especially in the fermentation process, is undeniable (Sionek et al., 2023), During the fermentation, these microorganisms enhance the flavour, texture, aroma and digestibility of food. The appreciation of fermented foods enhances shelf life, unique flavour, texture, aroma and safety of fermented food is a testament to the vital role of microorganisms in our food supply (Aguirre-Garcia et al., 2024). The most prevalent microbes in both traditional and modern food production are fungi, such as filamentous fungi (multicellular) and yeast (unicellular eukaryotes). Traditional food manufacturing has made considerable use of the yeast *Saccharomyces cerevisiae.* The ability to convert sugar into CO_2 and ethanol is essential in all processes, including baking, winemaking and brewing.

Additionally, filamentous fungi have been used as food for a remarkably long time, fostering a deep and enduring relationship between humans and microorganisms. For over

10,000 years, Asians have employed traditional miso procedures by using *Rhizopus* and *Aspergillus oryzae* (the "koji" mould) to ferment rice, barley or soybeans. However, it was unknown that the "mould" was present in the inoculate employed (Machida et al., 2008; Nout & Aidoo, 2010; Yokotsuka & Sasaki, 1998). Over a century ago, the first koji isolates were identified as naturally occurring manufacturers of enzyme "cocktails" beneficial to the food sector. An enzyme complex produced by *A. oryzae* was first manufactured in 1894 (Stone & Wright, 1898). In submerged fermentation, *A. oryzae* was engineered to produce a range of enzyme products, with a primary focus on amylases and proteases. Glucoamylase was also primarily produced by other *Aspergillus* species, including the black aspergilli, *A. awamori*, *A. niger*, *A. japonicus* and *A. aculeatus* (Cairns et al., 2018).

Acrylamide, a neurotoxin, is naturally produced as a by-product when food is fried or baked at temperatures exceeding 120 °C and with a minimal moisture content. When starchy foods are toasted, fried and baked, these factors determine their flavour and colour. The potential health risks of acrylamide, identified as a possible human carcinogen by the International Agency for Research on Cancer, should be a cause for concern (Zhivagi et al., 2019). It is mostly created when common foods are processed at high temperatures using free asparagine and reducing sugars. Coffee, cereals, bakery products and fried potatoes are the primary dietary sources of exposure (Pedreschi et al., 2014). Lipases, also known as triacylglycerol acyl hydrolases (EC: 3.1.1.3), are ubiquitous enzymes found in all living things, including bacteria, fungi, plants, and mammals. Their main function is to catalyse the hydrolysis of lipid into glycerol and free fatty acids at the aqueous-organic solvent interface. Expanding their utilisation in a variety of industries (Negi, 2019). They can react with various substrates, such as synthetic triglycerides, natural oils and fatty acid esters (Mehta et al., 2021). Trans fatty acids are solids fats found in margarine and baking fats, or they are fatty acid that have been eliminated by a chemical interesterification process. As a result, there are safety and environmental issues with the low-quality oil. Manufacturers are not allowed to use partially hydrogenated oil (PHOs) because of their high trans-fat content. Since the consumption of PHOs and trans fats is linked to an increased risk of coronary heart disease, the World Health Organisation started a global initiative to minimise these substances (Shin et al., 2009).

Furthermore, the policy domain of food safety is complex because it involves numerous stakeholders, encompasses a wide range of topics and disciplines, and has a significant global component. While nations can establish their standards, several issues limit their ability to regulate food safety through policy autonomy (Lindner, 2008). Every year, millions of diseases are caused by eating unsanitary food, making food safety a major global public health concern. The inclusion of chemicals (such as heavy metals and pesticides), physical hazards (including glass, metal and plastic) and microorganisms (like viruses and bacteria) makes contaminated food hazardous. Salmonellosis, campylobacteriosis, listeriosis, botulism, cholera, norovirus infection, *E. coli* infection and other foodborne infections are frequently caused by issues with food hygiene (Singh & Puniya, 2024). Biomaterials are now crucial in many fields, including biomedical engineering and regenerative medicine. However, using biological materials in clinical practice requires strict regulatory control and a commitment to standard ethics (Suhag, 2024).

8 Conclusion

Fungal metabolites demonstrate the inventiveness of nature with their different chemical structures and biological activity. The remarkable ability of fungi to create these chemicals serves as a foundation for synthetic biology techniques, and their impact extends to human creativity and natural ecosystems. This potential is not limited to medicinal research, sustainable agriculture and industrial biotechnology to include complex interaction with other bacteria, plants and animals that enable survival in a variety of situations. These metabolites have contributed to the development of vital agrochemicals and life-saving pharmaceuticals such as antibiotics and significant agrochemicals. However, due to their dual nature, some metabolites can also be dangerous as they can produce mycotoxins that endanger the health and safety of food.

Advancements in genome sequencing and analytic method have provided new opportunities for understanding and using fungal metabolites. In addition to enhancing our understanding of how these compounds are produced, their identification and elucidation of these gene clusters that produce them has paved the way for the advancement of synthetic biology techniques. These methods have the potential to produce vital molecules in regulated environments, thereby revolutionising pharmaceutical development, industrial biotechnology and sustainable agriculture.

The dual nature of fungal metabolites, yielding both beneficial as well as adverse effects emphasises the significance and necessity of future study. Researchers can improve the production of targeted compounds and develop approaches to reduce the hazards related to harmful metabolites by analysing biosynthesis pathways. This comprehension is essential for food security, environmental safety and public health. It is a call to action for all of us working in this field to continue our work and changing the world.

Fungal metabolites research is thriving and quickly expanding field with significant implications for science, commerce and society. We explore the synthesis solution for global problems through the research and application of fungi incredible metabolites capabilities, encompassing sustainable technologies and novel therapies derived from their enormous chemical repertoire.

Acknowledgements Gulafsha Parveen expresses gratitude to the Department of Biotechnology, Graphic Era (Deemed to be) University, for providing the financial support in the form of an institutional Research fellowship. The authors would like to express their gratitude to the Department of Biotechnology, Graphic Era (Deemed to be) University, and the Department of Biosciences, Graphic Era (Deemed to be) University, Department of Microbiology, Graphic Era (Deemed to be) University, Marwadi University Research Centre, Marwadi University, for their kind collaboration with this review paper.

References

Aguirre-Garcia, Y. L., Nery-Flores, S. D., Campos-Muzquiz, L. G., Flores-Gallegos, A. C., Palomo-Ligas, L., Ascacio-Valdés, J. A., et al. (2024). Lactic acid fermentation in the food industry and biopreservation of food. *Fermentation, 10*(3), 168.

Alizadeh, A. M., Hashempour-Baltork, F., Khaneghah, A. M., & Hosseini, H. (2021). New perspective approaches in controlling fungi and mycotoxins in food using emerging and green technologies. *Current Opinion in Food Science, 39*, 7–15.

Alonso, D. M., Hakim, S. H., Zhou, S., Won, W., Hosseinaei, O., Tao, J., et al. (2017). Increasing the revenue from lignocellulosic biomass: Maximizing feedstock utilization. *Science Advances, 3*(5), Article e1603301.

Altemimi, A., Lakhssassi, N., Baharlouei, A., Watson, D. G., & Lightfoot, D. A. (2017). Phytochemicals: Extraction, isolation, and identification of bioactive compounds from plant extracts. *Plants, 6*(4), 42.

Alves, V., Zamith-Miranda, D., Frases, S., & Nosanchuk, J. D. (2025). Fungal metabolomics: A comprehensive approach to understanding pathogenesis in humans and identifying potential therapeutics. *Journal of Fungi, 11*(2), 93.

Amani Machiani, M., Javanmard, A., Morshedloo, M. R., Janmohammadi, M., & Maggi, F. (2021). Funneliformis mosseae application improves the oil quantity and quality and eco-physiological characteristics of soybean (*Glycine max* L.) under water stress conditions. *Journal of Soil Science and Plant Nutrition, 21*, 3076–3090.

Amani Machiani, M., Javanmard, A., Habibi Machiani, R., & Sadeghpour, A. (2022). Arbuscular mycorrhizal fungi and changes in primary and secondary metabolites. *Plants, 11*(17), 2183.

Arnau, J., Yaver, D., & Hjort, C. M. (2020). Strategies and challenges for the development of industrial enzymes using fungal cell factories. *Grand challenges in fungal biotechnology* (pp. 179–210). Springer International Publishing.

Ashokkumar, V., Chandramughi, V. P., Kumar, G., Ngamcharussrivichai, C., Piechota, G., Igliński, B., et al. (2024). Advancements in lignocellulosic biomass: A critical appraisal of fourth-generation biofuels and value-added bioproduct. *Fuel, 365*, Article 130751.

Avalos, J., Pardo-Medina, J., Marente, J., & Limón, M. C. (2025). Biotechnological production and applications of fungal carotenoids. In *Biotechnology of yeasts and filamentous fungi* (pp. 229–262).

Avalos, J., & Limón, M. C. (2021). Fungal secondary metabolism. *Encyclopedia, 2*(1), 1–13.

Baran, M. T., Miziak, P., & Bonio, K. (2020). Characteristics of carotenoids and their use in the cosmetics industry. *Journal of Education, Health and Sport, 10*(7), 192–196.

Behera, B. C. (2020). Citric acid from *Aspergillus niger*: A comprehensive overview. *Critical Reviews in Microbiology, 46*(6), 727–749.

Bigliardi, B., & Galati, F. (2013). Innovation trends in the food industry: The case of functional foods. *Trends in Food Science & Technology, 31*(2), 118–129.

Bischof, R. H., Ramoni, J., & Seiboth, B. (2016). Cellulases and beyond: The first 70 years of the enzyme producer *Trichoderma reesei. Microbial Cell Factories, 15*, 1–13.

Boeriu, C. G. (2015). *Plants4Cosmetics: Perspectives for plant ingredients in cosmetics.* Wageningen UR Food & Biobased Research.

Begum, N., Qin, C., Ahanger, M. A., Raza, S., Khan, M. I., Ashraf, M., et al. (2019). Role of arbuscular mycorrhizal fungi in plant growth regulation: Implications in abiotic stress tolerance. *Frontiers in Plant Science, 10*, 1068.

Cai, P., & Dimopoulos, G. (2025). Microbial biopesticides: A one health perspective on benefits and risks. *One Health, 20*, Article 100962.

Cataldo, V., Arenas, N., López, J., Camilo, C., & Agosin, E. (2018). Sustainable Production of β-Xanthophylls in *Saccharomyces cerevisiae.* In Mic Eng. USA, https://dc.engconfintl.org/microbial/11/

Chai, K. F., Ng, K. R., Samarasiri, M., & Chen, W. N. (2022). Precision fermentation to advance fungal food fermentations. *Current Opinion in Food Science, 47*, Article 100881.

Devi, R., Kaur, T., Kour, D., Rana, K. L., Yadav, A., & Yadav, A. N. (2020a). Beneficial fungal communities from different habitats and their roles in plant growth promotion and soil health. *Microbial Biosystems, 5*(1), 21–47.

Dhevagi, P., Ramya, A., Priyatharshini, S., Geetha Thanuja, K., Ambreetha, S., & Nivetha, A. (2021). Industrially important fungal enzymes: Productions and applications. *Recent trends in mycological research: Volume 2: Environmental and industrial perspective* (pp. 263–309). Springer International Publishing.

dos Reis Celestino, J., de Carvalho, L. E., da Paz Lima, M., Lima, A. M., Ogusku, M. M., & de Souza, J. V. B. (2014). Bioprospecting of Amazon soil fungi with the potential for pigment production. *Process Biochemistry, 49*(4), 569–575.

Devi, R., Kaur, T., Guleria, G., Rana, K. L., Kour, D., Yadav, N., et al. (2020). Fungal secondary metabolites and their biotechnological applications for human health. In *New and future developments in microbial biotechnology and bioengineering* (pp. 147–161). Elsevier.

Downham, A., & Collins, P. (2000). Colouring our foods in the last and next millennium. *International Journal of Food Science & Technology, 35*(1), 5–22.

Duarte, L., Matte, C. R., Bizarro, C. V., & Ayub, M. A. Z. (2020). Review transglutaminases: Part II—Industrial applications in food, biotechnology, textiles and leather products. *World Journal of Microbiology and Biotechnology, 36*, 1–20.

Durán, N., Teixeira, M. F., De Conti, R., & Esposito, E. (2002). Ecological-friendly pigments from fungi. *Critical Reviews in Food Science and Nutrition, 42*(1), 53–66.

El-Baky, N. A., & Amara, A. A. A. F. (2021). Recent approaches towards control of fungal diseases in plants: An updated review. *Journal of Fungi, 7*(11), 900.

Elhalis, H. (2025). Exploring fungal mycelium for sustainable food solutions: From biomass utilization to byproduct innovation. In *Food reviews international* (pp. 1–33).

El-Shora, H. M., El-Sharkawy, R. M., Khateb, A. M., & Darwish, D. B. (2021). Production and immobilization of β-glucanase from *Aspergillus niger* with its applications in bioethanol production and biocontrol of phytopathogenic fungi. *Scientific Reports, 11*(1), 21000.

Fouillaud, M., Venkatachalam, M., Llorente, M., Magalon, H., Cuet, P., & Dufossé, L. (2017). Biodiversity of pigmented fungi isolated from marine environment in La Réunion Island, Indian Ocean: New resources for colored metabolites. *Journal of Fungi, 3*(3), 36.

Froese, A. (2021). *Towards increasing lignocellulose to biofuel conversion by Clostridium thermocellum: Co-culturing for increased hydrolysis and characterization of pyruvate phosphate dikinase for understanding atypical metabolism.* The Faculty of Graduate Studies of The University of Manitoba. https://mspace.lib.umanitoba.ca/items/e2f9bcf3-e3a3-4d24-b3d0-fcff67feca3d

Fuglie, K. O. (2018). Is agricultural productivity slowing? *Global Food Security, 17*, 73–83.

Gavahian, M., Sheu, S. C., Magnani, M., & Mousavi Khaneghah, A. (2022). Emerging technologies for mycotoxins removal from foods: Recent advances, roles in sustainable food consumption, and strategies for industrial applications. *Journal of Food Processing and Preservation, 46*(10), Article e15922.

Ghasemifar, E., Saeidian, S., & Setareh, P. (2013). Stability of seedless red grape anthocyanin under the effect of chlorogenic acid copigment, heating and UV irradiation. *Journal of Novel Applied Sciences, 2*, 594–597.

González-Medina, M., Owen, J. R., El-Elimat, T., Pearce, C. J., Oberlies, N. H., Figueroa, M., et al. (2017). Scaffold diversity of fungal metabolites. *Frontiers in Pharmacology, 8*, 180.

He, J., Jia, M., Li, W., Deng, J., Ren, J., Luo, F., et al. (2022). Toward improvements for enhancement the productivity and color value of Monascus pigments: A critical review with recent updates. *Critical Reviews in Food Science and Nutrition, 62*(26), 7139–7153.

Heredia-Bátiz, J. M., Manjarrez-Quintero, J. P., Valdez-Baro, O., Rivera-Salas, M. M., Bayardo-Rosales, H., Jiménez-Ortega, L. A., et al. (2025). Role of bioactive compounds in plant disease management. In Biotechnological intervention in production of bioactive compounds: Biosynthesis, characterization and applications (pp. 141–162). Springer Nature Switzerland.

Hridoy, M., Gorapi, M. Z. H., Noor, S., Chowdhury, N. S., Rahman, M. M., Muscari, I., et al. (2022). Putative anticancer compounds from plant-derived endophytic fungi: A review. *Molecules, 27*(1), 296.

Hobbs, D. H., Lu, C., & Hills, M. J. (1999). Cloning of a cDNA encoding diacylglycerol acyltransferase from *Arabidopsis thaliana* and its functional expression. *FEBS Letters, 452*(3), 145–149.

Hong, K. K., & Nielsen, J. (2012). Metabolic engineering of *Saccharomyces cerevisiae*: A key cell factory platform for future biorefineries. *Cellular and Molecular Life Sciences, 69*, 2671–2690.

Hushpulian, D. M., Gaisina, I. N., Nikulin, S. V., Chubar, T. A., Savin, S. S., Gazaryan, I. G., et al. (2024). High throughput screening in drug discovery: Problems and solutions. *Moscow University Chemistry Bulletin, 79*(2), 93–104.

Jakubowski, W., & Bartosz, G. (1997). Estimation of oxidative stress in *Saccharomyces cerevisae* with fluorescent probes. *The International Journal of Biochemistry & Cell Biology, 29*(11), 1297–1301.

Jouzani, G. S., Najafi, G. H., Gholami, M., & Ardjmand, M. (2013). Lignocellulosic biomass to bioethanol, a comprehensive review with a focus on pretreatment. *Renewable and Sustainable Energy Reviews, 27*, 77–93.

Jullesson, D., David, F., Pfleger, B., & Nielsen, J. (2015). Impact of synthetic biology and metabolic engineering on industrial production of fine chemicals. *Biotechnology Advances, 33*(7), 1395–1402.

Kalra, R., Conlan, X. A., & Goel, M. (2020). Fungi as a potential source of pigments: Harnessing filamentous fungi. *Frontiers in Chemistry, 8*, 369.

Kallscheuer, N., Classen, T., Drepper, T., & Marienhagen, J. (2019). Production of plant metabolites with applications in the food industry using engineered microorganisms. *Current Opinion in Biotechnology, 56*, 7–17.

Kamaraj, R., & Rao, Y. K. (2022). Biodiesel blends: a comprehensive systematic review on various constraints. In *Environmental science and pollution research* (pp. 1–16).

Kango, N., Jana, U. K., & Choukade, R. (2019). Fungal enzymes: Sources and biotechnological applications. In *Advancing frontiers in mycology & mycotechnology: Basic and applied aspects of fungi* (pp. 515–538).

Khan, S. U., Alsuhaibani, A., Alabduljabbar, A., Almarshad, F., Altherwy, Y. N., & Akram, T. (2025). A review on automated plant disease detection: Motivation, limitations, challenges, and recent advancements for future research. *Journal of King Saud University Computer and Information Sciences, 37*(3), 34.

Kour, D., Rana, K. L., Yadav, N., Yadav, A. N., Singh, J., Rastegari, A. A., et al. (2019). Agriculturally and industrially important fungi: Current developments and potential biotechnological applications. In *Recent advancement in white biotechnology through fungi: Volume 2: Perspective for value-added products and environments* (pp. 1–64).

Kumar, A., Dhiman, S., Krishan, B., Samtiya, M., Kumari, A., Pathak, N., et al. (2024). Microbial enzymes and major applications in the food industry: A concise review. *Food Production, Processing and Nutrition, 6*(1), 85.

Kumar, A., Verma, V., Dubey, V. K., Srivastava, A., Garg, S. K., Singh, V. P., et al. (2023). Industrial applications of fungal lipases: A review. *Frontiers in Microbiology, 14*, 1142536.

Lagashetti, A. C., Dufossé, L., Singh, S. K., & Singh, P. N. (2019). Fungal pigments and their prospects in different industries. *Microorganisms, 7*(12), 604.

Latz, M. A., Jensen, B., Collinge, D. B., & Jørgensen, H. J. (2018). Endophytic fungi as biocontrol agents: Elucidating mechanisms in disease suppression. *Plant Ecology & Diversity, 11*(5–6), 555–567.

Leclère, D., Obersteiner, M., Barrett, M., Butchart, S. H., Chaudhary, A., De Palma, A., et al. (2020). Bending the curve of terrestrial biodiversity needs an integrated strategy. *Nature, 585*(7826), 551–556.

Lindner, L. F. (2008). Regulating food safety: The power of alignment and drive towards convergence. *Innovation: The European Journal of Social Science Research, 21*(2), 133–143.

Linnemannstöns, P., Schulte, J., del Mar Prado, M., Proctor, R. H., Avalos, J., & Tudzynski, B. (2002). The polyketide synthase gene pks4 from Gibberella fujikuroi encodes a key enzyme in the biosynthesis of the red pigment bikaverin. *Fungal Genetics and Biology, 37*(2), 134–148.

Lopes, F. C., & Ligabue-Braun, R. (2021). Agro-industrial residues: Eco-friendly and inexpensive substrates for microbial pigments production. *Frontiers in Sustainable Food Systems, 5*, Article 589414.

Machida, M., Yamada, O., & Gomi, K. (2008). Genomics of *Aspergillus oryzae*: Learning from the history of Koji mold and exploration of its future. *DNA Research, 15*(4), 173–183.

Mapari, S. A., Thrane, U., & Meyer, A. S. (2010). Fungal polyketide azaphilone pigments as future natural food colorants? *Trends in Biotechnology, 28*(6), 300–307.

Masi, M., Freda, F., Sangermano, F., Calabrò, V., Cimmino, A., Cristofaro, M., et al. (2019). Radicinin, a fungal phytotoxin as a target-specific bioherbicide for invasive buffelgrass (*Cenchrus ciliaris*) control. *Molecules, 24*(6), 1086.

McKelvey, S. M., & Murphy, R. A. (2017). Biotechnological use of fungal enzymes. In *Fungi: Biology and applications* (pp. 201–225).

Mehta, A., Guleria, S., Sharma, R., & Gupta, R. (2021). The lipases and their applications with emphasis on food industry. In *Microbial biotechnology in food and health* (pp. 143–164). Academic Press.

Meruvu, H., & Dos Santos, J. C. (2021). Colors of life: A review on fungal pigments. *Critical Reviews in Biotechnology, 41*(8), 1153–1177.

Meyer, V. (2008). Genetic engineering of filamentous fungi—Progress, obstacles and future trends. *Biotechnology Advances, 26*(2), 177–185.

Mishra, S., Dash, D., & Das, A. P. (2022). Detection, characterization and possible biofragmentation of synthetic microfibers released from domestic laundering wastewater as an emerging source of marine pollution. *Marine Pollution Bulletin, 185*, Article 114254.

Mood, S. H., Golfeshan, A. H., Tabatabaei, M., Hawksworth, D. L., & Lücking, R. (2017). Fungal diversity revisited: 2.2 to 3.8 million species. *Microbiology Spectrum, 5*(4), 10–1128.

Moore, D., Robson, G. D., & Trinci, A. P. (2020). *21st century guidebook to fungi*. Cambridge University Press.

Mukherjee, G., Mishra, T., & Deshmukh, S. K. (2017). Fungal pigments: an overview. In *Developments in fungal biology and applied mycology* (pp. 525–541).

Muro Urista, C., Gracida Rodriguez, J., Abreu Corona, A., Arana Cuenca, A., & Tellez Jurado, A. (2016). Pigments from fungi, an opportunity of production for diverse applications. *Biologia, 71*(10), 1067–1079.

Nabi, B. G., Mukhtar, K., Ahmed, W., Manzoor, M. F., Ranjha, M. M. A. N., Kieliszek, M., et al. (2023). Natural pigments: Anthocyanins, carotenoids, chlorophylls, and betalains as colorants in food products. *Food Bioscience, 52*, Article 102403.

Naz, T., Ullah, S., Nazir, Y., Li, S., Iqbal, B., Liu, Q., et al. (2023). Industrially important fungal carotenoids: Advancements in biotechnological production and extraction. *Journal of Fungi, 9*(5), 578.

Nelson, G. (2002). Application of microencapsulation in textiles. *International Journal of Pharmaceutics, 242*(1–2), 55–62.

Newman, D. J., & Cragg, G. M. (2020). Plant endophytes and epiphytes: Burgeoning sources of known and "unknown" cytotoxic and antibiotic agents? *Planta Medica, 86*(13/14), 891–905.

Nielsen, A. A., Segall-Shapiro, T. H., & Voigt, C. A. (2013). Advances in genetic circuit design: Novel biochemistries, deep part mining, and precision gene expression. *Current Opinion in Chemical Biology, 17*(6), 878–892.

Nout, M. R., & Aidoo, K. E. (2010). Asian fungal fermented food. In *Industrial applications* (pp. 29–58). Springer.

Østergaard, L. H., & Olsen, H. S. (2010). Industrial applications of fungal enzymes. In *Industrial applications* (pp. 269–290). Springer.

Oviya, R., Sobanbabu, G., Mehetre, S. T., Kannan, R., Theradimani, M., & Ramamoorthy, V. (2023). Impact of persistence and movement of gliotoxin produced by *Trichoderma virens* in agricultural soil and crop plants. In *Microbial biocontrol: Molecular perspective in plant disease management* (pp. 129–142).

Pagano, M. C., & Dhar, P. P. (2015). Fungal pigments: An overview. In *Fungal biomolecules: Sources, applications and recent developments* (pp. 173–181).

Patil, A. G., Khan, K., Aishwarya, S., Padyana, S., Huchegowda, R., Reddy, K. R., et al. (2022). Fungal amylases and their industrial applications. *Industrially important fungi for sustainable development: Volume 2: Bioprospecting for biomolecules* (pp. 407–434). Springer International Publishing.

Pearce, T. R., Antonelli, A., Brearley, F. Q., Couch, C., Campostrini Forzza, R., Gonçalves, S. C., et al. (2020). International collaboration between collections-based institutes for halting biodiversity loss and unlocking the useful properties of plants and fungi. *Plants, People, Planet, 2*(5), 515–534.

Pedreschi, F., Mariotti, M. S., & Granby, K. (2014). Current issues in dietary acrylamide: Formation, mitigation and risk assessment. *Journal of the Science of Food and Agriculture, 94*(1), 9–20.

Pela, F. (2004). Biological activities of fungal metabolites. In *Handbook of industrial mycology* (pp. 68–111). CRC Press.

Pouris, J., Kolyva, F., Bratakou, S., Vogiatzi, C. A., Chaniotis, D., & Beloukas, A. (2024). The role of fungi in food production and processing. *Applied Sciences, 14*(12), 5046.

Pusztahelyi, T., Holb, I. J., & Pócsi, I. (2015). Secondary metabolites in fungus-plant interactions. *Frontiers in Plant Science, 6*, 573.

Qvirist, L., Mierke, F., Vazquez Juarez, R., & Andlid, T. (2022). Screening of xylose utilizing and high lipid producing yeast strains as a potential candidate for industrial application. *BMC Microbiology, 22*(1), 173.

Ramesh, C., Vinithkumar, N. V., Kirubagaran, R., Venil, C. K., & Dufossé, L. (2019). Multifaceted applications of microbial pigments: Current knowledge, challenges and future directions for public health implications. *Microorganisms, 7*(7), 186.

Rather, L. J., Mir, S. S., Ganie, S. A., & Li, Q. (2023). Research progress, challenges, and perspectives in microbial pigment production for industrial applications—A review. *Dyes and Pigments, 210*, Article 110989.

Ravindran, R., Hassan, S. S., Williams, G. A., & Jaiswal, A. K. (2018). A review on bioconversion of agro-industrial wastes to industrially important enzymes. *Bioengineering, 5*(4), 93.

Sajid, S., & Akbar, N. (2018). Applications of fungal pigments in biotechnology. *Pure and Applied Biology (PAB), 7*(3), 922–930.

Salaheldeen, M., Mariod, A. A., Aroua, M. K., Rahman, S. A., Soudagar, M. E. M., & Fattah, I. R. (2021). Current state and perspectives on transesterification of triglycerides for biodiesel production. *Catalysts, 11*(9), 1121.

Saloheimo, M. (2025). Filamentous fungi in biotechnology. In *Forest microbiology* (pp. 221–232). Academic Press.

Salvatore, D., Mok, K., Garrett, K. K., Poudrier, G., Brown, P., Birnbaum, L. S., et al. (2022). Presumptive contamination: A new approach to PFAS contamination based on likely sources. *Environmental Science & Technology Letters, 9*(11), 983–990.

Samtiya, M., Aluko, R. E., Dhewa, T., & Moreno-Rojas, J. M. (2021). Potential health benefits of plant food-derived bioactive components: An overview. *Foods, 10*(4), 839.

Sarica, D. (2024). Evaluating and forecasting pesticide usage in agricultural production in Türkiye. In *Environment, development and sustainability* (pp. 1–18).

Santra, H. K., & Banerjee, D. (2020). Natural products as fungicide and their role in crop protection. In *Natural bioactive products in sustainable agriculture* (pp. 131–219).

Sawant, A., & Sajwan, H. (2022). Fungal metabolites: Industrial applications and challenges. In *Sustainable utilization of fungi in agriculture and industry* (pp. 159–201).

Saxena, R. K., Gupta, R., Saxena, S., & Gulati, R. (2001). Role of fungal enzymes in food processing. In *Applied mycology and biotechnology* (Vol. 1, pp. 353–386). Elsevier.

Saye, L. M., Navaratna, T. A., Chong, J. P., O'Malley, M. A., Theodorou, M. K., & Reilly, M. (2021). The anaerobic fungi: Challenges and opportunities for industrial lignocellulosic biofuel production. *Microorganisms, 9*(4), 694.

Schmalhofer, M., Vagstad, A. L., Zhou, Q., Bode, H. B., & Groll, M. (2024). Polyketide trimming shapes dihydroxynaphthalene-melanin and anthraquinone pigments. *Advanced Science, 11*(22), 2400184.

Shankar, A., & Sharma, K. K. (2022). Fungal secondary metabolites in food and pharmaceuticals in the era of multi-omics. *Applied Microbiology and Biotechnology, 106*(9), 3465–3488.

Shin, J. A., Akoh, C. C., & Lee, K. T. (2009). Production and physicochemical properties of functional-butterfat through enzymatic interesterification in a continuous reactor. *Journal of Agricultural and Food Chemistry, 57*(3), 888–900.

Singh, N., & Gaur, S. (2021). GRAS fungi: A new horizon in safer food product. In *Fungi in sustainable food production* (pp. 27–37).

Singh, R., & Puniya, A. K. (2024). Role of food safety regulations in protecting public health. *Indian Journal of Microbiology, 64*(3), 1376–1378.

Singh, S., Kumar, V., Datta, S., Dhanjal, D. S., & Singh, J. (2020). Plant disease management by bioactive natural products. In *Natural bioactive products in sustainable agriculture* (pp. 15–29).

Sionek, B., Szydłowska, A., Küçükgöz, K., & Kołożyn-Krajewska, D. (2023). Traditional and new microorganisms in lactic acid fermentation of food. *Fermentation, 9*(12), 1019.

Solano, F. (2014). Melanins: Skin pigments and much more—Types, structural models, biological functions, and formation routes. *New Journal of Science, 2014*(1), Article 498276.

Stone, W. E., & Wright, H. E. (1898). Notes on Taka-diastase. *Journal of the American Chemical Society, 20*(9), 639–647.

Suhag, D. (2024). Regulatory and ethical considerations. In *Handbook of biomaterials for medical applications, Volume 2: Applications* (pp. 355–372). Springer Nature Singapore.

Takahashi, J. A., Contigli, C., Martins, B. A., & Lima, M. T. N. S. (2022). Fungal metabolites advances in contemporary industrial scenario. In *Fungal biotechnology* (pp. 87–129). CRC Press.

Talati, J. (2023). The role of fungal enzymes in microorganisms, biotechnology and chemical catalysts. *Fungal Genom Biol, 13*, 231.

Tao, B. Y. (2007). Industrial applications for plant oils and lipids. In *Bioprocessing for value-added products from renewable resources* (pp. 611–627). Elsevier.

Thakur, R. (2021). Use of *Trichoderma* spp. as biocontrol for disease management. *Indian Farmer, 8*(01), 108–115.

Umar, A., Mubeen, M., Ali, I., Iftikhar, Y., Sohail, M. A., Sajid, A., et al. (2024). Harnessing fungal bio-electricity: A promising path to a cleaner environment. *Frontiers in Microbiology, 14*, 1291904.

UNION, P. (2009). Regulation (EC) No 1223/2009 of the European parliament and of the council. *Official Journal of the European Union L, 342*, 59.

Usman, M. M., Idi, A., Aisami, A., & Maigari, F. U. (2023). A review on endophytic fungi: A natural source of industrial enzymes. *Asian Journal of Plant Biology, 5*(1), 7–11.

Vasić, K., Knez, Ž, & Leitgeb, M. (2021). Bioethanol production by enzymatic hydrolysis from different lignocellulosic sources. *Molecules, 26*(3), 753.

Venil, C. K., Velmurugan, P., Dufossé, L., Renuka Devi, P., & Veera Ravi, A. (2020). Fungal pigments: Potential coloring compounds for wide ranging applications in textile dyeing. *Journal of Fungi, 6*(2), 68.

Wadhwa, K., Kapoor, N., Kaur, H., Abu-Seer, E. A., Tariq, M., Siddiqui, S., et al. (2024). A comprehensive review of the diversity of fungal secondary metabolites and their emerging applications in healthcare and environment. In *Mycobiology* (pp. 1–53).

Xu, X., Sharma, P., Shu, S., Lin, T. S., Ciais, P., Tubiello, F. N., et al. (2021). Global greenhouse gas emissions from animal-based foods are twice those of plant-based foods. *Nature Food, 2*(9), 724–732.

Yadav, S. K. (2017). Technological advances and applications of hydrolytic enzymes for valorization of lignocellulosic biomass. *Bioresource Technology, 245*, 1727–1739.

Yadav, V. G., De Mey, M., Lim, C. G., Ajikumar, P. K., & Stephanopoulos, G. (2012). The future of metabolic engineering and synthetic biology: Towards a systematic practice. *Metabolic Engineering, 14*(3), 233–241.

Yokotsuka, T., & Sasaki, M. (1998). Fermented protein foods in the Orient: Shoyu and miso in Japan. *Microbiology of fermented foods* (pp. 351–415). Springer, US.

Zhivagui, M., Ng, A. W., Ardin, M., Churchwell, M. I., Pandey, M., Renard, C., et al. (2019). Experimental and pan-cancer genome analyses reveal widespread contribution of acrylamide exposure to carcinogenesis in humans. *Genome Research, 29*(4), 521–531.

Fungal Metabolomics and Analytical Techniques

Krystle Angelique A. Santiago, Roberth Riggs L. Rondilla, Jeane A. dela Cruz, Lloyd Christian J. Llames, and Thomas Edison E. dela Cruz

Abstract

Fungal metabolites have received so much attention due to the immense potential of these fungal natural products in health, agriculture and the environment. Now, like other metabolites in other organisms, fungal metabolites can also serve as indicators of the shifts in the biological and physiological states of fungal cell, and this has fuelled the rise of metabolomics. Metabolomics is a rapidly advancing field dedicated to the comprehensive analysis of the entire metabolome of an organism and combines biology, microbiology, chemistry and bioinformatics to provide a holistic understanding of the biochemical processes of an organism. This chapter listed analytical tools used in metabolomics and the application of metabolomics to fungal research. Key analytical techniques used to detect, characterize and isolate metabolites, such as nuclear magnetic resonance (NMR) and mass spectrometry (MS), will be emphasized. Also, the combination of metabolomics with other technologies in the field of omics including genomics, transcriptomics and proteomics will be presented. Web-based and software-based data analysis tools used in metabolomics research will also be discussed. The chapter concludes with the current trends in fungal metabolomics studies to know the fundamental role this research will have in future.

Keywords

Fungal metabolites · Analytical techniques · Metabolomes · Omics technologies · Biotechnology

1 Introduction

The high rate of antimicrobial resistance (AMR) is one of the most acute health issues in the modern world. AMR is accelerated by antibiotic abuse and overuse (Ahmed et al., 2024; Salam et al., 2023) that reduces the functional efficacy of the available drugs and medications (Bloom & Cadarette, 2019; Ferri et al., 2017; Tang et al., 2023). Moreover, the biology and natural expeditive nature of the pathogens enable them to develop resistance quickly that adds to the further emergence of AMR (Ahmad et al., 2021).

The war against AMR is not an easy task. First, it is complicated by the large variety of microbial pathogens, each having complex resistance mechanisms, thus complicating any therapeutic innovations (Baquero et al., 2021; Coque et al., 2023). Moreover, low-resource areas can increase data gaps through limited AMR surveillance systems, reflecting an understudied area thus impeding the formulation of possible interventions (Rony et al., 2023). Continuous efforts are being developed to identify resistance mechanisms and provide novel therapeutic targets (Bano et al., 2024; Bartelo et al., 2024; Ramzan et al., 2024; Wang et al., 2024). Hence, many organisms are now being explored to produce bioactive secondary metabolites that can fight these infectious agents (Linhares et al.,

K. A. A. Santiago
School of Science, Monash University Malaysia, Subang Jaya, Selangor Darul Ehsan, Malaysia

R. R. L. Rondilla
Department of Mathematics and Natural Sciences, College of Arts and Sciences, Southern Luzon State University, Lucban, Quezon, Philippines

J. A. dela Cruz · L. C. J. Llames · T. E. E. dela Cruz (✉)
Department of Biological Sciences, College of Science, University of Santo Tomas, Manila, Philippines
e-mail: tedelacruz@ust.edu.ph

J. A. dela Cruz · T. E. E. dela Cruz
The Graduate School, University of Santo Tomas, Manila, Philippines

J. A. dela Cruz · L. C. J. Llames · T. E. E. dela Cruz
Fungal Biodiversity, Ecogenomics and Systematics-Metabolomics (FBeS) Group, Research Center for the Natural and Applied Sciences, University of Santo Tomas, Manila, Philippines

© The Author(s), under exclusive license to Springer Nature Switzerland AG 2026

A. Prakash and M. K. Dubey (eds.), *Advances in Fungal Secondary Metabolites: Bioactive Attributes and Biotechnological Potential of Industrial Interest*, Sustainable Economy and Ecotechnology, https://doi.org/10.1007/978-3-032-23507-7_10

2023). One example of a technology proving huge impact is metabolomics.

Metabolomics is a comprehensive study of the metabolome, within any biological system (Oliver et al., 1998). The technique analyses complex biochemical networks and identifies unique metabolic markers of a disease. Therefore, exploring this technique could help enhance particularly the discovery of novel antimicrobial agents (Panter et al., 2021; Venkataraman et al., 2024). One can use the application either in a targeted or non-targeted approach (Kok et al., 2022; Tigistu-Sahle et al., 2023).

Targeted metabolomics is concerned with the measurement of a predetermined number of metabolites or a particular metabolic pathway and provides an opportunity to accurately analyse biochemical alterations under defined conditions (Fortuin & Soares, 2022; Kumar et al., 2022). It is beneficial to use it in tracking certain biomarkers and hypothesis testing. Nevertheless, the targeted quality of this methodology can fail to consider new or unexpected metabolites that may provide new therapeutic information. In the meantime, non-targeted metabolomics is a study of the entire metabolome without any form of predetermination that generates a more discovery-oriented and larger approach (Overton et al., 2022). This method is capable of identifying previously unknown metabolic pathways and the discovery of new biomarkers, compared to the specific metabolomics, and the resultant data are very complex and large and can only be reliably interpreted through the application of sophisticated analytical tools. A synthesis of the two of these approaches, including advanced analytical capabilities such as powerful bioinformatics software and cross-functional synergy, will enable metabolomics to have a role to play in dealing with the complex issues of AMR and in the identification of new therapeutic compounds.

This has been used in various biological systems in approaches to drug discovery. These techniques have discovered pharmacologically active plant compounds, such as alkaloids, flavonoids, and terpenoids, and many of them are antimicrobial (Marchev et al., 2021; Pandohee et al., 2023; Sieniawska & Georgiev, 2022). Other marine organisms such as sponges also are reported to synthesize antimicrobial compounds some of which have been found with the help of metabolomics (Patry, 2021). In addition, metabolomics, which were done using liquid chromatography-mass spectrometry (LC-MS), allowed the identification of structurally diverse metabolites in fungi. Some examples include the works on endophytic fungi by LC-HRMS (liquid chromatography-high-resolution mass spectrometry) and nuclear magnetic resonance (NMR)-based metabolomics (Alhadrami et al., 2021; Attia et al., 2020; Ibrahim et al., 2020), which have explained various fungal metabolites towards important drug discovery. Tang et al. (2020) also shed more light on Basidiomycota through the use of untargeted metabolomics through liquid chromatography-mass spectrometry (LC-MS) that defined 236 different metabolites under different development stages. The use of metabolomics to identify new bioactive compounds was also used in Flores et al. (2022), which further contributes to the usefulness of the method in screening antimicrobial metabolites of various fungal origins. These papers confirm the possibility of metabolomics in the advancement of the discovery of new therapeutic agents in different sources to fight multidrug-resistant pathogens. Also presented are different analytical tools and other studies in fungal metabolomics such as platforms and an assortment of softwares utilized to examine metabolomics data. The use of metabolomics in other omics technology is also addressed. The trend of research in fungal metabolomics is presented at the end of this book chapter.

2 Analytical Techniques as Applied to Fungal Metabolomics

Metabolomics gives a comprehensive and actual 'picture' of metabolic pathways being expressed by a particular organism. In microbial secondary metabolism, metabolomics allows for a rapid and extensive analysis of crude extracts, including the identification of compounds associated with pharmacological activities (Mishra et al., 2022). Exhaustive metabolic data were fundamentally brought about by recent advances in analytical instrumentation used to detect these metabolites. The two main analytical platforms applied in metabolomics studies are NMR spectroscopy and HRMS. Several separation techniques were coupled to MS, such as capillary electrophoresis (CE), gas chromatography (GC), and LC. MS mainly offers excellent sensitivity over NMR. A single fungal extract could possess numerous amounts of complex and diverse secondary metabolites, and consequently, there is no single analytical platform that can provide complete metabolic information (Stuart et al., 2020). Two or more of these techniques can be used in parallel to obtain a wider coverage of the fungal metabolome. Herein, we discuss the applicability of each instrument as it is applied to fungal metabolomics. The disadvantages and advantages of the different analytical platforms are presented in Table 1.

2.1 NMR-Based Metabolomics

NMR is a spectroscopic method widely used for elucidating the chemical structure of small organic molecules. This technique involves applying a magnetic field and radiofrequency to a sample, causing resonances at different frequen-

Table 1 Disadvantages and advantages of different analytical platforms which are used or applied in fungal metabolomics research

Parameters	NMR	GC-MS	LC-MS
Sample amount	At least 5 mg	1 mg	1 mg
Sample preparation	Minimal	Often needs derivatization	None
Nature of sample/ metabolites	Polar to non-polar	Volatile polar and non-polar	Non-polar (using a reverse phase column) Polar (using HILIC—Hydrophilic Interaction Liquid Chromatography)
Run time	5 min (for ^{1}H NMR)	> 10 min	> 10 min
Sensitivity	Low	Very high	Very high
Reproducibility	High	Low	Low
Resolution expressed as separation of metabolites	Low	High	High
Identification/ Annotation of metabolites	Structural information only	Easy (spectral libraries)	Database needed
Quantification	Yes	Yes (needs reference standards)	

Source Authors

cies for various atomic nuclei, such as ^{1}H. The quantification of specified molecules or biomarkers can be successfully done but only above micromolar concentrations and thus results in its low sensitivity. This means that minor components can often be missed in NMR metabolomic analysis. Although NMR has lower sensitivity as compared to the MS techniques (Table 1), it remains crucial in metabolomics research because of its high reproducibility, simplicity in sample preparation, non-destructive nature, and ease of quantification (Gowda & Djukovic, 2014). Due to its simplicity and rapid data acquisition time, NMR has gained interest in metabolomics research and is often used for high-throughput studies.

Given that crude extracts contain numerous metabolites, the detection and identification of individual molecules is impractical in NMR. Instead, NMR-based metabolomics provides an overview of the metabolites present in the extract, such as primary metabolites (amino acids, sugars, lipids, etc.) and secondary metabolites (alkaloids, flavonoids, phenolics, terpenes, etc.). These compounds can be differentiated based on the ^{1}H chemical shifts of specific functional groups. In other words, it offers a chemical fingerprint or metabolite profile of a given microbial sample or extract. As a case in point, the characteristic signals of terpenes were eventually correlated with the pronounced anti-inflammatory effects of a crude extract obtained from the fungal endophyte *Colletotrichum* sp. (Flores-Vallejo et al., 2020). In addition, the geographical origin (Korea vs China) of the medicinal mushroom *Ganoderma lucidum* were distinguished through NMR-based metabolomics approach with choline and propionic amino acids as discriminating metabolites (Wen et al., 2010). Through NMR metabolite profiling, secondary metabolite variations in relation to collection sites and parts (caps and stipes) of two edible mushroom species, *Hypholoma capnoides* and *Kuehneromyces mutabilis* were recognized (Alanne et al., 2019). NMR-based metabolomics also differentiates several groups of primary and secondary metabolites that were induced and regulated in plant roots which were inoculated with the ectomycorrhizal fungus *Pisolithus tinctorius* (Sebastiana et al., 2021). A computational method using principal component analysis (PCA) based on ^{1}H NMR data was developed to distinguish *Fusarium solani* from *F. oxysporum*, all of which were isolated from the rhizospheric soil of *Senna spectabilis* (Selegato et al., 2016). The method, when combined with an in-house metabolite library, can successfully identifies specific mycotoxins, including fusaric acid and beauveracin, thereby enhancing the dereplication process. NMR metabolomics in tandem with the one-strain many compounds (OSMAC) strategy was an efficient and rapid method to select microbial fractions with excellent chemical diversity (Liu et al., 2017).

Now, the metabolomics studies predominantly utilize one-dimensional ^{1}H NMR spectroscopy. This owes to the fact that ^{1}H nuclei have very high isotopic abundance (99%) as compared to other atomic nuclei such ^{13}C (1.1%) and ^{15}N (0.365%). Fortunately, there are other NMR experiments that can be applied to further enhance metabolomic data coverage and resolution. Two-dimensional (2D) NMR spectroscopy is usually used for structure elucidation, but in metabolomics, overlapping peaks can be further resolved. It potentially allows the identification of metabolites present within a mixture. Examples of 2D-NMR experiments applied in metabolomics are ^{13}C single quantum coherence (^{1}H-^{13}C-HSQC), 2D J-resolved spectroscopy (J-Res), correlation spectroscopy (COSY), nuclear overhauser effect (NOESY), total correlation spectroscopy (TOCSY), and ^{1}H-and heteronuclear multiple bond correlation (HMBC) (Emwas et al., 2019; Martineau et al., 2020). However, these experiments will take a much longer acquisition time as compared to one-dimensional proton NMR spectroscopy (1D-^{1}H NMR). 2D-NMR-based metabolomics has gained attention in plant sciences (Mahrous & Farag, 2015) and other natural product research (Robinette et al., 2012). For instance, the geographical origin of several pistachio samples was successfully examined by applying ^{1}H and 2D TOCSY (two-dimensional total correlation spectroscopy) experiments. Different metabolites were featured as markers for the certified origin and quality control of pistachio nuts

(Sciubba et al., 2014). There are little to no 2D-NMR metabolomics pursued for fungal metabolites, leaving a gap that future researchers can fill up.

Besides the innovations in 2D-NMR experiments, several techniques have been incorporated in recent metabolomics studies. These include high-resolution magic-angle sample spinning (HRMAS), hybrid NMR methods, hyperpolarization methods, pure shift NMR techniques, and ultrafast 2D NMR approaches (Emwas et al., 2019). HRMAS can analyse semisolid or gel-like samples without prior sample extraction steps (Garrod et al., 1999). Hence, intact samples like biological tissues can be directly subjected to the instrument and placed at the 'magic angle' (54.74°). HRMAS techniques were mainly applied in clinical research, particularly metabolic markers that can aid in the diagnosis of several types of cancer (Fuss & Cheng, 2016). It has also gained interest in plant metabolomics (Augustijn et al., 2021) and food and agricultural applications (Corsaro et al., 2016; Mazzei & Piccolo, 2017). Only a single study involving fungi can be found in the literature. In this paper, HRMAS metabolomics offered a fast and accurate method to examine metabolic changes in tomato plants treated with two metabolites from the fungus *Trichoderma* (Mazzei et al., 2016).

The characteristic limited detectability of NMR instruments can be eluded with hyperpolarization techniques. In this process, polarization of nuclear spins is significantly increased that helps to detect and analysis of minute compounds (Ribay et al., 2023). A number of methodologies are available to achieve hyperpolarization in metabolomics, including dissolution dynamic nuclear polarization (d-DNP) and para-hydrogen (p-H_2) hyperpolarization. Hyperpolarization NMR metabolomics is an emerging technique with only several research papers utilizing it over the past decade (Ribay et al., 2023).

Another innovative methodology is the ultrafast NMR technique. This strategy drastically reduces data collection time as compared to conventional 2D NMR experiments (Giraudeau & Frydman, 2014). It can provide multidimensional NMR data in a single scan. Ultrafast NMR results from a spatial-encoding procedure inspired from imaging (Tal & Frydman, 2010). There are only a few studies that employed ultrafast NMR in metabolomics. One article discussed the investigation of serum from colorectal cancer subjects. Acquisition time, sensitivity, and resolution were dramatically improved with the use of the said NMR technique (Guennec et al., 2014).

Resolution of overlapped peaks coming from several numbers of metabolites in a complex mixture is indeed challenging. Pure shift NMR converts multiplet peaks to singlets by relying on pulse sequences intended to let the interactions of nuclear spins with the main magnetic field (chemical shifts) evolve while suppressing spin–spin interactions or couplings (Foroozandeh et al., 2018). This procedure found its way in metabolite profiling of complex biological samples (Chen et al., 2023). One study demonstrated the practical benefits of pure shift NMR by applying the pure shift yielded by chirp excitation (PSYCHE) experiment to enhance resolution in the context of food quality and authenticity. In that study, tea extracts and honey samples were analysed to probe concerns related to adulteration and food origin. Their findings showed that the identification of major components in these samples was significantly improved using PSYCHE pure shift spectra (Bo et al., 2019). A statistically valid correlation between certain metabolites of soft coral extracts and its observed cytotoxicity was achieved with pure shift PSYCHE NMR tools (Santacruz et al., 2020).

NMR remains an invaluable analytical platform in metabolomics, including the study of fungal secondary metabolites. Though there are several downsides as compared to MS-based metabolomics, ingenuine advanced NMR experiments have been developed over the years to enhance resolution, sensitivity, and acquisition times. Certainly, NMR-based metabolomics will still be an essential instrument to bring a detailed understanding of the fungal metabolome.

2.2 MS-Based Metabolomics

Mass spectrometry has superior sensitivity and a wide dynamic range versus NMR spectroscopy (Table 1). Because of its sensitivity, thousands of compounds can be detected in a single crude extract, even at picomolar concentrations (Marshall & Powers, 2017). However, due to this complexity, it often necessitates prior separation of metabolites of interest. Hence, hyphenated techniques such as LC-MS (with liquid chromatography), GC-MS (with gas chromatography), and CE-MS (with capillary electrophoresis) methods have found their applications in numerous metabolomics studies. GC methods offer better separation and reduced matrix effects than LC (Mastrangelo et al., 2015). Another problem in mass spectrometry is the proper ionization of metabolites and ion suppression. GC-MS generally prevents ion suppression since its analytes are already in the gaseous phase. However, GC can only detect volatile and low-molecular weight compounds (ca. 50–600 Da) (Beale et al., 2018). That is why samples for GC typically need to be chemically derivatized to be analysed (Table 1). In this case, LC-MS has the advantage of detecting larger and intact metabolites. As a result, LC-MS and GC-MS are often combined to widen the spectrum of identified metabolites. CE-MS are mainly used for polar and charged compounds (amino acids, nucleotides, small organic acids, and sugar phosphates) without the need for sample pretreatment, very little to no use of organic solvents, and cheap capillary columns (Ramautar et al., 2009). However, CE-MS is less sensitive and reproducible than GC-MS or LC-MS.

Detection of the mass of the compounds is dependent on the MS instrumentation. Analytes should be first ionized to be able to measure its mass-to-charge ratio (m/z). There are a wide range of available ionization techniques, including atmospheric pressure chemical ionization (APCI), electron impact ionization (EI), electrospray ionization (ESI), and matrix-assisted laser desorption ionization (MALDI). All these have different mechanisms to ionize individual metabolites, which impacts its detectability in the MS detector. ESI is a general choice for metabolomics applications (Challen & Cramer, 2022).

Mass accuracy and resolution are critical performance characteristics of MS instruments. These are mainly dependent on the type of mass analysers or detectors inside the equipment. Numerous types of mass detectors are available, each offering its own unique benefits. This includes Fourier transform ion cyclotron resonance (FTICR), linear ion trap (LIT), quadruple (Q), quadrupole ion trap (QIT), Orbitrap, and time of flight (TOF). These are either operated as a single configuration or as tandem, e.g. MS/MS, albeit MS/MS methods cause fragmentation of ions, which gives additional structural information to correctly identify metabolites. Other common tandem MS/MS instruments include linear-quadrupole ion trap-Orbitrap (LTQ-Orbitrap), quadrupole-TOF (Q-TOF), triple quadrupole (TQ), and triple-quadrupole ion trap (QTrap). LC-Qtrap or LC-TQ is often used for targeted metabolomic studies due to their higher sensitivity and selectivity, while Q-TOF, LTQ-Orbitrap, and FTICR are better used in untargeted metabolomics studies because of higher mass accuracy (Demetrowitsch & Schwarz, 2014; Gowda & Djukovic, 2014). The types of mass analysers and ionization techniques are combinedly matched to efficiently bring out accurate metabolite profiling and quantification in metabolomics research.

2.2.1 Gas Chromatography-Mass Spectrometry

GC-MS are primarily used for volatile compounds. If compounds are not volatile, these need a prior derivatization step. Thermal stability of metabolites must also be considered in GC-MS studies. However, despite the concerns on the nature of the samples that can be analysed by the mass spectrometry, GC-MS is still a useful platform for metabolomics (Beale et al., 2018). Commonly, EI is used to ionize analytes in a GC-MS instrument. EI is a hard-ionization method as compared to the normally used ESI for LC-MS. Though it causes sample destruction, EI is much more reproducible than ESI. This is because the m/z data and its fragments can be compared to readily available mass spectral databases such as the NIST database which brings a highly accurate identification of compounds (Vinaixa et al., 2016).

There are numerous GC-MS studies in the regard to fungal metabolomics. *Phanerochaete chrysosporium*, the model white-rot fungus was exposed to different oxygen concentrations. Through GC-MS, differences in metabolite levels (erythronate, threonate, and veratryl alcohol) between the oxygen conditions provided insights into how *P. chrysosporium* adapts its metabolism in response to varying oxygen availability (Miura et al., 2004). GC-TOF MS detected phenylpropanoid compounds 4-hydroxystyrene, piceatannol, sinapate, and taxifolin under starvation of the grape endophytic fungi but not under normal conditions (Lu et al., 2020). The peptide ε-poly-l-lysine (ε-PL) increased the ability of industrial yeast to withstand freeze–thaw tolerance. Untargeted metabolomics based on GC-MS analysis revealed that the mechanism behind improved freeze–thaw tolerance was associated with alterations in intracellular metabolites and metabolic processes (Lu et al., 2022). The biochemical compounds during growth stages of the well-known East Asian medicinal mushroom *Cordyceps militaris* were investigated through GC-MS. Cordycepin, a major bioactive compound, was highly enriched during the later stages of fungal growth and thus unravelled optimal time of harvest (Oh et al., 2019). A GC was used in a selected ion monitoring mass spectrometry in conjunction with GC to study the metabolites in the isoprenoid pathway of a genetically modified Saccharomyces cerevisiae in a targeted metabolomics (Huang et al., 2011). Threitol was especially enhanced in *Eucalyptus grandis* root tip colonization by *Armillaria luteobubalina*, and its further GC-MS metabolomics analysis showed that the metabolite was of pathogenic origin. The observation highlights the role of fungal-derived signalling in the multifaceted nature of the plant-fungal interactions (Wong et al., 2020). It was possible to resolve the complex of *Rhizoctonia solani* species using various multivariate and hierarchical cluster analysis of its GC-MS metabolome (Aliferis et al., 2013).

Two-dimensional gas chromatography (GC × GC)–mass spectrometry (MS) is another strong advancement in the field. Two successive GC columns were employed in this arrangement, and the first column was a normal long GC column and the second one a shorter column. This gives extra separation of the compounds which improves the resolution of the co-eluting peaks and increases the peak capacity (Biedermann & Grob, 2019). It is an innovation that has begun to be of interest in metabolomics research (Feizi et al., 2021). As an illustration, the metabolome of the fungal endophytes of *Celtis africana* was comprehensively studied with its help. The endophytic fungi grew in isolation were capable of producing specific metabolites, and some compounds also were isolated by the host plant (Tapfuma et al., 2020).

2.2.2 Liquid Chromatography-Mass Spectrometry

LC-MS is often the technique of fungal metabolomics. The fact that in a review published recently (Li et al., 2021), 29 of 49 mentioned articles (approximately 60%) used this analytical platform to characterize fungal metabolites supports its prevalence. Sample pretreatment in LC-MS is also less complicated than GC-MS. At this point, two possible modes of separation methods are commonly applied in the context of LC-MS-based metabolomics method: (1) reverse-phased columns that are utilized to separate non-polar and semi-polar compounds, and (2) hydrophilic interaction liquid chromatography (HILIC) that provides better separation capabilities in the context of highly polar compounds (Tang et al., 2016). Ultra-performance liquid chromatography (UPLC) columns have been recently used to achieve a better resolution and high-quality MS as well as a faster analysis and more sensitivity (Plumb et al., 2023).

One of the most common ionization methods employed in the LC-MS metabolomics is electrospray ionization (ESI) (Agin et al., 2016; Schmid et al., 2023). Eluted samples of the LC system are sprayed directly to a vacuum environment in this technique with the voltage applied, and this produces nebulized charged droplets. It forms adducts of the generated ions based on the mode applied, i.e. positive or negative. The most common adducts used in the positive mode include $[M + H]^+$ and $[M + Na]^+$, whereas in the negative ESI mode, it includes $[M - H]^-$ (Manohar & Marzinke, 2016).

These ions were further subjected to fragmentation in tandem MS/MS instruments which allows substructuring matching in compound databases with high confidence (Suntivich et al., 2022). Thus, LC-MS/MS-based metabolomics offers highly accurate metabolite annotation.

Application of LC-MS in fungal metabolomics has been widely reported (Li et al., 2021; Nagarajan et al., 2022). As an example, anti-trypanosomal compounds were obtained from the endophytic fungus *Lasidiplodia theobromae* through LC-MS metabolomics-guided isolation strategy (Kamal et al., 2017). In another report, metabolites that can chemotaxonomically distinguish *Penicillium setosum* from its related species were identified through LC-MS metabolomics (George et al., 2019). Through UPLC-QTOF-MS, metabolites caused the inhibition of *Zygosaccharomyces rouxii* when it was co-cultured with *Aspergillus oryzae* (Liu et al., 2022). In another study, the same instrument was used to analyse the molecular and metabolic mechanisms of the browning of *Agaricus bisporus* (Cai et al., 2022). Additionally, HILIC-QTOF-MS could detect dopa melanin and other phenolics as the causal organisms of the browning of another edible mushroom *Flammulina filiformis* (Fu et al., 2021).

It has also been identified that metabolomics can be used as a secondary method of fungal identification (Frisvad et al., 2008; Hyde et al., 2023; Maciá-Vicente et al., 2018). As an example, seven Trichoderma species were characterized by chemotaxonomic characteristics by use of LC-ESI-MS/MS (Kang et al., 2011). The discriminatory biomarker identification using lipidomics was successfully implemented on various edible mushrooms (Yang et al., 2021).

A number of computational instruments were devised in order to optimize the interpretation and analysis of often complex data of tandem MS/MS instruments. Molecular networking (MN) is one of the finest algorithms that represents clusters of compounds that have similar substructures (Quinn et al., 2017). MN may be implemented within the web-based environment like global natural product social molecular networking (GNPS) (Aron et al., 2020). In the article of Oppong-Danquah et al. (2020), MN was applied to create networks of marine fungi metabolites cultivated as mono or co-cultures with phytopathogenic fungi. MN was also used with the bioactivity-guided isolation strategy to recover two new a-pyrones and one known derivative and three known indole diketopiperazines by the Antarctic fungi *Acrostalagmus luteoalbus* (Shi et al., 2022). The use of MN was considered the guiding tool towards the targeted isolation of bioactive compounds of *Aspergillus versicolor* (Han et al., 2020). The greater part of the instruments and programs applied in metabolomics is introduced in the subsequent section of this chapter.

2.3 Other MS-Based Techniques

2.3.1 Maldi-TOF-MS

The development of MS instrumentation also was further used to enhance metabolomics analysis. MALDI-TOF-MS is a soft ionization method that is similar to ESI but with a laser to ablate molecules in a surface to form ions in the gaseous phase. Over the past few decades, MALDI-TOF-MS competency to identify bacterial pathogens is accurate and time-saving (Haider et al., 2023; Hou et al., 2019). Moreover, it has already been used in several microbial metabolomics analyses (Santos et al., 2016). The MALDI-TOF-MS has been used to identify *Aspergillus* species and various yeast strains with high-resolution (Hettick et al., 2008; Qian et al., 2008). These are the best examples of how MALDI-TOF may be broadly used in fungal taxonomy and agriculture.

2.3.2 Ion Mobility Spectrometry (IMS)

The IMS technique is a separation method in which the separation of ions occurs according to their mobility in a gaseous medium under an applied electric field. The speed of IMS is very high relative to other chromatographic techniques because ions separation occurs within milliseconds (Lapthorn et al., 2013). MS is applicable in IMS that enhances selectivity, isomer resolution, detection limit, and accuracy in annotating the metabolic. Hence, IMS-MS has been successfully applied to plethora of metabolomics studies (Levy et al., 2019; Luo et al., 2024; Zhang et al., 2018). IMS facilitated in situ analysis of fungi-contaminated barley with a comparable detection limit with GC-APCI-MS (Erler et al., 2020). Real-time monitoring of wood fungal colonization/infestation was feasible, rapid, and sensitive through IMS-MS. (Hübert et al., 2011). Through IMS-MS coupled with UPLC, eight nucleosides and the biomarker cordycepin within *Ophiocordyceps sinensis* samples could be simultaneously quantified in (Joshi et al., 2018). Finally, GC-IMS is used for real-time monitoring of fungal volatile compounds to understand the influence of culture conditions on metabolite production by the plant pathogenic fungus *Fusarium oxysporum* (Speckbacher et al., 2021).

2.4 Combination of Techniques

The chemical characteristics and nature of metabolites of interest mainly dictate which will be the most suitable analytical instrument to answer specific research aims. As shown in Table 1, NMR and MS analytical platforms offer different advantages, but also have limitations. No single technique can fully measure the metabolome of samples. Hence, many researchers explored the idea of combining several analytical platforms to improve metabolite coverage (Jeppesen & Powers, 2023). For example, GC-MS and LC-MS were integrated to understand key metabolites and biochemical pathways influenced by methyl jasmonate in *Ganoderma lucidum* (Jiang et al., 2019). NMR and GC metabolomics analysis paved the way for the identification of affected metabolic pathways, such as trehalose biosynthesis in the mutant fungi (Liu et al., 2019). Combining NMR and GC-MS profiling led to the identification of significant metabolites, such as α-trehalose, D-glucose, and various fatty acids, which are important in distinguishing and classifying the developmental stages of sclerotia in the soilborne plant pathogen *Rhizoctonia solani* AG-3 (Aliferis & Jabaji, 2010). Applying both ^{1}H-NMR and direct injection MS allowed for the extensive investigation of intracellular metabolome of various *Fusarium* species under both toxin-producing and non-producing. It reveals that the external nutritional environment significantly influences the metabolome more than the genotype, with conserved responses in toxin-producing species, including increased levels of glycerol, mannitol, and γ-aminobutyric acid (Lowe et al., 2010). LC-QTOF-MS together with ^{1}H NMR was employed to compare the endo- and exo-metabolomic profiles of *Fusarium graminearum* cultured under in toxin-inducing and repressing conditions. Exposure of the said fungus to caffeic acid led to certain metabolic changes without affecting fungal growth (Atanasova-Penichon et al., 2018).

3 Integrating Metabolomics Data with Other 'Omics'

Metabolomics generates vast amounts of data (i.e. 'big data') from complex chemical, biological, and physiological interactions. Specifically, metabolomics focuses on the qualitative and quantitative analysis of these metabolites, providing insights into changes within biological systems in response to various biotic or abiotic factors (Cambiaghi et al., 2017). However, metabolite profiling alone is insufficient to fully understand the complexity of biological systems. Metabolomics can be integrated with other omics approaches (i.e. genomics, proteomics, and transcriptomics), leading to what is referred now as 'multi-omics' or 'cross-omics' technologies, and these generate more comprehensive and insightful data about biological systems and their metabolic interactions (Jendoubi, 2021). These integrative strategies are particularly crucial to advance metabolite discovery and understand complex biological processes. The 'multi-omics' or 'cross-omics' framework is illustrated in Fig. 1.

Metabolomics coupled with other omics technologies have been proven invaluable in the study of natural products. Nature-derived metabolites remain an unparalleled source of bioactive lead compounds in drug discovery owing to the unique structures and varied bioactivities of these metabolites (Zhang et al., 2021). Microorganisms, particularly bacteria and fungi, have been proven as valuable resource in drug discovery efforts (Palazzotto & Weber, 2018), and these have been targeted by pharmaceutical companies over the years. For example, in 1928 penicillin discovery from the green mould (*Penicillium notatum*) by Alexander Fleming has paved the way for the development of the antibiotic era (Bhattarai et al., 2021). With the application of the multi-omics approach, studies on chemical interactions and gene expressions under specific conditions are now possible. The holistic perspective provided by multi-omics studies has therefore revealed the intricate interplay between metabolites, genes, proteins, and other biomolecules. Exemplified below are some reports of multi-omics approaches.

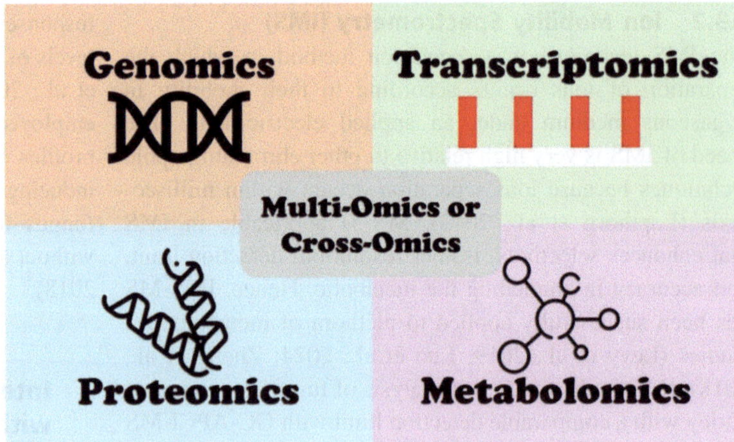

Fig. 1 'Multi-omics' or 'cross-omics' framework combines genomics, proteomics, and transcriptomics with metabolomics for a comprehensive understanding and better insights into biological processes or systems. *Source* Authors

The interaction between metabolites and proteins were conducted to advance fungal research. For instance, the integration of metabolomics and proteomics has successfully predicted 461 protein-metabolite interactions in the thermophilic fungus *Chaetomium thermophilum* (Li et al., 2021). Such an approach provided insights into the metabolic regulation of *C. thermophilum* and offered different applications in metabolic engineering and synthetic biology. Equally, integrating metabolomics with proteomics revealed the effect of the secondary metabolites of *Trichoderma harzianum* to the physiology of the phytopathogen *Fusarium culmorum* (Mironenka et al., 2021). The *T. harzianum* metabolites resulted in serious oxidative stress to *F. culmorum* and interfered with its expression of zearalenone (ZEN) gene that produces mycotoxins. Besides the fungi-pathogen relations, the combination of metabolomics and proteomics was also critical to the fermentation process, especially in fermentation by fungi. In another recent trial by Ma et al. (2021), the fermentation of tea leaves using the fungus *Aspergillus niger, A. tamarii* or *A. fumigatus,* and the integrated omics method disclosed that enzymes generated by *Aspergillus* species played a significant role in the metabolite concentration and composition of tea leaves.

Transcriptomics can also be used together with metabolomics. As an example, this combined type of omics method demonstrated the key response metabolites and the molecular mechanisms of *Aspergillus fumigatus* in response to heavy metal cadmium (Cd) (Tian et al., 2024). In the case, 358 differentials were found to be affected on exposure to Cd at varying incubation times, and a few of the genes were significantly up or down-regulated. Significant metabolic pathways, which comprise energy metabolism, amino acid metabolism, and the system of ABC transport, were observed to make significant contributions towards the process of the Cd tolerance and the detoxification of heavy metal in fungi. Alongside, the antifungal mechanisms of the cinnamic acid derivative of phenylcinnamic acid (PA) on the fungal plant pathogen *Colletotrichum gloeosporioides* have also been reported by the metabolomic and transcriptomic analysis (Zhang et al., 2022). It was found that the *C. gloeosporioides* can be affected by the *C. gloeosporioides* as the decrease of amino acid metabolism and interference with secondary metabolite biosynthesis of the pathogen. Lastly, the integrated metabolomic and transcriptomic method resulted in the observation of the metabolic and genetic differences in wild-type and cultivated *Ophiocordyceps sinensis*, a medicinal fungus native to Qinghai-Tibet Plateau (Zhang et al., 2023). The metabolomic and transcriptomic studies showed that wild type *O. sinensis* had more nucleosides and nucleotides, but cultivated strains had more amino acid derivatives, carbohydrates and derivatives, and phenolic acids (Zhang et al., 2023).

Finally, fungal studies have also been positively reflected in the discovery of important secondary metabolites by metabolomics and the analysis of gene clusters involved in their syntheses by genomics. The technique has been found especially helpful in the identification of new fungal metabolites. As an example, *Beauveria felina* SYSU-MS7908 marine-derived fungus was discovered through the use of this combined method to isolate and characterize new cyclopeptides (i.e. isaridins I-N) (Jiang et al., 2023). On the same note, 110 Ascomycetes have been successfully assessed to match their fungal metabolism to their biosynthetic gene clusters (BGCs) on the basis of joint analysis of genomic and metabolomic datasets (Caesar et al., 2023). Moreover, new metabolites have also been identified as a result of the combined approach, e.g. the chemical diversity of terpenes among the ectomycorrhizal fungi *S. cothurnatus* VC1858, *S. decipiens* EM49 and *Suillus hirtellus* EM16 were discovered (Mudbhari et al., 2024).

The use of metabolomics, and other omics technologies, has certainly opened the door to more detailed and informative studies of fungi. Since the multi-omics methodology is designed to make a total, unbiased, and accurate

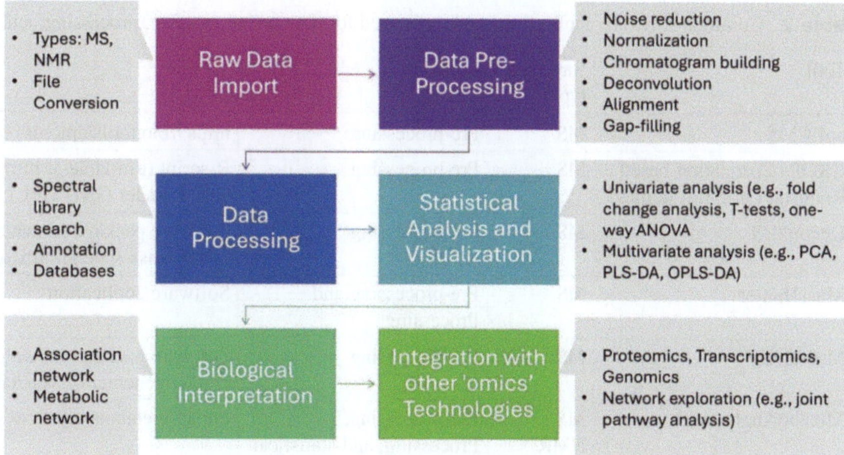

Fig. 2 General workflow for metabolomics analysis. These steps can be performed using a single or combination of software tools. MS: Mass Spectrometry, NMR: Nuclear Magnetic Resonance, ANOVA: Analysis of Variance, PCA: Principal Component Analysis, PLS-DA: Partial Least Squares-Discriminant Analysis, OPLS-DA: Orthogonal Partial Least Squares-Discriminant Analysis. *Source* Authors

apprehension of the molecules and interactions of interest, inherent in the inter personalization of these disparate omics datasets, the software, databases as well as analytical tools are advanced, robust, and sophisticated in quality to obtain an accurate analysis and interpretation.

4 Software Tools and Platforms Used in Metabolomics Analysis

Big data requires advanced and robust analytical tools and platforms. With the field of metabolomics drastically expanding over the last few years, distinct workflow, including data analysis methods, software tools and databases, to deliver comprehensive datasets, visualization, and interpretation, has been carefully developed. The general workflow for metabolomics analysis is illustrated in Fig. 2.

When conducting any metabolomics studies, it is essential to first determine the appropriate analytical platform, i.e. MS versus NMR, to be used or is appropriate for the study. While both MS and NMR platforms are widely used in metabolomics research (Sarkar et al., 2025; Wishart et al., 2022), each has its own strengths and limitations (Table 1). For instance, MS, particularly liquid chromatography-mass spectrometry (LC-MS), is widely popular due to its high sensitivity and excellent chromatographic resolution (Liu et al., 2019). However, MS has certain limitations, including the inability to perform without the availability of authentic standards. Additionally, MS struggles to analyse compounds that are not easily ionized or those lacking polar moieties, the MS-inactive components (Liu et al., 2019). NMR, in contrast, is a non-destructive, unbiased, and highly quantifiable analytical technique (Liu et al., 2019) and has earned a reputation as the 'gold standard' in metabolomics research due to its reliability in compound identification (Wishart et al., 2022). Nevertheless, NMR is relatively less sensitive than MS, and its sensitivity is usually 10–100 times lower (Emwas et al., 2019), which may restrict its use to identify low-abundance metabolites.

After the selection of the suitable analytical platform, it is necessary to pay much attention to the selection of suitable software tools to be used in the data pre-processing, processing, and statistical analysis. The common software tools used in metabolomic studies are listed in Table 2. These tools are either freely available from the web or may require a subscription or licensing fee. Nevertheless, data pre-processing is a crucial step and involves converting the raw data into a machine-readable format which is suitable for specific machine learning algorithms (Abram & McCloskey, 2022). This step ensures that the raw data are clean, organized and standardized for the necessary analysis. Among the key pre-processing tasks are noise reduction, normalization, peak picking/detection, deconvolution, alignment, and gap-filling. It is followed by the pre-processing and the data processing that is aimed at spectral library searches and annotation to detect the metabolites. This process is based on a number of known metabolomic databases and is necessary to find metabolites accurately which is a prerequisite to significant biological understanding (Hendriks et al., 2011). The common databases used in metabolomics research include Human Metabolome Database (HMDB) (https://hmdb.ca), METLIN Metabolite searching (https://metlin.scripps.edu/), Kyoto Encyclopedia of Genes and Genomes (KEGG) (https://www.genome.jp/kegg/), LIPID MAPS (https://www.lipidmaps.org), Metabolomics Workbench (https://www.metabolomicsworkbench.org), Chemical Entities of Biological Interest (ChEBI) (https://www.ebi.ac.uk), MassBank (https://massbank.eu), MetaboLights (https://www.ebi.ac.uk/metabolights/), and BioCyc (https://www.biocyc.org). While these databases are highly versatile and widely used for various applications, most, except for HMDB, are employed in fungal metabolomics research,

Table 2 Common software applications and tools used for data pre-processing, processing, and statistical analysis in metabolomics research

Tool	Raw data type[a]	Category	Website	Reference
apLCMS	MS	Pre-processing	https://bio.tools/aplcms	Liu et al. (2020)
CROP (Correlation-based Removal Of multiPlicities)	MS	Pre-processing	R-script (available at https://github.com/rendju/CROP under GNU GPL)	Kouřil et al. (2020)
Decon2LS	MS	Pre-processing	Software package (download from https://pnnl-comp-mass-spec.github.io/DeconTools/)	Jaitly et al. (2009)
MassHunter	MS	Pre-processing and Processing	Software application	Agilent Technologies (2021)
MAVEN2	MS	Pre-processing	Software package (download from https://github.com/eugenemel/maven?tab=readme-ov-file)	Seitzer et al. (2022)
MetaboAnalyst 6.0	MS, NMR	Pre-processing, Processing, and statistical analysis	https://metaboanalyst.ca/	Pang et al. (2024)
MSeasy	MS	Pre-processing and Processing	Software package (download from https://rdrr.io/cran/MSeasy/)	Nicolè et al. (2012)
MS-DIAL 4	MS	Pre-processing	https://systemsomicslab.github.io/en/	Tsugawa et al. (2020)
mzMatch	MS	Pre-processing	Software package (download from https://mzmatch.sourceforge.net)	Scheltema et al. (2011)
MZmine 4.0	MS	Pre-processing and processing	Software package (download from https://mzio.io/mzmine/)	Schmid et al. (2023)
OpenMS 3.3.0	MS	Pre-processing and processing	https://www.openms.de	Sturm et al. (2008)
SIMCA-P	Multiple[b]	Statistical analysis	Software application	Wu et al. (2010)
SIRIUS 4	MS	Processing	https://bio.informatik.uni-jena.de/sirius/	Dührkop et al. (2019)
XCMS Online	MS	Pre-processing and processing	https://xcmsonline.scripps.edu	Tautenhahn et al. (2012)

Source Authors compilation
[a] *MS* Mass spectrometry, *NMR* Nuclear magnetic resonance
[b] Multiple refers to spectroscopic, chemical and analytical, biological, sensory, and financial data

particularly providing valuable information on metabolite identities.

Another one of the critical steps in any metabolomics research is the statistical analysis of the metabolomic data. The statistical method depends on the nature and design of the study. Univariate analysis is typically applied when comparing two groups, both paired and not paired datasets (Chen et al., 2022). This method allows easy statistical techniques using T-tests, fold change analysis, and creation of volcano plots and offers valuable information and meaningful differences between groups. In the meantime, multivariate analysis is an appropriate method of research including various groups or variables that enables the interpretation of data more thoroughly. These are analysis of variance (ANOVA), principal component analysis (PCA), and partial least squares-discriminant analysis (PLS-DA) (Chen et al., 2022). Also, machine learning models like the random forest (RF) and support vector machine (SVM) are also becoming commonly used in metabolomics studies to perform pattern recognition, classification, and predictive modelling (Chen et al., 2022; Worley & Powers, 2013). These statistical methods are essential in deriving meaning out of metabolomics data, hence useful in assisting researchers to detect meaningful secondary metabolites (i.e. biomarkers), distinguish sample groups, and reduce complexity of biological systems to be easily understood.

There also exist software that can integrate metabolomic datasets to other omics technologies, as shown in Table 3. The work with big datasets that are produced by every single omics experiment is a computationally expensive procedure that requires sophisticated software to reveal meaningful interactions, correlations, and biological findings (Misra et al., 2019). These tools vary in the ease of use and interfaces that are intermediary and easy to use by non-specialists to more sophisticated platforms that need knowledge in programming languages. Due to this, one has to develop a comprehensive knowledge on every step of the analysis process in order to be able to process and interpret the data correctly. The reliability and reproducibility of the

Table 3 Available softwares commonly used for analysing and interpreting integrated omics datasets

Tool	Interface	Function	Website
Cytoscape	Software installation	Integration of metabolomics, transcriptomics, proteomics, and genomics datasets	https://cytoscape.org
IMPaLA (Integrated Molecular Pathway-Level Analysis)	Web-based	Integration of metabolomics and transcriptomics datasets	https://impala.molgen.mpg.de
MetaboAnalyst 6.0	Web-based	Integration of metabolomics, transcriptomics, proteomics, and genomics datasets	https://metaboanalyst.ca
PaintOmics 4	Web-based	Integration of metabolomics and transcriptomics datasets	https://paintomics.org
Paired Omics Data Platform	Software installation	Integration of metabolomics and genomics datasets	Download from https://zenodo.org
ProMeTra	Software installation	Integration of metabolomics, transcriptomics, and proteomics datasets	https://prometra.cebitec.uni-bielefeld.de
XCMS Online	Web-based	Integration of metabolomics, transcriptomics, and proteomics datasets	https://xcmsonline.scripps.edu

Source Authors compilation

results of any multi-omics studies can be greatly increased with proper training and knowledge of the selected tools.

These sophisticated software tools on how to undertake multi-omics research are beneficial to offer assistance to researchers who are interested in such research. Nevertheless, even with the availability of these technologies, it is always necessary to advance and improve the various methods of computation. This is necessary to improve data analysis, metabolite annotation and identification, storage, quantification, and smooth integration of the various datasets, which will remain to be the main role of bioinformatics in promoting various domains in metabolomics.

5 Trends in Fungal Metabolomics

Fungi are highly exuberant in the production of secondary metabolites that have potential bioactive applications such as antibiotics, antioxidants, and anti-inflammatory (Attia et al., 2020; Chávez et al., 2015; Jin et al., 2016; Tawfike et al., 2019). Metabolomics is an effective analytical instrument, which contributes to the identification of these bioactive metabolites. Metabolomics through the use of high-resolution mass spectrometry (HRMS), nuclear magnetic resonance (NMR) spectroscopy, Fourier transform ion cyclotron resonance mass spectrometry (FTIR), and high-performance liquid chromatography (HPLC), and/or their combinations has revolutionized fungal biology by accelerating fungal natural product research and revealing unknown metabolic pathways. In contemporary biological sciences, metabolomics, alongside other omics technology, has been at the core of novel bioactive compounds discovery and understanding of pathogenesis and drug resistance mechanisms. Moreover, optimization of industrial processes such as fermentation, understanding host–pathogen dynamics, organism stress response, and food safety have been significantly improved through metabolomics. Thus, it has emerged as a key party in overcoming international medicine, agriculture, and green issues. Here, the forms of metabolomics to be used on a few studies conducted on fungi are presented and are what result in their diverse usage.

5.1 Fungal Metabolomics in Agriculture

In agriculture, metabolomics has been instrumental in understanding plant-fungal interactions. This was achieved through metabolic profiling of different fungi, particularly those that colonize plants like symbionts (Pelo et al., 2021; Sebastiana et al., 2021; Uehling et al., 2019) and pathogens (Gotthardt et al., 2020; Liu et al., 2019; Ren et al., 2021). Metabolomics recognizes biochemical changes and novel metabolites that serve as biomarkers crucial in the early detection of pathogens and mycotoxin-producing fungi in plants (Chervin et al., 2022; Maia et al., 2020; Seybold et al., 2020). This approach facilitates the mitigation of infection before symptoms appear and significant damage happen particularly in economically-important crops. Mycotoxin detection and control is enhanced by combining metabolomics with machine learning tools, predicting aflatoxin contamination risks up to 47 days in advance with 87.2% prediction accuracy (Xie et al., 2022).

Global environmental changes due to natural and anthropogenic factors also affect plant productivity. It is imperative that more resilient crops and soil are developed and can persist amidst a cocktail of global change factors such as temperature, salinity, pH, and pesticide residues. Fungal symbionts are known to confer stress resistance to plants (Begum et al., 2019; Singh et al., 2011). Through metabolomics using liquid chromatography time-of-flight mass spectrometry (LC/TOF-MS), insights into the molecular stress tolerance mechanisms of plants in the presence of fungal symbionts are now available. For example, changes in metabolites were detected in arbuscular mycorrhizal (AM)-inoculated plants, and these interactions between fungus and plants rendered the crops more tolerant to drought, heavy metal, and soil alkalization (Yang et al., 2020; Zou et al., 2023). Such mechanisms include the production of certain amino acids, organic acids, and sterols to maintain and improve cell membrane stability (Yang et al., 2020). Likewise, regulation of metabolic pathways such as oxidative phosphorylation, phenylalanine metabolism, and biosynthesis was also observed in walnuts under drought stress (Zou et al., 2023). Volatile organic compounds increased in trifoliate orange under drought stress, and this was identified as an initiated plant defence response (Liang et al., 2021). Jia et al. (2024) also found that amino acid metabolism, phytohormones, and glycolysis pathway are regulated under heavy metal and salt stress. All of these are made known through metabolomics-based research.

Pests and pathogens are also increasingly becoming more resistant to existing agrochemicals. Metabolomic studies have been instrumental to unravel the mechanisms of fungicide resistance. This is achieved by identifying unique and distinct metabolites produced during the normal biological activities of the pest or pathogen. Resistance development may induce changes in the metabolic profiles which can be easily detected by metabolomics. Interestingly, as an alternative to synthetic fungicides, biological control agents are used due to safety and environmental issues associated with the use of synthetic chemicals. Metabolomic studies have revealed the suppressive ability of volatile organic compounds of metabolites from formulated biological control agents against pathogenic and mycotoxigenic fungi (Nguyen et al., 2020; Silva-Campos et al., 2022). Another strategy to reduce dependence on pesticide use is the development of more disease-resistant crop varieties. Using Fourier transform ion cyclotron resonance mass spectrometer (FT-ICR-MS), Maia et al. (2020), for instance, detected discovered possible biomarkers of susceptibility to fungal-related disease in grapevine. Identifying markers linked to disease-susceptibility genotypes can aid selection for and breeding of more disease-tolerant varieties. Furthermore, applications of metabolomics provide mechanistic insights to soil health. For example, the response of soil microbes and the consequent changes in the soil metabolite spectrum under localized fertilization regimes is elucidated using metabolomic approach (Cheng et al., 2022). Lastly, metabolomics is also able to provide a complete picture of the metabolic status of a plant. Agricultural practices are improved by identifying important biomarkers of stress tolerance that are caused by environmental changes and pollution, and knowing how nutrients are obtained and used. These insights would therefore also contribute to coming up with crops that are more resilient to environmental issues and less dependent on pesticides and fertilizers leading to sustainable farming.

5.2 Fungal Metabolomics in Food

Another area that metabolomics has changed is in improving the utility of fungi in food production, quality, and safety. The biomarkers of fungi growth in the foods are specific chemical patterns that are generated by the fungi and used to detect early growth of fungi and assess the food decay. As an illustration, untargeted metabolomic methods based on gas chromatography-mass spectrometry (GC-MS) have allowed characterizing volatile profiles of grape juice that has been infected by fungus like *Botrytis cinerea*, *Aspergillus* species, or *Penicillium expansum* (Schueuermann et al., 2019). The identification of discriminatory metabolites, such as 1,5-dimethylnaphthalene, a number of sesquiterpenes, 2-(4-hexyl-2,5-dioxo-2,5-dihydrofuran-3-yl) acetic acid, γ-nonalactone, and *m*-cresol, was subsequently made as invaluable biomarkers that can be used to determine the change of the quality, maturity, and safety of this food commodity.

As the fungi are also a part of fermentation, the metabolomics is also applied to demonstrate the dependence of the microbial community succession and the production of metabolites at different fermentation steps (Mu et al., 2020; Zhang et al., 2023). This could aid the development of products with improved pharmacological properties by facilitating strain selection and optimizing the fermentation process to enhance the production of targeted beneficial metabolites (Du et al., 2023; Ma et al., 2021). Fermented tea leaves, for example, exhibited increased levels of baclofen, a compound used to treat muscle spasticity, and glucosamine, a precursor for synthesizing glycosylated proteins and lipids (Ma et al., 2021). Therefore, by carefully selecting fungal inoculants, producers can maximize the medicinal properties of fermented products while optimizing quality, taste, and flavour. In addition, rice fermentation with basidiomycetes of known medicinal value has been demonstrated to upregulate metabolic pathways related to amino acid, lipid, and carbohydrate metabolism (Luo et al., 2024). Such insights offer an advantage in functional food preparation.

Functional foods contribute to health improvement apart from supplying fundamental nutrition. Temple (2022) further described them as '*novel foods that have been formulated so that they contain substances or live microorganisms that have a possible health-enhancing or disease-preventing value, and at a concentration that is both safe and sufficiently high to achieve the intended benefit*'. Mushrooms are thus considered good sources of functional foods. They are low in fat, high in proteins, dietary fibre, and essential nutrients (Ayimbila & Keawsompong, 2023). However, despite the popularity of edible mushrooms including morels, white button mushroom, maitake, and shiitake, very little is known about their chemical composition (Uffelman et al., 2023; Zhao et al., 2022). Only recently, with the aid of metabolomics, over 10,000 compounds were detected from these fungi paving way to more research on their health-promoting properties.

5.3 Fungal Metabolomics in Medicine

In medicine, fungal metabolomics underpins advancements in disease diagnostics and therapeutic discovery. Metabolic profiling has facilitated the identification of biomarkers for fungal infections, enhancing diagnostic accuracy (Lima et al., 2020). Metabolomic profiling was instrumental in the detection of the most common fungal pathogen, *Candida albicans*, including laboratory-induced azole-resistant strains. Oliver et al. (2020) also used metabolomics to characterize clinical isolates of various *Candida* species originating from different infection sources such as blood, vagina, and respiratory tract. They further observed that specific metabolic fingerprints such as those associated with biofilm formation, antifungal resistance (i.e. caspofungin and fluconazole), and the secreted aspartyl proteinase (Sap) production could be detected and be used to deepen our understanding of *Candida*'s virulence and pathogenicity. Furthermore, the identification of these unique metabolic fingerprints produced by pathogenic microorganisms by metabolomics and extending these to the infection-associated changes to the metabolism of the host can provide a more complete impression of the infection process (Brandt et al., 2021). Zhu et al. (2013) studied *Candida albicans* during the biofilm development phases and identified thirty-one differentially produced metabolites between the biofilm and planktonic specimens. Multi-omics analysis also revealed indicator features of host-microbe interactions during *C. albicans* colonization and subsequent infection, which can be utilized to identify clinical risk factors (Zhang et al., 2024).

Metabolomic approaches have also provided mechanistic insights into the biosynthesis of pharmacologically active compounds, which can pave way for new drug development. Microorganisms including fungal endophytes and microbiota that are associated with marine organisms have been identified as important natural sources of bioactive compounds (Jin et al., 2016). These compounds can have antibacterial, anticancer, and anti-trypanosomal properties, among the many pharmacological features known for these metabolites (Tawfike et al., 2019; Utermann et al., 2020). Examples of studies that employed metabolomics to discover novel compounds were from root endophytic fungi (Maciá-Vicente et al., 2018), mangrove-associated endophytic fungi (Mazlan et al., 2020), and the endophytic fungi *Aspergillus flocculus* (Tawfike et al., 2019), and *Lasiodiplodia theobromae* (Kamal et al., 2017). Likewise, fungi from extreme environments, such as deserts, have emerged as a rich source of metabolites with significant bioactivities (Chávez et al., 2015; Mohammed et al., 2021). These exemplified the power of metabolomics as applied to health and medicine, particularly in drug discovery.

6 Concluding Remarks

Research on fungi has continued to increase in the last 3–4 decades. With the advancement of knowledge, development of new technologies and/or the application of available tools to fungi, it is expected that fungal research will gain further momentum and take the CenterStage in scientific discoveries. Hyde et al. (2024) provided a run through of trends, limitations, and future directions of 15 important research areas with fungi as its centrepiece. Among these is the natural product-based drug discovery where fungi historically played a key role. As summarized by Bills and Gloer (2016), twenty-four key metabolites from fungi and more as its derivates were already processed into agrochemical, cosmetic, and pharmaceutical formulations. But with the range of chemically diverse natural products that fungi can produce and the high number of fungal taxa that remained untested for bioactivities, they remain as one of the most promising sources of new lead compounds for any drug discovery program. Integral to the fungal drug research is natural product chemistry, with the intent of isolating and identifying novel fungal compounds. An initial step to this critical stage is the metabolomics-based screening which can shorten the drug discovery process. Hence, fungal metabolomics has cemented its position within the bigger scheme of drug biodiscovery. Fungal metabolomics has also the potential to lead future research directions and revolutionize multiple knowledge domains in the next decades. In food science, fungal metabolomics can enhance our understanding of the chemical process of fermentation and aid in developing and producing safe and high-quality functional foods. Fungal metabolomics can help in detection of pathogen, surveillance and control, understanding the mechanism of stress tolerance in agricultural industry, and produce eco-friendly agrochemicals in a way that leads to sustainable agricultural practices. Fungal metabolomics in health and medicine is a field that has been used in innovations in disease diagnostics and development of therapeutics, especially the identification of bioactive compounds that were previously unknown. And since it is an evolved method of analysis, the possibilities of fungal metabolomics are infinite and can be innovated in future overcoming the limitation of existing methods, therefore highlighting the importance of this technique in solving most of our world, regional, and local problems.

References

Abram, K. J., & McCloskey, D. (2022). A comprehensive evaluation of metabolomics data preprocessing methods for deep learning. *Metabolites, 24*, 202. https://doi.org/10.3390/metabo12030202

Agilent Technologies. (2021). Agilent MassHunter. Agilent Technologies. Available from https://www.agilent.com

Agin, A., Heintz, D., Ruhland, E., de La Barca, J. C., Zumsteg, J., Moal, V., et al. (2016). Metabolomics—An overview: From basic principles to potential biomarkers (part 1). *Médecine Nucléaire, 40*(1), 4–10. https://doi.org/10.1016/j.mednuc.2015.12.006

Ahmad, I., Malak, H. A., & Abulreesh, H. H. (2021). Environmental antimicrobial resistance and its drivers: A potential threat to public health. *Journal of Global Antimicrobial Resistance, 27*, 101–111. https://doi.org/10.1016/j.jgar.2021.08.001

Ahmed, S. K., Hussein, S., Qurbani, K., Ibrahim, R. H., Fareeq, A., Mahmood, K. A., et al. (2024). Antimicrobial resistance: Impacts, challenges, and future prospects. *Journal of Medicine, Surgery, and Public Health, 2*, Article 100081. https://doi.org/10.1016/j.glmedi.2024.100081

Alanne, A.-L., Issakainen, J., Pihlaja, K., Jokioja, J., & Sinkkonen, J. (2019). Metabolomic discrimination of the edible mushrooms *Kuehneromyces mutabilis* and *Hypholoma capnoides* (Strophariaceae, Agaricales) by NMR spectroscopy. *Zeitschrift Für Naturforschung C, 74*(7–8), 201–210. https://doi.org/10.1515/znc-2018-0214

Alhadrami, H. A., Sayed, A. M., El-Gendy, A. O., Shamikh, Y. I., Gaber, Y., Bakeer, W., et al. (2021). A metabolomic approach to target antimalarial metabolites in the *Artemisia annua* fungal endophytes. *Scientific Reports, 11*(1), 2770. https://doi.org/10.1038/s41598-021-82201-8

Aliferis, K. A., Cubeta, M. A., & Jabaji, S. (2013). Chemotaxonomy of fungi in the *Rhizoctonia solani* species complex performing GC/MS metabolite profiling. *Metabolomics, 9*(1), 159–169. https://doi.org/10.1007/s11306-011-0340-1

Aliferis, K. A., & Jabaji, S. (2010). 1H NMR and GC-MS metabolic fingerprinting of developmental stages of *Rhizoctonia solani* sclerotia. *Metabolomics, 6*(1), 96–108. https://doi.org/10.1007/s11306-009-0180-4

Aron, A. T., Gentry, E. C., McPhail, K. L., Nothias, L. F., Nothias-Esposito, M., Bouslimani, A., et al. (2020). Reproducible molecular networking of untargeted mass spectrometry data using GNPS. *Nature Protocols, 15*(6), 1954–1991. https://doi.org/10.1038/s41596-020-0317-5

Atanasova-Penichon, V., Legoahec, L., Bernillon, S., Deborde, C., Maucourt, M., Verdal-Bonnin, M. N., et al. (2018). Mycotoxin biosynthesis and central metabolism are two interlinked pathways in *Fusarium graminearum*, as demonstrated by the extensive metabolic changes induced by caffeic acid exposure. *Applied and Environmental Microbiology, 84*(8), e01705-e1717. https://doi.org/10.1128/AEM.01705-17

Attia, E. Z., Farouk, H. M., Abdelmohsen, U. R., & Mo'men, H. (2020). Antimicrobial and extracellular oxidative enzyme activities of endophytic fungi isolated from alfalfa (*Medicago sativa*) assisted by metabolic profiling. *South African Journal of Botany, 134*, 156–162. https://doi.org/10.1016/j.sajb.2019.12.003.

Augustijn, D., de Groot, H. J., & Alia, A. (2021). HR-MAS NMR applications in plant metabolomics. *Molecules, 26*(4), 931. https://doi.org/10.3390/molecules26040931

Ayimbila, F., & Keawsompong, S. (2023). Nutritional quality and biological application of mushroom protein as a novel protein alternative. *Current Nutrition Reports, 12*(2), 290–307. https://doi.org/10.1007/s13668-023-00468-x

Bano, N., Mohammed, S. A., & Raza, K. (2024). Integrating machine learning and multitargeted drug design to combat antimicrobial resistance: A systematic review. *Journal of Drug Targeting*, 1–13. https://doi.org/10.1080/1061186X.2024.2428984

Baquero, F., Martinez, J. L., Lanza, V. F., Rodríguez-Beltrán, J., Galán, J. C., San Millán, A., et al. (2021). Evolutionary pathways and trajectories in antibiotic resistance. *Clinical Microbiology Reviews, 34*(4), e00050-e119. https://doi.org/10.1128/CMR.00050-19

Bartelo, N., Mishra, S., Singh, P. R., Padmanarayana, M., & Soni, V. (2024). Future prospective of omics-system biology to control AMR: Recommendations and directions. In *Antimicrobial resistance: factors to findings: omics and systems biology approaches* (pp. 415–449). Springer International Publishing. https://doi.org/10.1007/978-3-031-65986-7_13

Beale, D. J., Pinu, F. R., Kouremenos, K. A., Poojary, M. M., Narayana, V. K., Boughton, B. A., et al. (2018). Review of recent developments in GC–MS approaches to metabolomics-based research. *Metabolomics, 14*, 1–31. https://doi.org/10.1007/s11306-018-1449-2

Begum, N., Qin, C., Ahanger, M. A., Raza, S., Khan, M. I., Ashraf, M., et al. (2019). Role of arbuscular mycorrhizal fungi in plant growth regulation: Implications in abiotic stress tolerance. *Frontiers in Plant Science, 10*, 1068. https://doi.org/10.3389/fpls.2019.01068

Bhattarai, K., Kabir, M. E., Bastola, R., & Baral, B. (2021). Fungal natural products galaxy: Biochemistry and molecular genetics toward blockbuster drugs discovery. *Advances in Genetics, 107*, 193–284. https://doi.org/10.1016/bs.adgen.2020.11.006

Biedermann, M., & Grob, K. (2019). Advantages of comprehensive two-dimensional gas chromatography for comprehensive analysis of potential migrants from food contact materials. *Analytica Chimica Acta, 1057*, 11–17. https://doi.org/10.1016/j.aca.2018.10.046

Bills, G. F., & Gloer, J. B. (2016). Biologically active secondary metabolites from the fungi. *Microbiology Spectrum, 4*(6). https://doi.org/10.1128/microbiolspec.funk-0009-2016

Bloom, D. E., & Cadarette, D. (2019). Infectious disease threats in the twenty-first century: Strengthening the global response. *Frontiers in Immunology, 10*, 549. https://doi.org/10.3389/fimmu.2019.00549

Bo, Y., Feng, J., Xu, J., Huang, Y., Cai, H., Cui, X., et al. (2019). High-resolution pure shift NMR spectroscopy offers better metabolite discrimination in food quality analysis. *Food Research International, 125*, Article 108574. https://doi.org/10.1016/j.foodres.2019.108574

Brandt, P., Garbe, E., & Vylkova, S. (2021). Catch the wave: Metabolomic analyses in human pathogenic fungi. In *Advances in clinical immunology, medical microbiology, COVID-19, and Big Data* (pp. 413–424). Jenny Stanford Publishing. https://doi.org/10.1371/journal.ppat.1008757

Caesar, L. K., Butun, F. A., Robey, M. T., Ayon, N. J., Gupta, R., Dainko, D., et al. (2023). Correlative metabologenomics of 110 fungi reveals metabolite–gene cluster pairs. *Nature Chemical Biology, 19*(7), 846–854. https://doi.org/10.1038/s41589-023-01276-8

Cai, Z. X., Chen, M. Y., Lu, Y. P., Guo, Z. J., Zeng, Z. H., Liao, J. H., et al. (2022). Metabolomics and transcriptomics unravel the mechanism of browning resistance in *Agaricus bisporus. PLoS ONE, 17*(3), Article e0255765. https://doi.org/10.1371/journal.pone.0255765

Cambiaghi, A., Ferrario, M., & Masseroli, M. (2017). Analysis of metabolomic data: Tools, current strategies and future challenges for omics data integration. *Briefings in Bioinformatics, 18*(3), 498–510. https://doi.org/10.1093/bib/bbw031

Challen, B., & Cramer, R. (2022). Advances in ionisation techniques for mass spectrometry-based omics research. *Proteomics, 22*(15–16), 2100394. https://doi.org/10.1002/pmic.202100394

Chávez, R., Fierro, F., García-Rico, R. O., & Vaca, I. (2015). Filamentous fungi from extreme environments as a promising source of novel bioactive secondary metabolites. *Frontiers in Microbiology, 6*, 903. https://doi.org/10.3389/fmicb.2015.00903

Chen, X., Bertho, G., Caradeuc, C., Giraud, N., & Lucas-Torres, C. (2023). Present and future of pure shift NMR in metabolomics. *Magnetic Resonance in Chemistry, 61*(12), 654–673. https://doi.org/10.1002/mrc.5356

Chen, Y., Li, E. M., & Xu, L. Y. (2022). Guide to metabolomics analysis: A bioinformatics workflow. *Metabolites, 12*(4), 357. https://doi.org/10.3990/metabo12040357

Cheng, H., Yuan, M., Tang, L., Shen, Y., Yu, Q., & Li, S. (2022). Integrated microbiology and metabolomics analysis reveal responses of soil microorganisms and metabolic functions to phosphorus fertilizer on semiarid farm. *Science of the Total Environment, 817*, Article 152878. https://doi.org/10.1016/j.scitotenv.2021.152878

Chervin, J., Romeo-Oliván, A., Fournier, S., Puech-Pages, V., Dumas, B., Jacques, A., et al. (2022). Modification of early response of *Vitis vinifera* to pathogens relating to esca disease and biocontrol agent Vintec revealed by untargeted metabolomics on woody tissues. *Frontiers in Microbiology, 13*, Article 835463. https://doi.org/10.3389/fmicb.2022.835463

Coque, T. M., Cantón, R., Pérez-Cobas, A. E., Fernández-de-Bobadilla, M. D., & Baquero, F. (2023). Antimicrobial resistance in the global health network: Known unknowns and challenges for efficient responses in the 21st century. *Microorganisms, 11*(4), 1050. https://doi.org/10.3390/microorganisms11041050

Corsaro, C., Cicero, N., Mallamace, D., Vasi, S., Naccari, C., Salvo, A., et al. (2016). HR-MAS and NMR towards foodomics. *Food Research International, 89*, 1085–1094. https://doi.org/10.1016/j.foodres.2016.09.033

Demetrowitsch, T., & Schwarz, K. (2014). Metabolomics: New analytical methods for metabolome research. *Ernährungs-Umschau, 61*, 102–111. https://doi.org/10.4455/eu.2014.019

Du, J., Wu, X., Sun, S., Qin, Y., Liao, K., Liu, X., et al. (2023). Study on inoculation fermentation by fungi to improve the taste quality of summer green tea. *Food Bioscience, 51*, Article 102321. https://doi.org/10.1016/j.fbio.2022.102321

Dührkop, K., Fleischauer, M., Ludwig, M., Aksenov, A. A., Melnik, A. V., Meusel, M., et al. (2019). SIRIUS 4: A rapid tool for turning tandem mass spectra into metabolite structure information. *Nature Methods, 16*(4), 299–302. https://doi.org/10.1038/s41592-019-0344-8

Emwas, A. H., Roy, R., McKay, R. T., Tenori, L., Saccenti, E., Gowda, G. N., et al. (2019). NMR spectroscopy for metabolomics research. *Metabolites, 9*(7), 123. https://doi.org/10.3390/metabo9070123

Erler, A., Riebe, D., Beitz, T., Löhmannsröben, H.-G., Grothusheitkamp, D., Kunz, T., et al. (2020). Characterization of volatile metabolites formed by molds on barley by mass and ion mobility spectrometry. *Journal of Mass Spectrometry, 55*(5), Article e4501. https://doi.org/10.1002/jms.4501

Feizi, N., Hashemi-Nasab, F. S., Golpelichi, F., Saburouh, N., & Parastar, H. (2021). Recent trends in application of chemometric methods for GC-MS and GC×GC-MS-based metabolomic studies. *Trends in Analytical Chemistry, 138*, Article 116239. https://doi.org/10.1016/j.trac.2021.116239

Ferri, M., Ranucci, E., Romagnoli, P., & Giaccone, V. (2017). Antimicrobial resistance: A global emerging threat to public health systems. *Critical Reviews in Food Science and Nutrition, 57*(13), 2857–2876. https://doi.org/10.1080/10408398.2015.1077192

Flores, G. A., Girometta, C. E., Cusumano, G., Angelini, P., Tirillini, B., Ianni, F., et al. (2022). Untargeted metabolomics used to describe the chemical composition, antioxidant and antimicrobial effects of extracts from *Pleurotus* spp. mycelium grown in different culture media. *Antibiotics, 11*(11), 1468. https://doi.org/10.3390/antibiotics11111468

Flores-Vallejo, R. del C., Folch-Mallol, J. L., Sharma, A., Cardoso-Taketa, A., Alvarez-Berber, L., & Villarreal, M. L. (2020). ITS2 ribotyping, in vitro anti-inflammatory screening, and metabolic profiling of fungal endophytes from the Mexican species *Crescentia alata* Kunth. *South African Journal of Botany, 134*, 213–224. https://doi.org/10.1016/j.sajb.2019.12.030

Foroozandeh, M., Morris, G. A., & Nilsson, M. (2018). PSYCHE pure shift NMR spectroscopy. *Chemistry—A European Journal, 24*(53), 13988–14000. https://doi.org/10.1002/chem.201800524

Fortuin, S., & Soares, N. C. (2022). The integration of proteomics and metabolomics data paving the way for a better understanding of the mechanisms underlying microbial acquired drug resistance. *Frontiers in Medicine, 9*, Article 849838. https://doi.org/10.3389/fmed.2022.849838

Frisvad, J. C., Andersen, B., & Thrane, U. (2008). The use of secondary metabolite profiling in chemotaxonomy of filamentous fungi. *Mycological Research, 112*(Pt 2), 231–240. https://doi.org/10.1016/j.mycres.2007.08.018

Fu, Y., Yu, Y., Tan, H., Sun, C., Shi, J., Li, Y., et al. (2021). The use of metabolomics in assessing the quality and safety of functional foods: A review. *Comprehensive Reviews in Food Science and Food Safety, 20*(5), 4229–4252. https://doi.org/10.1111/1541-4337.12688

Fuss, T. L., & Cheng, L. L. (2016). Evaluation of cancer metabolomics using ex vivo high resolution magic angle spinning (HRMAS) magnetic resonance spectroscopy (MRS). *Metabolites,* 6(1), Article 1. https://doi.org/10.3390/metabo6010011

Garrod, S., Humpfer, E., Spraul, M., Connor, S., Polley, S., Connelly, J., Lindon, J., Nicholson, J., & Holmes, E. (1999). High-resolution magic angle spinning 1H NMR spectroscopic studies on intact rat renal cortex and medulla. *Magnetic Resonance in Medicine: An Official Journal of the International Society for Magnetic Resonance in Medicine, 41*(6), 1108–1118. https://doi.org/10.1002/(SICI)1522-2594(199906)41:63.0.CO;2-M

George, T. K., Devadasan, D., & Jisha, M. S. (2019). Chemotaxonomic profiling of *Penicillium setosum* using high-resolution mass spectrometry (LC-Q-ToF-MS). *Heliyon, 5*(9), Article e02484. https://doi.org/10.1016/j.heliyon.2019.e02484

Giraudeau, P., & Frydman, L. (2014). Ultrafast 2D NMR: An emerging tool in analytical spectroscopy. *Annual Review of Analytical Chemistry, 7*, 129–161. https://doi.org/10.1146/annurev-anchem-071213-020208

Gotthardt, M., Kanawati, B., Schmidt, F., Asam, S., Hammerl, R., Frank, O., et al. (2020). Comprehensive analysis of the *Alternaria mycobolome* using mass spectrometry-based metabolomics. *Molecular Nutrition and Food Research, 64*(3), 1–16. https://doi.org/10.1002/mnfr.201900558

Gowda, G. A. N., & Djukovic, D. (2014). Overview of mass spectrometry-based metabolomics: Opportunities and challenges.

Methods in Molecular Biology (Clifton, N.J.), 1198, 3–12. https://doi.org/10.1007/978-1-4939-1258-2_1

Guennec, A. L., Giraudeau, P., & Caldarelli, S. (2014). Evaluation of Fast 2D NMR for metabolomics. *Analytical Chemistry, 86*(12), 5946–5954. https://doi.org/10.1021/ac500966e

Haider, A., Ringer, M., Kotroczó, Z., Mohácsi-Farkas, C., & Kocsis, T. (2023). The current level of MALDI-TOF MS applications in the detection of microorganisms: A short review of benefits and limitations. *Microbiology Research, 14*(1), Article 1. https://doi.org/10.3390/microbiolres14010008

Han, X., Tang, X., Luo, X., Sun, C., Liu, K., Zhang, Y., Li, P., & Li, G. (2020). Isolation and identification of three new sterigmatocystin derivatives from the fungus *Aspergillus versicolor* guided by molecular networking approach. *Chemistry & Biodiversity, 17*(6), Article e2000208. https://doi.org/10.1002/cbdv.202000208

Hendriks, M. M. W. B., van Eeuwijk, F. A., Jellema, R. H., Westerhuis, J. A., Reijmers, T. H., Hoefsloot, H. C. J., & Smilde, A. K. (2011). Data-processing strategies for metabolomics studies. *Trends in Analytical Chemistry, 30*, 1685–1698. https://doi.org/10.1016/j.trac.2011.04.019

Hettick, J. M., Green, B. J., Buskirk, A. D., Kashon, M. L., Slaven, J. E., Janotka, E., Blachere, F. M., Schmechel, D., & Beezhold, D. H. (2008). Discrimination of Aspergillus isolates at the species and strain level by matrix-assisted laser desorption/ionization time-of-flight mass spectrometry fingerprinting. *Analytical Biochemistry, 380*(2), 276–281. https://doi.org/10.1016/j.ab.2008.05.051

Hou, T.-Y., Chiang-Ni, C., & Teng, S.-H. (2019). Current status of MALDI-TOF mass spectrometry in clinical microbiology. *Journal of Food and Drug Analysis, 27*(2), 404–414. https://doi.org/10.1016/j.jfda.2019.01.001

Huang, B., Zeng, H., Dong, L., Li, Y., Sun, L., Zhu, Z., Chai, Y., & Chen, W. (2011). Metabolite target analysis of isoprenoid pathway in Saccharomyces cerevisiae in response to genetic modification by GC-SIM-MS coupled with chemometrics. *Metabolomics, 7*(1), 134–146. https://doi.org/10.1007/s11306-010-0240-9

Hübert, T., Tiebe, C., & Stephan, I. (2011). Detection of fungal infestations of wood by ion mobility spectrometry. *International Biodeterioration & Biodegradation, 65*(5), 675–681. https://doi.org/10.1016/j.ibiod.2011.03.008

Hyde, K. D., Abdel-Wahab, M. A., Abdollahzadeh, J., Abeywickrama, P. D., Absalan, S., Afshari, N., et al. (2023). Global consortium for the classification of fungi and fungus-like taxa. *Mycosphere, 14*(1), 1960–2012. https://doi.org/10.5943/mycosphere/14/1/23

Hyde, K. D., Baldrian, P., Chen, Y., Thilini Chethana, K. W., De Hoog, S., Doilom, M., et al. (2024). Current trends, limitations and future research in the fungi? *Fungal Diversity, 125*(1), 1–71. https://doi.org/10.1007/s13225-023-00532-5

Ibrahim, A., Tanney, J. B., Fei, F., Seifert, K. A., Cutler, G. C., Capretta, A., et al. (2020). Metabolomic-guided discovery of cyclic nonribosomal peptides from *Xylaria ellisii* sp. nov., a leaf and stem endophyte of *Vaccinium angustifolium. Scientific Reports, 10*(1), 4599. https://doi.org/10.1038/s41598-020-61088-x

Jaitly, N., Mayampurath, A., Littlefield, K., Adkins, J. N., Anderson, G. A., & Smith, R. D. (2009). Decon2LS: An open-source software package for automated processing and visualization of high-resolution mass spectrometry data. *BMC Bioinformatics, 10*, 1–15. https://doi.org/10.1186/1471-2105-10-87

Jendoubi, T. (2021). Approaches to integrating metabolomics and multi-omics data: A primer. *Metabolites, 11*, 184. https://doi.org/10.3390/metabo11030184

Jeppesen, M. J., & Powers, R. (2023). Multiplatform untargeted metabolomics. *Magnetic Resonance in Chemistry, 61*(12), 628–653. https://doi.org/10.1002/mrc.5350

Jia, B., Cui, X., Zhang, Z., Li, X., Hou, Y., Luo, J., et al. (2024). Arbuscular mycorrhizal fungi regulate amino acid metabolism, phytohormones and glycolysis pathway to promote the growth of *Suaeda salsa* under combined Cd and NaCl stresses. *Plant Physiology and Biochemistry, 214*, Article 108921. https://doi.org/10.1016/j.plaphy.2024.108921

Jiang, L., Li, J., Zhang, Y., Li, X., & Li, Z. (2019). Investigating metabolic responses to stress in *Pseudomonas syringae* using liquid chromatography–mass spectrometry. *Metabolomics, 16*(12), 127. https://doi.org/10.1007/s11306-020-01492-7

Jiang, M., Chen, S., Lu, X., Guo, H., Chen, S., Yin, X., et al. (2023). Integrating genomics and metabolomics for the targeted discovery of new cyclopeptides with antifungal activity from a marine-derived fungus *Beauveria felina. Journal of Agricultural and Food Chemistry, 71*(25), 9782–9795. https://doi.org/10.1021/acs.jafc.3c02415

Jin, L., Quan, Hou, X., & Fan, S. (2016). Potential pharmacological resources: natural bioactive compounds from marine-derived fungi. *Marine Drugs, 14*(4). https://doi.org/10.3390/md14040076

Joshi, R., Sharma, A., Thakur, K., Kumar, D., & Nadda, G. (2018). Metabolite analysis and nucleoside determination using reproducible UHPLC-Q-ToF-IMS in Ophiocordyceps sinensis. *Journal of Liquid Chromatography & Related Technologies, 41*(15–16), 927–936. https://doi.org/10.1080/10826076.2018.1541804

Kamal, N., Viegelmann, C. V., Clements, C. J., & Edrada-Ebel, R. (2017). Metabolomics-guided isolation of anti-trypanosomal metabolites from the endophytic fungus *Lasiodiplodia theobromae. Planta Medica, 234*(06), 565–573. https://doi.org/10.1055/s-0042-118601

Kang, D., Kim, J., Choi, J. N., Liu, K.-H., & Lee, C. H. (2011). Chemotaxonomy of *Trichoderma* spp. using mass spectrometry-based metabolite profiling. *Journal of Microbiology and Biotechnology, 21*(1), 5–13. https://doi.org/10.4014/jmb.1008.08018

Kok, M., Maton, L., van der Peet, M., Hankemeier, T., & van Hasselt, J. C. (2022). Unravelling antimicrobial resistance using metabolomics. *Drug Discovery Today, 27*(6), 1774–1783. https://doi.org/10.1016/j.drudis.2022.03.015

Kouřil, Š, de Sousa, J., Václavík, J., Friedecký, D., & Adam, T. (2020). CROP: Correlation-based reduction of feature multiplicities in untargeted metabolomic data. *Bioinformatics, 36*(9), 2941–2942. https://doi.org/10.1093/bioinformatics/btaa012

Kumar, A., Singh, S., Gupta, S. K., Kumar, S., Kumar, S., Singh, R., et al. (2022). Identification of metabolite extraction method for targeted exploration of antimicrobial resistance associated metabolites of *Klebsiella pneumoniae. Scientific Reports, 12*(1), 8939. https://doi.org/10.1038/s41598-022-12153-0

Lapthorn, C., Pullen, F., & Chowdhry, B. Z. (2013). Ion mobility spectrometry-mass spectrometry (IMS-MS) of small molecules: Separating and assigning structures to ions. *Mass Spectrometry Reviews, 32*(1), 43–71. https://doi.org/10.1002/mas.21349

Levy, A. J., Oranzi, N. R., Ahmadireskety, A., Kemperman, R. H., Wei, M. S., & Yost, R. A. (2019). Recent progress in metabolomics using ion mobility-mass spectrometry. *Trends in Analytical Chemistry, 116*, 274–281. https://doi.org/10.1016/j.trac.2019.05.001

Li, Y., Kuhn, M., Zukowska-Kasprzyk, J., Hennrich, M. L., Kastritis, P. L., O'Reilly, F. J., et al. (2021). Coupling proteomics and metabolomics for the unsupervised identification of protein–metabolite interactions in *Chaetomium thermophilum. PLoS ONE, 16*(7), Article e0254429. https://doi.org/10.1371/journal.pone.0254429

Liang, S. M., Zhang, F., Zou, Y. N., Kuča, K., & Wu, Q. S. (2021). Metabolomics analysis reveals drought responses of trifoliate orange by arbuscular mycorrhizal fungi with a focus on terpenoid profile. *Frontiers in Plant Science, 12*, Article 740524. https://doi.org/10.3389/fpls.2021.740524

Lima, E. D. O., Navarro, L. C., Morishita, K. N., Kamikawa, C. M., Rodrigues, R. G. M., Dabaja, M. Z., et al. (2020). Metabolomics and machine learning approaches combined in pursuit for more accurate paracoccidioidomycosis diagnoses. *Msystems, 5*(3), 10–1128. https://doi.org/10.1128/msystems.00258-20

Linhares, Y., Kaganski, A., Agyare, C., Kurnaz, I. A., Neergheen, V., Kolodziejczyk, B., et al. (2023). Biodiversity: The overlooked source of human health. *Trends in Molecular Medicine, 29*(3), 173–187. https://doi.org/10.1016/j.molmed.2022.12.002

Liu, Z., Kang, B., Duan, X., Hu, Y., Li, W., Wang, C., Li, D., & Xu, N. (2022). Metabolomic profiles of the liquid state fermentation in co-culture of *A. oryzae* and *Z. rouxii*. *Food Microbiology, 103*, 103966. https://doi.org/10.1016/j.fm.2021.103966

Liu, C., Chen, F., Zhang, J., Liu, L., Lei, H., Li, H., et al. (2019). Metabolic changes of *Fusarium graminearum* induced by TPS gene deletion. *Journal of Proteome Research, 18*(9), 3317–3327. https://doi.org/10.1021/acs.jproteome.9b00259

Liu, M., Grkovic, T., Liu, X., Han, J., Zhang, L., & Quinn, R. J. (2017). A systems approach using OSMAC, Log P and NMR fingerprinting: An approach to novelty. *Synthetic and Systems Biotechnology, 2*(4), 276–286. https://doi.org/10.1016/j.synbio.2017.10.001

Liu, Q., Walker, D., Uppal, K., Liu, Z., Ma, C., Tran, V., et al. (2020). Addressing the batch effect issue for LC/MS metabolomics data in data preprocessing. *Scientific Reports, 10*(1), 13856. https://doi.org/10.1038/s41598-020-70850-0

Lowe, R. G., Allwood, J. W., Galster, A. M., Urban, M., Daudi, A., Canning, G., Ward, J. L., Beale, M. H., & Hammond-Kosack, K. E. (2010). A combined 1H nuclear magnetic resonance and electrospray ionization–mass spectrometry analysis to understand the basal metabolism of plant-pathogenic fusarium spp. *Molecular Plant-Microbe Interactions, 23*(12), 1605–1618. https://doi.org/10.1094/MPMI-04-10-0092

Lu, Y., Che, J., Xu, X., Pang, B., Zhao, X., Liu, Y., & Shi, J. (2020). Metabolomics reveals the response of the phenylpropanoid biosynthesis pathway to starvation treatment in the grape endophyte Alternaria sp. MG1. *Journal of Agricultural and Food Chemistry, 68*(4), 1126–1135. https://doi.org/10.1021/acs.jafc.9b05302

Lu, L., Zhu, K.-X., Yang, Z., Guo, X.-N., & Xing, J.-J. (2022). Metabolomics analysis of freeze–thaw tolerance enhancement mechanism of ε-poly-l-lysine on industrial yeast. *Food Chemistry, 382*, Article 132315. https://doi.org/10.1016/j.foodchem.2022.132315

Luo, X., Jiang, J. H., Liu, S. L., Gao, J. Y., & Zhou, L. W. (2024). Metabolomics analysis of rice fermented by medicinal fungi providing insights into the preparation of functional food. *Food Chemistry, 459*, Article 140372. https://doi.org/10.1016/j.foodchem.2024.140372

Ma, Y., Ling, T. J., Su, X. Q., Jiang, B., Nian, B., Chen, L. J., et al. (2021). Integrated proteomics and metabolomics analysis of tea leaves fermented by *Aspergillus niger*, *Aspergillus tamarii* and *Aspergillus fumigatus*. *Food Chemistry, 334*, Article 127560. https://doi.org/10.1016/j.foodchem.2020.127560

Maciá-Vicente, J. G., Shi, Y. N., Cheikh-Ali, Z., Grün, P., Glynou, K., Kia, S. H., et al. (2018). Metabolomics-based chemotaxonomy of root endophytic fungi for natural products discovery. *Environmental Microbiology, 20*(3), 1253–1270. https://doi.org/10.1111/1462-2920.14072

Mahrous, E. A., & Farag, M. A. (2015). Two dimensional NMR spectroscopic approaches for exploring plant metabolome: A review. *Journal of Advanced Research, 6*(1), 3–15. https://doi.org/10.1016/j.jare.2014.10.003

Maia, M., Ferreira, A. E., Nascimento, R., Monteiro, F., Traquete, F., Marques, A. P., et al. (2020). Integrating metabolomics and targeted gene expression to uncover potential biomarkers of fungal/oomycetes-associated disease susceptibility in grapevine. *Scientific Reports, 10*(1), 15688. https://doi.org/10.1038/s41598-020-72781-2

Manohar, M., & Marzinke, M. A. (2016). Chapter 3—Application of chromatography combined with mass spectrometry in therapeutic drug monitoring. In W. Clarke & A. Dasgupta (Eds.), *Clinical challenges in therapeutic drug monitoring* (pp. 45–70). Elsevier. https://doi.org/10.1016/B978-0-12-802025-8.00003-9

Marchev, A. S., Vasileva, L. V., Amirova, K. M., Savova, M. S., Balcheva-Sivenova, Z. P., & Georgiev, M. I. (2021). Metabolomics and health: From nutritional crops and plant-based pharmaceuticals to profiling of human biofluids. *Cellular and Molecular Life Sciences, 78*(19–20), 6487–6503. https://doi.org/10.1007/s00018-021-03918-3

Marshall, D. D., & Powers, R. (2017). Beyond the paradigm: Combining mass spectrometry and nuclear magnetic resonance for metabolomics. *Progress in Nuclear Magnetic Resonance Spectroscopy, 100*, 1–16. https://doi.org/10.1016/j.pnmrs.2017.01.001

Martineau, E., Dumez, J.-N., & Giraudeau, P. (2020). Fast quantitative 2D NMR for metabolomics and lipidomics: A tutorial. *Magnetic Resonance in Chemistry, 58*(5), 390–403. https://doi.org/10.1002/mrc.4899

Mastrangelo, A., Ferrarini, A., Rey-Stolle, F., García, A., & Barbas, C. (2015). From sample treatment to biomarker discovery: A tutorial for untargeted metabolomics based on GC-(EI)-Q-MS. *Analytica Chimica Acta, 900*, 21–35. https://doi.org/10.1016/j.aca.2015.10.001

Mazlan, N. W., Tate, R., Yusoff, Y. M., Clements, C., & Edrada-Ebel, R. (2020). Metabolomics-guided isolation of anti-trypanosomal compounds from endophytic fungi of the mangrove plant *Avicennia lanata*. *Current Medicinal Chemistry, 27*(11), 1815–1835. https://doi.org/10.2174/0929867326666190704130105

Mazzei, P., & Piccolo, A. (2017). HRMAS NMR spectroscopy applications in agriculture. *Chemical and Biological Technologies in Agriculture, 4*(1), 11. https://doi.org/10.1186/s40538-017-0093-9

Mazzei, P., Vinale, F., Woo, S. L., Pascale, A., Lorito, M., & Piccolo, A. (2016). Metabolomics by proton high-resolution magic-angle-spinning nuclear magnetic resonance of tomato plants treated with two secondary metabolites isolated from Trichoderma. *Journal of Agricultural and Food Chemistry, 64*(18), 3538–3545. https://doi.org/10.1021/acs.jafc.6b00801

Mironenka, J., Różalska, S., Soboń, A., & Bernat, P. (2021). *Trichoderma harzianum* metabolites disturb *Fusarium culmorum* metabolism: Metabolomic and proteomic studies. *Microbiological Research, 249*, Article 126770. https://doi.org/10.1016/j.micres.2021.126770

Mishra, S., Priyanka, & Sharma, S. (2022). Metabolomic insights into endophyte-derived bioactive compounds. *Frontiers in Microbiology*, 13. https://doi.org/10.3389/fmicb.2022.835931

Misra, B. B., Langefeld, C., Olivier, M., & Cox, L. A. (2019). Integrated omics: Tools, advances and future approaches. *Journal of Molecular Endocrinology, 62*(1), R21–R45. https://doi.org/10.1530/JME-18-0055

Miura, D., Tanaka, H., & Wariishi, H. (2004). Metabolomic differential display analysis of the white-rot basidiomycete *Phanerochaete chrysosporium* grown under air and 100% oxygen. *FEMS Microbiology Letters, 234*(1), 111–116. https://doi.org/10.1111/j.1574-6968.2004.tb09521.x

Mohammed, A. E., Sonbol, H., Alwakeel, S. S., Alotaibi, M. O., Alotaibi, S., Alothman, N., et al. (2021). Investigation of biological activity of soil fungal extracts and LC/MS-QTOF based metabolite profiling. *Scientific Reports, 11*(1), 4760. https://doi.org/10.1038/s41598-021-83556-8

Mu, Y., Su, W., Mu, Y., & Jiang, L. (2020). Combined application of high-throughput sequencing and metabolomics reveals metabolically active microorganisms during panxian ham processing. *Frontiers in Microbiology, 10*, 1–12. https://doi.org/10.3389/fmicb.2019.03012.

Mudbhari, S., Lofgren, L., Appidi, M. R., Vilgalys, R., Hettich, R. L., & Abraham, P. E. (2024). Decoding the chemical language of *Suillus* fungi: Genome mining and untargeted metabolomics uncover terpene chemical diversity. *Msystems, 9*(4), e01225-e1323. https://doi.org/10.1128/msystems.01225-23

Nagarajan, K., Ibrahim, B., Ahmad Bawadikji, A., Lim, J.-W., Tong, W.-Y., Leong, C.-R., Khaw, K. Y., & Tan, W.-N. (2022). Recent developments in metabolomics studies of endophytic fungi. *Journal of Fungi, 8*(1), Article 1. https://doi.org/10.3390/jof8010028

Nguyen, P., Strub, C., Lagree, M., Bertrand-Michel, J., Schorr-Galindo, S., & Fontana, A. (2020). Study of in vitro interaction between *Fusarium Verticillioides* and *Streptomyces* Sp. using metabolomics. *Folia Microbiologica, 65*(2), 303–314. https://doi.org/10.1007/s12223-019-00725-z

Nicole, F., Guitton, Y., Courtois, E. A., Moja, S., Legendre, L., & Hossaert-McKey, M. (2012). MSeasy: Unsupervised and untargeted GC-MS data processing. *Bioinformatics, 28*(17), 2278–2280. https://doi.org/10.1093/bioinformatics/bts427

Oh, J., Yoon, D.-H., Shrestha, B., Choi, H.-K., & Sung, G.-H. (2019). Metabolomic profiling reveals enrichment of cordycepin in senescence process of *Cordyceps militaris* fruit bodies. *Journal of Microbiology, 57*(1), 54–63. https://doi.org/10.1007/s12275-019-8486-z

Oliver, J. C., Laghi, L., Parolin, C., Foschi, C., Marangoni, A., Liberatore, A., et al. (2020). Metabolic profiling of *Candida* clinical isolates of different species and infection sources. *Scientific Reports, 10*(1), 16716. https://doi.org/10.1038/s41598-020-73889-1

Oliver, S. G., Winson, M. K., Kell, D. B., & Baganz, F. (1998). Systematic functional analysis of the yeast genome. *Trends in Biotechnology, 16*(9), 373–378. https://doi.org/10.1016/s0167-7799(98)01214-1

Oppong-Danquah, E., Budnicka, P., Blümel, M., & Tasdemir, D. (2020). Design of fungal co-cultivation based on comparative metabolomics and bioactivity for discovery of marine fungal agrochemicals. *Marine Drugs, 18*(2). https://doi.org/10.3390/md18020073

Overton, J. M., Linke, L., Magnuson, R., Broeckling, C. D., & Rao, S. (2022). Metabolomic profiles of multidrug-resistant *Salmonella typhimurium* from humans, bovine, and porcine hosts. *Animals, 12*(12), 1518. https://doi.org/10.3390/ani12121518

Palazzotto, E., & Weber, T. (2018). Omics and multi-omics approach to study the biosynthesis of secondary metabolites in microorganisms. *Current Opinion in Microbiology, 45*, 109–116. https://doi.org/10.1016/j.mib.2018.03.004

Pandohee, J., Kyereh, E., Kulshrestha, S., Xu, B., & Mahomoodally, M. F. (2023). Review of the recent developments in metabolomics-based phytochemical research. *Critical Reviews in Food Science and Nutrition, 63*(19), 3734–3749. https://doi.org/10.1080/10408398.2021.1993127

Pang, Z., Lu, Y., Zhou, G., Hui, F., Xu, L., Viau, C., et al. (2024). MetaboAnalyst 6.0: Towards a unified platform for metabolomics data processing, analysis and interpretation. *Nucleic Acids Research*, gkae253. https://doi.org/10.1093/nar/gkae253

Panter, F., Bader, C. D., & Müller, R. (2021). Synergizing the potential of bacterial genomics and metabolomics to find novel antibiotics. *Chemical Science, 12*(17), 5994–6010. https://doi.org/10.1039/D0SC06919A

Patry, S. A. D. (2021). *Omics based approaches for the identification of novel bioactive secondary metabolites from marine sponge derived bacterial isolates*. Ph.D. Thesis, University College Cork.

Pelo, S. P., Adebo, O. A., & Green, E. (2021). Chemotaxonomic profiling of fungal endophytes of *Solanum mauritianum* (alien weed) using gas chromatography high resolution time-of-flight mass spectrometry (GC-HRTOF-MS). *Metabolomics, 17*(5), 1–13. https://doi.org/10.1007/s11306-021-01790-7

Plumb, R. S., Gethings, L. A., Rainville, P. D., Isaac, G., Trengove, R., King, A. M., & Wilson, I. D. (2023). Advances in high throughput LC/MS based metabolomics: A review. *TrAC Trends in Analytical Chemistry, 160*, Article 116954. https://doi.org/10.1016/j.trac.2023.116954

Qian, J., Cutler, J. E., Cole, R. B., & Cai, Y. (2008). MALDI-TOF mass signatures for differentiation of yeast species, strain grouping and monitoring of morphogenesis markers. *Analytical and Bioanalytical Chemistry, 392*(3), 439–449. https://doi.org/10.1007/s00216-008-2288-1

Quinn, R. A., Nothias, L.-F., Vining, O., Meehan, M., Esquenazi, E., & Dorrestein, P. C. (2017). Molecular networking as a drug discovery, drug metabolism, and precision medicine strategy. *Trends in Pharmacological Sciences, 38*(2), 143–154. https://doi.org/10.1016/j.tips.2016.10.011

Ramautar, R., Somsen, G. W., & de Jong, G. J. (2009). CE-MS in metabolomics. *Electrophoresis, 30*(1), 276–291. https://doi.org/10.1002/elps.200800512

Ramzan, M., Raza, A., Nisa, Z., Abdel-Massih, R. M., Al Bakain, R., Cabrerizo, F. M., et al. (2024). Detection of antimicrobial resistance (AMR) and antimicrobial susceptibility testing (AST) using advanced spectroscopic techniques: A review. *Trends in Analytical Chemistry, 172*, Article 117562. https://doi.org/10.1016/j.trac.2024.117562

Ren, Z., Fang, M., Muhae-Ud-Din, G., Gao, H., Yang, Y., Liu, T., Chen, W., & Gao, L. (2021). Metabolomics analysis of grains of wheat infected and noninfected with Tilletia controversa Kühn. *Scientific Reports, 11*(1), 18876. https://doi.org/10.1038/s41598-021-98283-3

Ribay, V., Praud, C., Letertre, M. P. M., Dumez, J.-N., & Giraudeau, P. (2023). Hyperpolarized NMR metabolomics. *Current Opinion in Chemical Biology, 74*, Article 102307. https://doi.org/10.1016/j.cbpa.2023.102307

Robinette, S. L., Brüschweiler, R., Schroeder, F. C., & Edison, A. S. (2012). NMR in metabolomics and natural products research: Two sides of the same coin. *Accounts of Chemical Research, 45*(2), 288–297. https://doi.org/10.1021/ar2001606

Rony, M. K. K., Sharmi, P. D., & Alamgir, H. M. (2023). Addressing antimicrobial resistance in low and middle-income countries: Overcoming challenges and implementing effective strategies. *Environmental Science and Pollution Research, 30*(45), 101896–101902. https://doi.org/10.1007/s11356-023-29434-4

Salam, M. A., Al-Amin, M. Y., Salam, M. T., Pawar, J. S., Akhter, N., Rabaan, A. A., & Alqumber, M. A. (2023). Antimicrobial resistance: A growing serious threat for global public health. *Healthcare, 11*(13), 1946. https://doi.org/10.3390/healthcare11131946

Santacruz, L., Hurtado, D. X., Doohan, R., Thomas, O. P., Puyana, M., & Tello, E. (2020). Metabolomic study of soft corals from the Colombian Caribbean: PSYCHE and 1H-NMR comparative analysis. *Scientific Reports, 10*(1), 5417. https://doi.org/10.1038/s41598-020-62413-0

Santos, D. C., Lima, M. L., Toledo, J. S., Fernandes, P. A., Aguiar, M. M., López-Gonzálvez, Á., et al. (2016). Metabolomics as a tool to evaluate the toxicity of formulations containing amphotericin B, an antileishmanial drug. *Toxicology Research, 5*(6), 1720–1732. https://doi.org/10.1039/c6tx00253f

Sarkar, J., Singh, R., & Chandel, S. (2025). Understanding LC/MS-based metabolomics: A detailed reference for natural product analysis. *PROTEOMICS–Clinical Applications, 19*(1), e202400048. https://doi.org/10.1002/prca.202400048

Scheltema, R. A., Jankevics, A., Jansen, R. C., Swertz, M. A., & Breitling, R. (2011). PeakML/mzMatch: A file format, Java library, R library, and tool-chain for mass spectrometry data analysis. *Analytical Chemistry, 83*, 2786–2793. https://doi.org/10.1021/ac2000994

Schmid, R., Heuckeroth, S., Korf, A., Smirnov, A., Myers, O., Dyrlund, T. S., et al. (2023). Integrative analysis of multimodal mass spectrometry data in MZmine 3. *Nature Biotechnology, 41*(4), 447–449. https://doi.org/10.1038/s41587-023-01690-2

Schueuermann, C., Steel, C. C., Blackman, J. W., Clark, A. C., Schwarz, L. J., Moraga, J., et al. (2019). A GC–MS untargeted metabolomics approach for the classification of chemical differences in grape juices based on fungal pathogen. *Food Chemistry, 270*, 375–384. https://doi.org/10.1016/j.foodchem.2018.07.057

Sciubba, F., Capuani, G., Di Cocco, M. E., Avanzato, D., & Delfini, M. (2014). Nuclear magnetic resonance analysis of water soluble metabolites allows the geographic discrimination of pistachios (*Pistacia vera*). *Food Research International, 62*, 66–73. https://doi.org/10.1016/j.foodres.2014.02.039

Sebastiana, M., Gargallo-Garriga, A., Sardans, J., Pérez-Trujillo, M., Monteiro, F., Figueiredo, A., et al. (2021). Metabolomics and transcriptomics to decipher molecular mechanisms underlying ectomycorrhizal root colonization of an oak tree. *Scientific Reports, 11*(1), 8576. https://doi.org/10.1038/s41598-021-87886-5

Seitzer, P., Bennett, B., & Melamud, E. (2022). MAVEN2: An updated open-source mass spectrometry exploration platform. *Metabolites, 12*(8), 684. https://doi.org/10.3390/metabo12080684

Selegato, D. M., Freire, R. T., Tannús, A., & Castro-Gamboa, I. (2016). New dereplication method applied to NMR-based metabolomics on different fusarium species isolated from Rhizosphere of Senna spectabilis. *Journal of the Brazilian Chemical Society, 27*(8), 1421–1431. https://doi.org/10.5935/0103-5053.20160139

Seybold, H., Demetrowitsch, T., Amine, H. M., Szymczak, S., Reim, E., Haueisen, J., et al. (2020). A fungal pathogen induces systemic susceptibility and systemic shifts in wheat metabolome and microbiome composition. *Nature Communications, 11*(1), 1–12. https://doi.org/10.1038/s41467-020-15633-x

Shi, T., Li, X.-Q., Wang, Z.-M., Zheng, L., Yu, Y.-Y., Dai, J.-J., & Shi, D.-Y. (2022). Bioactivity-guided screening of antimicrobial secondary metabolites from antarctic cultivable fungus *Acrostalagmus luteoalbus* CH-6 combined with molecular networking. *Marine Drugs, 20*(5), Article 5. https://doi.org/10.3390/md20050334

Sieniawska, E., & Georgiev, M. I. (2022). Metabolomics: Towards acceleration of antibacterial plant-based leads discovery. *Phytochemistry Reviews, 21*(3), 765–781. https://doi.org/10.1007/s11101-021-09762-4

Silva-Campos, M., Callahan, D., & Cahill, D. (2022). Metabolites derived from fungi and bacteria suppress in vitro growth of *Gnomoniopsis smithogilvyi*, a major threat to the global chestnut industry. *Metabolomics, 18*(9), 1–14. https://doi.org/10.1007/s11306-022-01933-4

Singh, L. P., Gill, S. S., & Tuteja, N. (2011). Unraveling the role of fungal symbionts in plant abiotic stress tolerance. *Plant Signaling & Behavior, 6*(2), 175–191. https://doi.org/10.4161/psb.6.2.14146

Speckbacher, V., Zeilinger, S., Zimmermann, S., Mayhew, C. A., Wiesenhofer, H., & Ruzsanyi, V. (2021). Monitoring the volatile language of fungi using gas chromatography-ion mobility spectrometry. *Analytical and Bioanalytical Chemistry, 413*(11), 3055–3067. https://doi.org/10.1007/s00216-021-03242-6

Stuart, K. A., Welsh, K., Walker, M. C., & Edrada-Ebel, R. (2020). Metabolomic tools used in marine natural product drug discovery. *Expert Opinion Drug Discovery, 15*(4), 499–522. https://doi.org/10.1080/17460441.2020.1722636

Sturm, M., Bertsch, A., Gröpl, C., Hildebrandt, A., Hussong, R., Lange, E., et al. (2008). OpenMS—An open-source software framework for mass spectrometry. *BMC Bioinformatics, 9*, 1–11. https://doi.org/10.1186/1471-2105-9-163

Suntivich, R., Songjang, W., Jiraviriyakul, A., Ruchirawat, S., & Chatwichien, J. (2022). LC-MS/MS metabolomics-facilitated identification of the active compounds responsible for anti-allergic activity of the ethanol extract of *Xenostegia tridentata*. *PLoS ONE, 17*(4), Article e0265505. https://doi.org/10.1371/journal.pone.0265505

Tal, A., & Frydman, L. (2010). Single-scan multidimensional magnetic resonance. *Progress in Nuclear Magnetic Resonance Spectroscopy, 57*(3), 241–292. https://doi.org/10.1016/j.pnmrs.2010.04.001

Tang, D. Q., Zou, L., Yin, X. X., & Ong, C. N. (2016). HILIC-MS for metabolomics: An attractive and complementary approach to RPLC-MS. *Mass Spectrometry Reviews, 35*(5), 574–600. https://doi.org/10.1002/mas.21445

Tang, K. W. K., Millar, B. C., & Moore, J. E. (2023). Antimicrobial resistance (AMR). *British Journal of Biomedical Science, 80*, 11387. https://doi.org/10.3389/bjbs.2023.11387

Tang, L., Shang, J., Song, C., Yang, R., Shang, X., Mao, W., et al. (2020). Untargeted metabolite profiling of antimicrobial compounds in the brown film of *Lentinula edodes* mycelium via LC–MS/MS analysis. *ACS Omega, 5*(13), 7567–7575. https://doi.org/10.1021/acsomega.0c00398

Tapfuma, K. I., Nchabeleng, E. K., Adebo, O. A., Hussan, R., Williams, R. D., Ravuluvulu, A. B., Ndinteh, D. T., Gan, R.-Y., Habimana, O., Niemann, N., Muganza, F. M., Mekuto, L., & Mavumengwana, V. (2020). Antibacterial activity and gas chromatography mass spectrometry (GC–MS)-based metabolite profiles of *Celtis africana* and its endophytic extracts. *Industrial Crops and Products, 157*, Article 112933. https://doi.org/10.1016/j.indcrop.2020.112933

Tautenhahn, R., Patti, G. J., Rinehart, D., & Siuzdak, G. (2012). XCMS Online: A web-based platform to process untargeted metabolomics data. *Analytical Chemistry, 84*, 5035–5039. https://doi.org/10.1021/ac300698c

Tawfike, A. F., Romli, M., Clements, C., Abbott, G., Young, L., Schumacher, M., et al. (2019). Isolation of anticancer and anti-trypanosome secondary metabolites from the endophytic fungus *Aspergillus flocculus* via bioactivity guided isolation and MS based metabolomics. *Journal of Chromatography, B, 1106*, 71–83. https://doi.org/10.1016/j.jchromb.2018.12.032

Temple, N. (2022). A rational definition for functional foods: A perspective. *Frontiers in Nutrition, 9*. https://doi.org/10.3389/fnut.2022.957516

Tian, Q., Wang, J., Shao, S., Zhou, H., Kang, J., Yu, X., et al. (2024). Combining metabolomics and transcriptomics to analyze key response metabolites and molecular mechanisms of *Aspergillus fumigatus* under cadmium stress. *Environmental Pollution*, 124344. https://doi.org/10.1016/j.envpol.2024.124344

Tigistu-Sahle, F., Mekuria, Z. H., Satoskar, A. R., Sales, G. F., Gebreyes, W. A., & Oliveira, C. J. (2023). Challenges and opportunities of molecular epidemiology: Using omics to address complex One Health issues in tropical settings. *Frontiers in Tropical Diseases, 4*, 1151336. https://doi.org/10.3389/fitd.2023.1151336

Tsugawa, H., Ikeda, K., Takahashi, M., Satoh, A., Mori, Y., Uchino, H., et al. (2020). A lipidome atlas in MS-DIAL 4. *Nature Biotechnology, 38*(10), 1159–1163. https://doi.org/10.1038/s41587-020-0531-2

Uehling, J., Entler, M., Meredith, H., Millet, L., Timm, C., Aufrecht, J., et al. (2019). Microfluidics and metabolomics reveal symbiotic bacterial–fungal interactions between *Mortierella elongata* and *Burkholderia* include metabolite exchange. *Frontiers in Microbiology, 10*, 1–13. https://doi.org/10.3389/fmicb.2019.02163

Uffelman, C. N., Doenges, K. A., Armstrong, M. L., Quinn, K., Reisdorph, R. M., Tang, M., et al. (2023). Metabolomics profiling of white button, crimini, portabella, lion's mane, maitake, oyster, and shiitake mushrooms using untargeted metabolomics and targeted amino acid analysis. *Foods, 12*(16), 2985. https://doi.org/10.3390/foods12162985

Utermann, C., Echelmeyer, V. A., Blümel, M., & Tasdemir, D. (2020). Culture-dependent microbiome of the *Ciona intestinalis* tunic: Isolation, bioactivity profiling and untargeted metabolomics. *Microorganisms, 8*(11), 1732. https://doi.org/10.3390/microorganisms8111732

Venkataraman, R., Nakedi, K., Akhade, A. S., & Soni, V. (2024). Microbial metabolomics to understand antimicrobial resistance and drug metabolism. In *Antimicrobial resistance: Factors to findings: omics and systems biology approaches* (pp. 151–177). Springer International Publishing. https://doi.org/10.1007/978-3-031-65986-7_5

Vinaixa, M., Schymanski, E. L., Neumann, S., Navarro, M., Salek, R. M., & Yanes, O. (2016). Mass spectral databases for LC/MS- and GC/MS-based metabolomics: State of the field and future prospects. *TrAC Trends in Analytical Chemistry, 78*, 23–35. https://doi.org/10.1016/j.trac.2015.09.005

Wang, Z., Xie, W., Luo, D., Wang, L., Xu, Y., Rosenn, E. H., et al. (2024). Omics-based approaches in developing treatments against antimicrobial resistance. In *Antimicrobial resistance: factors to findings: Omics and systems biology approaches* (pp. 387–413). Springer International Publishing. https://doi.org/10.1007/978-3-031-65986-7_12

Wen, H., Kang, S., Song, Y., Song, Y., Sung, S. H., & Park, S. (2010). Differentiation of cultivation sources of *Ganoderma lucidum* by NMR-based metabolomics approach. *Phytochemical Analysis, 21*(1), 73–79. https://doi.org/10.1002/pca.1166

Wishart, D. S., Cheng, L. L., Copié, V., Edison, A. S., Eghbalnia, H. R., Hoch, J. C., et al. (2022). NMR and metabolomics—A roadmap for the future. *Metabolites, 12*(8), 678. https://doi.org/10.3390/metabo12080678

Wong, J.W.-H., Plett, K. L., Natera, S. H. A., Roessner, U., Anderson, I. C., & Plett, J. M. (2020). Comparative metabolomics implicates threitol as a fungal signal supporting colonization of *Armillaria luteobubalina* on eucalypt roots. *Plant, Cell & Environment, 43*(2), 374–386. https://doi.org/10.1111/pce.13672

Worley, B., & Powers, R. (2013). Multivariate analysis in metabolomics. *Current Metabolomics, 1*(1), 92–107. https://doi.org/10.1016/B978-0-12-818607-7.00023-2

Wu, Z., Li, D., Meng, J., & Wang, H. (2010). Introduction to SIMCA-P and its application. In *Handbook of partial least squares: Concepts, methods and applications* (pp. 757–774). https://doi.org/10.1007/978-3-540-32827-8_33

Xie, H., Wang, X., van der Hooft, J. J., Medema, M. H., Chen, Z. Y., Yue, X., et al. (2022). Fungi population metabolomics and molecular network study reveal novel biomarkers for early detection of aflatoxigenic *Aspergillus* species. *Journal of Hazardous Materials, 424*, Article 127173. https://doi.org/10.1016/j.jhazmat.2021.127173

Yang, C., Zhao, W., Wang, Y., Zhang, L., Huang, S., & Lin, J. (2020). Metabolomics analysis reveals the alkali tolerance mechanism in *Puccinellia tenuiflora* plants inoculated with arbuscular mycorrhizal fungi. *Microorganisms, 8*(3), 327. https://doi.org/10.3390/microorganisms8030327

Yang, X., Zhan, J., & Ma, Y. (2021). Targeted and untargeted metabolomics for analyzing mycotoxins in food. *TrAC Trends in Analytical Chemistry, 151*, Article 116537. https://doi.org/10.1016/j.trac.2022.116537

Zhang, B., Gao, X., Wang, Q., Li, Y., He, C., Luo, H., et al. (2022). Integrated application of transcriptomics and metabolomics provides insights into the antifungal activity of α-phenylcinnamic acid against *Colletotrichum gloeosporioides. Postharvest Biology and Technology, 186*, Article 111834. https://doi.org/10.1016/j.postharvbio.2022.111834

Zhang, H., Song, D., Luo, Q., Yu, J., Wei, Y., Chen, D., et al. (2024). Multi-omics analysis reveals indicator features of microbe-host interactions during *Candida albicans* colonization and subsequent infection. *Frontiers in Microbiology, 15*, 1476429. https://doi.org/10.3389/fmicb.2024.1476429

Zhang, X., Quinn, K., Cruickshank-Quinn, C., Reisdorph, R., & Reisdorph, N. (2018). The application of ion mobility mass spectrometry to metabolomics. *Current Opinion in Chemical Biology, 42*, 60–66. https://doi.org/10.1016/j.cbpa.2017.11.001

Zhang, Y., Yang, Y., Ding, L., Wang, Z., Xiao, Y., & Xiao, W. (2021). Emerging applications of metabolomics to assess the efficacy of traditional Chinese medicines for treating type 2 diabetes mellitus. *Frontiers in Pharmacology, 12*, 735410. https://doi.org/10.3389/fphar.2021.735410

Zhang, Y. Y., Zhang, P., Le, M. M., Qi, Y., Yang, Z., Hu, F. L., et al. (2023). Improving flavor of summer *Keemun* black tea by solid-state fermentation using *Cordyceps militaris* revealed by LC/MS-based metabolomics and GC/MS analysis. *Food Chemistry, 407*, Article 135172. https://doi.org/10.1016/j.foodchem.2022.135172

Zhao, X., Hengchao, E., Dong, H., Zhang, Y., Qiu, J., Qian, Y., et al. (2022). Combination of untargeted metabolomics approach and molecular networking analysis to identify unique natural components in wild *Morchella* sp. by UPLC-Q-TOF-MS. *Food Chemistry, 366*, 130642. https://doi.org/10.1016/j.foodchem.2021.130642

Zhu, Z., Wang, H., Shang, Q., Jiang, Y., Cao, Y., & Chai, Y. (2013). Time course analysis of *Candida albicans* metabolites during biofilm development. *Journal of Proteome Research, 12*(6), 2375–2385. https://doi.org/10.1021/pr300447k

Zou, Y. N., Qin, Q. Y., Ma, W. Y., Zhou, L. J., Wu, Q. S., Xu, Y. J., et al. (2023). Metabolomics reveals arbuscular mycorrhizal fungi-mediated tolerance of walnut to soil drought. *BMC Plant Biology, 23*(1), 118. https://doi.org/10.1186/s12870-023-04111-3

The Field of Biotechnology Concerning Fungal Secondary Metabolites

Noorul Samsoon Maharifa Haja Mohaideen, Aparna Lakshmi Narasimhan, Anupam Prakash, and Hemalatha Srinivasan

Abstract

The fungal secondary metabolites have become potent biotechnological weapons, and they provide a fantastic diversity of bioactive compounds which are utilized in medicine, agriculture, and industry. In this chapter, the extensive metabolic potential of fungi, especially the filamentous Ascomycota and Basidiomycota will be discussed, along with the advancements in genetics, systems biology,and fermentation engineering that enable us to reach their full potential. Fungi have a historical lifesaving drug record, including penicillin and lovastatin. These are now being re-engineered with advanced technologies including CRISPR/Cas9 and synthetic biology to improve yields, unlock silent gene clusters, or synthesize new molecules. The present chapter explains how early methods such as random mutagenesis and protoplast fusion paved the way to modern strain improvement, which is currently being transformed by targeted genome editing and pathway rewiring. Co-culturing, promoter swapping, and synthetic regulatory circuits enable fine control of metabolite production. Furthermore, the fungal biosynthesis genes are becoming more applicable to scalable and economical hosts due to their heterologous expression in microbial hosts as *Pichia pastoris* and *Bacillus subtilis*. Genetic manipulation aside, this chapter also discusses the problems of laboratory-to-industrial scale-up—particular in viscous fungal fermentations. Parameter optimization, such as oxygen transfer, shear stress, nutrient availability, and incorporation of bioprocess modelling have been found vital to repeatable yields at scale. Case studies illustrate how designed strains of *Fusarium heterosporum*, *Trichoderma reesei*, and *Aspergillus oryzae* are opening the route to sustainable biomanufacturing with inexpensive raw materials. In whole, this chapter shows that biology and engineering are coming together in contemporary fungal biotechnology. By utilizing sophisticated instruments and carefully thought-out plans for expansion, fungi are in a strong position to lead the next phase of sustainable innovation in the biosynthesis of natural products.

Keywords

Biosynthetic gene clusters (BGCs) · CRISPR/Cas9 genome editing · Synthetic biology · Metabolic engineering · Bioprocess optimization

1 Introduction

Fungal secondary metabolites are various chemical compounds that help fungi to live in their environments, interact with other organisms, and defend themselves (Brakhage, 2013). Metabolites are produced by fungi at points in their life cycles and consist of antibiotics, mycotoxins, pigments, and growth regulators that greatly assist in medicine, agriculture, and industry (Keller, 2019). Sophisticated metabolic pathways in fungi as eukaryotes make them important contributors to biotechnology (Corbu et al., 2023). The fungal kingdom's Ascomycota and Basidiomycota are remarkable for having highly evolved secondary metabolisms which are backed by the large numbers of their species and the continuous discovery of new BGCs (Panstruga et al., 2023; Wijayawardene et al.,

N. S. M. H. Mohaideen · A. L. Narasimhan · H. Srinivasan (✉)
School of Life Sciences, B. S. Abdur Rahman Crescent Institute of Science & Technology, Chennai, Tamil Nadu, India
e-mail: hemalatha.sls@bsauniv.ac.in

N. S. M. H. Mohaideen
Department of Microbiology, Thassim Beevi Abdul Kader College for Women, Kilakarai, Tamil Nadu, India

A. Prakash
Department of Life Sciences, School of Biosciences and Technology, Galgotias University, Greater Noida, India

© The Author(s), under exclusive license to Springer Nature Switzerland AG 2026

A. Prakash and M. K. Dubey (eds.), *Advances in Fungal Secondary Metabolites: Bioactive Attributes and Biotechnological Potential of Industrial Interest*, Sustainable Economy and Ecotechnology, https://doi.org/10.1007/978-3-032-23507-7_11

2024). Because of their metabolic flexibility, they may hold great unexplored potential for discovering useful new compounds in both therapies and industries (Xu et al., 2024).

Fungi produce secondary metabolites according to their nutrient supply, the temperature, the pH, and the presence of different elicitors (Xue et al., 2023). Positive compounds like penicillin and lovastatin have enhanced medical care, while harmful ones such as mycotoxins are a danger to food safety and health (Hashem et al., 2023). For these compounds to be applied on a large scale, challenges with their yields, stability, and purification must be solved which has encouraged significant inventions within biotechnology (Shenouda et al., 2022). Approaches like random mutagenesis and protoplast fusion have established a base for improving strains (Zhgun, 2023a, 2023b). Thankfully, new approaches such as genetic engineering, metabolic engineering, and synthetic biology have made it possible to alter fungal genomes and metabolism confidently (Garg et al., 2025). With the help of tools like CRISPR/Cas9, scientists have now managed to activate inactive BGCs, revealing additional metabolites and synthetic regulatory circuits and heterologous systems have made it easier to produce more (Jiang et al., 2021; Li et al., 2019).

Such progress is crucial since more and more sectors are looking for fungal secondary metabolites. Still, transferring cell production to industry is difficult and calls for optimizing fermentation methods, bioreactors, and growing conditions (Yang et al., 2023). Both systems biology and predictive modelling contribute to better designing and optimizing, dealing with issues related to metabolic inconsistency and contamination risks (Verma et al., 2023). The blend of science and technology in this area highlights how fungal biotechnology can create better solutions for the world.

It describes the many areas of fungal secondary metabolites, outlines how to classify them, highlights their functions in nature, and mentions biotechnological steps aimed at boosting production. It uses various traditional and cutting-edge genetic techniques as well as CRISPR/Cas9 editing to modify fungi and discusses new pathways and engineering of metabolic networks for improved biosynthesis. It also covers using yeasts and bacteria to create proteins from foreign genes, describes the issues involved, and defines important steps for scaling-up fermentation in industrial bioreactors. The study points out the progress already made and the prospects for the future of using fungi in making secondary metabolites.

2 Secondary Metabolite Producers

Fungi represent crucial and varied elements within our planet's ecosystems (Shi et al., 2024). Their functional diversity is notable, accompanied by a range of dispersal mechanisms that categorize them as both macro and microorganisms (Zhao, 2023). As our comprehension of microbial biogeography has evolved, it has become evident that the fungal assembly patterns and processes differ significantly from those of other microorganisms, such as bacteria and plants. The remarkable adaptability of fungi, allowing them to overcome barriers of distance, time, and biological interactions, elucidates their success as organisms and their substantial impact on the environment in ways that are not achievable by other groups of organisms (Wijayawardene et al., 2024).

Among the three principal kingdoms of life—bacteria, fungi, and plants. Fungi exhibit a notably advanced secondary metabolism (Panstruga et al., 2023). While the Zygomycota, Glomeromycota, and Chytridiomycota display less sophisticated secondary metabolic pathways, the filamentous Ascomycota and Basidiomycota are characterized by their highly developed secondary metabolisms, as illustrated in Fig. 1 (Wijayawardene et al., 2024). The extensive variety of fungal species, particularly within the Ascomycota and Basidiomycota, along with the corresponding diversification of genes and gene clusters responsible for biosynthesis, indicates that the potential for metabolic variation is virtually limitless. Recent genome sequencing efforts have revealed that many fungal species harbour dozens of BGCs, most of which remain uncharacterized, suggesting a wealth of undiscovered metabolites (Xu et al., 2024).

The biological functions of secondary metabolites are remarkably varied. While some have contributed positively to societal advancements, others are associated with significant and often complex challenges. The array of natural compounds synthesized by fungi can be classified as either beneficial, such as Penicillin, Lovastatin, Cyclosporine, Griseofulvin, Ergotamine, and Ergometrine, or harmful, including Ochratoxin produced by *Aspergillus* and *Penicillium* species, Fumonisins generated by *Fusarium verticillioides* and *Fusarium proliferatum* and Trichothecenes from various *Fusarium* species. A range of Indole-terpene, tremor-inducing compounds, such as cytochalasins, is produced by various genera of fungi (Corbu et al., 2023). All these compounds are categorized into primary and secondary metabolites. The primary metabolites are directly linked to central metabolic processes and are essential for the growth and development of fungi; however, secondary metabolites are nonessential (Corbu et al., 2023).

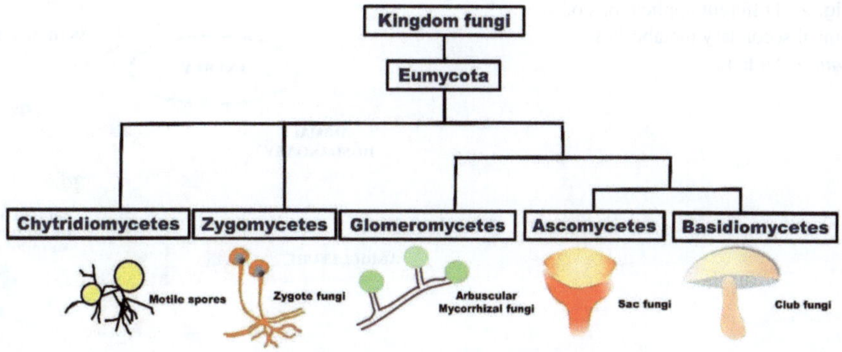

Fig. 1 Classification of Eumycota. *Source* Authors

2.1 Engineering Fungi for Enhanced Metabolite Production

Fungi, a eukaryotic organism with well-developed cell organelles and defined nucleus. They produce both primary metabolites, which are essential for growth, development, reproduction, and secondary metabolites, which are supplement for their survival having advantages like antibiotics, mycotoxins, pigments, growth hormones, etc. (Fig. 2) (Corbu et al., 2023; Wadhwa et al., 2024). Recent studies have demonstrated the efficacy of co-culturing strategies, where inter-microbial interactions activate silent biosynthetic gene clusters, yielding novel metabolites (Wang et al., 2022). The demand for these secondary metabolites in medicine, agriculture, and industry has driven the development of various methods to improve fungal strains, including genetic engineering, metabolic engineering, synthetic biology, and optimization of environmental conditions (Garg et al., 2025). These approaches not only enhance yield but also enable the production of metabolites with improved bioactivity and reduced production costs (Rodrigues et al., 2021).

Fungi are known to produce various pharmaceutical, food, agricultural, and industrial products using pathways such as polyketide pathways, non-ribosomal peptide, terpenoid, and hybrid pathways (Keller, 2019). The secondary metabolites are the end product of these pathways including penicillin, immunosuppressant cyclosporine, cholesterol-lowering drug lovastatin, blue cheese, soy sauce, amylase, protease etc., (Wijayawardene et al., 2024). These secondary metabolites are administered by a regulatory mechanism that includes global, cluster-specific, and epigenetic regulation. Epigenetic modifiers, such as histone deacetylase inhibitors, have been shown to activate silent BGCs, leading to the discovery of new compounds (Verma et al., 2023).

The generation of new secondary metabolites can be significantly facilitated by multiple strategic approaches as shown in Fig. 3. These are co-culture, molecular-based strategies, chemical elicitors, and systems biology strategies. Co-culturing strategies involve inter-microbial interactions of different microbial strains; for example, fungal-fungal, fungal-bacterial, and host-fungal co-cultures might activate silent biosynthetic gene clusters. Molecular-based strategies include gene editing, promoter swapping, site-directed mutagenesis, and rRNA gene substitution to activate or upregulate metabolite biosynthesis (Tamano et al., 2023). CRISPR/Cas9-mediated gene modification is an example of modifying a single strain genome, allowing many compounds to be produced from a single bacterial or fungal strain (Garg et al., 2025). In addition, methylation histone acetylation and de-acetylation can also chemically prime and epigenetically modify gene expression that can stimulate secondary metabolite production (Zhgun, 2023a, 2023b). Systems biology strategies including sequence alignment, primer design, molecular docking, and in silico prediction can help enable the rational design and discovery of secondary metabolites (Verma et al., 2023). Collectively, these strategies coupled with combinations of fermentation and cultivation conditions have the potential to create a powerful platform for the discovery and production of bioactive secondary metabolites (Fig. 3).

2.1.1 Techniques in Strain Improvement and Genetic Modification

With fungi producing important secondary metabolites, their demand in industry, agriculture, and medicine expands (Hashem et al., 2023). Due to their insufficient production in their host organism and to meet the growing demands, there is a need for continuous modification of the fungal strains to enhance their metabolite production, functional properties, cost reduction, improved stress tolerance, and enhanced biocontrol capabilities (Garg et al., 2025). Before the advent of molecular techniques, traditional methods were used for strain improvement aiming for better production of metabolites (Coca-Ruiz et al., 2024). These laid the foundation for enhancement of widely used methods such as random mutagenesis, protoplast fusion, and parasexual reproduction (Zhgun, 2023a, 2023b) (Fig. 4). These traditional approaches, while effective, often yielded unpre-

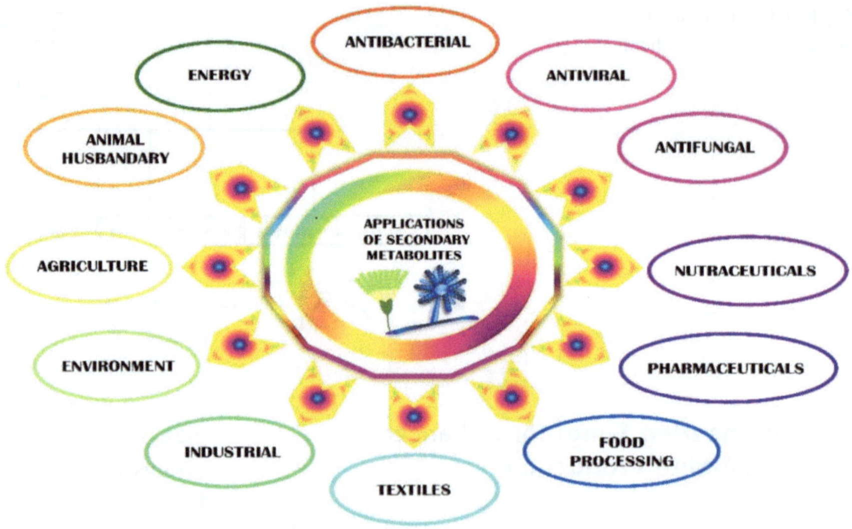

Fig. 2 Different applications of fungal secondary metabolites. *Source* Authors

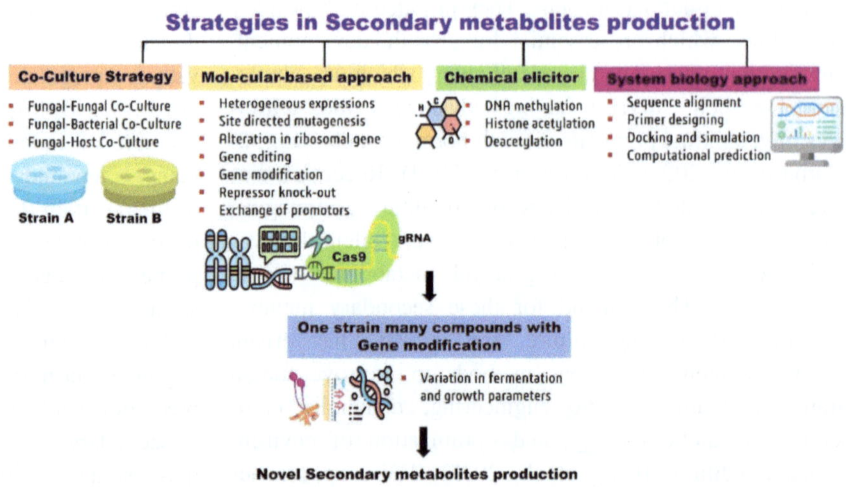

Fig. 3 Strategies in secondary metabolites production. *Source* Authors

Fig. 4 Process involved in traditional molecular methods. *Source* Authors

dictable outcomes due to their reliance on random genetic alterations (Coca-Ruiz et al., 2024).

Random Mutagenesis and Screening

This involves creating genetic variation in the organism by inducing mutations in the DNA; this is achieved by mutagens—agents that cause genetic variation (Zhgun, 2023a, 2023b). The primary goal of mutagenesis is to produce varying mutants with varying traits that enhance the production of metabolites (Wang et al., 2022). Screening is the process of identifying and evaluating the microbial strains that show the desired traits and the best performance under specified conditions (Fig. 5) (Nevalainen, 2001).

Protoplast Fusion

Protoplast fusion is a method that allows the fusion of strains, by removing the cell walls. This can be done by either using polyethylene glycol or using electric fields, and electrofusion, resulting in the formation of a heterokaryon that carries the genetic material (Coca-Ruiz et al., 2024). The fusion results in the formation of mixed genetic materials that show enhanced traits and improved metabolic capabilities (Nevalainen, 2001). Protoplast fusion has been particularly useful in combining desirable traits from different fungal species, although it remains technically challenging (Coca-Ruiz et al., 2024).

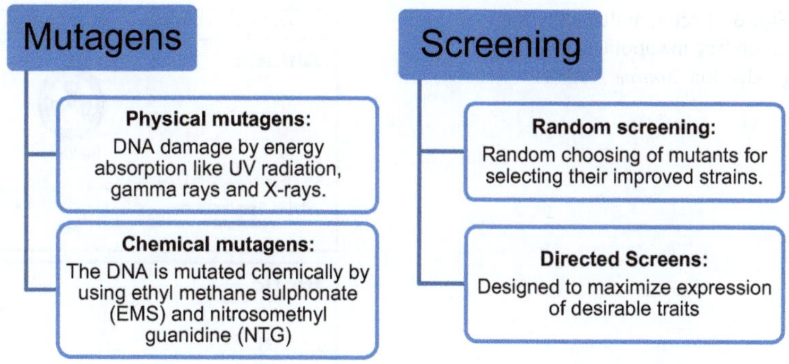

Fig. 5 Methods involved in modern secondary metabolite production. *Source* Authors

Sexual and Parasexual Reproduction

The fungi can reproduce through sexual and parasexual reproduction. In sexual reproduction, special cells fuse forming diploid cells (Michaliski et al., 2023). These diploid cells undergo meiosis, resulting in a blend of genetic materials from the different fungi, which results in boosting yield and increases therapeutic potential. In the case of parasexual reproduction, the fungus connects with each other and shares their genetic material without forming diploid cells. This leads to the formation of a heterokaryon, which further divides and gives new characteristic traits to the offspring. Fungi that cannot reproduce sexually can use parasexuality to exchange genes and enhance their strains, instead (Zhgun, 2023a, 2023b).

Nowadays, genetic alterations are mostly done using CRISPR/Cas9 and similar methods, due to their better accuracy and speed. Overall, utilizing CRISPR/Cas9 can help make fungal genome modification more exact, so researchers can activate silent BGCs or strengthen existing pathways (Jiang et al., 2021). However, applying traditional techniques, even now, can be useful when used along with modern strategies.

2.1.2 Synthetic Pathways and Metabolic Engineering

Fungal secondary metabolites are of low molecular weight with various chemical structures and biological activities (Yang et al., 2023). They play a crucial role in their defence mechanism, protection against environmental stress caused by both biotic and abiotic, can influence a wide array of microbial reactions, and also support Quorum sensing (Brakhage, 2013). Synthetic biology has emerged as a transformative tool, enabling the redesign of these pathways to produce novel metabolites with enhanced bioactivity (Cravens et al., 2019). Elicitors, which are considered stressors, can induce or amplify the production of secondary metabolites within a biological system. These elicitors are classified into distinct categories based on their nature and origin, which include biotic or abiotic factors, as well as physical or chemical influences (Xue et al., 2023) (Fig. 6). They also come with several advantages like antibiotics, immunosuppressants, anticancer agents, and agrochemicals. These secondary metabolites are produced in fungi during post exponential phase (Brakhage, 2013), but these are also governed by factors like nutrient availability, temperature, and shift in pH.

The production mechanism of these secondary metabolites is completely different from the primary metabolites. The clusters of genes that were modulated rationally are called biosynthetic gene clusters (BGCs). The activation of the gene clusters is regulated by environmental conditions or epigenetic changes. The production of the metabolites is synthesized through various synthetic pathways (Nevalainen, 2001). The most prominent pathways include the non-ribosomal pathway, polyketide pathway, terpenoid, alkaloid, and hybrid pathway with their enzymatic machinery, as presented in Table 1. By using metabolic engineering, researchers can now improve these routes and increase secretion of wanted chemicals, decreasing the amount of by-products in the same process (Tamano et al., 2023). It is possible to design synthetic pathways by using metabolic engineering to combine genes from different organisms into a fungi host using heterologous expression. Researchers have been able to use this method to make complex metabolites that chemical synthesis finds difficult, which increased the range of uses for fungal biotechnology (Shenouda et al., 2022).

2.2 Pathway Optimization Using Genetic Tools

2.2.1 CRISPR/Cas9 and Gene Editing

CRISPR/Cas9 is a cutting-edge genetic engineering tool in comparison with traditional gene editing tools like zinc finger nucleases and transcription activator-like effector nucleases. The Cas 9 system, which uses simple gRNA to direct the genome editing, is more efficient, faster, and relatively lower cost compared to the traditional use of costume-designed proteins specific to the DNA sequence. The CRISPR/Cas9 technology helps in targeting multiple genes at once and has a higher potential for off-target effects

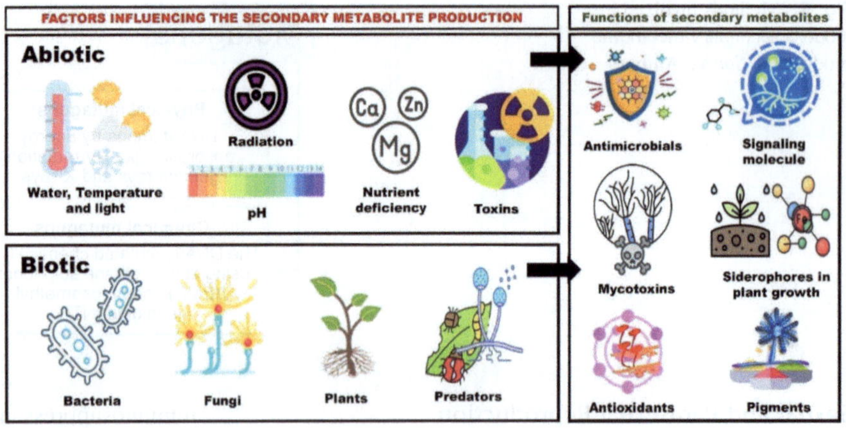

Fig. 6 Factors influencing the secondary metabolites production. *Source* Authors

Table 1 Pathway connections in secondary metabolite production of fungi

S. No	Pathway	Key enzyme	Mechanism	Secondary metabolite	References
1	Polyketide pathway	Polyketide synthases	The multi-domain enzyme uses acetyl CoA and malonyl CoA. They help in adding 2 units of carbon molecules and removing CO_2. Once the molecule is fully modified, it is released by thioesterase	Lovastatin, Aflatoxins, and Tetracycline	Kakule et al. (2015), Ridley and Khosla (2009)
2	Non-ribosomal peptide pathway	Non-ribosomal peptide synthetase	The enzyme is independent of mRNA and ribosomes; each functional module of the enzyme adds amino acid. The final products are the small cyclic peptides with modification. The functional modules are adenylation, thiolation, and condensation	Penicillin and Cyclosporin	Zobel et al. (2015), Shenouda et al. (2022)
3	Terpenoid pathway	Terpene synthases	Constructed by isoprene units—IPP (isopentenyl pyrophosphate) and DMAPP. This follows 2 routes; the mevalonate and non-mevalonate pathway. The enzyme helps in catalysing the formation or arrangement of prenyl diphosphatase	Trichothecenes and Fumonisins	Shi et al. (2024); Wadhwa et al. (2024)
4	Alkaloid pathway	This involves a series of enzymes like cytochrome P450s, methyltransferases, and oxidases	The reaction starts with the addition of amino acids, followed by enzymatic activities like decarboxylation, oxidation methylation, and cyclization	Ergot alkaloids	Xue et al. (2023)
5	Hybrid pathway	Fusion of enzymes from the different pathways	This pathway involves a combination of different pathways like PKS + NRPS, PKS + Terpenes, etc. This allows the fungi to synthesize diverse secondary molecules	Mycotoxins and antibiotics	Shenouda et al. (2022)

Source Authors compilation

which is challenging in traditional methods. CRISPR/Cas9 provides a more flexible, accessible, and powerful platform for editing genes in the fungal genome producing higher secondary metabolites (Doudna & Charpentier, 2014; Gaj et al., 2013; Joung & Sander, 2013).

Filamentous fungi with their secondary metabolites having diverse bioactivity remain undiscovered due to their silent BGCs (Xu et al., 2024). CRISPR/Cas9 genome editing helped in the discovery of new BGCs. The Cas 9 system, a molecular scissor, is highly specific and efficient, and it is a versatile genome-editing tool. The guide RNA (gRNA) facilitates the Cas 9 to the target site cutting the double-stranded DNA and repairing cells via non-homologous end joining by introducing mutations and homology-directed repair that helps in precise editing (Doudna & Charpentier, 2014). However, its application in fungi raises ethical and regulatory considerations, particularly regarding environmental release and safety (Garg et al., 2025).

A new gene responsible for polyketide non-ribosomal peptide in species *T. atroroseus* was identified using CRISPR/Cas9 technology. The researchers primarily focused on deleting gene UA08_00425, responsible for green pigment, and it was then confirmed in fungus with the production of white spores (Nielsen et al., 2017). They then deleted another gene UA08_04451, suggesting that the compounds ZG-1494α and talaroconvolutin-A were synthesized from the same origin by polyketide non-ribosomal peptide fusion enzyme encoded by talA gene. Here the CRISPR/Cas9 system developed for *Aspergillus* fungal species was used for polyketide non-ribosomal peptide in species (Nielsen et al., 2017).

A new gene editing kit was developed using CRISPR/Cas9 to enhance the production of gibberellin, which is an important plant hormone used industrially (Jiang et al., 2021). The new strategy was developed with CRISPR/Cas9 for the fungus *Fusarium fujikuroi*, by proving its efficiency with the addition of HTB_NLS, a signal from protein histone H2B, which helped in entering the nucleus (Shi et al., 2019). To increase the production of GA, two genes—Cps/Ks and tHmgR—were introduced stably into the cell that precisely engineered the fungus genome, boosting the production of GA4 and GA7 (Jiang et al., 2021).

2.2.2 Synthetic Regulatory Circuits for Metabolite Overproduction

Regulatory circuits are complex organized networks of molecular components that sense specific input and give the determined output to control gene expression and cellular behaviour (Peng et al., 2022). Synthetic regulatory circuits, which mimic natural genetic control mechanisms, give us an advantage of predictability and tunability. The components of the synthetic circuit include ribosome binding sites, activators, repressors, promoters, and riboswitches; these components are structured in such a way that they activate in specified conditions (Teixeira & Fussenegger, 2024). These circuits help in sensing and controlling metabolite concentrations in real time, rerouting metabolic pathways, and stimulating or repressing the BGCs (Cravens et al., 2019).

Synthetic regulatory circuits afford metabolic engineers the precision needed to control the trade-off between metabolite production and growth (Müller et al., 2025). Inductive promoters and modular sensor technology are examples of synthetic biology approaches to control gene expression in complex metabolic networks (Cravens et al., 2019). This allows microorganisms to maximize metabolite production while maintaining homeostasis, even in nutrient-limiting and stress (Garg et al., 2025).

The titres of bioactive organic products have been considerably enhanced in *Actinobacteria* by rerouting precursor availability and harmonizing the expression of pathway genes using synthetic regulatory networks (Li et al., 2019). In the same vein, the cofactors engage biosensor-derived regulatory systems, regulating the production of secondary metabolites through signal-responsive components (BGCs) (Jia et al., 2024). The addition of pathway-specific activators into synthetic regulatory circuits greatly enhanced the production of non-ribosomal peptide and polyketide profiles (Li et al., 2019). Synthetic circuits for the temporal regulation of expression of genes clustered in BGCs were developed in *Aspergillus* species showing a greater abundance of metabolites while maintaining cell viability (Jia et al., 2024). The CsdA/RsdA RNA binding protein complex regulates transcript stability and expression to control the metabolism of fungi. Synthetic analogues of both CsdA and RsdA RNA components provide a new template for building circuits for research and new products (Song et al., 2023). Using synthetic circuits that combine real-time sensors, microorganisms can self-correct their gene expression in response to changes in nutrients or metabolite concentrations, allowing maximum output during fermentation (Zhang et al., 2024).

3 Heterologous Expression of Fungal Biosynthetic Pathways

3.1 Techniques for Expressing Fungal Genes in Yeasts and Bacteria

3.1.1 Host Selection and Expression Systems

Expressing fungal genes in heterologous hosts like yeast and bacteria in facilitating the production of enzymes, proteins, and secondary metabolites is a unique trend in the field of biotechnology (Shenouda et al., 2022). This approach is instrumental in overcoming challenges associated with fungal secondary metabolites production, by enhancing yield and scalability. The selection of the host system needs to be tailored to target the protein's complexity and intended application. Gene editing tools like CRISPR/Cas9 have enabled the expression of fungal genes in heterologous host systems (Xu et al., 2024).

Prominent yeast hosts, such as *Saccharomyces cerevisiae* and *Komagataella phaffii* (*Pichia pastoris*), are widely used to express fungal genes because of their eukaryotic nature which facilitates essential post-translational modifications such as proper protein folding and glycosylation which are often impossible in prokaryotic systems (Karbalaei et al., 2020). They are also known to produce a high yield of recombinant proteins, thereby simplifying downstream processing. The yeast system has well-established markers and promotors enabling the expression of complex biosynthetic gene clusters, providing novel secondary metabolite products. Fungal laccases, especially those from white-rot basidiomycetes such as *Trametes trogii*, are multi-copper oxidases that can oxidatively polymerize a wide variety of phenolic compounds, aromatic amines, and xenobiotics, including many synthetic dyes and pesticides, which makes them very useful for bioremediation, textile processing, and organic syntheses (Colao et al., 2006). Nevertheless, the industrial application of these enzymes has been limited due to the complexity of culturing fungi and have multiple isoenzymes, which complicate attempts to purify an active enzyme. To address these obstacles, we heterologous lcc1 gene expressed from *T. trogii*encoding the major phenol oxidase in the yeast *Pichia pastoris*, known for high-level expression of proteins because *P. pastoris* linearly grows in the presence of methanol which controllably drives carbon to the alcohol oxidase promoter (P_AOX1) (Colao et al., 2006). The total RNA was isolated from clone lcc1 cDNA into expression vectors pHIL-D2 and pPIC9 with either native signal sequence or *S. cerevisiae* α-mating factor signal peptide and transformed into *P. pastoris* GS115 cells. The majority of the transformants generated exhibited successful secretion of active laccase from the yeast and found that expression in a fermenter raised the expression level. Finally, the secreted recombinant Lcc1 was purified by several chromatographic stages, obtaining a 64 kDa protein according to SDS-PAGE, consistent with that of the native enzyme (Colao et al., 2006).

Bacteria are widely used expression systems due to their rapid growth and well-organized genome structure. However, these expression systems come with limitations as they lack post-translational modifications (Teixeira & Fussenegger, 2024). The successful heterologous expression of the *esyn* gene from *Fusarium oxysporum*,which encodes enniatin synthetase (ESYN) and a non-ribosomal peptide synthetase (NRPS) enzyme (Zobel et al., 2015). This study describes the first instance of heterologous expression of a eukaryotic NRPS gene in *B. subtilis*, more impressive given that the esyn gene encodes a large, multifunctional biosynthetic enzyme. By using an acetoin-inducible promoter system, the enniatin was produced at an approximate concentration of 1.1 mg/L. This is a remarkable achievement since *B. subtilis* is a well-characterized, fast-growing, and generally-recognized-as-safe organism, thus making it an excellent chassis for biosynthesis of fungal metabolites at an industrial scale. This study confirmed *B. subtilis* as a suitable host for complex fungal biosynthetic gene expression while developing potential ways to increase yield through metabolic engineering, the optimization of transcriptional and translational controls, and/or feeding strategies with native or synthetic precursors. This study demonstrates an effective approach to using bacterial systems to produce pharmaceutical compounds derived from fungi in a more controlled and scalable manner (Zobel et al., 2015).

The decision to use yeast versus bacterial systems will depend on the requirements of the fungal gene of interest, such as the requirement for post-transitional modifications, the ability to secrete, and the complexity of the protein. The continued innovation of genetic engineering and synthetic biology grows the tools available to optimize heterologous expression for fungi, thus improving the ability to produce compounds from fungi in other systems more readily.

3.2 Overcoming Challenges in Heterologous Production

Issues of Yield, Stability, and Compound Purification

Advances in fungal synthetic biology have recently detailed the strategic deployment of native regulatory components of biosynthetic gene clusters (BGCs) in order to enhance yields of secondary metabolites. An outstanding example is *Fusarium heterosporum*, in which improved growth conditions led to the discovery of high-titre equisetin production (> 2 g/L)—an inherent regulatory network ready to be overproduced (Kakule et al., 2015). Based on this, investigators took advantage of two important native components: the pathway-specific transcription factor EqxR and the BGC promoter eqxS. This toolkit regulatory has been effective in heterologous expression of varied biosynthetic genes of fungi in *F. heterosporum* achieving engineered strains capable of producing several secondary metabolites. These included a selective antituberculosis compound greater than 800 mg/L. More importantly, this strategy also induced a silent biosynthetic pathway of an endophytic fungus source. Using a related approach, the fast-growing and enzyme-producing industrial workhorse *Trichoderma reesei* was heterologous fungal gene expression engineered to degrade agricultural waste (Shenouda et al., 2022). To reduce metabolic interference and create a simplified metabolic chassis, the native sorbicillinoid pathway was deleted. Then BGCs (aspks1 of *Acremonium strictum* and tenS/tenC of *Beauveria bassiana*) of large size were introduced. Growth on low-cost residues (orange peel, banana peel, and barley straw) led to the generation of new and known metabolites—pretenellin A and 3-methylorcinaldehyde—proving once more that *T. reesei* is an effective bioconversion platform of waste-derived metabolites. Besides that, the model heterologous expression host *Aspergillus oryzae* NSAR1 underwent targeted genetic optimization (Dao et al., 2021). Deletion of global repressor mcrA and kojic acid biosynthetic gene kojA generated a derivative strain that had lost the ability to produce kojic acid and decreased native metabolite interference. Background noise was also reduced by inhibition of the mcrA-dependant stress response, faithfully detecting heterologously expressed metabolites. This designed *A. oryzae* NSAR1 strain has turned into a high-precision cellular factory to natural product discovery and pathway elucidation, and the analytical resolution is higher.

4 Bioprocessing and Fermentation Technology

4.1 Scale-Up Strategies for Industrial Production

4.1.1 From Lab-Scale to Industrial Fermentation

Scaling up of laboratory-scale fermentation processes to commercial production is a complicated series of issues, especially in microbial and fungal biotechnology. The most important goal of the scale-up process is to demonstrate that the ideal conditions that were discovered under the controlled conditions in the laboratory can be consistently replicated at the larger scales without loss in yield, product quality, or process effectiveness (Fig. 7). That takes careful consideration of numerous operational conditions, such as pH, temperature, oxygen transfer, agitation, nutrient supply, and shear stress, which can all be erratic as the scale increases, whether it is shake flasks or bench-scale bioreactors to pilot- or commercial-scale fermenters (Yang et al., 2023). When these scale-up parameters are not carefully controlled, key features of the process including biomass development, metabolite production, and substrate utilization may be adversely impacted. Typical problems are oxygen limitation, nutrient gradients, and excessive foaming, any of which can affect the consistency or scalability of products. Therefore, clearly-defined scale-up paths are required not just to sustain the robustness of a bioprocess, but also to assure economic viability and compliance with regulation—both of which are the focal points of successful industrial biotechnology projects.

The filamentous nature of the fungi is one of the greatest challenges that need to be addressed urgently during the scale-up of fungal fermentation as it may result in very viscous broths that may hamper both mixing and oxygen transfer. An interesting case study, conducted by Pollard et al. (2007), examined this problem, and it was found that the traditional scale-up methods are frequently unsuccessful because of the unavailability of tools that could identify essential constraints during the fermentation process. To counter this, the researchers used regime analysis approach—adding oxygen to the fermenter at the point at which oxygen transfer constraints would be expected to limit further oxygen transfer even without direct measurements.

Further development of this approach has been made by Zhang et al. (2024) who proposed an idea of operating boundary conditions to determine the acceptable limits of main physical parameters, i.e. reactor geometry, specific volume, and dimensional ratios. The objective in standardizing these conditions at the various scales was to reproduce hydrodynamic and mass transfer environments between bench-scale systems and industrial reactors. This structured approach made it easier to consistently replicate and predict outcomes during the scale-up of viscous fungal fermentations, which are often difficult to manage due to their complex nature.

In conclusion, physical and biological aspects are crucial to the success of industrial fungal bioprocesses. Although the behaviour of fluids and the gas exchange, as well as the design of the reactor, can be described using engineering principles, the biological factors, namely the growth behaviour of fungi, their morphology, and metabolic requirements have to be considered in the design of the process. Consideration of these interdependencies in a

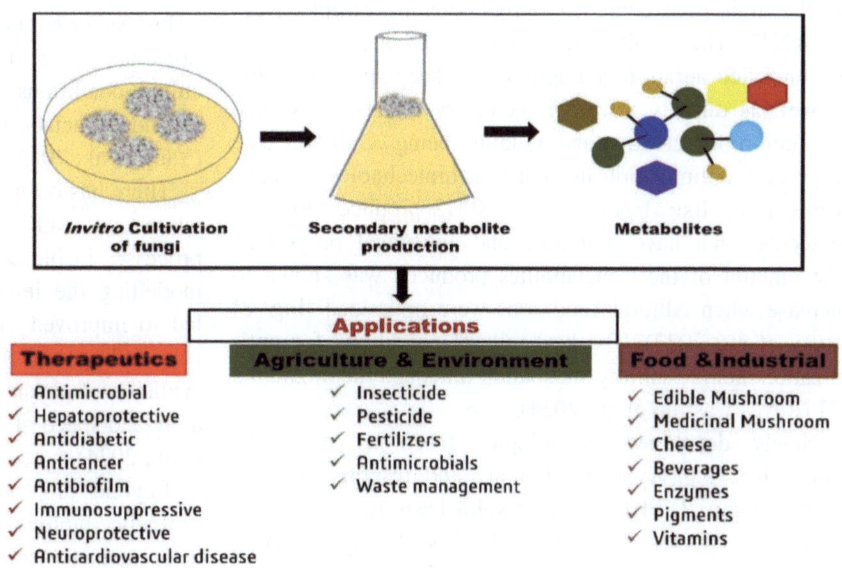

Fig. 7 Various applications of the fungal secondary metabolites, Wadhwa et al. (2024)

balanced strategy is key to optimization of industrial-scale production of fungal-derived biochemicals and pharmaceuticals, to ensure high performance and economic feasibility.

4.1.2 Optimization of Growth Conditions and Media

The composition of the growth medium, including carbon, nitrogen and trace elements, is critical in determining fungal secondary metabolism. The carbon source, i.e. glucose and starch, is the main factor correlated with biomass growth, though organic nitrogen sources, i.e. peptone and yeast extract, have a higher probability of influencing the production of metabolites (Yang et al., 2025; Zhang et al., 2024). Elsewhere, the response surface methodology and medium design optimization 30-fold increase in physcion production in *Aspergillus chevalieri* (Zhang et al., 2024). These results put the significance of nutrient optimization into perspective, allowing small scale, focused production of metabolites.

Temperature, pH, aeration, and moisture are important parameters driving microbial growth and metabolite production. Solid-state fermentation (SSF) by *Trichoderma* species can be an advantage due to the possibilities they provide for downstream processing and higher products/concentrations (Mulatu et al., 2021). Recently, mathematic modelling methodologies have now able to provide an approach to minimize experimental costs and estimate the parameters for optimal fermentation.

The rate of agitation and dissolved oxygen concentration in submerged fermentation processes are important factors that need to be manipulated to balance the providing of substrate to the microbial metabolism and the damage caused to the sensitive bacterial and fungal cells due to shear forces. Recently, a study by Yang et al. (2025) revealed that optimization of these parameters had a significant effect on the antifungal compound production by Streptomyces sp. KN37. The results have demonstrated the importance of controlled aeration and agitation in keeping cells viable as well as directly affecting secondary metabolite yields, with controlled aeration and agitation being of significance to process optimization in industrial biotechnology. Endophytic fungi like *Aspergillus sp.* SPH2 produce bioactive molecules that have acaricidal and antifungal properties. The amount of these metabolites produced was shown to increase when cultural conditions were optimized (Fig. 8) (Rusu et al., 2023). The key cultural conditions for optimization include salinity, incubation time, and the medium's pH (Reyes Castillo et al., 2024).

Newly developed technologies (such as predictive approaches, e.g., synthetic biology and response surface methodology (RSM)) can lay a solid basis for optimizing production systems (Li et al., 2019). Specifically, synthetic biology can enable informing biosynthetic pathways around the selective manipulation of biosynthetic gene clusters, as shown with *Actinobacteria* and *Aspergillus oryzae* (Jia et al., 2024; Li et al., 2019). In addition to simple optimization actions, these tools can provide scientific ways for increasing target metabolite levels.

4.1.3 Fermentation Optimization for High Yield

Bioreactor Design and Control

Yield results are very sensitive to the specific bioreactor type, especially with filamentous fungi and actinomycetes, which are the primary sources of secondary metabolites. The stirred-tank bioreactor remains user-preferred due to effective mixing, oxygen transfer, and so on, but designs like airlift and bubble column bioreactors. This provides suitable environments for certain oxygen-sensitive processes or shear-sensitive microbes (such as *Streptomyces* and *Aspergillus* spp.), which may experience lower levels of oxygen transfer restrictions or gradients that lead to scale-up problems due to proper control of agitation, aeration, and nutrient supply (Reyes Castillo et al., 2025).

Advanced control strategies for preserving optimal growth conditions are becoming increasingly important in fermentation systems. Key factors, including real-time pH, temperature, dissolved oxygen (DO), and redox potential analysis, should be monitored. The feedback control system based on this analysis promotes uniformity in products and yield by facilitating automatic adjustments of airflow, agitation speed, and nutrient input rates (Mulatu et al., 2021).

Batch and fed-batch fermentations that employ dissolved oxygen-stat, pH–stat, or alkaline scavengers have been freedom-partitioned to maintain a physiologically recognizable habitat for the production of secondary metabolites (Rusu et al., 2023). The importance of the dynamic control of bioreactor cultures wherein the use of aeration and agitation rates were monitored and changed based on current culture conditions, maximizing the production of antifungal secondary metabolites from the *Streptomyces* sp. KN37 cells (Yang et al., 2025).

There has been an increase in the use of computer simulation and mathematical modelling, to plan and improve processes in bioreactors (Xu et al., 2024). Through RSM, modelling the interactions of various process factors has led to improved yields successfully modelled the various surroundings of *Aspergillus chevalieri* to maximize the synthesis of physicians by changing factors such as temperature and speed of agitation to allow for good yields (Zhang et al., 2024).

Beyond bioprocess engineering practices, the use of synthetic biology has introduced alternatives to bioreactor management of the environment and an organism's

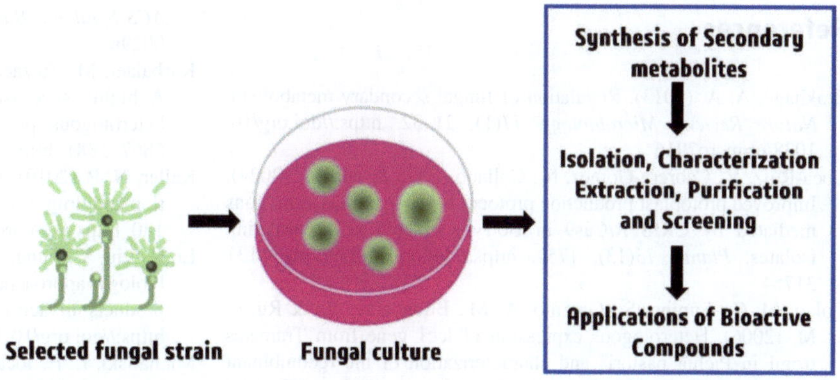

Fig. 8 Steps in preliminary screening and optimization of secondary metabolite production. *Source* Authors

understanding of biological mechanisms. External circuit regulators and biosensors have now been constructed to respond to fermentation microenvironments and to subsequently activate independent modifications in gene expression according to the different parameters of the bioreactor (Jia et al., 2024; Li et al., 2019). This enables two-way feedback between the environmental response of the bioreactor and the metabolism of the microbe allowing for potential intelligent fermentation-control systems.

The design of bioreactors for solid-state fermentation presents unique challenges due to the lower moisture content associated with solid-state fermentation and lower convection (Sentís-Moré et al., 2023). SSF systems have an established history for the development of bio-fungicides with *Trichoderma* species under regulated conditions of temperature, moisture, and aeration (Mulatu et al., 2021). The goal is to further increase control in SSF conditions with custom-designed tray or pack-bed reactors with sensors and actuators.

4.1.4 Limitations

Due to the complexity and interconnectedness of fungal metabolic networks, predicting the effects of changing a single pathway is challenging. Genetic engineering presents difficulties in introducing foreign DNA into fungi can be unstable and inefficient, and our understanding of fungal biology remains limited. Moreover, the concerns about the potential health and environmental effects of genetically modified fungi and their metabolites in developing and scaling up the production of fungal metabolites can be expensive and complex. Microbial strain restrictions, media formulation challenges, and contamination control are all factors in creating and executing economical and successful fermentation processes for industrial output. Strategies for industrial production scale-up include preserving uniformity, investing in expensive equipment, and ensuring process verification. Additionally, intricate relationships between variables, metabolic engineering constraints, and erratic microbial behaviour complicate matters. The high cost and energy requirements of low-scale production pose risks such as contamination, natural fluctuations over time, product impurities, the need for additional treatment, unexpected and undesired products, and the growth of undesirable bacteria while favourable microbes may die off.

5 Conclusion

The potential in this fascinating field of fungal secondary metabolites is enormous. Researchers are using synthetic biology to transform fungi into true factories that generate useful compounds. To enhance metabolite synthesis, researchers are rewiring fungal pathways through techniques such as metabolic engineering, strain improvement, and CRISPR-Cas9 gene editing. As synthetic biology unleashes the full potential of these microbial maestros, biotechnology is expected to undergo a revolution in the future, leading to breakthroughs in sectors like agriculture and medicine. Fungal biosynthesis pathway heterologous expression is a complex area rich in potential discoveries. Even though enduring challenges with yield optimization and purification, the progress made thus far instils hope that this technology will eventually transform biotechnology, agriculture, and medicine. The foundations of biotechnology enable the production of numerous products that enhance our quality of life through fermentation and bioprocessing technologies. The evolution from laboratory-scale experimentation to industrial-scale manufacturing highlights human creativity consistently improving growth conditions, media composition, and bioreactor design, which can maximize the capacity of microbes to produce valuable chemicals precisely and effectively. In the future, biotechnological techniques could significantly transform industrial operations, making them more cost-effective and environmentally sustainable secondary metabolites for various applications.

References

Brakhage, A. A. (2013). Regulation of fungal secondary metabolism. *Nature Reviews Microbiology, 11*(1), 21–32. https://doi.org/10.1038/nrmicro2916

Coca-Ruiz, V., Cabrera-Gómez, N., Collado, I. G., & Aleu, J. (2024). Improved protoplast production protocol for fungal transformations mediated by CRISPR/Cas9 in Botrytis cinerea non-sporulating isolates. *Plants, 13*(13), 1754. https://doi.org/10.3390/plants13131754

Colao, M. C., Lupino, S., Garzillo, A. M., Buonocore, V., & Ruzzi, M. (2006). Heterologous expression of lcc1 gene from Trametes trogii in Pichia pastoris and characterization of the recombinant enzyme. *Microbial Cell Factories, 5*, 1–11. https://doi.org/10.1186/1475-2859-5-31

Corbu, V. M., Gheorghe-Barbu, I., Dumbravă, A. Ş, Vrâncianu, C. O., & Şesan, T. E. (2023). Current insights in fungal importance—A comprehensive review. *Microorganisms, 11*(6), 1384. https://doi.org/10.3390/microorganisms11061384

Cravens, A., Payne, J., & Smolke, C. D. (2019). Synthetic biology strategies for microbial biosynthesis of plant natural products. *Nature Communications, 10*(1), 2142. https://doi.org/10.1038/s41467-019-09848-w

Dao, T. T., de Mattos-Shipley, K. M. J., Prosser, I. M., Williams, K., Zacharova, M. K., Lazarus, C. M., Willis, C. L., & Bailey, A. M. (2021). Cleaning the Cellular Factory–Deletion of McrA in Aspergillus oryzae NSAR1 and the Generation of a Novel Kojic Acid Deficient Strain for Cleaner Heterologous Production of Secondary Metabolites. *Frontiers in Fungal Biology, 2*, 1–12. https://doi.org/10.3389/ffunb.2021.632542

Doudna, J. A., & Charpentier, E. (2014). The new frontier of genome engineering with CRISPR-Cas9. *Science, 346*(6213), 1258096. https://doi.org/10.1126/science.1258096

Gaj, T., Gersbach, C. A., & Barbas, C. F. (2013). ZFN, TALEN, and CRISPR/Cas-based methods for genome engineering. *Trends in Biotechnology, 31*(7), 397–405. https://doi.org/10.1016/j.tibtech.2013.04.004

Garg, S., Kim, M., & Romero-Suarez, D. (2025). Current advancements in fungal engineering technologies for sustainable development goals. *Trends in Microbiology, 33*(3), 285–301. https://doi.org/10.1016/j.tim.2024.11.001

Hashem, A. H., Attia, M. S., Kandil, E. K., Fawzi, M. M., Abdelrahman, A. S., Khader, M. S., & Abdelaziz, A. M. (2023). Bioactive compounds and biomedical applications of endophytic fungi: A recent review. *Microbial Cell Factories, 22*(1), 107. https://doi.org/10.1186/s12934-023-02118-x

Jia, X., Song, J., Wu, Y., Feng, S., Sun, Z., Hu, Y., Yu, M., Han, R., & Zeng, B. (2024). Strategies for the enhancement of secondary metabolite production via biosynthesis gene cluster regulation in *Aspergillus oryzae. Journal of Fungi, 10*(5), 312. https://doi.org/10.3390/jof10050312

Jiang, C., Lv, G., Tu, Y., Cheng, X., Duan, Y., Zeng, B., & He, B. (2021). Applications of CRISPR/Cas9 in the synthesis of secondary metabolites in filamentous fungi. *Frontiers in Microbiology, 12*, Article 638096. https://doi.org/10.3389/fmicb.2021.638096

Joung, J. K., & Sander, J. D. (2013). TALENs: A widely applicable technology for targeted genome editing. *Nature Reviews Molecular Cell Biology, 14*(1), 49–55. https://doi.org/10.1038/nrm3486.TALENs

Nevalainen, K. M. H. (2001). Strain improvement in filamentous fungi-an overview. *Applied Mycology and Biotechnology, 1*, 289–304. https://doi.org/10.1016/S1874-5334(01)80013-7

Kakule, T. B., Jadulco, R. C., Koch, M., Janso, J. E., Barrows, L. R., & Schmidt, E. W. (2015). Native promoter strategy for high-yielding synthesis and engineering of fungal secondary metabolites. *ACS Synthetic Biology, 4*(5), 625–633. https://doi.org/10.1021/sb500296p

Karbalaei, M., Rezaee, S. A., & Farsiani, H. (2020). Pichia pastoris: A highly successful expression system for optimal synthesis of heterologous proteins. *Journal of Cellular Physiology, 235*(9), 5867–5881. https://doi.org/10.1002/jcp.29583

Keller, N. P. (2019). Fungal secondary metabolism: Regulation, function and drug discovery. *Nature Reviews Microbiology, 17*(3), 167–180. https://doi.org/10.1038/s41579-018-0121-1.Fungal

Li, L., Liu, X., Jiang, W., & Lu, Y. (2019). Recent advances in synthetic biology approaches to optimize production of bioactive natural products in actinobacteria. *Frontiers in Microbiology, 10*, 2467. https://doi.org/10.3389/fmicb.2019.02467

Michaliski, L. F., Ióca, L. P., Oliveira, L. S., Crnkovic, C. M., Takaki, M., Freire, V. F., & Berlinck, R. G. (2023). Improvement of targeted fungi secondary metabolite production using a systematic experimental design and chemometrics analysis. *Methods and Protocols, 6*(5), 77. https://doi.org/10.3390/mps6050077

Mulatu, A., Alemu, T., Megersa, N., & Vetukuri, R. R. (2021). Optimization of culture conditions and production of bio-fungicides from *Trichoderma species* under solid-state fermentation using mathematical modeling. *Microorganisms, 9*(8), 1675. https://doi.org/10.3390/microorganisms9081675

Müller, M. M., Arndt, K. M., & Hoffmann, S. A. (2025). Genetic circuits in synthetic biology: Broadening the toolbox of regulatory devices. *Frontiers in Synthetic Biology, 3*, 1548572. https://doi.org/10.3389/fsybi.2025.1548572

Nielsen, M. L., Isbrandt, T., Rasmussen, K. B., Thrane, U., Hoof, J. B., Larsen, T. O., & Mortensen, U. H. (2017). Genes linked to production of secondary metabolites in *Talaromyces atroroseus* revealed using CRISPR-Cas9. *PLoS ONE, 12*(1), Article e0169712. https://doi.org/10.1371/journal.pone.0169712

Panstruga, R., Antonin, W., & Lichius, A. (2023). Looking outside the box: A comparative cross-kingdom view on the cell biology of the three major lineages of eukaryotic multicellular life. *Cellular and Molecular Life Sciences, 80*(8), 198. https://doi.org/10.1007/s00018-023-04843-3

Peng, B., Bandari, N. C., Lu, Z., Howard, C. B., Scott, C., Trau, M., Dumsday, G., & Vickers, C. E. (2022). Engineering eukaryote-like regulatory circuits to expand artificial control mechanisms for metabolic engineering in *Saccharomyces cerevisiae. Communications Biology, 5*(1), 135. https://doi.org/10.1038/s42003-022-03070-z

Pollard, D. J., Kirschner, T. F., Hunt, G. R., Tong, I. T., Stieber, R., & Salmon, P. M. (2007). Scale up of a viscous fungal fermentation: Application of scale-up criteria with regime analysis and operating boundary conditions. *Biotechnology and Bioengineering, 96*(2), 307–317. https://doi.org/10.1002/bit

Reyes Castillo, N., Díaz, C. E., Andres, M. F., Imperial, J., Valcárcel, F., & González Coloma, A. A. (2024). Optimization of fungicidal and acaricidal metabolite production by endophytic fungus *Aspergillus sp.* SPH2. *Bioresources and Bioprocessing*, 11(1), 28. https://doi.org/10.1186/s40643-024-00745-9

Reyes Castillo, N., Rojas López-Menchero, J., Pacheco Useche, W. A., Díaz, C. E., Andres, M. F., & González-Coloma, A. (2025). Advanced fermentation techniques enhance dioxolanone type biopesticide production from Phyllosticta capitalensis. *Scientific Reports, 15*(1), 7989. https://doi.org/10.1038/s41598-025-92626-0

Ridley, C. P., & Khosla, C. (2009). Polyketides. In M. Schaechter (Ed.), *Encyclopedia of microbiology* (3rd ed., pp. 472–481). Oxford: Academic Press.

Rodrigues, F., Ries, L., Wang, C., & Wong, K. H., (Eds.) (2021). *Fungal primary and secondary metabolism and its importance for virulence and biomedical applications.* Lausanne: Frontiers Media SA. https://doi.org/10.3389/978-2-88966-889-2

Rusu, A. V., Trif, M., & Rocha, J. M. (2023). Microbial secondary metabolites via fermentation approaches for dietary supplementation formulations. *Molecules, 28*(16), 6020. https://doi.org/10.3390/molecules28166020

Sentís-Moré, P., Romero-Fabregat, M. P., Rodríguez-Marca, C., Guerra-Sánchez, A. J., & Ortega-Olivé, N. (2023). Design optimization of a tray bioreactor for solid-state fermentation: Study of process parameters through protein modification of by-products. *Fermentation, 9*(10), 921. https://doi.org/10.3390/fermentation9100921

Shenouda, M. L., Ambilika, M., Skellam, E., & Cox, R. J. (2022). Heterologous expression of secondary metabolite genes in *Trichoderma reesei* for waste valorization. *Journal of Fungi, 8*(4), 355. https://doi.org/10.3390/jof8040355

Shi, T. Q., Gao, J., Wang, W. J., Wang, K. F., Xu, G. Q., Huang, H., & Ji, X. J. (2019). CRISPR/Cas9-based genome editing in the filamentous fungus *Fusarium fujikuroi* and its application in strain engineering for gibberellic acid production. *ACS Synthetic Biology, 8*(2), 445–454. https://doi.org/10.1021/acssynbio.8b00478

Shi, Y., Ji, M., Dong, J., Shi, D., Wang, Y., Liu, L., Feng, S., & Liu, L. (2024). New bioactive secondary metabolites from fungi: 2023. *Mycology, 15*(3), 283–321. https://doi.org/10.1080/21501203.2024.2354302

Song, Z., Zhou, S., Zhang, H., Keller, N. P., Oakley, B. R., Liu, X., & Yin, W. B. (2023). Fungal secondary metabolism is governed by an RNA-binding protein CsdA/RsdA complex. *Nature Communications, 14*(1), 7351. https://doi.org/10.1038/s41467-023-43205-2

Tamano, K., Brown, D. W., & Yoshimi, A. (2023). The use of metabolic engineering techniques to increase the productivity of primary and secondary metabolites within filamentous fungi. *Frontiers in Fungal Biology, 4*, 1178290. https://doi.org/10.3389/ffunb.2023.1178290

Teixeira, A. P., & Fussenegger, M. (2024). Synthetic gene circuits for regulation of next-generation cell-based therapeutics. *Advanced Science, 11*(8), 2309088. https://doi.org/10.1002/advs.202309088

Verma, A., Tiwari, H., Singh, S., Gupta, P., Rai, N., Kumar Singh, S., & Gautam, V. (2023). Epigenetic manipulation for secondary metabolite activation in endophytic fungi: Current progress and future directions. *Mycology, 14*(4), 275–291. https://doi.org/10.1080/21501203.2023.2241486

Wadhwa, K., Kapoor, N., Kaur, H., Abu-Seer, E. A., Tariq, M., Siddiqui, S., Yadav, V. K., Niazi, P., Kumar, P., & Alghamdi, S. (2024). A comprehensive review of the diversity of fungal secondary metabolites and their emerging applications in healthcare and environment. *Mycobiology, 52*(6), 335–387. https://doi.org/10.1080/12298093.2024.2416736

Wang, G., Ran, H., Fan, J., Keller, N. P., Liu, Z., Wu, F., & Yin, W. B. (2022). Fungal-fungal cocultivation leads to widespread secondary metabolite alteration requiring the partial loss-of-function VeA1 protein. *Science Advances, 8*(17), eabo6094. https://doi.org/10.1126/sciadv.abo6094

Wijayawardene, N. N., Hyde, K. D., Mikhailov, K. V., Péter, G., Aptroot, A., Pires-Zottarelli, C. L. A., Goto, B. T., Tokarev, Y. S., Haelewaters, D., Karunarathna, S. C., Kirk, P. M., André, A. L. C. M., Saxena, R. K., Schoutteten, N., Wimalasena, M. K., Aleoshin, V. V., Al-Hatmi, A. M. S., Ariyawansa, K. G. S. U., & Assunção, A. R., … Karpov, S. Y. (2024). Classes and phyla of the kingdom Fungi. *Fungal Diversity, 128*(1), 1–165. https://doi.org/10.1007/s13225-024-00540-z

Xu, X., Sun, Y., Zhang, A., Li, S., Zhang, S., Chen, S., Lou, C., Cai, L., Chen, Y., Luo, C., & Yin, W. B. (2024). Quantitative characterization of gene regulatory circuits associated with fungal secondary metabolism to discover novel natural products. *Advanced Science, 11*(47), 2407195. https://doi.org/10.1002/advs.202407195

Xue, M., Hou, X., Fu, J., Zhang, J., Wang, J., Zhao, Z., Xu, D., Lai, D., & Zhou, L. (2023). Recent advances in search of bioactive secondary metabolites from fungi triggered by chemical epigenetic modifiers. *Journal of Fungi, 9*(2), 172. https://doi.org/10.3390/jof9020172

Yang, J., Yue, H.R., Pan, L. Y., Feng, J. X., Zhao, S., Suwannarangsee, S., Champreda, V., Liu, C. G., & Zhao, X. Q. (2023). Fungal strain improvement for efficient cellulase production and lignocellulosic biorefinery: Current status and future prospects. *Bioresource Technology, 385*, 129449. https://doi.org/10.1016/j.biortech.2023.129449

Yang, X., Yuan, L., Zeeshan, M., Yang, C., Gao, W., Zhang, G., & Wang, C. (2025). Optimization of fermentation conditions to increase the production of antifungal metabolites from Streptomyces sp. KN37. *Microbial Cell Factories, 24*(1), 26. https://doi.org/10.1186/s12934-025-02652-w

Zhang, S., Jiang, Z., An, S., Jiang, X., & Zhang, Y. (2024). Optimization of fermentation conditions for physcion production of *Aspergillus chevalieri* BYST01 by response surface methodology. *PeerJ, 12*, Article e18380. https://doi.org/10.7717/peerj.18380

Zhao, R. L. (2023). Edible and medicinal macrofungi. *Journal of Fungi, 9*(9), 908. https://doi.org/10.3390/jof9090908

Zhgun, A. A. (2023a). Fungal BGCs for production of secondary metabolites: Main types, central roles in strain improvement, and regulation according to the piano principle. *International Journal of Molecular Sciences, 24*(13), 11184. https://doi.org/10.3390/ijms241311184

Zhgun, A. A. (2023b). Industrial production of antibiotics in fungi: Current state, deciphering the molecular basis of classical strain improvement and increasing the production of high-yielding strains by the addition of low-molecular weight inducers. *Fermentation, 9*(12), 1027. https://doi.org/10.3390/fermentation9121027

Zobel, S., Kumpfmüller, J., Süssmuth, R. D., & Schweder, T. (2015). *Bacillus subtilis* as heterologous host for the secretory production of the non-ribosomal cyclodepsipeptide enniatin. *Applied Microbiology and Biotechnology, 99*(2), 681–691. https://doi.org/10.1007/s00253-014-6199-0

Fungal Secondary Metabolites in Drug Discovery and Development

Abrar Hussain, Amina Zia, Muhammad Waseem Akram, and Syed Abid Ali

Abstract

Fungi represent a diverse class of microorganisms with promising applications in vast areas. Fungi are considered a rich source of thousands of secondary metabolites with potential therapeutic properties. Besides other applications of secondary metabolites, their role in the discovery and development of drugs remains of utmost importance. Fungal secondary metabolites (SMs) have various biological properties and variations in chemical and biological properties, thus becoming an important lead substance to be used as drug molecules. Currently, a good number of drug candidates are identified from fungal sources. The advantages of using fungal metabolites as lead drug candidates include easy production, low cost, greater bioactivity, and greater availability. Advances in genomics also added to the identification of fungal secondary metabolites as drug precursor molecules and thus enhanced the accuracy, lowered the screening time, and elucidated the drug-related properties at the genetic level. Nevertheless, fungal metabolites have greater drug-like properties, but still, challenges and limitations exist. For instance, antibiotic resistance, difficulty in fungal growth, and extensive clinical trials can limit the potential of fungal SMs as drugs. While intellectual property rights, patenting, and lengthy approval process can limit their commercialization. In the future, the usage of advanced technologies, computational tools, and artificial intelligence tools can overcome these limitations and will add more to the fungal secondary metabolites drug properties.

Keywords

Fungi · Metabolites · Drug · Therapeutics · Penicillin

1 Introduction

Natural products are thought to be the primary source of novel drugs and therapeutic agents. According to the International Pharmacopoeia, which listed 371 drugs, more than 80 drugs are natural products or their derivatives (Zhang et al., 2020). Humans used biologically active secondary metabolites, mostly derived from plants, as traditional medicines and occasionally as natural poisons since ancient times (Larsen et al., 2005). In the fields of infectious disease and cancer, 60 and 75% of new pharmaceutical substances come from natural sources (Gullo et al., 2006). Berdy (2005) reported that microorganisms are the source of over 20,000 biologically active secondary metabolites. Microorganisms have ubiquitous distributions and are found in all possible places of existence. These are tiny creatures and are comprised of bacteria, fungi, viruses, and protist (Hussain et al. 2024a). Among the ocean of these microorganisms, fungi are the largest eukaryotic groups with great potential and applications. Fungi are the most significant groups of being widely known for the synthesis of new secondary metabolites. These fungal-derived metabolites can be used directly as drugs or serve as lead structures for other bioactive products. The advanced technology has brought in new strategies to drug discovery from natural products (Berdy, 2005). Currently, new technologies like 3D bioprinting are also employed in the drug developmental

A. Hussain (✉) · S. A. Ali
Third World Center for Science and Technology, HEJ Research Institute of Chemistry, International Center for Chemical and Biological Sciences, University of Karachi, Karachi, Pakistan
e-mail: abrarhussain0307@gmail.com

A. Zia
Institute of Microbiology, Faculty of Veterinary Science, University of Agriculture Faisalabad, Faisalabad, Pakistan

M. W. Akram
Industrial Biotechnology, National Institute for Biotechnology and Genetic Engineering (NIBGE), Pakistan Institute of Engineering and Applied Sciences (PIEAS), Islamabad, Pakistan

© The Author(s), under exclusive license to Springer Nature Switzerland AG 2026

A. Prakash and M. K. Dubey (eds.), *Advances in Fungal Secondary Metabolites: Bioactive Attributes and Biotechnological Potential of Industrial Interest*, Sustainable Economy and Ecotechnology, https://doi.org/10.1007/978-3-032-23507-7_12

process that can reduce the screening time and enhance identification (Hussain et al., 2024b).

Prior works have shown that higher fungi have had a long history of applications in traditional medicines, especially within the Asian region, and their research has attracted important interest (Lindequist et al., 2010). Fungal-derived compounds have been applied in the management of many chronic infections and autoimmune diseases. Over the last few years, small molecules from fungi, which are classified under secondary metabolites, have been under trials for the treatment of chronic diseases like drug-resistant depression and cancer (Prescott et al., 2023). The main aim of research into fungal-derived metabolites is to find active substances that may be used in the synthesis of new pharmaceutical substances and products for crop protection and cosmetics (Aly et al., 2011). For example, since ancient times, *Ganoderma lucidum* is a medicinal higher fungi that has been utilized for the treatment of gastric cancer, hepatitis, high blood pressure, and excessive cholesterol (Tang et al., 2006). Another example is *Cordyceps*, which is used in Traditional Chinese Medicine (TCM) for the treatment of various diseases affecting the heart, liver, kidneys, and lungs, as well as hyposexuality and hyperlipidemia (Khan et al., 2010). There have been recent reports of their usage as an adjuvant to modern cancer therapies and in the treatment of immunological diseases (Holliday & Cleaver, 2008). Additionally, extracts of *Cordyceps militaris* and *Cordyceps sinensis* showed strong anticancer effects by triggering cell death and modulating the immune system (Khan et al., 2010). Drug development is a lengthy, costly, and time-consuming process, requiring more developments to reduce these shortcomings. The general schematic illustration of the drug development process is shown in Fig. 1.

Currently, drugs or therapeutic agents are identified from diverse sources, each with its advantages and disadvantages. Fungi offer a promising set of metabolites for various applications. Some of their metabolites are also used as drugs or therapeutic agents (Hussain et al., 2025a). Despite being a heap of work has done to explore the fungal metabolites, properties, classifications, and potential applications, there is a vacant place to decipher their role in the development of drug molecules. Hence, this chapter aims to explore the role of fungal metabolites in drug developmental processes, their importance, and exceptional properties elucidation. This chapter also explores the basic terminology and auxiliary information that help the audience to understand the subject from the basics.

2 Fungal-Derived Drugs: A Historical Perspective

The concept that fungi contain biologically active substances is not new; it existed before the popularity of scientific methods, like compound isolation methods dating back to 1893, while fungi bioactive compounds were known earlier (Prescott et al., 2023). However, the concept of fungal bioactivity was known early from the discovery of penicillin in 1928 by Sir Alexander Fleming, which is largely considered the beginning of the history of fungal drug discovery. Additionally, the Second World War's outbreak increased the urgency of novel treatment options, thus more studies have been conducted to achieve the therapeutic agents. By 1940, penicillin's production had grown, and enough of the drug was utilized for the first human trial in February 1941, which created a major advancement in treating bacterial infections. Due to the discovery of penicillin, the scientific community became aware of the idea that fungi could harbor numerous small therapeutic molecules that have significant potential as therapeutic agents. This discovery made a huge impact on the scientific community and thus researchers are devoted to exploring such compounds in detail (Butler, 2008).

In the following years, many more antimicrobial drugs of fungal origin were isolated including griseofulvin (commercialized in 1959) (Aris et al., 2022) and cephalosporin C (developed in the 1950s) (Hamilton-Miller, 2000). These were followed by the successful isolation of some novel natural products that act as lead compounds and thus open potential sources of therapeutic compounds (Butler, 2004). Wild-type or genetically modified fungi can be used in fermentation to synthesize many of these compounds at a large scale and in a cost-effective manner. Microfungi were predominantly found in soil samples in the early days of fungal-derived drug discovery. Studies on soil fungi reveal a lower hit rate of new compounds, thus focus was shifted to more conventional sources, such as marine microbes as well as plants-associated fungi (endophytic fungi). Exploring fungi from novel or understudied ecological habitats is expected to eventually yield new lead structures for drugs or agrochemicals, as described (Aly et al., 2011). In order to find effective treatments for numerous old and newly emerging diseases that require treatment through various therapeutic approaches such as chemotherapy, immunotherapy, and gene therapy, it is still imperative to fund new therapeutic candidates like anticancer agents, antimicrobial, antiparasitics agents, and agents that combat resistant microorganisms and parasitic

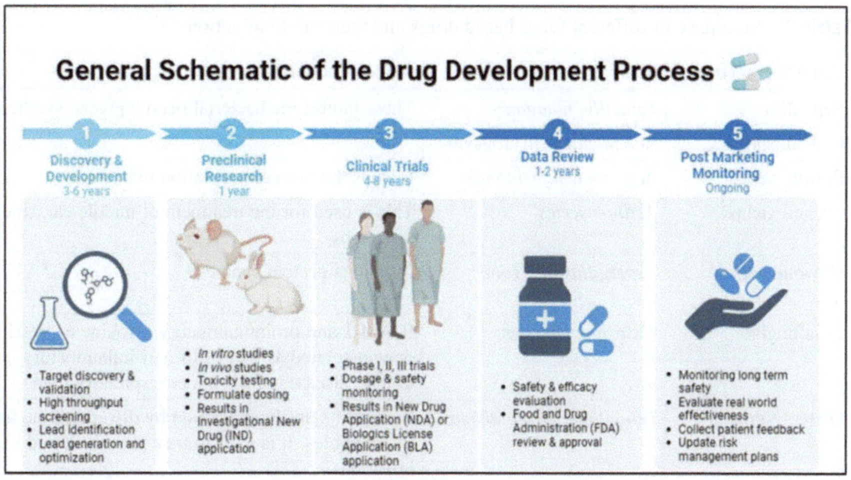

Fig. 1 General schematic of the drug developmental process. *Source* Authors

protozoans (Prescott et al., 2023) The different fungi-based drugs or therapeutic agents are identified are summarized in Table 1.

3 Screening Methods for Fungal-Based Lead Drug Molecules

The growing need for effective treatments against fungal infections as well as bacterial and other infectious agents has accelerated the investigation of compounds originating from fungi for their potential as therapeutic agents. To find these bioactive substances, a number of screening techniques have been developed, each with unique benefits and uses. Important among them are discussed below and summarized in Fig. 2.

3.1 Agar Disk-Diffusion Method and Variations

Agar disk-diffusion assay is the most common method used for the screening of antimicrobial susceptibility in microbiological laboratories (Lage et al., 2018). In this assay, the test compound is used to check whether they inhibit the growth of selected microorganisms or not on the agar plates. The test substance is impregnated onto filter paper discs, which are then set on the medium surface that has been inoculated with selected microorganisms. After incubation, the zone of inhibition is measured, which reflects their potential. This is an easy and inexpensive method that is capable of testing a large number of compounds and microorganisms; however, it has certain limitations. For instance, it cannot distinguish between bactericidal and bacteriostatic effects and is not suitable for the determination of minimum inhibitory concentration (MIC) of microorganisms (Lage et al., 2018).

Agar well diffusion assay can be modified according to the experimental design. For instance, it is modified to a method in which agar plates are inoculated with microorganisms of interest, followed by the loading of compound/extract into the tiny holes of the medium. If antimicrobial agents are present, they diffuse into the agar medium and prevent microbial growth. Similarly, in the gradient method (E-test), a strip with increasing concentration of the test compound/extract is placed over the agar surface, and the inhibition of the compound is measured. This gradient approach is used to determine the MIC interaction between two different antimicrobial drugs (Lage et al., 2018).

3.2 Poisoned Food Method

In this method, the required final concentration of the test compound is added to the melted agar medium. Next, the medium is poured into the Petri plates and upon solidification, the target fungal mycelium is introduced into the agar surface. Consequently, the fungal growth diameters are then measured after incubation and compared with the control, which does not include the extract/test compound. This technique is commonly used in the study of antifungal efficacy as well as in evaluating the percent inhibitory effect of mycelial growth, minimum inhibitory concentration, and minimum fungicidal concentration (Li et al., 2016).

3.3 Thin-Layer Chromatography (TLC)-Bioautography

This method combines contact bioautography and thin-layer chromatography (TLC). Several extracts from natural sources are screened for bioactivity-guided fractionation using this quick, easy, efficient, and affordable method (Dewanjee et al., 2015). This method localizes active

Table 1 Summary of different fungi-based drugs and their mode of action

Drugs/therapeutics	Sources	Mode of action	References
Penicillin	*Penicillin notatum*	These inhibit the bacterial peptidoglycan synthesis	Prescott et al. (2023)
Cephalosporin C	*Acremonium chrysogenum*		Bo (2000)
Fusidic acid	*Acremonium fusidioides*	It stops the protein translation in bacteria	Eid et al. (2019)
Co-amoxiclav	*Actinomycetes*	This is used for the treatment of middle ear, sinus, and respiratory tract infections	Fromage (2013)
Helvolic acid	*Aspergillus fumigotus*	It inhibits protein synthesis	Al-Fakih and Almaqtri (2019)
Fusafungine	*Fusarium laterium*	ICAM-1 and proinflammatory cytokine expression are believed to be downregulated as part of its anti-inflammatory mode of action. It is used to treat diseases of the upper respiratory tract	Sy-Cordero et al. (2012)
Griseofulvin	*Penicillium griseofulvum*	It exerts its antifungal action by disrupting the function of mitotic spindle microtubules. It is used to treat skin, nail, and hair dermatophyte infections	Petersen et al. (2014)
Strobilurins	*Strobilurus tenacellus*	It inhibits the respiratory systems of fungal cells	Balba (2007)
Mycophenolic acid	*Penicillium brevicompactum*	It prevents the production of guanine by inhibiting inosine monophosphate dehydrogenase, which stops the proliferation of lymphocytes It is also employed in surgery for organ transplantation	Freedman et al. (2020)
Mizoribine	*Penicillium dodge*	Mizoribine blocks the proliferation of lymphocytes by inhibiting inosine monophosphate dehydrogenase in a similar way to mycophenolic acid. It has been used to reduce rejection in kidney transplants in Japan since 1984	Kawasaki (2009)
Cyclosporin A	*Tolypocladium inflatum*	It acts by blocking the calcineurin pathway, which prevents human T-cell activation help in organ transplantation	Bushley et al. (2013)
Gliotoxin	*Aspergillus fumigotus*	It inhibits a number of thiol-requiring enzymes	Coméra et al. (2007)
Mevastatin	*Penicillium citrinum*	They act as competitive inhibitors of HMG-CoA reductase, an enzyme necessary for the biosynthesis of cholesterol	Stossel (2008)
Lovastatin	*Monascus ruber* and *Aspergillus terreus*		Hajjaj et al. (2001)
Pravachol	*Penicillium compactum*		Squibb and Sons (2007)
Terrain	*A. terreus*	It induces apoptosis in breast cancer	(Liao et al., 2012)
Brefeldin A	*P. brefeldin*	It induces apoptosis in leukemia	Gnägi et al. (2019)
Aspirin	*A. nidulans*	It induces G2/M cell cycle arrest	He et al. (2011)
Austocystin D	*A. pseudoustus*	Inhibit the growth of human colon carcinoma	Wang et al. (2008)
Norsolorinic acid	*A. parasiticus* and *A. nidulans*	Induce Go/G1 cell cycle arrest	Yamada et al., (2012)
Chondrosterin	*Chondrostereum* spp.	Cytotoxic effects on tumor cell lines	Li et al. (2014)
PF1022A	*Mycelia sterile*	It demonstrates significant anthelmintic activity against *Ascaridia galli* in chickens	Jeschke et al. (2005)

Source Authors compilation

compounds that have antimicrobial activity in addition to separating a complex mixture or extracts. Some laboratories utilize the most advanced and updated version of this procedure with high-performance liquid chromatography (HPLC), which is now the method of choice for the final clean-up of active fractions to obtain pure compounds, along with bioautography (Favre-Godal et al., 2013).

3.4 Dilution Methods

These methods are used mostly for the determination of MIC of bacterial and fungal strains. During the experiment, the amount or concentration of the tested antimicrobial agent, which is toxic to the test microorganism is determined. The MIC endpoint is measured by the least concentration of an antimicrobial substance that effectively inhibits growth after incubation (Lage et al., 2018).

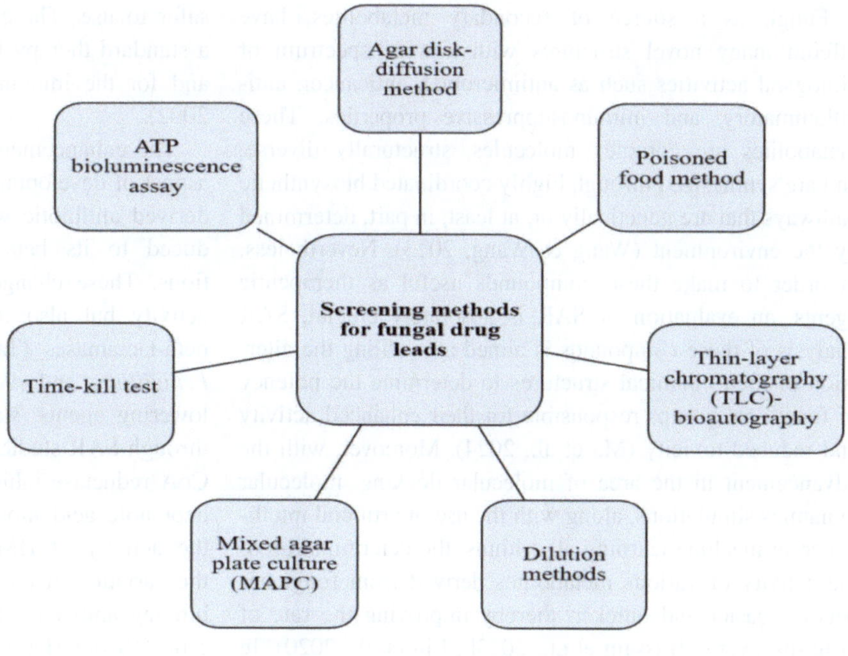

Fig. 2 Summary of various methods used in the screening of fungal metabolites for drug discovery. *Source* Authors

3.5 Mixed Agar Plate Culture (MAPC)

The MAPC approach enhances the screening process by using agar plates with fungi and bacterial cultures. The incorporation of mixed cultures from the onset of the screening process increases the pace of screening. It facilitates the screening of a large number of filamentous fungi naturally sampled within a given period. This method enhances the probability of identifying the novel compound that preferably possesses antibacterial and antifungal effects. Many laboratories use this method because of its convenient and simple nature (Guohua Chen, 2012).

3.6 Time-Kill Test

This method is applied to know whether bactericidal or fungicidal medicines work synergistically or antagonistically. It is the best technique for figuring out antimicrobial activity that is dependent on both time and concentration. After testing each drug in broth culture media containing bacterial suspension, the live cells are counted using the agar plate count method (Franconi & Lupetti, 2023).

3.7 Adenosine Triphosphate (ATP) Bioluminescence Assay

The chemical form of cell energy, adenosine triphosphate (ATP), can be measured by luminescence. This method has traditionally been used for drug screening as well as for cytotoxicity tests, but it can also be used to estimate the growth of microbial populations. Briefly, when luciferase is present, ATP generated by bacteria or fungi transforms D-luciferin into oxyluciferin, which generates light. A luminometer measures the amount of light emitted, and the result is given as a relative light unit (RLU). The observed luminescence and cell viability have a linear relationship. This method has been applied to test antifungal, antibacterial, and antimycobacterial agents. Rapidity is one of its primary benefits; in contrast to the conventional dilution method, which takes three or four weeks, antimycobacterial test results are obtained in just three to five days (Rakotonirainy et al., 2003).

4 Structure–Activity Relationship (SAR) Studies

A step in designing and optimizing effective drugs derived from fungi is the structure–activity relationship (SAR) analysis studies. These studies enable us to establish a relationship between the structure of a given chemical compound and its efficacy and toxicity, thus showing which part of the compound may be credited to the effectiveness of the drug. Identifying SAR is useful for small molecules in fungal metabolites targeted for optimal pharmaceutical properties and so suitable for human usage (Masi et al., 2018). The huge structural diversities of fungal metabolites offer a good target in drug discovery. These studies are of particular importance for fungal metabolites since these compounds are characterized by a high chemical diversity and bioactivity of metabolites produced by fungi.

Fungi, as a source of secondary metabolites, have offered many novel structures with a wide spectrum of biological activities such as antimicrobial, anticancer, anti-inflammatory, and immunosuppressive properties. These metabolites are complex molecules, structurally diverse, and are synthesized through highly coordinated biosynthetic pathways that are genetically or, at least, in part, determined by the environment (Wang & Wang, 2023). Nevertheless, in order to make these compounds useful as therapeutic agents, an evaluation of SAR is absolutely crucial. SAR analysis of these compounds is aimed at profiling the alteration of their chemical structures to determine the potency of functional groups responsible for their enhanced activity and reduced toxicity (Ma et al., 2024). Moreover, with the advancement in the area of molecular docking, molecular dynamics simulations, along with the use of artificial intelligence in machine learning algorithms, the determination of the activity of various metabolites derived from fungi has become easier and quicker, thereby improving the rate of drug discovery (Hussain et al., 2025b; Lin et al., 2020). In this way, it is possible to create desired derivatives against the target disease.

4.1 Designing Potent Fungal-Derived Drugs by Modifications to Enhance Efficacy and Reduce Toxicity

The two broad stages in the bioactivity-guided design of drugs from fungal metabolites include the discovery of active metabolites and the deployment of efficient optimization strategies. Since such drugs are derived from fungi, the design of such compounds encompasses natural product chemistry, medicinal chemistry, and biotechnology. The bioactive and structurally diverse metabolites from fungi make them a source of choice for biologically active compounds. However, the transition from natural metabolite to a clinically proven drug involves a highly randomized approach (Xu et al., 2023). It is necessary to define the so-called pharmacophore, which denotes the structural elements in molecules responsible for biological activity, as one of the key features of SAR research. From these altered compounds, researchers can understand which regions of a fungal metabolite are needed for it to bind to a target biomolecule (Huang et al., 2024). For instance, the antifungal agent echinocandins, which is isolated from *Aspergillus nidulans*, pertains to the class of drugs that affect the synthesis of β-(1,3)-D-glucan in the fungal cell walls. Reports with SAR on echinocandins analogs show that even a slight change in the lipophilic side chain enhances effects on antifungal and pharmacokinetic properties. This as a result brought out better echinocandins like caspofungin and micafungin that are more effective and safer to use. These refinements have made echinocandins a standard therapy for systemic fungal infections generally and for the immunocompromised in particular (Denning, 2002).

The enhancement of antimicrobial properties is another aspect of developing fungal-derived drugs. The first fungal-derived antibiotic was penicillin. Acyl groups were introduced to its beta-lactam ring through SAR investigations. These changes not only advanced its spectrum of activity but also increased its ability to resist bacterial beta-lactamases (Tahlan & Jensen, 2013). Similarly, the *Penicillium* and *Aspergillus* species produce cholesterol-lowering agents, statins, which have also been optimized through SAR studies to enhance their selectivity for HMG-CoA reductase inhibition. It was found that the dihydroxy heptanoic acid moiety was essential for the inhibition of the activity of HMG-CoA reductase. At the same time, the variation of the hydrophobic ring system enhanced the binding ability of the molecules and pharmaceutical properties (Sadowska et al., 2023). The first naturally derived statin, i.e., lovastatin, provides guidelines for the development of synthetic-based statins such as atorvastatin, which has better pharmacokinetic characteristics. Statins such as atorvastatin and rosuvastatin were subsequently synthesized, and these are more potent and have longer half-lives than those of natural statins (Endo, 2017). These developments have put statins among the most used drug classes globally, shaping the future of healthcare.

SAR studies also play a key role in optimizing fungal metabolites in terms of their toxicity. For example, cyclosporine, an immunosuppressant drug obtained from the soil fungus *Tolypocladium inflatum*, is an ideal example of how studying SAR works effectively to compete for toxicities with efficacy. The cyclic peptide structure of cyclosporine was studied in detail to determine specific amino acid positions important in calcineurin inhibition, which is the primary mode of immunosuppressive action. Altering the cyclic structure of SAR minimized nephrotoxicity while retaining immunosuppressive efficacy. It progressed into cyclosporine A, which has significantly changed the sphere of organ transplantation as it prevents graft rejection (Corbett et al., 2021). Such advancements, therefore, call for the expansion of SAR studies in the quest to bring fungal metabolites into the drug development stream.

Compared with the previous studies on SAR of fungal metabolites, new strategies in current investigations have included higher throughput techniques and metabolomics investigations. Advanced molecular biology screening techniques allow the selection and testing of multiple fungal metabolites and their modified compounds with candidates for subsequent optimization and development (Ayon, 2023). Metabolomics profiling complemented with genomic data has made it possible to discover biosynthetic gene clusters and their complex regulatory frames, giving SAR studies new prospects (Gaudêncio et al., 2023).

4.2 In Silico Tools for SAR Analysis

In the Structure–Activity Relationship (SAR) investigation of fungal metabolites, in silico methods have become essential because they enable us to predict and optimize natural compounds' biological activities before physical testing. Notably, these computational methods aid in the grand challenge of recognizing interaction trends and transforming structures and biological parts in a faster way than other methods. This development has been greatly boosted by the integration of SAR with computational techniques in the development of fungus-derived drugs (De Araújo et al., 2013). Molecular docking and QSAR modeling are two in silico strategies that provide an opportunity to anticipate how changes in structure will influence activity. These tools are immensely helpful for drug design, which enables rapid screening of multiple sets of fungal metabolites and their analogs (Brogi et al., 2020). Furthermore, the delicate regulation of biosynthesis of fungal secondary metabolites by molecular biology enhances the production of new bioactive molecules or enhances the production of known compounds, thus expanding the chemical space to be searched for SAR (Trosset & Carbonell, 2015).

4.2.1 Molecular Docking

Among the several in silico techniques, molecular docking successfully emulates the binding of fungal metabolites to their target molecules. This technique indicates how strongly the protein target's active site will bind to the ligand and to the relevant position. Concretely, docking analysis carried out on proteasome inhibitors synthesized by fungi, such as epoxomicin, has led to the development of the bortezomib drug, used in the treatment of multiple myeloma (Brogi et al., 2020). In the same way, the compound aspergillomarasmine A, derived from *Aspergillus fumigatus,* was investigated by molecular docking study for the possibility of it serving as an inhibitor of metalloproteinase enzymes associated with cancer metastases. Biochemical and in vivo validation still required for proper validation, as the in silico studies only reflect the binding affinities (Isaac et al., 2021). Advanced studies in molecular docking have allowed researchers to predict, on the in silico level, the binding constant of metabolites with enzymes or receptors responsible for disease progression in fungi. For example, AutoDock and GROMACS were used to evaluate the binding characteristics of fungal secondary metabolites with other targets, such as bacterial ribosomes or fungal enzymes (Jacob et al., 2022).

4.2.2 Quantitative Structure–Activity Relationship (QSAR) Modeling

QSAR employs mathematical models to assess the correlations of the descriptors of chemicals to the known biological activity; this leads to a likelihood of the activity of the compounds that are not yet tested. QSAR analysis has been actively involved in the friendly modulation of fungal metabolite production with special emphasis on polyketides and non-ribosomal peptides. QSAR models allow it by applying the AI algorithms to the large data set in an attempt to identify the structural fragments that define potency and selectivity (González-Medina et al., 2017). The new investigation in the field of medicinal chemistry studies has adopted the QSAR models applied in the recent research in the medicinal chemistry domain on the general probable antifungal and anticancer activities of the compounds derived from *Fusarium* and *Trichoderma* (Borrego-Muñoz et al., 2024).

4.2.3 High-Throughput Virtual Screening

The high-throughput virtual screening (HTVS) molecular docking in conjugation with QSAR modeling is used to screen the large database of fungal metabolites and related analogs with comparable efficiency. This approach arranges compounds in order of their expected activity so as to facilitate the identification of the lead compounds. For example, HTVS has been applied in the search for new inhibitors of fungal cell wall synthesis, thus increasing drug development and the effectiveness of antifungal drugs (Weinberger et al., 2020). Again, using HTVS with the support of experimental validation allows us to select the possible candidates for additional optimization effectively.

4.2.4 Profiling and Biosynthetic Pathway Analysis

As mentioned above, metabolomic profiling, when complemented with in silico techniques, enriches SAR analysis. Thus, analyzing the metabolic networks in fungi and recognizing biosynthetic gene clusters will help to find new metabolites and their products. Utilizing data from the genomics of the source organism, antiSMASH, and PRISM allow for the generation of the chemical structures of secondary metabolites that can be investigated in SAR research (Blin et al., 2019). It has been notably useful in the identification of active fungal metabolites that are not expressed in baseline conditions in a laboratory setting.

4.2.5 Synthetic Biology and CRISPR-Cas Systems

Developments in synthetic biology have allowed the generation of engineered fungal strains specifically for metabolite generation for SAR analysis. Modern CRISPR-Cas systems and other strategies enable the targeted manipulation of biosynthetic pathways for the synthesis of libraries of derivatives with structural diversity. These engineered metabolites can then be used in silico SAR analysis for better lead identification with desired characteristics. Synthetic biology also helps in getting a high quantity of the promising metabolites of matter and, in the same process, helps solve yield and scalability issues that are normally associated with the process. Further, synthetic biology enables the generation of a large amount of the promising metabolites of matter and simultaneously solves yield and scalability problems that are usually linked to this process (Brooks & Alper, 2021).

However, major issues persist with regard to the design of highly effective drugs from fungal sources. Due to the structural intricacy of many fungal metabolites, the synthesis and alteration of these products may be quite challenging. Further, the regulatory networks controlling secondary metabolite synthesis are frequently unknown in fungi, thus hindering the control of biosynthetic pathways. These challenges need to be approached through collaboration in a systems biology fashion, considering both computational biology and wet-labs biology and synthetic biology and medicinal chemistry to harness the abundant therapeutic assets discovered among fungal metabolites efficiently (Esheli et al., 2022; Shankar & Sharma, 2022).

4.3 Modifying Metabolite Structures for Enhanced Activity

Conformational and chemical changes in the structures of fungal metabolites are often used to increase their biological activity. Given improved knowledge of fungal genomics and the fact that many more biosynthetic gene clusters are yet to be identified, researchers are now focusing on regulating fungal SM using techniques such as genome editing, enzyme engineering, and synthetic biology. These techniques enable the molecular composition of fungal metabolites to enhance their pharmacological profile and upstream problems such as drug resistance, poor solubility, or toxicity (Gerke & Braus, 2014). The general modifications methods that are thought to help in the advancement of metabolite activates are shown in Fig. 3.

CRISPR/Cas9 system: One advancement in this area is genome editing through the CRISPR/Cas9 system, which allows for the use of gene editing to knock out genes or alter genes that are involved in the biosynthesis of special metabolites. This way, biosynthetic genes can be upregulated or downregulated in order to force the metabolic pathways in the creation of new molecules or to enhance the production of existing molecules that have better biological activity (Jiang et al., 2021).

Enzyme engineering: The use of enzyme engineering approaches like directed evolution along with computational enzyme design is used to optimize enzymes for the biosynthesis of fungal-derived natural products, giving rise to novel chemical entities with enhanced potency against highly structured diseases like cancer and multi-drug resistance infections (Skellam, 2019).

Artificial intelligence (AI): The use of machine learning and artificial intelligence (AI) in fungal metabolite engineering is also assisting in the determination of structure–activity relationships (SAR) so that more effective and even more potent resultant derivatives with minimal side effects can be synthesized. Artificial intelligence systems are then used to forecast the optimal chemical changes from the libraries of known fungal metabolites and related activities (Riedling et al., 2024). These emerging technologies are innovations in the methods used with the production of drugs obtained from fungi because they enable the creation of a new generation of bioforms with required pharmacokinetics features.

4.4 Chemical Derivatization for Activity Enhancement

The most frequent preparative technique that is used to improve the properties of fungal metabolites is chemical derivatization. This suggests that if the functional groups are somehow altered or if the molecules frames are changed, then the pharmacological properties such as potency, selectivity, solubility, and many others can be boosted (Ding & Xue, 2024). Chemical modification includes derivatization, which provides new functions on the chain of the parent compound whereby solubility, total effectiveness, and specificity of a drug derived from fungi can be improved. In addition, optimization can also involve altering other aspects of the pharmacokinetics of metabolites, including the extension of half-life or improving the ability of the metabolite to cross over the blood–brain barrier (Markovic et al., 2020). Structural variation and the subsequent chemical modification of fungal natural products have emerged as formidable tactics or tools in drug discovery campaigns. Such chemical modifications tried to improve the pharmacology properties of the natural products, biological activity, stability, and solubility for the intended medical application. The fungal metabolites are routinely utilized as the lead, and by a process known as structure optimization, their activity is further enhanced (Ding & Xue, 2024).

Some successful attempts with chemical derivatization procedures to enhance the bioactivity of fungal metabo-

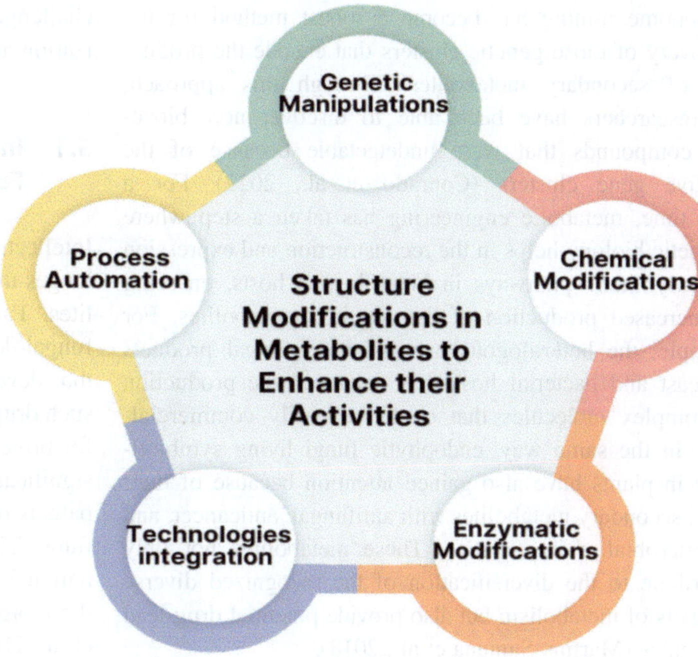

Fig. 3 General strategies that help to enhance the metabolites activities. *Source* Authors

lites have also been reported earlier. For instance, modifications in the structure of penicillamine have been carried out with the intention of developing new pen derivatives that are more effective against resistant *Staphylococcus aureus* (Pinar & Rodríguez-Couto, 2024). Similarly, derivatives of trichothecene toxins from the species of *Fusarium* have also fostered the creation of non-toxic compounds that can effectively cure fungal infections (Wachowska et al., 2017).

4.5 Examples of Chemical Derivatization

4.5.1 Halogenation

The incorporation of atoms like chloride or bromide also improves the permeability of the membrane as well as increases receptor affinities. This approach was successfully applied to modify antifungal agents obtained from *Aspergillus niger* (Faleye et al., 2024).

4.5.2 Hydroxylation

Poly-hydroxylation affords improved water solubility, which is a virtue in drug design for improving oral bioavailability. A few fungal alkaloids were enhanced through hydroxylation, for instance ergotamine for use as anti-migraine drugs (Plazas & Faraone, 2023). Similarly, antitumor compounds such as epothilones derived from *Sorangium cellulosum* and modified to improve their efficacy further as cancer therapies have been optimized extensively using this technique (Majhi & Das, 2021).

4.5.3 Prodrugs

Prodrug strategies have been employed on fungal metabolites such as fumagillin by converting inactive precursors to active drugs on metabolism for improved therapeutic windows when used in cancer treatments (Markovic et al., 2020; Pela, 2004).

4.6 Optimization Through Semi-synthesis

Another strategy for achieving higher levels of regeneration of natural products with the desired properties is semi-synthesis, which includes the chemical conversion of a fungal metabolite. A clear case is the semi-synthesis of taxol, which is isolated from the *Taxus* plant but also available from strains of fungi. Through semi-synthesis, such modifications enhance the anticancer properties of the molecule as well as provide the feasibility for large-scale production (Yan-Hua et al., 2020). The semi-synthetic synthesis of complex natural compounds from fungal strains is also being investigated extensively. These products can subsequently be altered to improve their pharmacological characteristics. Another prominent example is the chemical modification of penicillin derivatives, which led to the development of semi-synthetic antibiotics such as amoxicillin and methicillin. These derivatives contain improved resistance to β-lactamase enzymes and enhanced activity against a broader range of bacteria (Volpato et al., 2010).

Genome mining has become a robust method for the discovery of those genetic clusters that encode the production of secondary metabolites. Through this approach, the researchers have been able to discover new bioactive compounds that were undetectable because of the inactive gene clusters (Conrado et al., 2022). For a long time, metabolic engineering has taken a step where synthetic biology helps in the reconstruction and expression of biosynthetic pathways in heterologous hosts, ensuring an increased production of the required metabolites. For example, the heterologous expression of fungal products in yeast and bacterial hosts has enhanced the production of complex molecules that could be easily commercialized. In the same way, endophytic fungi living symbiotically in plants have also gained attention because of their novel secondary metabolites with antifungal, anticancer, and antimicrobial characteristics. These metabolites not only contribute to the diversification of the recognized diverse products of metabolism but also provide potential drug lead molecules (Martins-Santana et al., 2018).

Nevertheless, some obstacles persist in realizing the potential of fungal metabolites as follows. A significant challenge is the induction of previously dormant biosynthetic gene clusters that need to be addressed by novel strategies and technologies, including epigenetic modification and co-culturing. Other strategies, such as the One Strain Many Compounds (OSMAC) approach, have also been used to maximize the production of metabolite diversity under different cultural conditions (Nevalainen, 2020).

In addition, the issue of sustainability is still an area of concern because sustainable production has not been developed to offer efficient and environmentally friendly production solutions in terms of cost. Protein and metabolomics technologies globally and high-throughput screening techniques are poised to enhance and expedite the discovery of novel fungal metabolites for optimization in response to current-day health concerns.

5 Challenges in Commercialization

The pathway from discovering a compound with an origin that originated from fungi to converting it into the product is filled with challenges. These challenges, therefore, are complex and encompass intellectual property (IP) barriers, legal and regulatory requirements, and questions of bioethical principles. Nevertheless, there are significant opportunities for fungal metabolites to be used in drug discovery, and these challenges need to be overcome to ensure that discoveries can be realized as genuine treatments. The major challenges associated with fungal metabolites commercialization are summarized in Fig. 4.

5.1 Intellectual Property and Patenting Fungal Compounds

Intellectual property (IP) is one of the most critical challenges that inhibit the commercialization of fungal metabolites. Patents are important in the commercialization of fungal-derived drugs since they make sure that the entity that developed the drugs has the exclusive right to market such drugs. Among all the intellectual property rights (IPRs) for protection solutions, patents are proving to be the most significant, and the landscape reveals a growing body of patents on process and product innovations associated with fungi. This is, in fact, a very convincing reason that international markets have discovered the great potential that these organisms have a leading role in the industry (Hüttner et al., 2020; Hyde et al., 2024). However, patenting natural compounds, specifically those derived from microorganisms such as fungi, presents exceptional challenges.

Legal protection of fungi and other microorganisms through patents and other forms of IPRs is complicated in two ways. The first intensified risk relates to inventions made out of biological material to disclose them in other fields. It also provides that for a creation to be patented, it should be described in a manner sufficiently elaborated for a person with average skill in the field. Where inventions concern microorganisms, such disclosure is normally improbable. The second reason is that a large part of the legislation governing the patenting of microorganisms is bureaucratized (Meyer & Mengel, 2024).

Clear legal frameworks are necessary to guarantee equal access to fungal biodiversity, as demonstrated by the recent case studies, such as those involving *Penicillium* strains used to produce antifungal medications. For example, the patent laws about the extraction of natural products have changed in the United States. However, it is still difficult to get patents for naturally occurring fungal metabolites that do not require significant synthetic alteration (Ayaz et al., 2023). Many patent laws hold that only altered or synthesized natural products can be patented in most countries around the world. This limitation poses a challenge to the commercialization of fungal metabolites since their structures are originally abstracted from nature (Weissman, 2017). Also, international agreements like the Budapest Treaty promote the deposit of biological materials but also bring in procedural measures that must be followed throughout the patent application process (Hüttner et al., 2020).

Fig. 4 Major challenges associated with fungal metabolites commercialization. *Source* Authors

5.2 Issues in Patent Protection for Natural Compounds

Naturally occurring fungal-derived compounds are complex and often contentious to patent. It is important to understand that natural substances, although the synthetic derivatives of the natural compounds may be patented, are not easily patented. The first and perhaps the greatest difficulty that accompanies the filing of patents for fungal metabolites is the problem of novelty, where fungal metabolites, already described in literature or other patents, are likely to be patented as prior arts. For instance, antifungal or antimicrobial compounds from a sample of a fungal species, new to experts, can encounter challenges such as synthesized or identified compounds with similarities in structures and activities already existing (Harrison, 2014).

Taxol (generally known as Paclitaxel), originally extracted from the bark of the Pacific yew tree, is an approved anticancer agent by the Food and Drug Administration (FDA). However, its production has also been associated with endophytic fungi such as *Taxomyces andreanae.* The "prior art doctrine" made it difficult for the patenting process because the use of taxol was prior to the application of the fungal sources (Torres-Mendoza et al., 2020). Similarly, lovastatin is a product of a cholesterol-lowering drug that comes from a fungus known as *Aspergillus terreus.* Even though the synthetic derivatives of lovastatin are still patentable, problems are observed concerning the natural compound because of its previous usage in folk medicine and existing information about its effects. This case highlights the problems that arise when there is an attempt to procure patents on natural products that have existed throughout history (Srinivasan et al., 2022). Griseofulvin, an antifungal agent of natural source and obtained from *Penicillium griseofulvum*; griseofulvin has also suffered from problems of patents faced by most natural products (Aris et al., 2022). These examples illustrate the difficulties in ensuring that metabolites obtained from fungi meet the strict requirements for regulatory approval.

The main difficulties stem from the necessity to prove the originality and uniqueness of the compound at stake, especially if one or another similar compound has been described in the literature or used prior to the priority date (Hüttner et al., 2020). In certain cases, a metabolite's usage in a therapeutic application may be denied even if it is innovative because it overlaps with pre-existing patents on the general class of substances (Rai et al., 2008). Indeed, the very nature of natural products suggests that the chemical composition of natural products may well not be new enough to meet patent policies in terms of invention.

Another problem is connected with the process of natural product extraction and isolation, which, in fact, can be considered as the most evident way of getting a natural product from a fungal source. The transition from extraction to isolation seems conventional, which makes it possible to patent the compound in question (Sarker & Nahar, 2012). Patenting of natural products is even more problematic where there is traditional knowledge, especially medicinal, as many indigenous people have utilized certain fungal metabolites. This situation has raised concerns about biopiracy.

The marketing of spinosad, an insecticidal metabolite of *Saccharopolyspora spinosa*, presented a number of controversies. Critics said that companies were not doing enough to ensure the benefits of the organism to the region where the organism was first discovered (Maroufpoor et al., 2010). Also, there are disparities in the patenting approaches across the regions. In Europe, for instance, the European Patent

Office (EPO) has very tough requirements when it comes to patenting natural compounds, meaning it could be harder to get a patent for fungal-based drugs. On the other hand, the United States of America has been relatively flexible, especially with the patenting of biotechnological inventions, especially where synthetic alterations have been made (Gates, 2014; Sharma & Jain, 2019).

Certain practical difficulties can, nonetheless, arise from benefit-sharing agreements. They are extremely long and hierarchical and may add several parties of interest, such as governments, indigenous populations, and major private companies. At other times, legal uncertainty can also dissuade bioprospecting, thus postponing the identification of novel fungal metabolites (Bhatti et al., 2009).

5.3 Legal and Ethical Considerations in Bioprospecting

Bioprospecting, the use of biological resources, such as fungi, to determine if any valuable commercial compounds (e.g., drugs) can be extracted, is associated with legal and ethical problems pertaining to fair access and the sharing of benefits. Bioprospecting of fungal metabolites, notably for pharmaceuticals, cosmetics, or agrochemical products, is one of the largest sources. The rights to these natural resources and the benefits to be derived from them have, however, been discussed for decades (Efferth et al., 2016).

In December 1992, the Convention on Biological Diversity (CBD) established important rules for regulating bioprospecting, which consist of specifying rules for access to genetic resources and the obligation that the benefits of these resources be shared fairly with the country of origin (Herrera Izaguirre, 2008). The present framework is of particular interest for fungal metabolites, which often originate from the species naturally occurring in specific geographical locations or countries. Where there are no defined access and benefit-sharing regulations in compliance with the CBD, the pharmaceutical companies may take full advantage of such resources without compensation to the supplying nations of the natural resource or the indigenous people who initially guarded the biodiversity for the world's use (Geng et al., 2023). Further, bioprospecting also involves the use of an assay of natural compounds, which has some legal and ethical implications. It is built on the Convention On Biological Diversity (CBD) and the more recent Nagoya Protocol that came with the objective of guaranteeing that the achievement of this potential of genetic resources, associated with fungi genetic resources in this case, is reasonable, and fair and that must be accompanied by prior informed consent from the country of origin with regard to access to genetic resources and benefit sharing (Beato & Veneroso, 2023; Keller, 2019). However, the commercialization of fungal metabolites has been hindered by the development of a very complicated structure in the areas of intellectual property rights (IPRs) and treaties. Moreover, such protocols have also urged researchers and companies to consider the legal steps to gain access to fungal resources in an ethically and legally sound way.

5.4 Regulatory Challenges in Drug Approval

Once a fungal metabolite has gone through synthesis and has been analyzed and optimized as a drug, the resultant molecule goes through a myriad of tests to reach the regulatory requirement as per the Food and Drug Administration, European Medicines Agency, or any other healthcare authority across the globe. Safety and efficacy trials are basic prerequisites to establish fungal metabolites for their therapeutic applications, as preclinical and clinical investigations are mandatory to establish the efficacy of a drug. Obtaining approval of a drug or a metabolite is an essential step because it guarantees the drug in terms of safety, efficacy, and quality. However, it also provides a set of challenges that can hinder or even entirely stop the commercialization of fungal-derived metabolites (Keller, 2019).

5.5 Safety, Efficacy, and Quality Control Measures

Although fungal-derived drugs are expected to demonstrate efficacy in the course of their development, they also need to be safe for human use. Fungal metabolites are batch-to-batch variables like most of the other natural products because of the different sources, i.e., the strain of fungus, the conditions under which that fungus is grown, and the manner in which it is harvested. To obtain results that can be trusted by replicating the same experiment, there is a requirement for rigorous manageability and consistency (Vrabl et al., 2019). Fungal-derived drugs have to be first demonstrated to be safe and effective as drugs by rigorous examination through preclinical and clinical trials. Since the fungal metabolites are extremely complex and numerous, these drugs are ordinarily highly effective with multiple biological effects; in other words, the full spectrum of these effects is still unknown at this point and requires further clarification for patients' safety (Prescott et al., 2023).

Fungal metabolites should be carefully evaluated for cytotoxicity, hepatotoxicity, and other adverse effects. For instance, the antifungal drug amphotericin B, derived from *Streptomyces nodosus*, is associated with nephrotoxicity and requires close monitoring during treatment (Esheli et al., 2022). Likewise, the anticancer drug L-asparaginase, isolated from *Escherichia coli* but produced using fungal strains in bioreactors, has been used in the treatment of acute lymphoblastic leukemia (Rath et al., 2024). However, the drug presents severe side effects such as hypersensitivity reactions (Mireles et al., 2024). So, managing side effects associated with fungal-derived drugs requires execution, and failure to do so results in time-consuming and expensive consequences such as obtaining approval.

Other critical elements to be considered involve pharmacokinetics, including bioavailability, half-life, and metabolism in drug development. Targeted fungal-derived drugs have to be designed so that they will be sufficiently absorbed and have enough distribution within the body to reach the therapeutic effect without toxic concentrations in the body (Bellmann & Smuszkiewicz, 2017). This process is often challenging due to the fact that fungal fermentation produces products that differ in composition with respect to the strain used as well as the conditions under which the fermentation process is undertaken. This also requires detailed in vivo testing and formulation optimization and can easily add several years to the drug approval process. Metabolites derived from fungi also claim to face the challenge of many patient's response variabilities to therapy, hence causing unequal flowing of the drug's therapeutic effects (Vena et al., 2020).

The medicine agencies legal regulating bodies, e.g., the FDA in the USA or the EMA in the EU, have very specific criteria for drug approval, and it cannot be easy to meet some of these criteria when it comes to fungal-derived compositions. Novel approaches to drug formulation and administration may be required because of suspended enzymes and the intricate chemical structures of fungal metabolites, which introduce additional regulatory levels of complexity (Tillotson & Tillotson, 2015). Nonetheless, a crucial challenge in the development of drugs from fungi is that there is a deficiency of harmonization among regulatory standards across the world. For a long time, the FDA/EMA had been concerned with standardizing the operating procedures for the research and marketing of drugs. Still, standards differ from region to region and often delay the introduction of new promising drugs to the market. Additionally, the regulatory process becomes more difficult because of the absence of standardized processes for analyzing natural compounds, such as fungus metabolites. Consequently, these substances can be subjected to more scrutiny or have to wait longer for approval (Verma, 2016). In this regard, the use of probiotics as medicine is not governed by the FDA.

5.6 Safety and Toxicity Concerns

Concerns about the toxicity and safety of drugs derived from fungi have become important problems that need attention. Well-known antifungal compounds, as well as anticancer agents, are contained in the list of existing fungal drugs, which may have various negative side effects on patients. Knowledge of safety profiles is essential for clinicians and researchers because it defines possibilities for treatment and the establishment of safer therapeutic ways.

A significant number of fungal metabolites are ambivalent; that is, they may act, at the same time, a bioactive agent or a toxin. For example, among mycotoxins, aflatoxins generated by *Aspergillus flavus* have a very extensive application in therapeutic practice and can also have severe adverse effects on human health. Although they have been considered for their anticancer properties, aflatoxins, especially AFB1, are known to produce serious toxic effects. These are hepatotoxicity, nephrotoxicity, and immune suppression, to mention but a few, which may result in complications such as liver failure and increased susceptibility to developing infection (Abrehame et al., 2023). Similarly, amphotericin B is a potent antifungal drug, but it is also associated with nephrotoxicity. Thus, it requires reformulations like liposomal delivery systems for minimal side effects (Safdar et al., 2010). Before being authorized for use in organ transplantation, cyclosporine A also required completing a significant amount of clinical research to lessen its nephrotoxic effects. The main reason for regulatory delays was often the need to evaluate the immunosuppressive advantages against any possible negative consequences (Wu et al., 2018). In the case of voriconazole, a triazole antifungal drug derived from fungi, it faced approval challenges due to concerns over liver toxicity and its narrow therapeutic window (Pata et al., 2021). These cases demonstrate the challenges in ensuring the safety of fungal-derived compounds.

6 Conclusion

Drug development is a long way process that spans longer time and has a huge cost. Different natural and synthetic or semi-synthetic compounds, molecules, etc., are widely screened for their therapeutic potential and drug-like actions. The use of metabolites from different sources also got attention and currently various metabolite-based therapeutic agents are available commercially. Fungal metabolites offer unique properties and good therapeutic potential; hence, they are extensively targeted for this potential.

Recently, the use of derived and modified fungal metabolites is under consideration in exploring their drug-like potential. Various modifications have been made to enhance their effectiveness as prompt agents for treating various diseases. Structure activity relationship methods are widely employed in such processes, while following the in silico and screening approaches. Unfortunately, there are several hurdles existing that poses challenges to commercialize the fungal metabolites. Intellectual property rights, the urge of patenting, and tiresome approval process slow down the approval of fungal metabolites as therapeutic agents. In the future, the use of molecule techniques, integrating artificial intelligence tools, and applying collaborations can enhance the screening, identification, and commercialization process.

References

Abrehame, S., Manoj, V. R., Hailu, M., Chen, Y.-Y., Lin, Y.-C., & Chen, Y.-P. (2023). Aflatoxins: Source, detection, clinical features and prevention. *Processes, 11*(1), 204.

Al-Fakih, A. A., & Almaqtri, W. Q. A. (2019). Overview on antibacterial metabolites from terrestrial *Aspergillus* spp. *Mycology, 10*(4), 191–209.

Aly, A. H., Debbab, A., & Proksch, P. (2011). Fifty years of drug discovery from fungi. *Fungal Diversity, 50*, 3–19. https://doi.org/10.1007/s13225-011-0116-y

Aris, P., Wei, Y., Mohamadzadeh, M., & Xia, X. (2022). Griseofulvin: An updated overview of old and current knowledge. *Molecules, 27*(20), 7034.

Ayaz, M., Li, C.-H., Ali, Q., Zhao, W., Chi, Y.-K., Shafiq, M., Ali, F., Yu, X.-Y., Yu, Q., & Zhao, J.-T. (2023). Bacterial and fungal biocontrol agents for plant disease protection: Journey from lab to field, current status, challenges, and global perspectives. *Molecules, 28*(18), 6735.

Ayon, N. J. (2023). High-throughput screening of natural product and synthetic molecule libraries for antibacterial drug discovery. *Metabolites, 13*(5), 625.

Balba, H. (2007). Review of strobilurin fungicide chemicals. *Journal of Environmental Science and Health Part B, 42*(4), 441–451.

Beato, M. S., & Veneroso, V. (2023). The Nagoya Protocol on access and benefit sharing: The neglected issue of animal health. *Frontiers in Microbiology, 14*, 1124120.

Bellmann, R., & Smuszkiewicz, P. (2017). Pharmacokinetics of antifungal drugs: Practical implications for optimized treatment of patients. *Infection, 45*, 737–779.

Berdy, J. (2005). *Journal of Antibiotics. Antibiotics, 58*(1), 1–26.

Bhatti, S., Young, T., Carrizosa, S., & McGuire, P. (2009). *Contracting for ABS: The legal and scientific implications of bioprospecting contracts* (Vol. 4). IUCN.

Blin, K., Shaw, S., Steinke, K., Villebro, R., Ziemert, N., Lee, S. Y., Medema, M. H., & Weber, T. (2019). antiSMASH 5.0: Updates to the secondary metabolite genome mining pipeline. *Nucleic Acids Research, 47*(W1), W81–W87.

Bo, G. (2000). Giuseppe Brotzu and the discovery of cephalosporins. *Clinical Microbiology and Infection, 6*, 6–8.

Borrego-Muñoz, P., Coy-Barrera, E., & Quiroga, D. (2024). New strategies in the chemical control of *Fusarium oxysporum* using synthetic bioisosteres of secondary metabolites: A review of the synthetic methods for novel compounds with potential antifungal activity. *Mini-Reviews in Organic Chemistry, 21*(8), 794–810.

Brogi, S., Ramalho, T. C., Kuca, K., Medina-Franco, J. L., & Valko, M. (2020). In silico methods for drug design and discovery. *Frontiers in Chemistry, 8*, 612.

Brooks, S. M., & Alper, H. S. (2021). Applications, challenges, and needs for employing synthetic biology beyond the lab. *Nature Communications, 12*(1), 1390.

Bushley, K. E., Raja, R., Jaiswal, P., Cumbie, J. S., Nonogaki, M., Boyd, A. E., Owensby, C. A., Knaus, B. J., Elser, J., & Miller, D. (2013). The genome of Tolypocladium inflatum: Evolution, organization, and expression of the cyclosporin biosynthetic gene cluster. *PLoS Genetics, 9*(6), Article e1003496.

Butler, M. S. (2004). The role of natural product chemistry in drug discovery. *Journal of Natural Products, 67*(12), 2141–2153.

Butler, M. S. (2008). Natural products to drugs: Natural product-derived compounds in clinical trials. *Natural Product Reports, 25*(3), 475–516.

Chen, G. (2012). A new screening method for discovering antibacterial agents from filamentous fungi. *African Journal of Biotechnology, 11*(34), 8600–8605. https://doi.org/10.5897/ajb11.2000

Coméra, C., André, K., Laffitte, J., Collet, X., Galtier, P., & Maridonneau-Parini, I. (2007). Gliotoxin from *Aspergillus fumigatus* affects phagocytosis and the organization of the actin cytoskeleton by distinct signalling pathways in human neutrophils. *Microbes and Infection, 9*(1), 47–54.

Conrado, R., Gomes, T. C., Roque, G. S. C., & De Souza, A. O. (2022). Overview of bioactive fungal secondary metabolites: Cytotoxic and antimicrobial compounds. *Antibiotics, 11*(11), 1604.

Corbett, K. M., Ford, L., Warren, D. B., Pouton, C. W., & Chalmers, D. K. (2021). Cyclosporin structure and permeability: From A to Z and beyond. *Journal of Medicinal Chemistry, 64*(18), 13131–13151.

De Araújo, R. S. A., Guerra, F. Q. S., de O. Lima, E., De Simone, C. A., Tavares, J. F., Scotti, L., Scotti, M. T., De Aquino, T. M., De Moura, R. O., & Mendonça Jr, F. J. B. (2013). Synthesis, structure-activity relationships (SAR) and in silico studies of coumarin derivatives with antifungal activity. *International Journal of Molecular Sciences, 14*(1), 1293–1309.

Denning, D. W. (2002). Echinocandins: A new class of antifungal. *Journal of Antimicrobial Chemotherapy, 49*(6), 889–891.

Dewanjee, S., Gangopadhyay, M., Bhattacharya, N., Khanra, R., & Dua, T. K. (2015). Bioautography and its scope in the field of natural product chemistry. *Journal of Pharmaceutical Analysis, 5*(2), 75–84.

Ding, Y., & Xue, X. (2024). Medicinal chemistry strategies for the modification of bioactive natural products. *Molecules, 29*(3), 689.

Efferth, T., Banerjee, M., Paul, N. W., Abdelfatah, S., Arend, J., Elhassan, G., Hamdoun, S., Hamm, R., Hong, C., & Kadioglu, O. (2016). Biopiracy of natural products and good bioprospecting practice. *Phytomedicine, 23*(2), 166–173.

Eid, A. M., Istateyeh, I., Salhi, N., & Istateyeh, T. (2019). Antibacterial activity of Fusidic acid and sodium Fusidate nanoparticles incorporated in pine oil Nanoemulgel. *International Journal of Nanomedicine*, 9411–9421.

Endo, A. (2017). Discovery and development of statins. *Natural Product Communications, 12*(8). https://doi.org/10.1177/1934578X1701200801

Esheli, M., Thissera, B., El-Seedi, H. R., & Rateb, M. E. (2022). Fungal metabolites in human health and diseases—An overview. *Encyclopedia, 2*(3), 1590–1601. https://doi.org/10.3390/encyclopedia2030108

Faleye, O. S., Boya, B. R., Lee, J.-H., Choi, I., & Lee, J. (2024). Halogenated antimicrobial agents to combat drug-resistant pathogens. *Pharmacological Reviews, 76*(1), 90–141.

Favre-Godal, Q., Queiroz, E. F., & Wolfender, J.-L. (2013). Latest developments in assessing antifungal activity using TLC-bioautography: A review. *Journal of AOAC International, 96*(6), 1175–1188.

Franconi, I., & Lupetti, A. (2023). In vitro susceptibility tests in the context of antifungal resistance: Beyond minimum inhibitory concentration in *Candida spp. Journal of Fungi, 9*(12), 1188.

Freedman, R., Yu, R., Sarkis, A. W., & Hedstrom, L. (2020). A structural determinant of mycophenolic acid resistance in eukaryotic inosine 5′-monophosphate dehydrogenases. *Protein Science, 29*(3), 686–694.

Fromage, G. (2013). Structure and medical uses of antibiotics: Erythromycin and co-amoxiclav. *Journal of Aesthetic Nursing, 2*(4), 184–188.

Gates, C. (2014). Patenting the life sciences at the European patent office. *Cold Spring Harbor Perspectives in Medicine, 4*(12), Article a020792.

Gaudêncio, S. P., Bayram, E., Lukić Bilela, L., Cueto, M., Díaz-Marrero, A. R., Haznedaroglu, B. Z., Jimenez, C., Mandalakis, M., Pereira, F., & Reyes, F. (2023). Advanced methods for natural products discovery: Bioactivity screening, dereplication, metabolomics profiling, genomic sequencing, databases and informatic tools, and structure elucidation. *Marine Drugs, 21*(5), 308.

Geng, Y., Li, Z., & Tian, Y. (2023). Convention on biological diversity: The current status, ongoing challenges, and future prospects of marine biodiversity conservation. *Biodiversity Science, 31*(4), 22645.

Gerke, J., & Braus, G. H. (2014). Manipulation of fungal development as source of novel secondary metabolites for biotechnology. *Applied Microbiology and Biotechnology, 98*(20), 8443–8455.

Gnägi, L., Martz, S. V., Meyer, D., Schärer, R. M., & Renaud, P. (2019). A short synthesis of (+)-brefeldin C through enantioselective radical hydroalkynylation. *Chemistry—A European Journal, 25*(50), 11646–11649.

González-Medina, M., Owen, J. R., El-Elimat, T., Pearce, C. J., Oberlies, N. H., Figueroa, M., & Medina-Franco, J. L. (2017). Scaffold diversity of fungal metabolites. *Frontiers in Pharmacology, 8*, 180.

Gullo, V. P., McAlpine, J., Lam, K. S., Baker, D., & Petersen, F. (2006). Drug discovery from natural products. *Journal of Industrial Microbiology and Biotechnology, 33*(7), 523–531. https://doi.org/10.1007/s10295-006-0107-2

Hajjaj, H., Niederberger, P., & Duboc, P. (2001). Lovastatin biosynthesis by *Aspergillus terreus* in a chemically defined medium. *Applied and Environmental Microbiology, 67*(6), 2596–2602.

Hamilton-Miller, J. (2000). Sir Edward Abraham's contribution to the development of the cephalosporins: A reassessment. *International Journal of Antimicrobial Agents, 15*(3), 179–184.

Harrison, C. (2014). Patenting natural products just got harder. *Nature Biotechnology, 32*(5), 403–405.

He, L., Nan, M.-H., Oh, H. C., Kim, Y. H., Jang, J. H., Erikson, R. L., Ahn, J. S., & Kim, B. Y. (2011). Asperlin induces G2/M arrest through ROS generation and ATM pathway in human cervical carcinoma cells. *Biochemical and Biophysical Research Communications, 409*(3), 489–493.

Herrera Izaguirre, J. A. (2008). The 1992 United Nations convention on biological diversity. *Boletín Mexicano de Derecho Comparado*, 1023–1040. https://doi.org/10.22201/iij.24484873e.2008.122.4000

Holliday, J., & Cleaver, M. P. (2008). Medicinal value of the caterpillar fungi species of the genus Cordyceps (Fr.) Link (Ascomycetes). A review. *International Journal of Medicinal Mushrooms, 10*(3).

Huang, L., Li, Y., Pang, J., Lv, L., Zhou, J., Liang, L., He, X., Li, J., Xu, W., & Yang, R. (2024). Isolation and characterization of antimicrobial metabolites from the *Sophora tonkinensis*-associated fungus *Penicillium* sp. GDGJ-N37. *Molecules, 29*(2), 348.

Hussain, A., Rahman, M., Bibi, T., Fatima, R., Arif, I., Barwant, M. M., & Ali, S. A. (2024a). Prevalence of microorganisms in indoor household environments and their pathogenesis. *Journal of Antimicrobial Agents, 10*(1), 1–5.

Hussain, A., Rajpoot, T., & Rehman, M. (2024b). The role of 3d bioprinting in drug discovery: A review. *Journal of Advanced Biomedical and Pharmaceutical Sciences, 7*(3), 88–95.

Hussain, A., Rabbani, N., Areeb, M., Malik, S., & Ali, S. A. (2025a). Fungi as probiotics: Properties and applications. In *Fungal Biotechnology; Advancements in Drug Discovery, Healthcare, and Environmental Solutions* (1st ed., pp. 245–258). CRC Press. https://doi.org/10.1201/9781003594840-22

Hussain, A., Alvi, A., Qureshi, S. R., Parveen, S., Um-E-Habiba-U-Nisa, & Ali, S. A. (2025b). Exploring the role of artificial intelligence in microbial identification, environmental distribution, and ecological interactions. In *Artificial Intelligence and Animal Ecology* (pp. 221–243). CRC Press. https://doi.org/10.1201/9781003492788-10

Hüttner, S., Johansson, A., Gonçalves Teixeira, P., Achterberg, P., & Nair, R. B. (2020). Recent advances in the intellectual property landscape of filamentous fungi. *Fungal Biology and Biotechnology, 7*(1), 16.

Hyde, K. D., Baldrian, P., Chen, Y. et al. (2024). Current trends, limitations and future research in the fungi?. *Fungal Diversity 125*, 1–71.

Isaac, C., Narayanaswamy, R., & Vallivitan, K. (2021). *Molecular docking analysis of selected Clitoriaternatea constituents as matrix metalloproteinases (MMP 2 & MMP 9) inhibitors.*

Jacob, S., Bühring, S., & Bersching, K. (2022). Recent advances in research on molecular mechanisms of fungal signaling. *Encyclopedia, 2*(2), 840–863.

Jeschke, R., Iinuma, K., Harder, A., Schindler, M., & Murakami, T. (2005). Influence of the cyclooctadepsipeptides PF1022A and PF1022E as natural products on the design of semi-synthetic anthelmintics such as emodepside. *Parasitology Research, 97*(S1), S11–S16.

Jiang, C., Lv, G., Tu, Y., Cheng, X., Duan, Y., Zeng, B., & He, B. (2021). Applications of CRISPR/Cas9 in the synthesis of secondary metabolites in filamentous fungi. *Frontiers in Microbiology, 12*, Article 638096.

Kawasaki, Y. (2009). Mizoribine: A new approach in the treatment of renal disease. *Journal of Immunology Research, 2009*(1), Article 681482.

Keller, N. P. (2019). Fungal secondary metabolism: Regulation, function and drug discovery. *Nature Reviews Microbiology, 17*(3), 167–180.

Khan, M. A., Tania, M., Zhang, D., & Chen, H. (2010). Cordyceps mushroom: A potent anticancer nutraceutical. *The Open Nutraceutical Journal, 3*, 179.

Lage, O. M., Ramos, M. C., Calisto, R., Almeida, E., Vasconcelos, V., & Vicente, F. (2018). Current screening methodologies in drug discovery for selected human diseases. *Marine Drugs, 16*(8), 1–31. https://doi.org/10.3390/md16080279

Larsen, T. O., Smedsgaard, J., Nielsen, K. F., Hansen, M. E., & Frisvad, J. C. (2005). Phenotypic taxonomy and metabolite profiling in microbial drug discovery. *Natural Product Reports, 22*(6), 672–695.

Li, H.-J., Jiang, W.-H., Liang, W.-L., Huang, J.-X., Mo, Y.-F., Ding, Y.-Q., Lam, C.-K., Qian, X.-J., Zhu, X.-F., & Lan, W.-J. (2014). Induced marine fungus *Chondrostereum* sp. as a means of producing new sesquiterpenoids chondrosterins I and J by using glycerol as the carbon source. *Marine Drugs, 12*(1), 167–175. https://doi.org/10.3390/md12010167

Li, W.-R., Shi, Q.-S., Dai, H.-Q., Liang, Q., Xie, X.-B., Huang, X.-M., Zhao, G.-Z., & Zhang, L.-X. (2016). Antifungal activity, kinetics and molecular mechanism of action of garlic oil against Candida albicans. *Scientific Reports, 6*(1), 22805. https://doi.org/10.1038/srep22805

Liao, W.-Y., Shen, C.-N., Lin, L.-H., Yang, Y.-L., Han, H.-Y., Chen, J.-W., Kuo, S.-C., Wu, S.-H., & Liaw, C.-C. (2012). Asperjinone, a nor-neolignan, and terrein, a suppressor of ABCG2-expressing breast cancer cells, from thermophilic Aspergillus terreus. *Journal of Natural Products, 75*(4), 630–635.

Lin, X., Li, X., & Lin, X. (2020). A review on applications of computational methods in drug screening and design. *Molecules, 25*(6), 1375.

Lindequist, U., Rausch, R., Füssel, A., & Hanssen, H. P. (2010). Higher fungi in traditional and modern medicine. *Medizinische Monatsschrift fur Pharmazeuten, 33*(2), 40–48.

Ma, Y.-M., Miao, X., Jia, B., Sun, Z.-Y., Ma, S.-Y., & Yan, C. (2024). Design, synthesis, antifungal evaluation, structure-activity relationship (SAR) study, and molecular docking of novel spirotryprostatin A derivatives. *Molecules, 29*(4), 864.

Majhi, S., & Das, D. (2021). Chemical derivatization of natural products: Semisynthesis and pharmacological aspects—A decade update. *Tetrahedron, 78*, Article 131801.

Markovic, M., Ben-Shabat, S., & Dahan, A. (2020). Prodrugs for improved drug delivery: Lessons learned from recently developed and marketed products. *Pharmaceutics, 12*(11).

Maroufpoor, M., Safaralizadeh, M. H., Pourmirza, A. A., Allahvaisy, S., & Ghasemzadeh, S. (2010). Lethal effects of spinosad on *Chrysoperla carnea* larvae (Neuroptera: Chrisopidae) under laboratory conditions. *Journal of Plant Protection Research, 50*(2).

Martins-Santana, L., Nora, L. C., Sanches-Medeiros, A., Lovate, G. L., Cassiano, M. H. A., & Silva-Rocha, R. (2018). Systems and synthetic biology approaches to engineer fungi for fine chemical production. *Frontiers in Bioengineering and Biotechnology, 6*, 117.

Masi, M., Nocera, P., Reveglia, P., Cimmino, A., & Evidente, A. (2018). Fungal metabolites antagonists towards plant pests and human pathogens: Structure-activity relationship studies. *Molecules, 23*(4), 834.

Meyer, V., & Mengel, S. (2024). Patent landscape analysis for materials based on fungal mycelium: A guidance report on how to interpret the current patent situation. *Fungal Biology and Biotechnology, 11*(1), 11.

Mireles, J. A. G., Asseff, I. L., Espinoza, E. A. R., Hurtado, L. P. C., Gutiérrez, F. de M. R., Cabrera, A. S., Castañeda, V. L., González, C. D., Vázquez, A. G., & Rodríguez, A. R. N. (2024). L-asparaginase in the treatment of acute lymphoblastic leukemia, its adverse effects, and opportunities for treatment optimization. *Revista de Hematología, 24*(3), 145–162.

Nevalainen, H. (2020). *Grand challenges in fungal biotechnology*. Springer.

Pata, R., Dolkar, T., Patel, M., & Nway, N. (2021). Voriconazole-induced acute liver injury: A case report. *Cureus, 13*(12).

Pela, F. (2004). Biological activities of fungal metabolites. In *Handbook of industrial mycology* (pp. 68–111). CRC Press.

Petersen, A. B., Rønnest, M. H., Larsen, T. O., & Clausen, M. H. (2014). The chemistry of griseofulvin. *Chemical Reviews, 114*(24), 12088–12107. https://doi.org/10.1021/cr400368e

Pinar, O., & Rodríguez-Couto, S. (2024). Biologically active secondary metabolites from white-rot fungi. *Frontiers in Chemistry, 12*, 1363354.

Plazas, E., & Faraone, N. (2023). Indole alkaloids from psychoactive mushrooms: Chemical and pharmacological potential as psychotherapeutic agents. *Biomedicines, 11*(2), 461.

Prescott, T. A. K., Hill, R., Mas-Claret, E., Gaya, E., & Burns, E. (2023). Fungal drug discovery for chronic disease: history new discoveries and new approaches. *Biomolecules, 13*(6), 986. https://doi.org/10.3390/biom13060986

Rai, A. K., Reichman, J. H., Uhlir, P. F., & Crossman, C. (2008). Pathways across the valley of death: Novel intellectual property strategies for accelerated drug discovery. *Yale Journal of Health Policy, Law, and Ethics, 8*, 1.

Rakotonirainy, M., Hanus, J., Bonassies-Termes, S., Heraud, C., & Lavédrine, B. (2003). Detection of fungi and control of disinfection by ATP-bioluminescence assay. *AICCM Bulletin, 28*(1), 16–22.

Rath, G., Prusty, M., Nivedita, S., Behera, S. S., Ojha, P., Achary, K. G., Behera, S. K., Behera, H. T., & Ray, L. (2024). Unveiling the anticancer potential of L-Asparaginase (ASNase) from novel Streptomyces isolate HB2AG: Purification and characterization study. *The Microbe, 3*, 100076.

Riedling, O., Walker, A. S., & Rokas, A. (2024). Predicting fungal secondary metabolite activity from biosynthetic gene cluster data using machine learning. *Microbiology Spectrum, 12*(2), e03400-e3423.

Sadowska, A., Osiński, P., Roztocka, A., Kaczmarz-Chojnacka, K., Zapora, E., Sawicka, D., & Car, H. (2023). Statins—From fungi to pharmacy. *International Journal of Molecular Sciences, 25*(1), 466.

Safdar, A., Ma, J., Saliba, F., Dupont, B., Wingard, J. R., Hachem, R. Y., Mattiuzzi, G. N., Chandrasekar, P. H., Kontoyiannis, D. P., & Rolston, K. V. (2010). Drug-induced nephrotoxicity caused by amphotericin B lipid complex and liposomal amphotericin B: A review and meta-analysis. *Medicine, 89*(4), 236–244.

Sarker, S. D., & Nahar, L. (2012). *An introduction to natural products isolation*. Springer.

Shankar, A., & Sharma, K. K. (2022). Fungal secondary metabolites in food and pharmaceuticals in the era of multi-omics. *Applied Microbiology and Biotechnology, 106*(9–10), 3465–3488.

Sharma, A., & Jain, R. (2019). Patenting: Product of Nature. *International Journal of Legal Science and Innovation, 1*(2), 10.

Skellam, E. (2019). Strategies for engineering natural product biosynthesis in fungi. *Trends in Biotechnology, 37*(4), 416–427.

Squibb, B. M., & Sons, L. L. C. (2007). Pravachol (pravastatin sodium) tablet. *FDA-Approved Drug Product Labeling*.

Srinivasan, N., Thangavelu, K., & Uthandi, S. (2022). Lovastatin production by an oleaginous fungus, *Aspergillus terreus* KPR12 using sago processing wastewater (SWW). *Microbial Cell Factories, 21*(1), 22.

Stossel, T. P. (2008). The discovery of statins. *Cell, 134*(6), 903–905.

Sy-Cordero, A. A., Pearce, C. J., & Oberlies, N. H. (2012). Revisiting the enniatins: A review of their isolation, biosynthesis, structure determination and biological activities. *The Journal of Antibiotics, 65*(11), 541–549.

Tahlan, K., & Jensen, S. E. (2013). Origins of the β-lactam rings in natural products. *The Journal of Antibiotics, 66*(7), 401–410.

Tang, W., Gu, T., & Zhong, J. J. (2006). Separation of targeted ganoderic acids from *Ganoderma lucidum* by reversed phase liquid chromatography with ultraviolet and mass spectrometry detections. *Biochemical Engineering Journal, 32*(3), 205–210.

Tillotson, J., & Tillotson, G. S. (2015). The regulatory pathway for antifungal drugs: a US perspective. *Clinical Infectious Diseases, 61*(suppl_6), S678–S683.

Torres-Mendoza, D., Ortega, H. E., & Cubilla-Rios, L. (2020). Patents on endophytic fungi related to secondary metabolites and biotransformation applications. *Journal of Fungi, 6*(2), 58.

Trosset, J.-Y., & Carbonell, P. (2015). Synthetic biology for pharmaceutical drug discovery. *Drug Design, Development and Therapy*, 6285–6302.

Vena, A., Muñoz, P., Mateos, M., Guinea, J., Galar, A., Pea, F., Alvarez-Uria, A., Escribano, P., & Bouza, E. (2020). Therapeutic drug monitoring of antifungal drugs: Another tool to improve patient outcome? *Infectious Diseases and Therapy, 9*, 137–149.

Verma, N. (2016). Current regulatory challenges and approaches in the registration of herbal drugs in Europe. *Clinical Research and Regulatory Affairs, 33*(1), 9–24.

Volpato, G., C. Rodrigues, R., & Fernandez-Lafuente, R. (2010). Use of enzymes in the production of semi-synthetic penicillins and cephalosporins: Drawbacks and perspectives. *Current Medicinal Chemistry, 17*(32), 3855–3873.

Vrabl, P., Schinagl, C. W., Artmann, D. J., Heiss, B., & Burgstaller, W. (2019). Fungal growth in batch culture—What we could benefit if we start looking closer. *Frontiers in Microbiology, 10*, 2391.

Wachowska, U., Packa, D., & Wiwart, M. (2017). Microbial inhibition of Fusarium pathogens and biological modification of trichothecenes in cereal grains. *Toxins, 9*(12), 408.

Wang, W., & Wang, T.-T. (2023). Fungal secondary metabolites as valuable chemical entities for medicines and agrochemicals. *Frontiers in Microbiology, 14*, 1150023.

Wang, C. C. C., Chiang, Y., Kuo, P., Chang, J., & Hsu, Y. (2008). Norsolorinic acid inhibits proliferation of T24 human bladder cancer cells by arresting the cell cycle at the G0/G1 phase and inducing a FAS/membrane-bound FAS ligand-mediated apoptotic pathway. *Clinical and Experimental Pharmacology and Physiology, 35*(11), 1301–1308.

Weinberger, S., Beyer, R., Schüller, C., Strauss, J., Pellis, A., Ribitsch, D., & Guebitz, G. M. (2020). High throughput screening for new fungal polyester hydrolyzing enzymes. *Frontiers in Microbiology, 11*, 554.

Weissman, R. (2017). A long, strange TRIPS: The pharmaceutical industry drive to harmonize global intellectual property rules, and the remaining WTO legal alternatives available to third world countries. *Globalization and Intellectual Property*, 255–311.

Wu, Q., Wang, X., Nepovimova, E., Wang, Y., Yang, H., & Kuca, K. (2018). Mechanism of cyclosporine A nephrotoxicity: Oxidative stress, autophagy, and signalings. *Food and Chemical Toxicology, 118*, 889–907.

Xu, S., Li, M., Hu, Z., Shao, Y., Ying, J., & Zhang, H. (2023). The potential use of fungal co-culture strategy for discovery of new secondary metabolites. *Microorganisms, 11*(2), 464.

Yamada, T., Jinno, M., Kikuchi, T., Kajimoto, T., Numata, A., & Tanaka, R. (2012). Three new azaphilones produced by a marine fish-derived *Chaetomium globosum. The Journal of Antibiotics, 65*(8), 413–417.

Yan-Hua, Y., Jia-Wang, M. A. O., & Xiao-Li, T. A. N. (2020). Research progress on the source, production, and anti-cancer mechanisms of paclitaxel. *Chinese Journal of Natural Medicines, 18*(12), 890–897.

Zhang, L., Song, J., Kong, L., Yuan, T., Li, W., Zhang, W., Hou, B., Lu, Y., & Du, G. (2020). The strategies and techniques of drug discovery from natural products. *Pharmacology and Therapeutics, 216*, Article 107686.

Future Directions in Fungal Secondary Metabolite Research

Richa Choudhary, Upendra Kumar Soni, Mohini Soni, and Yash Vardhan Raj

Abstract

Fungal secondary metabolites (SMs) hold a wealth of bioactive compounds useful in medicine, agriculture and biotechnology. With new developments in genomics, mitogenomics, transcriptomics, metabolomics, and synthetic biology, it is a rapidly evolving field and contemporary methods to explore and exploit these molecules are the future. In this chapter, we identified some possible future directions in fungal secondary metabolite research by discussing the synergy between multi-omics, artificial intelligence processes, and CRISPR technology to unlock cryptic biosynthetic gene clusters (BGCs). Furthermore, I have reinforced the significance of fungal-fungal and ultimately fungal-host interactions and how they regulate SM production and raised the concept of the metabolite mining of unexplored fungal species and extreme environments. Furthermore, I highlighted the relevance of the ecological roles and evolutionary pathway of SM production and how it may ultimately improve sustainable production and applications. Apart from linking interacting disciplines to develop new knowledge, we also have the potential to broaden the exploration of finding new compounds of interest in terms of health and food security with urgent global impacts.

Keywords

Fungal secondary metabolites · Biosynthetic gene clusters (BGCs) · Bioactive compounds · Drug discovery · Artificial intelligence in bioscience

1 Introduction

Fungal secondary metabolites (SMs) are bioactive compounds having tremendous impact in medicine, agriculture, and biotechnology. SMs are usually low-molecular-weight compounds that may not be critically important for fundamental cellular processes but play vital ecological roles by facilitating the fungal adaptation to specific environmental conditions (Conrado et al., 2022; O'Brien & Wright, 2011). The discovery of penicillin by Fleming heightened scientific interest in fungal SMs as potential sources of pharmaceutical agents (Aly et al., 2011). Partially fuelled by a resurgence of interest from the pharmaceutical industry, studies in fungal SMs have increased significantly. Researchers are exploring their potential not only as therapeutic agents but also as tools to understand microbial ecology and host-pathogen interactions. Research on fungal SMs has revealed their exceptional bioactivity, highlighting their potential for applications in medical and agricultural benefit. These investigations have also spurred significant progress in fields such as analytical techniques, synthetic chemistry, and medicinal chemistry (Kaneria et al., 2024). Nevertheless, a comprehensive understanding of the ecological and biological functions of these compounds, particularly in relation to fungal survival and environmental adaptation, remains incomplete.

Our understanding of the functional roles of fungal SMs in plant disease systems is still evolving. Certain phytotoxic SMs produced by plant-pathogenic fungi have been found to act as photosensitizers that damage plant

R. Choudhary (✉) · Y. V. Raj
Department of Life Sciences, School of Biosciences and Technology, Galgotias University, Greater Noida, UP, India
e-mail: sinharicha11@gmail.com

U. K. Soni (✉)
Endocrinology Division, Council of Scientific and Industrial, Research (CSIR)-Central Drug Research Institute (CDRI), Lucknow, UP, India
e-mail: upendrasoni20@gmail.com

M. Soni
Biochemistry and Structural Biology Division, Council of Scientific and Industrial, Research (CSIR)-Central Drug Research Institute (CDRI), Lucknow, UP, India

© The Author(s), under exclusive license to Springer Nature Switzerland AG 2026

A. Prakash and M. K. Dubey (eds.), *Advances in Fungal Secondary Metabolites: Bioactive Attributes and Biotechnological Potential of Industrial Interest*, Sustainable Economy and Ecotechnology, https://doi.org/10.1007/978-3-032-23507-7_13

Fig. 1 Diverse roles of fungal secondary metabolites (SMs) in plant pathosystems. *Source* Authors

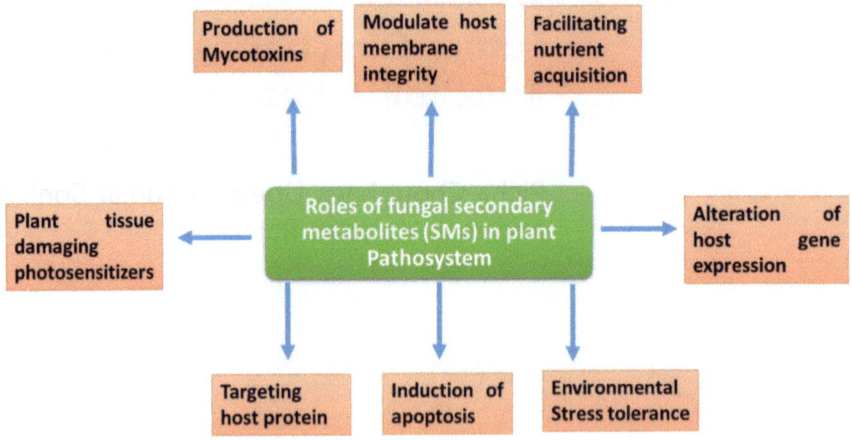

tissues, interfere with host proteins to hinder their function, modulate membrane integrity, or induce programmed cell death in plant cells (Möbius & Hertweck, 2009). Some of these metabolites function as host-specific toxins (HSTs), while others demonstrate activity across a wider spectrum of species. Beyond direct pathogenicity, fungal SMs contribute to diverse processes such as facilitating nutrient acquisition like iron through siderophore production, enhancing tolerance to environmental stresses, influencing interactions between species, and even modulating the host's gene expression (Scharf et al., 2014). Despite the known in vivo roles, these compounds can elicit physiological effects across a range of organisms. Notably, those SMs that are detrimental to humans and animals, irrespective of their biological function in fungi, are categorized as mycotoxins (Bennett & Klich, 2003; Reverberi et al., 2010) (Fig. 1).

Fungal SMs display extensive structural variation, yet they are generally classified into four principal groups according to their biosynthetic pathways: polyketides, non-ribosomal peptides (NRPs), terpenoids and tryptophan-derived compounds. These metabolite classes are synthesized by key enzymatic systems, including polyketide synthases (PKSs), non-ribosomal peptide synthetases (NRPSs), prenyltransferases or terpene synthases (PTs/TSs), and dimethylallyl tryptophan synthases (DMATSs) (Wiemann & Keller, 2014). One more emerging category of SMs comprises small peptides referred to as ribosomally synthesized and post-translationally modified peptides (RiPPs) and known to exhibit significant antimicrobial and antiviral properties, making them promising candidates for therapeutic applications (Zhong et al., 2022). Advancements in fungal genome sequencing have revolutionized our understanding of secondary metabolism. Genome mining efforts have revealed an unexpectedly large number of biosynthetic gene clusters (BGCs) within these fungi, underscoring their potential to produce a wide range of structurally diverse SMs. However, a significant gap exists between the number of predicted gene clusters and the actual SM compounds identified through traditional methods. This discrepancy suggests that much of the biosynthetic potential remains untapped, possibly due to the silent or cryptic nature of many gene clusters under standard laboratory conditions (Hur et al., 2023; Keller, 2019).

The advent of modern genomics has reshaped natural product research into a highly interdisciplinary field, where diverse methodologies are employed to explore various aspects of fungal SMs that range from their evolutionary development and genetic regulation to their biosynthetic pathways and ecological roles in nature. Moreover, the rapid advancement of omics technologies including genomics, transcriptomics, metabolomics, and mitogenomics alongside emerging tools like artificial intelligence (AI) and CRISPR-Cas-based genome editing is significantly enhancing the discovery and characterization of fungal SMs (Duncan et al., 2018; Zhou et al., 2022).

This chapter examines contemporary advances and future directions in fungal SM research, emphasizing the transformative role of integrative approaches combining multi-omics technologies (genomics, transcriptomics, metabolomics, and mitogenomics), artificial intelligence (AI)-driven data analysis, and synthetic biology platforms. The chapter also underscores the urgent need to explore the fungal SMs produced by neglected fungal lineages and fungus taxa from extreme or underexplored ecosystems. These studies are required to uncover novel bioactive metabolites and develop innovative biological solutions to pressing challenges in global health, food security, and environmental sustainability. Ultimately, this integrative framework provides an insight into the full metabolic potential of fungi in the modern post-genomic era.

2 Emerging Fungal Species and Novel Metabolites

Fungi are the most diverse creatures on Earth. Fungi synthesize SMs at various stages of their growth, encompassing groups such as terpenoids, alkaloids, polyketides, and non-ribosomal peptides. These compounds possess notable bioactivity, making them valuable for developing innovative pharmaceuticals, eco-friendly pesticides, and agents for environmental clean-up. In the recent years, fungal SMs have found extensive applications across the food and beverage sectors as fermentation agents, preservatives, alternative protein sources, and in dairy processing (Wadhwa et al., 2024). In the healthcare sector, it is utilized as an antibacterial, anticancer, anti-inflammatory, and immunosuppressive agent. In the recent years, scientists have found new fungus species that produce active metabolites. They are still finding new fungal species in new places, even in tough environments. Fungal organisms exhibit diverse bioactivities that contribute to plant protection, disease suppression, and various industrial applications (Giddings & Newman, 2022). The exploration of novel fungal taxa, particularly those originating from extreme or previously uncharted environments, has emerged as a significant strategy for the identification of structurally and functionally unique SMs. These compounds demonstrate considerable potential in biomedical research, particularly in the areas of antimicrobial, anticancer, antiviral, and antioxidant therapeutics (Zhang et al., 2018). This section underscores the biotechnological relevance of extremophilic fungi and reflects the growing scientific interest in fungal bioprospecting for the discovery of novel bioactive molecules.

2.1 Discovery of New Fungi in Extreme Environments

Extreme environments, including polar tundra, high-temperature zones, deep-sea ecosystems, and high-altitude regions, are increasingly recognized as promising reservoirs of novel fungal diversity. The distinctive ecological pressures in these habitats drive fungi to adapt through the production of specialized SMs essential for their survival. As a result, scientific efforts are now focused on systematically investigating such environments to uncover new fungal species and the unique bioactive compounds they synthesize.

Extremophiles, organisms that live in some of the harshest places on Earth, have caught the interest of researchers in the recent years. This includes certain types of fungi that thrive in places with extreme heat or cold, high salt levels, strong radiation, or crushing pressure. Fungi from these habitats are often discovered to produce unique metabolites that serve as defence mechanisms against environmental stress. Many of these compounds exhibit antimicrobial, anti-inflammatory, and antioxidant properties, making them valuable candidates for development in both pharmaceutical and industrial fields (Giddings & Newman, 2022).

2.1.1 Antarctic Fungi

Fungi found in Antarctica are especially fascinating because they manage to survive in one of the most extreme environments on Earth with freezing temperatures, intense UV radiation, and very little access to nutrients. Barreiro et al. (2024) showed that some *Penicillium* species from Antarctic soil produce new types of antimicrobial peptides that work well against drug-resistant bacteria. These compounds may also help protect the fungi from UV damage and oxidative stress. More recently, Cavalcante et al. (2024) discovered that Antarctic fungi create colourful pigments with antimicrobial and antiparasitic properties. In fact, compounds like Asterric acid, Violaceol, Mollicellin, Psegynamide A, Diorcinol, and Thailandolide A were found in fungi collected from Collins Glacier, and many of them were effective against Gram-positive bacteria and parasites.

2.1.2 Marine Fungi from Deep-Sea Hydrothermal Vents

Hydrothermal vent ecosystems, despite their mineral richness, present extreme environmental stressors such as elevated hydrostatic pressure and minimal light availability. Keeler et al. (2021) assessed fungal diversity within sediment samples collected from the Guaymas Basin, an underexplored deep-sea hydrothermal site with potential for biotechnological innovation. Using classical culture-dependent techniques, filamentous fungi and yeasts were isolated from 11 sediment samples. A total of 12 distinct fungal isolates were recovered and taxonomically classified within the phyla Ascomycota and Basidiomycota based on morphological and molecular analyses. Notably, seven of these isolates exhibited significant antibacterial activity against clinically relevant human pathogens, suggesting that hydrothermal vent-associated fungi may serve as a valuable source of novel bioactive metabolites with potential pharmaceutical applications.

A study by Tang et al. (2019) demonstrated the strong cytotoxic effects of two bioactive compounds produced by the deep-sea fungal strain *Penicillium citreonigrum* in cancer cell lines Bel7402 and HT1080. These compounds exhibited significant anticancer activity, highlighting their potential for cancer therapeutics. Moreover, such findings underscore the enormous untapped potential of fungi from extreme environments in the search for novel and effective medical treatments.

2.1.3 Desert Fungi

Despite the harsh conditions of deserts, many fungi have evolved critical to adapt and survive. Santiago et al. (2018) suggested that fungi found in the Atacama Desert are able to survive not only in high temperatures and drought but also in strong ultraviolet exposure. But many of these species residing in this extreme conditions are poorly characterized. Interestingly, a significant number of desert fungi are not rare but they are common one that have developed the ability to persist in such extreme environments. These fungal communities remain largely understudied and may contain previously unidentified species with potential biotechnological value. Murgia et al. (2019) further reported the economical importance of some such adaptable fungal classes such as Dothideomycetes, Pezizomycetes, and Sordariomycetes. Their research suggests that these habitat are rich in such fungus taxa that could enhance plant growth, offering promising applications in agriculture (Muria-Gonzalez et al., 2014).

2.2 Bioprospecting in Unexplored Habitats

In addition to extreme environments, the bioprospecting of unexplored habitats such as caves, volcanic regions, and rainforests continues to yield novel fungal species with unique SMs.

2.2.1 Cave Fungi

The search for novel fungal species in lesser-known environments such as caves continues to reveal fungi that produce distinct SMs. The cave ecosystems, though nutrient-limited but environmentally stable, have been found to contain fungi capable of generating bioactive compounds. A study by Nuehring (2023) sampled water from WTH Lake and Silent Lake in Wind Cave National Park, South Dakota, isolating eight pure fungal strains. These underwent chemical extraction, antimicrobial screening, and advanced techniques like NMR and HPLC. DNA sequencing of the ITS region revealed three fungal genera: Talaromyces, Linnemannia, and Gliomastix. Notably, Talaromyces was found to produce flavomannin A, an antimicrobial metabolite with potential implications for developing new antibiotics—an urgent need in the light of rising antimicrobial resistance (Zhai et al., 2016).

2.2.2 Volcanic Regions

Volcanic habitat, are rich in minerals and elements such as sulphur and copper, provide unique conditions for fungal growth. Recently, Yu et al. (2022) isolated as well as identified three novel phenalenone derivatives and four known derivatives from the Hawaiian soil fungus *Penicillium herquei FT729*. These compounds were found to exhibit anti-cancerous activity as it inhibits one of the enzyme, indoleamine 2,3-dioxygenase 1 (IDO1), which is associated with cancer progression. Several of the isolated metabolites demonstrated notable anti-cancerous activity, suggesting their potential applications in cancer immunotherapy.

2.2.3 Potential for Novel Bioactive Metabolites

Since the ground breaking discovery of penicillin by Alexander Fleming's in 1928, it has been well documented that fungal SMs have medicinal values and have the ability to treat major illness like cancer, infections, etc. They have been instrumental in producing therapeutic agents, including well-known anticancer drugs like Taxol. In the recent years, growing interest has been shown by microbiologists to discover new bioactive metabolites from a wide range of fungal species, especially those with antimicrobial, anticancer, antiviral, and antioxidant activities (Behar et al., 2024).

2.2.4 New Frontiers in Antimicrobial, Anticancer, and Antiviral Research

Antimicrobial resistance (AMR) is a raising alarmingly and mounting global health concern, driving the urgent requirement for new antimicrobial agent. This demand has sparked interest in fungal bioprospecting. It has been shown that extremophilic fungi, which thrive in harsh conditions, produce diverse range of antimicrobial compounds that may help to overcome AMR. Moreover, a study by Keeler et al. (2021) explored fungal diversity in hydrothermal sediments from the Guaymas Basin, Mexico (an area relatively unexplored in the search for novel microbial compounds). They have isolated 12 fungal strains from 11 hydrothermally affected sediment samples, which mainly belongs to the phyla Ascomycota and Basidiomycota. These strains were identified through morphological assessments and molecular techniques targeting ribosomal DNA and other genetic markers. Remarkably, 11 of these strains had previously been reported in other deep-sea environments, and seven exhibited antibacterial activity against human pathogens such as *Staphylococcus aureus* and *Escherichia coli*. This highlights the potential of hydrothermal ecosystems as untapped reservoirs of fungal biodiversity with promising biomedical applications. Further supporting this potential,

Lin et al. (2023) identified seven novel polyketides named fusarisolins F–K and fusarin I isolated from the marine-derived fungus *Fusarium solani* 8388. These compounds displayed potent antibacterial effects, particularly against methicillin-resistant *Staphylococcus aureus* (MRSA), indicating their potential as new antimicrobial agents.

Fungal SMs have long been explored for anticancer properties and new fungal strains continue to yield previously unknown bioactive molecules. One of the known SMs Fusidic acid which was first identified in 1962 from the fungus *Fusidium coccineum* is a steroidal antibiotic widely marketed under names such as Fucithalmic, Fucidin, Usidin, Fucicort, Fucibet, and Taksta. It is primarily used for the topical treatment of infections, including ocular conditions, caused by Gram-positive bacteria like *Staphylococcus aureus*, including methicillin-resistant strains (MRSA). Recently, ongoing research aims to develop fusidic acid analogues with potent anticancer properties (Huang et al., 2021). Similarly, Evidente (2024) recently reviewed advances in understanding the anticancer capabilities of fungal-derived compounds. Over the past decade, fungi from phyla including Ascomycota, Basidiomycota, and Zygomycota have emerged as important sources of such metabolites. These fungi produce a wide array of SMs such as alkaloids, terpenoids, polyketides, and peptides with strong anticancer potential. It has been reported that *Aspergillus* species generate bioactives like aspergillic acid and fumitremorgin, which have demonstrated abilities to inhibit cancer cell growth and trigger programmed cell death (apoptosis) (Kim et al., 2017). Similarly, *Ganoderma lucidum* (commonly known as Reishi mushroom) synthesizes ganoderic acids and triterpenoids with antitumor effects across various cancer models (Suárez-Arroyo et al., 2013). Meanwhile, *Phellinus linteus* is known for producing polysaccharides that enhance immune function and suppress tumour growth. These anticancer effects are often mediated through mechanisms such as induction of apoptosis, inhibition of angiogenesis, and modulation of signalling pathways like PI3K/AKT and MAPK (Ghogare & Pathan, 2025). Despite these promising findings, challenges remain in terms of bioavailability, clinical application, and large-scale production. Nonetheless, fungal metabolites continue to be a vibrant area of cancer drug discovery.

The antiviral properties of fungal metabolites have also gained momentum, particularly in response to emerging viral threats like COVID-19. Fungal SMs are often favoured over synthetic antivirals due to their potential to be effective at lower doses, their lower risk of inducing resistance, and their capacity for broad-spectrum activity (Salem et al., 2022). Moreover, fungal-derived coumarins and their derivatives have shown activity against a range of viruses, including influenza, HIV, dengue, and chikungunya. These compounds often interfere with multiple stages of the viral life cycle, offering broader mechanisms of action compared to traditional antivirals (Pal et al., 2023).

Endophytic fungi are another valuable source, producing compounds with antiviral, antibacterial, and immunomodulatory properties. Species such as *Pestalotiopsis* sp., *Schizophyllum commune*, and *Trichoderma* sp. have been demonstrated to exhibit antiviral potential (Aguilar-Pérez et al., 2020). The bioactive compounds from basidiomycete fungi, in particular, remain underexplored and offer immense potential for future antiviral discovery. Continued investigation into the bioactivity of fungal SMs is likely to yield novel therapeutic agents across a wide spectrum of diseases.

3 Omics Strategies in Research on SMs Derived from Fungi

Omics technologies have completely changed how we understand SMs, allowing for a complete and total understanding of the intricate biochemical pathways in a fungal system. Genomics, transcriptomics, proteomics, and metabolomics change the speed of discovering new SMs, especially in the context of fungi, due to their large metabolic diversity (Fig. 2). Figure 2 summarizes the application of various omics techniques (genomics, transcriptomics, proteomics, metabolomics, and integrative omics) in the characterization of secondary metabolite biosynthesis in fungi. It highlights key insights into the identification of biosynthetic gene clusters (BGCs), the regulation of secondary metabolite production, and the discovery of novel bioactive compounds. The integration of multi-omics data provides a comprehensive understanding of metabolic networks and the environmental factors influencing fungal secondary metabolite synthesis. This section discusses the role of these omics approaches, including the advances in next-generation sequencing (NGS) technology, as well as applications being developed to integrate these approaches for a comprehensive and complete metabolite profile (Table 1).

3.1 Role of Genomics, Transcriptomics, and Proteomics

Omics approaches will certainly be integral in the discovery of SMs through a pathway mapping approach, using genomics and transcriptomics to understand the genetics involved in the biosynthesis of these bioactive products, and using proteomics to understand the protein network in biosynthesis. This combination of these three high-throughput technologies will allow greater rates of identification and prediction of new fungal metabolites (Palazzotto & Weber, 2018).

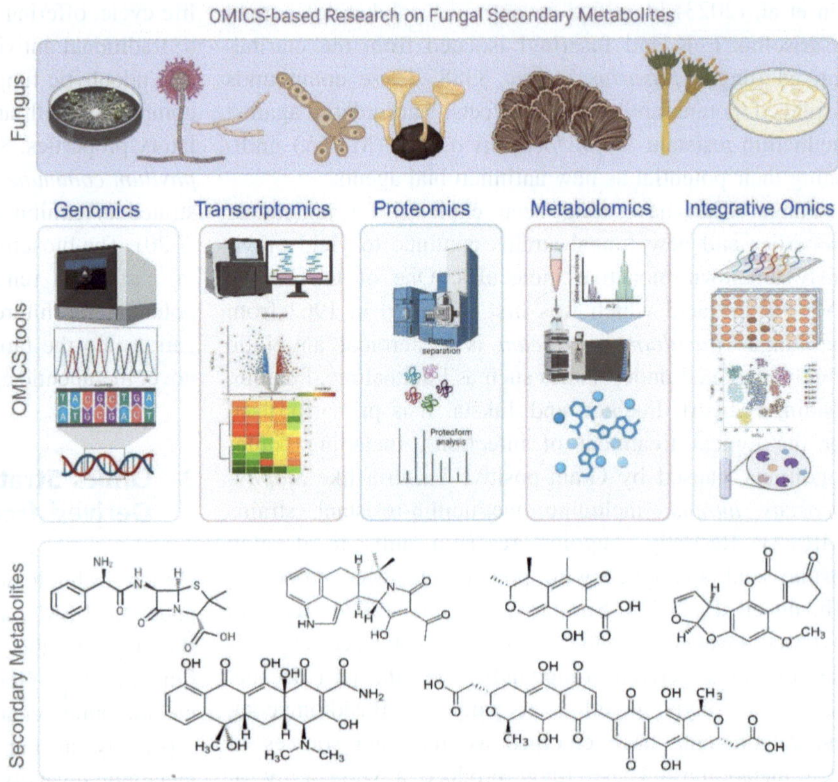

Fig. 2 Application of genomics, transcriptomics, proteomics, metabolomics, and integrative omics in study of fungal SMs. *Source* Authors

Table 1 'Omics' approaches in fungal-derived secondary metabolite research

Omics approach	Key insights and contributions	Examples of SMs	References
Genomics	Next-generation sequencing (NGS) facilitates the identification of biosynthetic gene clusters (BGCs) and novel SMs in fungi	Penicillin (*Penicillium chrysogenum*), Aspergillomarasmine A (*Aspergillus* spp.)	Drott et al. (2020)
	Mapping BGCs to corresponding metabolites enhances pathway elucidation		Cacho et al. (2015)
Transcriptomics	RNA-Seq identifies genes regulating secondary metabolite biosynthesis under various environmental conditions	Penicillin (*Penicillium chrysogenum*), Cyclopiazonic acid (*Aspergillus flavus*)	Jiang et al. (2022)
	Expression profiles provide insights into the regulation of metabolic pathways		Nielsen et al. (2019)
Proteomics	Mass spectrometry enables the identification of enzymes involved in the biosynthesis of SMs	Aflatoxin (*Aspergillus flavus*), Citrinin (*Penicillium citrinum*)	Yang et al. (2021)
	Proteomic analysis uncovers regulatory mechanisms and enzymatic functions in metabolic pathways		Li et al. (2017)
Metabolomics	Metabolomic profiling links BGCs with specific SMs, enhancing pathway elucidation	Antifungal metabolites (*Gaeumannomyces tritici*), Citrinin (*Penicillium citrinum*)	Geremia et al. (2024)
	Provides insights into the environmental influence on secondary metabolite production		Li et al. (2017)
Integrative omics	Integrating genomics, transcriptomics, proteomics, and metabolomics offers a comprehensive understanding of secondary metabolite biosynthesis	Actinorodins (*Streptomyces* spp.), Penicillin acid (*Penicillium* spp.)	Ramarajan et al. (2024)
	Predicts novel metabolites and enhances comprehension of regulatory networks under various conditions		Hautbergue et al. (2018b)
Metabolite discovery	Combining genomics and metabolomics facilitates the prediction of biosynthetic pathways and the discovery of novel metabolites. These integrated approaches uncover new biosynthetic routes for SMs	Antifungal metabolites (*Aspergillus fumigatus*), polyketides (*Streptomyces* spp.)	Gopalakrishnan Meena et al. (2023)

Source Authors compilation

Advanced high-throughput sequencing methods such as next-generation sequencing (NGS) technologies including Illumina, PacBio, and Oxford Nanopore, have significantly contributed to our understanding of the genomic architecture of fungi. By sequencing their entire genomes, researchers can explore the genetic composition of these organisms, uncover their capacity to synthesize SMs, and examine their molecular adaptations to symbiosis with host plants. Comparative genomic analyses across various endophytic species allow for the identification of shared genes, species-specific traits, and candidate genes involved in pathogenicity or mutualistic interactions. Next-generation sequencing (NGS) has changed the field of genomics by providing rapid and high-throughput sequencing of genomes, including those of fungi. In the area of SMs, NGS allows the identification of biosynthetic gene clusters (BGCs), which are groups of genes associated with secondary metabolite production (Kumar et al., 2018). During the initial phases of fungal genomics, model organisms like *Aspergillus nidulans* and *Saccharomyces cerevisiae* were pivotal in identifying biosynthetic gene clusters (BGCs) responsible for secondary metabolite (SM) production (Sagita, 2021). These findings facilitated the discovery of novel and, in some cases, previously uncharacterized SMs. Moreover, Drott et al. (2020) explored the genetic basis of SM biosynthesis in *Aspergillus*, shedding light on regulatory and biosynthetic pathways underlying metabolite diversity.

Traditionally, research on *A. nidulans* has been based on a single reference genome; however, emerging studies emphasize the significant intra-species variation in biosynthetic gene clusters (BGCs) and secondary metabolite (SM) profiles. The identification of six previously uncharacterized BGCs in the study by Drott et al. (2020) underscores this genomic and metabolic diversity within *A. nidulans*. The advent of next-generation sequencing (NGS) technologies has greatly accelerated the pace and precision of fungal genome sequencing. Park et al. performed whole-genome sequencing (WGS) of the strain, resulting in a 33.67 Mb assembled genome. This study demonstrates the effectiveness of WGS as a robust tool for identifying biosynthetic genes linked to the production of pharmacologically valuable SMs. For instance, Cacho et al. (2015) illustrated how NGS facilitated the discovery of new SMs through the localization of their associated BGCs in fungal genomes. This technological advancement has dramatically enhanced our ability to detect novel BGCs, including those responsible for producing bioactive compounds like penicillin and other antimicrobials in *Penicillium* species, thereby deepening our understanding of their biosynthetic pathways and supporting the development of innovative therapeutic agents.

NGS has also aided the discovery of novel BGCs for the direct identification of *Aspergillus* species that are linked to the production of mycotoxins and other SMs through NGS, which could assist in limiting toxic metabolites produced by harmful fungal species as well as encourage the production of beneficial metabolites. Additionally, NGS has discovered mycotoxin-encoding BGCs in *Fusarium* species responsible for producing SMs, including fumonisins. Similarly, Whole-Genome Sequencing of the Fungus *Penicillium citrinum* identified 29 biosynthetic gene clusters (BGCs) including 9 clusters encoding type I polyketide synthases (T1PKS), 3 clusters associated with non-ribosomal peptide synthetases (NRPS), 2 clusters representing PKS-NRPS hybrid clusters, and 1 clusters indole-NRPS hybrid cluster. Additionally, 2 clusters were linked to fatty acid biosynthesis, 1 cluster to terpene production, while 11 clusters are remained unclassified or not further characterized. Mapping the above genes allows for the characterization of regulation, safety, and therapeutic potential (Kawahara et al., 2012; Lin et al., 2023). These examples illustrate that NGS not only enhances our current understanding of previously explored metabolic pathways, but it also leads to novel metabolites and biosynthetic pathways that influence fungal ecology and biotechnology. On other hand, transcriptomic analyses underscore the diverse potential applications of fungi in various field. Transcriptomics primarily aims to investigate gene expression at the transcript level, focusing on the identification and quantification of various RNA types, including messenger RNA (mRNA), non-coding RNAs (ncRNAs), and small RNAs. Among the key tools in this field is RNA sequencing (RNA-Seq), which enables comprehensive profiling of the fungal transcriptome. This approach offers insights into gene expression dynamics, alternative splicing events, and the discovery of previously unannotated transcripts across varying conditions, particularly during host interactions. In addition, microarray technology serves as another valuable platform, allowing simultaneous assessment of expression levels for thousands of genes. Through differential gene expression analysis, transcriptomics reveals genes that are either upregulated or downregulated in response to specific stimuli or environmental changes, thereby enhancing our understanding of functional genomics in different biological contexts. Recently, Van Goethem et al. (2021) demonstrated that metagenomic sequencing of soil environments enables the identification of previously unrecognized fungal species along with their associated biosynthetic gene clusters (BGCs), thereby facilitating the discovery of novel SMs. This can potentially provide avenues for scouting new metabolites, without the need to undergo culturing of individuals. In the study of SMs, transcriptomics can go further to identify genes and identify how the genes' expression may correlate with the production of SMs. Transcriptomic studies have shown that *Aspergillus oryzae* adapts to temperature-induced stress by modulating the expression of genes involved in sugar and lipid metabolic pathways. Under thermal stress conditions, the fungus adjusts its carbohydrate metabolism to maintain energy balance and osmotic stability, while also altering lipid metabolism to preserve membrane integrity and fluidity. These molecular responses reflect a broader stress adaptation strategy that enables *A. oryzae* to survive and maintain functionality under fluctuating environmental temperatures (Jiang et al., 2022). While this research increased knowledge about the mechanisms that *A. oryzae* has to cope with temperature stress, and can also guide future genetic modifications to select for improved temperature tolerance. In addition, Nielsen et al. (2019) mentioned that transcriptomics is important to gain an understanding of the gene expression of different species of *Penicillium* in different growth conditions. By studying the transcriptomes, the researchers were able to identify gene expression levels that describe the metabolic processes involved in secondary metabolite production during a stationary phase of growth. The data were used to create coexpression networks that identified gene modules with similar expression patterns, which uncovered linkages between primary carbon metabolism and the biosynthesis of SMs. This study facilitated the identification of a metabolic signature that connected the precursor-generating pathways, i.e. fatty acids and branched-chain amino acid (BCAA) degradation, with the SMs. Furthermore, transcripts provided the ability to correlate the secondary metabolite backbone genes (e.g. PKS and NRPS) with the pathway-like precursor-generating pathways. Together, the incorporation of transcriptomics is an invaluable methodology for mapping out the multifaceted connections and interactions that exist between their metabolic pathways and help to understand the tailored functions of fungal metabolism to make energetically efficient secondary metabolite production, which could ideally generate ideas for metabolic engineering pathways and improve production in industry.

RNA-Seq analysis enabled the identification of secondary metabolite biosynthetic gene clusters (BGCs) that were transcriptionally active under infection-mimicking conditions. By measuring gene expression levels, the study identified several upregulated clusters particularly *GtPKS1* suggesting their potential involvement in the pathogenic process (Studt et al., 2012). This approach offered a comprehensive view of the transcriptional landscape related to secondary metabolism and facilitated the linkage of specific BGCs, like *GtPKS1*, to the biosynthesis of bioactive compounds such as dichlorodiaporthin, known for its antifungal and cytotoxic effects. Overall, RNA sequencing served as a foundational method for detecting and confirming the role of key genes involved in secondary metabolite production during host-pathogen interactions.

Proteomics refers to the comprehensive analysis of the entire protein complement expressed by a cell, tissue, or organism at a given time. In secondary metabolite research, proteomics allows the identification of key enzymes that are responsible for the biosynthesis of metabolites. A commonly used method in proteomics involves combining two-dimensional gel electrophoresis with mass spectrometry. Currently, mass spectrometry-based proteomics stands out as a powerful strategy for profiling the complete proteome of fungal endophytes, offering valuable insights into their roles in host-plant interactions, associated signalling pathways, and post-translational modifications but sometime it has certain limitation while analysing complex data. To address these difficulties, mass spectrometry (MS) is frequently paired with chromatographic techniques such as high-performance liquid chromatography (HPLC), gas chromatography (GC), or ultraviolet (UV) spectroscopy.

Moreover, Yuan et al. (2019) conducted a proteomic study on *Atractylodes lancea* inoculated with the endophytic fungus *Gilmaniella* sp. AL12. Their findings revealed an upregulation of proteins linked to carbon fixation, as well as carbohydrate and energy metabolism, suggesting that the endophyte may contribute to enhanced plant biomass production. Research has provided valuable understanding of the SMs synthesized by *Aspergillus fumigatus* and their involvement in disease development (Keller, 2017). Scientists have uncovered a variety of biologically active molecules such as gliotoxin, fumagillins, and pseurotins produced by *A. fumigatus* that play key roles in promoting virulence and weakening the host immune response. These substances enable *A. fumigatus* to thrive in different host environments, including low-oxygen areas like the lungs. Some metabolites, like hydroxy-(sulfooxy) benzoic acid, might also exhibit anti-inflammatory effects. Furthermore, a newly identified tetrapeptide called aspergitide, unique to *Aspergillus* species, shows promise as a diagnostic marker for aspergillosis. The wide range of SMs produced by *A. fumigatus* not only supports its adaptation to environmental challenges but also underpins its effectiveness as an opportunistic pathogen, emphasizing their relevance in both ecological and medical settings (Kim et al., 2017).

Additionally, the proteomic analysis using mass spectrometry of *Piriformospora indica* revealed the presence of proteins involved in key physiological processes such as reactive oxygen species (ROS) defence, photosynthesis, and energy transport under drought stress conditions (Nanjareddy et al., 2017). However, the limited scope of insights gained from proteomics alone reflects the intricate nature of plant-endophyte interactions. Therefore, integrating proteomics with other omics approaches offers a more comprehensive understanding of these complex biological systems.

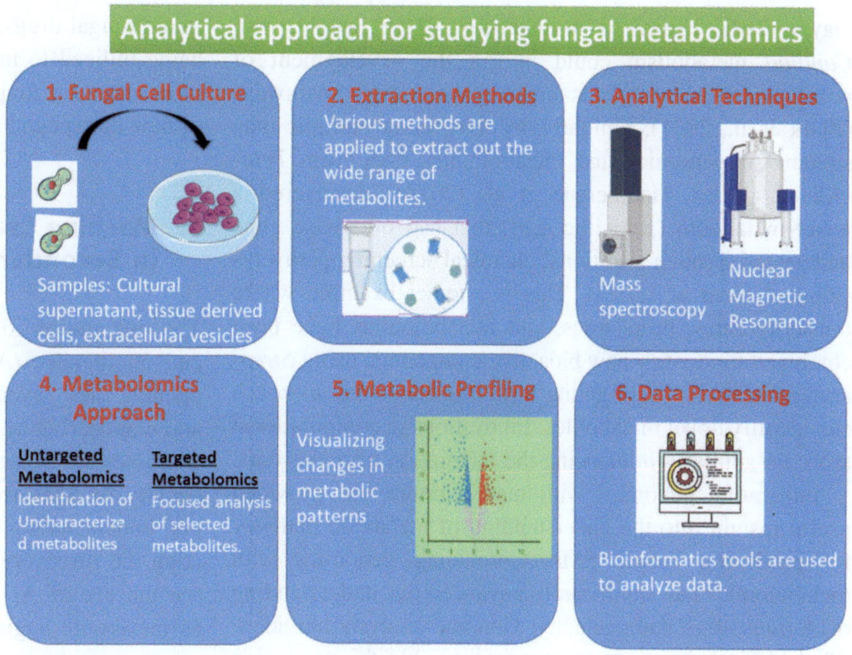

Fig. 3 Analytical workflow for fungal metabolomics, encompassing fungal cell culture (1), metabolite extraction (2), characterization through advanced analytical platforms (3), targeted or untargeted metabolomics approaches (4), metabolic profiling (5), bioinformatics-driven data processing, and (6) to elucidate metabolic patterns and functions. *Source* Authors

3.2 Role of Metabolomics

Metabolomics is an emerging field of omics technology that focus on detection and measurement of metabolites present in biological systems, as shown in Fig. 3. It includes uses of wide range of advanced technologies, particularly mass spectrometry (MS) and nuclear magnetic resonance (NMR). The recent progress in this area has been propelled by improvements in chromatography, cutting-edge instrumentation, and powerful bioinformatics tools. These developments have made it possible to identify and analyse hundreds of metabolites within a single analytical run (Kaneria et al., 2017; Kumar et al., 2018). Metabolomics plays a crucial role in exploring fungal SMs, particularly in fungi that thrive in extreme environments such as high salinity, temperature, or pH conditions. These stress-adapted fungi often produce unique bioactive compounds as part of their survival strategy. By profiling the complete set of metabolites under such harsh conditions, metabolomics enables researchers to uncover novel SMs with potential ecological or pharmaceutical relevance. Recently, *Aspergillus sydowii*, isolated from hypersaline environments, produces SMs like sydowinin A and B, which exhibit antimicrobial activity (Gostinčar et al., 2011). Similarly, *Penicillium* species from Arctic soils have been found to generate compounds with potential antifreeze or cryoprotective functions (Toscan et al., 2020). Through metabolomics profiling, researchers can analyse the effect of environmental stress on specific biosynthetic pathways, leading to the production of such novel metabolites. This not only enhances our understanding of fungal adaptation and resilience but also opens up possibilities for discovering bioactive compounds with pharmaceutical, agricultural, or industrial applications.

4 Integrative Omics for Comprehensive Metabolite Profiling

The combination of different omics techniques genomics, transcriptomics, proteomics, and metabolomics, has proven to be an effective way to provide an overall view of secondary metabolite biosynthesis. This can provide the means to integrate and have a detailed action of a metabolome pathway with characterization of potential metabolites and predict how the metabolite is biosynthesized under different conditions. The recent research has investigated the metabolic characteristics of different *Candida* species, such as *C. albicans*, *C. auris*, and *C. glabrata*, using gas chromatography-mass spectrometry (GC-MS). These studies have uncovered a wide range of metabolites linked to pathogenicity traits, including biofilm development and resistance to antifungal drugs. *C. auris* produces phenylethyl alcohol, a compound that suppresses hyphal growth and is also found in other *Candida* species. In *C. albicans*, metabolomics analysis of biofilm cells showed distinct metabolic profiles compared to free-floating (planktonic) cells, with notable differences in the TCA cycle, lipid metabolism, and amino acid biosynthesis. Moreover, azole-resistant *C. albicans* strains displayed changes in their amino acid pathways, TCA cycle activity, and phospholipid content, indicating a metabolic adaptation that may underlie resistance mechanisms. These insights into *Candida* metabolism could inform the development of novel antifungal treatments targeting these altered pathways (Zhou et al., 2021). Similarly, by integrating multiple data streams of genomics and metabolomics data from *Penicillium* species, Hautbergue et al. (2018a) discovered a novel metabolite (penicillic acid) that had otherwise gone undetected through traditional chemical screening pathways (Ramarajan et al., 2024). This is another example where transcriptomics, proteomics, and metabolomics have been combined to identify new bioactive compounds using *Streptomyces* species, highlighting how this integrated approach can identify novel metabolites. Innovations in computational tools are similarly influencing the integration of omics data.

Integrating genomics with metabolomics has provided useful insight into the understanding of metabolic pathway for producing fungal SMs. Combining genomics with metabolomics has significantly advanced our understanding of *A. fumigatus* pathogenicity. Genome analysis has identified biosynthetic gene clusters responsible for producing SMs like gliotoxin and fumitremorgin, both of which are linked to fungal virulence, especially in individuals with weakened immune systems (Keller, 2017).

Similarly, integrating transcriptomic and metabolomics approaches has been instrumental in uncovering about adaptation by *C. albicans* during infection. Changes in metabolism play a vital role in influencing its virulence, stress tolerance, and sensitivity to antifungal agents. Advanced techniques such as RNA-Seq and liquid chromatography-mass spectrometry (LC-MS) have been employed to study the transcriptional and metabolic shifts across diverse *C. albicans* strains, revealing intricate regulatory mechanisms in response to environmental pressures (Brown et al., 2020). Additionally, microarray analyses have helped identify genes linked to infection and provided comprehensive transcriptomic profiles under various conditions, highlighting pathways associated with biofilm development and resistance to antifungal drugs.

Moreover, combining transcriptomic and metabolomics approaches has been instrumental in uncovering the adaptation mechanism of *C. albicans* during infection. Changes in metabolism play a vital role in influencing its virulence, stress tolerance, and sensitivity to antifungal agents. Advanced techniques such as RNA-Seq and liquid chromatography-mass spectrometry (LC-MS) have been employed to study the transcriptional and metabolic shifts across diverse *C. albicans* strains, revealing intricate regulatory mechanisms in response to environmental pressures (Liu et al., 2021; Tebbji et al., 2020), Additionally, microarray analyses have helped identify genes linked to infection and provided comprehensive transcriptomic profiles under various conditions, highlighting pathways associated with biofilm development and resistance to antifungal drugs. Multi-omics strategies are increasingly being utilized to investigate antifungal resistance, discover biomarkers for fungal diseases, and aid in the development of new therapeutic drugs.

5 Machine Learning Approaches in Secondary Metabolite Research

The integration of machine learning (ML) and artificial intelligence (AI) in secondary metabolite research is opening new avenues for the prediction, identification, and engineering of bioactive compounds. Machine learning processes are becoming more frequently used to predict the activity of SMs from genomic data, therefore improving our understanding of biosynthetic pathways and accelerating the discovery of new metabolites. In future, we will see the use of AI with respect to drug discovery, as well as metabolite engineering, eventually allowing us to optimize compounds for therapeutics or industrial applications. This section details the current and future uses of machine learning and AI with respect to economic research, specifically fungal SMs (Table 2).

5.1 Predicting Secondary Metabolite Activity from Genomic Data

Machine learning is a powerful tool for predicting activity and function for SMs using genomic data. By examining datasets collectively, ML models can predict which biosynthetic gene clusters (BGCs) are derived and which metabolites will be produced by BGCs. This could provide a huge advantage by saving time and resources that are required by traditional methods for metabolite identification.

5.2 Role of Machine Learning in Identifying Biosynthetic Gene Clusters

Biosynthetic gene clusters (BGCs) produce SMs in fungi, and identification of these clusters is important for elucidating the biosynthesis of bioactive compounds. Machine learning has become a key component for the annotation and the prediction of the BGCs, which can be a cryptic feature of the microbial genome and often undetectable using traditional sequencing. For instance, the recent papers have demonstrated how ML algorithms can predict BGCs from fungal genomes. Microorganisms produce bioactive compounds through secondary metabolism, which have applications in medicine and agriculture, such as antimicrobial, anticancer and other bioactivities. Genome mining

Table 2 Machine learning approaches in secondary metabolite research

Research focus	ML/AI approach	Methodology	Examples and applications	Key references
Biosynthetic gene cluster (BGC) prediction	Supervised learning (random forest, SVM, XGBoost)	– Feature extraction from genomic sequences – Classification of biosynthetic genes	– antiSMASH 7.0: improved BGC classification for bacteria and fungi – DeepBGC: uses CNNs to predict cryptic BGCs	Blin et al. (2024) and Ologun et al. (2019)
Metabolite structure prediction	Deep learning (graph neural networks, transformers)	– SMILES-based molecular generation – Structure-activity relationship (SAR) modelling	– MolBERT: transformer-based model for predicting bioactivity – GROVER: self-supervised GNNs for compound property prediction	Li and Jiang (2021)
Activity prediction of SMs	Ensemble learning (random forest, gradient boosting)	– QSAR models for predicting antimicrobial, anticancer, and antiviral activity	– Bioactivity databases (ChEMBL, NPASS) integrated with AI to identify promising drug candidates	Paul et al. (2021)
Natural product virtual screening	AI-driven docking and molecular dynamics simulations	– Predicts binding affinity of metabolites to target proteins	– DeepDocking: AI-assisted docking for drug discovery – AlphaFold-Multimer: protein-ligand binding prediction	Scardino et al. (2023)
AI for drug repurposing using SMs	Transfer learning (pre-trained deep learning models)	– Reusing existing chemical knowledge for new bioactive molecule discovery	– DeepChem: AI framework for small molecule prediction – CANDLE: uses deep learning for personalized drug discovery	
Metabolic pathway prediction in fungi	Reinforcement learning and Bayesian networks	– Predicting metabolic flux and biosynthetic pathways	– PathPred: AI-based prediction of novel biosynthetic pathways – RetroPath2.0: automated retrosynthesis pathway generation	Delépine et al. (2018)
Fungal metabolite optimization for bioprocessing	AI-driven metabolic engineering	– Flux Balance Analysis (FBA) with AI for yield optimization	– OptKnock and CobraPy: ML-based strain optimization for enhanced metabolite production	Lee et al. (2020) and Ebrahim et al. (2013)
Integration of multi-omics data for secondary metabolite discovery	Multi-omics AI (deep learning + Bayesian models)	– Integrating genomics, transcriptomics, proteomics, and metabolomics for drug discovery	– OmicsIntegrator: AI-driven multi-omics analysis for fungal metabolites	Falster et al. (2021)

Source Authors compilation

has proven an effective bioactive secondary metabolite discovery approach. 'antiSMASH' was released in 2011 and has been widely used in the detection and characterization of biosynthetic gene clusters (BGCs) in microbes. The latest version, antiSMASH 7, builds on the previous version, adding additional features such as improved chemical structure prediction, visualization of enzymatic assembly lines and gene cluster regulation. It also now supports 81 cluster types from the previous 71 types of clusters (Blin et al., 2023). Riedling et al. (2024) used deep learning algorithms to predict that the genome of *Fusarium* species had cryptic BGCs, which discovered novel antifungal (cerulenin and hinokitiol) and anticancer metabolites. These advanced algorithms are capable of detecting BGCs that traditional genomic tools may overlook, offering an efficient method for discovering new SMs.

5.3 Computational Models for Metabolite Prediction

After BGCs are identified, the following task is to predict what metabolites they generate. ML models can support this effort by examining the genomic sequence BGCs and associating them with the previously known biosynthetic pathways. Thus, by utilizing databases of SMs and with some unknowns attached to a BGC, models can predict the chemical structure of unknown compounds. Sonah et al. (2016) advocated for the use of computational models to identify fungal effector proteins, and emphasized that the recent advances now allow for increased efficiency of computational prediction, particularly for the characterization of Avr genes, increasing the accuracy of effector protein identification. The review discusses the potential of several tools and pipelines that employ various computational methods to conduct a genome-wide identification of effector proteins. The screening approaches considered include conserved motifs, signature sequences, and structural features in proteins, providing a computational framework to identify potential effectors.

5.4 AI in Drug Discovery and Metabolite Engineering

Artificial intelligence (AI) and machine learning have also radically altered the process of drug discovery and metabolite engineering to optimize bioactive metabolites for pharmaceutical and industrial use. AI is speeding up identifying drug candidates and optimizing fungal metabolites for effective/safer/improved yield, all while utilizing vast data sets and computational power.

5.5 Application of AI Tools for Drug Optimization

AI works well to optimize SMs for their drug use. By studying the structure-activity relationships (SAR) of metabolites, AI can identify what structural modifications to a compound will enhance its bioactivity without increasing toxicity or improving pharmacokinetic characteristics. Advances in computing, data storage, natural language processing, and machine learning (ML) are completely changing the way we look at natural products, including medicinal products, and the first change is that the volume of research in the field is growing. Since 2010, there have been more the 600,000 scientific publications related to natural products release. AI is ultimately a major player in this area, especially with the identification, classification, and activity prediction of natural products. AI and ML algorithms will aid in the analysis of bioactive compounds found in plants and fungi, like alkaloids and flavonoids, which have health-promoting properties, including antiviral, anticancer, and therapeutic uses.

These technologies can expedite the drug discovery process (bioactivity predictions), ways to extract/drugs combined with these bioactive molecules, and give a new use for natural products, especially in mushrooms and other fungal species. AI would also be instrumental in metabolic engineering, whereby compounds and the production of SMs can be optimized. Wainaina and Taherzadeh (2023) suggested that automation and artificial intelligence (AI) have significant opportunities for fungi-based bioprocesses, in relation to the circular economy and bio-manufacturing. Filamentous fungi offer many applications, due to their metabolic versatility; their integration with automation and AI has the potential to improve process flexibility, data reliability, and productivity. Automation involves replacing human operator duties with mechanized tools, whereas AI allows machines to perform functions that require human intelligence.

However, the study highlights a gap in research regarding the use of these technologies in fungi-based food production, suggesting it is an area for further exploration. The potential for AI in drug discovery and metabolite engineering is vast, and future research will likely focus on the development of more sophisticated models that integrate multiple omics datasets (genomic, transcriptomic, proteomic, and metabolomics) for comprehensive drug optimization. Additionally, AI could play a crucial role in predicting the pharmacological profiles of fungal metabolites before they are synthesized, reducing the need for lengthy and costly experimental validation. AI is already being explored for multi-target drug discovery, where AI models predict the ability of SMs to interact with multiple disease pathways.

6 Challenges and Future Prospects in Fungal Secondary Metabolite Research

The increased attention to fungal SMs due to their many potential uses in pharmaceuticals, agriculture, and industry has raised some challenges involving sustainability, conservation, and ethics. Although the study of fungal SMs has advanced considerably but several core challenges persist. A major difficulty lies in the complex and highly variable nature of fungal metabolomes, which can vary significantly between species, strains, environmental settings, and host organisms. This high degree of variability complicates the development of standardized methods for detecting and quantifying metabolites, thereby affecting the consistency and comparability of research outcomes. Additionally, the vast chemical diversity of fungal SMs, which play crucial roles in virulence, host-pathogen interactions, and resistance to antifungal treatments, makes accurate identification and analysis difficult (Bennett & Klich, 2003; Keller, 2019). These compounds are often produced in minute amounts or closely resemble other molecules, limiting their detectability with current analytical technologies. These concerns are particularly acute in the context of bioprospecting, where fungi are collected for bioactive metabolites from their native/ecological environments.

A significant limitation in this area is the lack of well-developed, curated databases dedicated to fungal metabolites, particularly those linked to pathogenic traits. Unlike the extensive resources available for model organisms, databases for fungal species are still underdeveloped. This gap hinders a comprehensive understanding of the roles specific metabolites play in virulence, antimicrobial resistance, and biofilm development. Additionally, translating metabolomics findings into clinically actionable insights, such as identifying therapeutic targets or diagnostic biomarkers, remains a considerable challenge.

To overcome these hurdles, future studies must prioritize the refinement of analytical methodologies capable of detecting the broad spectrum of fungal metabolites, particularly those implicated in pathogenic mechanisms. Enhancements in high-resolution mass spectrometry and the integration of multi-omics strategies offer promising solutions. Equally important is the establishment of expansive, publicly accessible metabolomics databases tailored to specific fungi species, which will improve data consistency, cross-study comparisons, and real-world applicability. These developments are essential to bridge the current gap between foundational fungal SM research and clinical implementation, paving the way to find new therapeutic interventions for deadly diseases.

7 Conclusion

Fungal metabolomics is a rising field in the interdisciplinary bioscience, at the interface of mycology, analytical chemistry, and computational systems biology. This chapter has described the analysis framework and modern metabolomics tools employed to study fungal secondary metabolism that have transformed the possibility of deciphering fungal metabolic sophistication. In addition, these combined methods not only showed the biochemical complexity that regulates the formation of fungal secondary metabolites, ecological adjustment, but also stress on the industrial use of fungal secondary metabolites.

With the help of the modern omics approaches, nowadays microbiologists can encompass the entire second metabolic fingerprint of fungi in various environmental conditions. Combining both sophisticated analytical tools with data analysis based on bioinformatics has enabled both the rapid and massively identified and characterized metabolites as well as offered understanding of regulatory networks and metabolic pathways that regulate fungal responses to stress, host interactions, and developmental transitions.

Fungal metabolomics is not only important in basic research. Fungal secondary metabolites still hold a promising potential source of pharmacologically active products, such as antibiotics, antitumor agents, and immunomodulators. They are also significant in agricultural, food, textile among other sectors. In addition, analyses based on metabolomics can make contributions to the comprehension of pathogenicity mechanisms, symbiotic relationships, and ecological resilience, which is relevant to construct sustainable strategies to promote health, food security, and environmental biotechnology. Fungal metabolomics is a discovery tool, as well as a platform of translation between laboratory discovery and practical innovation.

In addition, combining metabolomics and genomics, transcriptomics, proteomics, and artificial intelligence will speed up the discovery of metabolites and their functional annotation. These integrative multi-omics and computational systems have a gigantic potential to explain cryptic biosynthetic gene clusters and design metabolic pathways, as well as predicting the synthesis of bioactive molecules with a precision never previously witnessed. Further development of analytical resolution and machine learning algorithms and data harmonization will be able to increase the predictive ability and reproducibility of fungal metabolomics.

To conclude, fungal metabolomics has developed into an invaluable staple of the contemporary bioscience, irreversibly changing how we study and capitalize on fungal biochemical variations. It blends precision of analysis and calculation intelligence with the possibilities of the future that will not only enlighten our knowledge about fungal biology but will also broaden natural product discovery, synthetic biology, and global sustainability.

References

Aguilar-Pérez, M. M., Torres-Mendoza, D., Vásquez, R., Rios, N., & Cubilla-Rios, L. (2020). Exploring the antibacterial activity of *Pestalotiopsis* spp. under different culture conditions and their chemical diversity, using LC-ESI-Q-TOF-MS. *Journal of Fungi (Basel, Switzerland), 6*(3), 140. https://doi.org/10.3390/jof6030140

Aly, A. H., Debbab, A., & Proksch, P. (2011). Fifty years of drug discovery from fungi. *Fungal Diversity, 50*(1), 3–19. https://doi.org/10.1007/s13225-011-0116-y

Barreiro, C., Albillos, S. M., & García-Estrada, C. (2024). *Penicillium chrysogenum*: Beyond the penicillin. In S. Sariaslani & G. M. Gadd (Eds.), *Advances in applied microbiology* (Vol. 123, pp. 143–221). Elsevier. https://doi.org/10.1016/bs.aambs.2024.01.004

Behar, N., Sharma, S., Parihar, R., Dubey, S. K., Mehta, S., & Pandey, V. (2024). Role of fungal metabolites in pharmaceuticals, human health, and agriculture. In *Fungal biotechnology* (pp. 69–93). Elsevier. https://doi.org/10.1016/b978-0-323-95241-5.00005-8

Bennett, J. W., & Klich, M. (2003). Mycotoxins. *Clinical Microbiology Reviews, 16*(3), 497–516. https://doi.org/10.1128/cmr.16.3.497-516.2003

Blin, K., Shaw, S., Medema, M. H., & Weber, T. (2024). The antiSMASH database version 4: Additional genomes and BGCs, new sequence-based searches and more. *Nucleic Acids Research, 52*(D1), D586–D589. https://doi.org/10.1093/nar/gkad984

Brown, A. J. P., Brown, G. D., Netea, M. G., & Gow, N. A. R. (2020). Metabolism impacts upon *Candida* immunogenicity and pathogenicity at multiple levels. *Trends in Microbiology, 28*(9), 795–807. https://doi.org/10.1016/j.tim.2020.03.003

Cacho, R. A., Tang, Y., & Chooi, Y.-H. (2015). Next-generation sequencing approach for connecting secondary metabolites to biosynthetic gene clusters in fungi. *Frontiers in Microbiology, 5.* https://doi.org/10.3389/fmicb.2014.00774

Cavalcante, S. B., da Silva, A. F., Pradi, L., Lacerda, J. W. F., Tizziani, T., Sandjo, L. P., Modesto, L. R., de Freitas, A. C. O., Steindel, M., Stoco, P. H., Duarte, R. T. D., & Robl, D. (2024). Antarctic fungi produce pigment with antimicrobial and antiparasitic activities. *Brazilian Journal of Microbiology, 55*(2), 1251–1263. https://doi.org/10.1007/s42770-024-01308-y

Conrado, R., Gomes, T. C., Roque, G. S. C., & de Souza, A. O. (2022). Overview of bioactive fungal secondary metabolites: Cytotoxic and antimicrobial compounds. *Antibiotics, 11*(11), 1604. https://doi.org/10.3390/antibiotics11111604

Delépine, B., Duigou, T., Carbonell, P., & Faulon, J. L. (2018). RetroPath2.0: A retrosynthesis workflow for metabolic engineers. *Metabolic Engineering, 45*, 158–170. https://doi.org/10.1016/j.ymben.2017.12.002

Drott, M. T., Bastos, R. W., Rokas, A., Ries, L. N. A., Gabaldón, T., Goldman, G. H., & Keller, N. P. (2020). Diversity of secondary metabolism in *Aspergillus nidulans* clinical isolates. *mSphere, 5*(3), e00156–20. https://doi.org/10.1128/mSphere.00156-20

Duncan, K. R., Hauptmann, M., et al. (2018). From genomics to metabolomics: Moving toward an integrated strategy for the discovery of fungal secondary metabolites. *Natural Product Reports, 35*(2), 147–173. https://doi.org/10.1039/C7NP00032D

Ebrahim, A., Lerman, J. A., Palsson, B. Ø., & Hyduke, D. R. (2013). COBRApy: Constraints-based reconstruction and analysis for Python. *BMC Systems Biology, 7*, 74. https://doi.org/10.1186/1752-0509-7-74

Evidente, A. (2024). Advances on anticancer fungal metabolites: Sources, chemical and biological activities in the last decade (2012–2023). *Natural Products and Bioprospecting, 14*(1), 4. https://doi.org/10.1007/s13659-024-00452-0

Falster, D., Gallagher, R., Wenk, E. H., Wright, I. J., Indiarto, D., Andrew, S. C., et al. (2021). AusTraits, a curated plant trait database for the Australian flora. *Scientific Data, 8*(1), 254. https://doi.org/10.1038/s41597-021-01006-6

Geremia, F., Paim, I., da Silva Camargo, M., Schrank, A., & Sbaraini, N. (2024). Secondary metabolite gene clusters from the phytopathogenic fungus *Gaeumannomyces tritici*. *Journal of Plant Pathology: An International Journal of the Italian Phytopathological Society, 106*(2), 657–669. https://doi.org/10.1007/s42161-024-01605-3

Ghogare, S. S., & Pathan, E. K. (2025). Intratumor fungi specific mechanisms to influence cell death pathways and trigger tumor cell apoptosis. *Cell Death Discovery, 11*(1), 188. https://doi.org/10.1038/s41420-025-02483-z

Giddings, L.-A., & Newman, D. J. (2022). Extremophilic fungi from marine environments: Underexplored sources of antitumor, anti-infective and other biologically active agents. *Marine Drugs, 20*(1), 62. https://doi.org/10.3390/md20010062

Gopalakrishnan Meena, M., Lane, M. J., Tannous, J., Carrell, A. A., Abraham, P. E., Giannone, R. J., Ané, J.-M., Keller, N. P., Labbé, J. L., Geiger, A. G., Kainer, D., Jacobson, D. A., & Rush, T. A. (2023). A glimpse into the fungal metabolomic abyss: Novel network analysis reveals relationships between exogenous compounds and their outputs. *PNAS Nexus, 2*(10).

Gostinčar, C., Grube, M., & Gunde-Cimerman, N. (2011). Evolution of fungal pathogens in salt stress environments. *Fungal Biology Reviews, 25*(2), 84–90. https://doi.org/10.1016/j.fbr.2011.01.001

Hautbergue, T., Jamin, E. L., Debrauwer, L., Puel, O., & Oswald, I. P. (2018a). From genomics to metabolomics: A mechanistic approach to mycotoxin production in *Penicillium* species. *Fungal Genetics and Biology, 115*, 14–23. https://doi.org/10.1016/j.fgb.2018.02.003

Hautbergue, T., Jamin, E. L., Debrauwer, L., Puel, O., & Oswald, I. P. (2018b). From genomics to metabolomics, moving toward an integrated strategy for the discovery of fungal secondary metabolites. *Natural Product Reports, 35*(2), 147–173. https://doi.org/10.1039/c7np00032d

Huang, W., Ge, X. Y., Huang, Y., Chai, X. T., Zhang, L., Zhang, Y. X., Deng, L. N., Liu, C. Q., Xu, H., & Gao, J. (2021). High-yield strain of fusidic acid obtained by atmospheric and room temperature plasma mutagenesis and the transcriptional changes involved in improving its production in fungus *Fusidium coccineum*. *Journal of Applied Microbiology, 130*(2), 581–596. https://doi.org/10.1111/jam.14797

Hur, J. Y., Jeong, E., Kim, Y. C., & Lee, S. R. (2023). Strategies for natural product discovery by unlocking cryptic biosynthetic gene clusters in fungi. *Separations, 10*(6), 333. https://doi.org/10.3390/separations10060333

Jiang, C., Qian, L., Liu, Y., Zhang, Y., & Chen, F. (2022). Transcriptomic analysis reveals *Aspergillus oryzae* responds to temperature stress by regulating sugar metabolism and lipid metabolism. *Food Microbiology, 102*, 103896. https://doi.org/10.1016/j.fm.2022.103896

Kaneria, M., Rakholiya, K., Jakasania, R., Dave, R., & Chanda, S. (2017). Metabolite profiling and antioxidant potency of Couroupita guianensis Aubl. Using LC-QTOF-MS based metabolomics. *Research Journal of Phytochemistry, 11*, 150–169. https://doi.org/10.3923/rjphyto.2017.150.169

Kaneria, M., Rakholiya, K., Bavaliya, K. R., et al. (2024). Untargeted metabolomics-based identification of bioactive compounds from *Mangifera indica* L. seed extracts in drug discovery through molecular docking and assessment of their anticancer potential. *Journal of the Science of Food and Agriculture, 104*(10), 5907–5920. https://doi.org/10.1002/jsfa.13421

Kawahara, Y., et al. (2012). Simultaneous RNA-Seq analysis of rice and blast fungus transcriptomes during infection reveals fungal effector gene expression and host defense responses. *PLoS ONE, 7*(4), e49423. https://doi.org/10.1371/journal.pone.0049423

Keeler, E., Burgaud, G., Teske, A., Beaudoin, D., Mehiri, M., Dayras, M., Cassand, J., & Edgcomb, V. (2021). Deep-sea hydrothermal vent sediments reveal diverse fungi with antibacterial activities. *FEMS Microbiology Ecology, 97*(8), fiab103. https://doi.org/10.1093/femsec/fiab103

Keller, N. P. (2017). Heterogeneity confounds establishment of *Aspergillus fumigatus* as a pathogen. *PLoS Pathogens, 13*(11), e1006711. https://doi.org/10.1371/journal.ppat.1006711

Keller, N. P. (2019). Fungal secondary metabolism: Regulation, function and drug discovery. *Nature Reviews Microbiology, 17*(3), 167–180. https://doi.org/10.1038/s41579-018-0121-1

Kim, Y.-S., Kim, S.-K., & Park, S. J. (2017). Apoptotic effect of demethoxyfumitremorgin C from marine fungus *Aspergillus fumigatus* on PC3 human prostate cancer cells. *Chemico-Biological Interactions, 269*, 18–24. https://doi.org/10.1016/j.cbi.2017.03.015

Kumar, A., et al. (2018). Genome sequencing and analyses of two marine fungi reveal a wealth of novel biosynthetic gene clusters. *Scientific Reports, 8*, 10762. https://doi.org/10.1038/s41598-018-28473-z

Lee, D. Y., Park, S. H., & Kim, H. U. (2020). OptKnock: A bilevel programming framework for identifying gene knockout strategies for microbial strain optimization. *Biotechnology and Bioengineering, 84*(6), 775–784. https://doi.org/10.1002/bit.10828

Li, J., & Jiang, X. (2021). Mol-BERT: An effective molecular representation with BERT for molecular property prediction. *Wireless Communications and Mobile Computing, 2021*(1), 1–7. https://doi.org/10.1155/2021/7181815

Li, T., Zhang, C., Du, L., & Ding, S. (2017). Proteomics: Insights into regulatory mechanisms and enzymatic functions in metabolic pathways. *Proteomics, 17*(3–4), 1600398. https://doi.org/10.1002/pmic.201600398

Lin, C., Huang, R., Liu, J., Li, H., Zhu, L., Huang, X., Ding, B., Liu, L., Huang, H., & Tao, Y. (2023). Antibacterial polyketides isolated from the marine-derived fungus *Fusarium solani* 8388. *Journal of Fungi (Basel, Switzerland), 9*(9), 875. https://doi.org/10.3390/jof9090875

Liu, Y., She, X., Chen, H., Chen, Z., Zhang, Y., & Pan, W. (2021). Integrated transcriptomic and metabolomic analyses reveal the adaptive response of *Candida albicans* to antifungal drugs. *mSystems, 6*(1), e01144–20. https://doi.org/10.1128/mSystems.01144-20

Möbius, N., & Hertweck, C. (2009). Fungal phytotoxins as mediators of virulence. *Current Opinion in Plant Biology, 12*(4), 390–398. https://doi.org/10.1016/j.pbi.2009.06.004

Murgia, M., Fiamma, M., Barac, A., Deligios, M., Mazzarello, V. L., Paglietti, B., Cappuccinelli, P., Al-Qahtani, A., Squartini, A., Rubino, S., & Al-Ahdal, M. N. (2019). Biodiversity of fungi in hot desert sands. *MicrobiologyOpen, 8*(1), e00595. https://doi.org/10.1002/mbo3.595

Muria-Gonzalez, M. J., Chooi, Y.-H., Breen, S., & Solomon, P. S. (2014). The past, present and future of secondary metabolite research in the Dothideomycetes. *Molecular Plant Pathology, 16*(1), 92–107. https://doi.org/10.1111/mpp.12162

Nanjareddy, K., Arthikala, M.-K., Gómez, B. M., Blanco, L., Lara, M., & Quinto, C. (2017). Proteomic analysis of *Piriformospora indica*-colonized chickpea (*Cicer arietinum*) roots reveals changes in host plant protein complement. *Plant Signaling & Behavior, 12*(4), e1297002. https://doi.org/10.1080/15592324.2017.1297002

Nielsen, J. C., Prigent, S., Grijseels, S., Workman, M., Ji, B., & Nielsen, J. (2019). Comparative transcriptome analysis shows conserved metabolic regulation during production of secondary metabolites in filamentous fungi. *mSystems, 4*(2). https://doi.org/10.1128/msystems.00012-19

Nuehring, J. (2023). *Bioactive secondary metabolites in cave dwelling fungi* [Master's thesis]. University of Northern Iowa. UNI ScholarWorks. https://scholarworks.uni.edu/etd/1342

O'Brien, J., & Wright, G. D. (2011). An ecological perspective of microbial secondary metabolism. *Current Opinion in Biotechnology, 22*(4), 552–558. https://doi.org/10.1016/j.copbio.2011.03.010

Ologun, G. O., Snyder, H., Hannigan, C., Njoku, D., & Aman, M. (2019). Celiac artery compression syndrome in a middle-age woman treated laparoscopically. *Cureus, 11*(9), Article e5582. https://doi.org/10.7759/cureus.5582

Pal, D., Bareth, K., Rani, P., Kandar, C. C., & Mishra, A. (2023). Coumarins as emerging anti-viral compounds from natural origins: Ethnopharmacology, chemistry, mechanism of action, clinical and preclinical studies, and future perspectives. In *Reference Series in Phytochemistry*. Springer. https://doi.org/10.1007/978-3-031-12199-9_29

Palazzotto, E., & Weber, T. (2018). Omics and multi-omics approaches to study the biosynthesis of secondary metabolites in microorganisms. *Current Opinion in Microbiology, 45*, 109–116. https://doi.org/10.1016/j.mib.2018.03.004

Paul, D., Sanap, G., Shenoy, S., Kalyane, D., Kalia, K., & Tekade, R. K. (2021). Artificial intelligence in drug discovery and development. *Drug Discovery Today, 26*(1), 80–93. https://doi.org/10.1016/j.drudis.2020.10.010

Ramarajan, M., Devilla, R., Dow, L., Walsh, N., Mead, O., Zakeel, M. C. M., Gallart, M., Richardson, A. E., & Thatcher, L. F. (2024). Genomic and untargeted metabolomic analysis of secondary metabolites in the *Streptomyces griseoaurantiacus* strain MH191 shows media-based dependency for the production of bioactive compounds with potential antifungal activity. *Journal of Agricultural and Food Chemistry, 72*(44), 24432–24448.

Reverberi, M., Punelli, F., Scarpari, M., Zjalic, S., Fanelli, C., & Fabbri, A. A. (2010). Lipoperoxidation affects ochratoxin A biosynthesis in *Aspergillus ochraceus* and its interaction with wheat seeds. *Applied Microbiology and Biotechnology, 85*(6), 1935–1946. https://doi.org/10.1007/s00253-009-2220-4

Riedling, O., Walker, A. S., & Rokas, A. (2024). Predicting fungal secondary metabolite activity from biosynthetic gene cluster data using machine learning. *Microbiology Spectrum, 12*(2). https://doi.org/10.1128/spectrum.03400-23

Sagita, R. (2021). Current state and future directions of genetics and genomics approaches for elucidating secondary metabolite biosynthetic pathways in endophytic fungi. *Frontiers in Bioengineering and Biotechnology, 9*, 649906. https://doi.org/10.3389/fbioe.2021.649906

Salem, M., El-Metwally, M., Saber, W., Negm, S., El-Kott, A., Mazroua, Y., Makhlouf, A., & Moustafa, M. (2022). Secondary antiviral metabolites from fungi with special reference to coronaviruses. *Biocell: Official Journal of the Sociedades Latinoamericanas de Microscopia Electronica, 46*(8), 1979–1988. https://doi.org/10.32604/biocell.2022.019301

Santiago, I. F., Gonçalves, V. N., Gómez-Silva, B., Galetovic, A., & Rosa, L. H. (2018). Fungal diversity in the Atacama Desert. *Antonie van Leeuwenhoek, 111*(8), 1345–1360. https://doi.org/10.1007/s10482-018-1060-6

Scardino, V., Di Filippo, J. I., & Cavasotto, C. N. (2023). How good are AlphaFold models for docking-based virtual screening? *iScience, 26*(1), 105920. https://doi.org/10.1016/j.isci.2022.105920

Scharf, D. H., Heinekamp, T., & Brakhage, A. A. (2014). Human and plant fungal pathogens: The role of secondary metabolites. *PLoS Pathogens, 10*(1), e1003859.

Sonah, H., Deshmukh, R. K., Bélanger, S., & Bélanger, R. R. (2016). Computational prediction of effector proteins in fungi: Opportunities and challenges. *Frontiers in Plant Science, 7*, 126. https://doi.org/10.3389/fpls.2016.00126

Studt, L., Wiemann, P., Kleigrewe, K., Humpf, H.-U., & Tudzynski, B. (2012). Biosynthesis of the cytotoxic diaporthin and orthosporin in *Glarea trichophora* and identification of the *GtPKS1* polyketide synthase gene. *Fungal Genetics and Biology, 49*(8), 495–505. https://doi.org/10.1016/j.fgb.2012.04.003

Suárez-Arroyo, I. J., Rosario-Acevedo, R., Aguilar-Perez, A., Clemente, P. L., Cubano, L. A., Serrano, J., Schneider, R. J., & Martínez-Montemayor, M. M. (2013). Anti-tumor effects of *Ganoderma lucidum* (Reishi) in inflammatory breast cancer in vivo and in vitro models. *PLoS ONE, 8*(2), e57431. https://doi.org/10.1371/journal.pone.0057431

Tang, X.-X., Liu, S.-Z., Yan, X., Tang, B.-W., Fang, M.-J., Wang, X.-M., Wang, H.-Y., Chen, J.-R., & Qiu, Y.-K. (2019). Two new cytotoxic compounds from a deep-sea *Penicillium citreonigrum* XT20-134. *Marine Drugs, 17*(9), 509. https://doi.org/10.3390/md17090509

Tebbji, F., Chen, Y., Richard, M., van Wijlick, L., Khemiri, I., O'Meara, T. R., & Whiteway, M. (2020). Transcriptomic analysis of *Candida albicans* biofilm cells reveals links between cell wall integrity and metabolism. *mBio, 11*(4), e01085-20. https://doi.org/10.1128/mBio.01085-20

Toscan, R., Blanchette, R. A., Held, B. W., & Gauthier, G. M. (2020). Fungal adaptation to extreme cold: Metabolite profiling of *Peni-*

cillium species from Arctic environments. *Fungal Biology, 124*(2), 140–150. https://doi.org/10.1016/j.funbio.2019.11.004

Van Goethem, M. W., Osborn, A. R., Bowen, B. P., Andeer, P. F., Swenson, T. L., Clum, A., Riley, R., He, G., Koriabine, M., Sandor, L., Yan, M., Daum, C. G., Yoshinaga, Y., Makhalanyane, T. P., Garcia-Pichel, F., Visel, A., Pennacchio, L. A., O'Malley, R. C., & Northen, T. R. (2021). Long-read metagenomics of soil communities reveals phylum-specific secondary metabolite dynamics. *Communications Biology, 4*, 1302. https://doi.org/10.1038/s42003-021-02809-4

Wadhwa, K., Kapoor, N., Kaur, H., Abu-Seer, E. A., Tariq, M., Siddiquie, S., Yadav, V. K., Niazi, P., Kumar, P., & Alghamdi, S. (2024). A comprehensive review of the diversity of fungal secondary metabolites and their emerging applications in healthcare and environment. *Mycobiology, 52*(6), 335–387. https://doi.org/10.1080/12298093.2024.2416736

Wainaina, S., & Taherzadeh, M. J. (2023). Automation and artificial intelligence in filamentous fungi-based bioprocesses: A review. *Bioresource Technology, 369*, 128421. https://doi.org/10.1016/j.biortech.2022.128421

Wiemann, P., & Keller, N. P. (2014). Strategies for mining fungal natural products. *Journal of Industrial Microbiology & Biotechnology, 41*(2), 301–313. https://doi.org/10.1007/s10295-013-1366-3

Yang, M., Zhu, Z., Zhuang, Z., Bai, Y., Wang, S., & Ge, F. (2021). Proteogenomic characterization of the pathogenic fungus *Aspergillus flavus* reveals novel genes involved in aflatoxin production. *Molecular & Cellular Proteomics, 20*, 100013. https://doi.org/10.1074/mcp.ra120.002144

Yu, J. S., Jeong, S. Y., Li, C., Oh, T., Kwon, M., Ahn, J. S., Ko, S. K., Kim, C. S., Cao, S., & Kim, K. H. (2022). New phenalenone derivatives from the Hawaiian volcanic soil-associated fungus *Penicillium herquei* FT729 and their inhibitory effects on indoleamine 2,3-dioxygenase 1 (IDO1). *Archives of Pharmacal Research, 45*(2), 105–113. https://doi.org/10.1007/s12272-022-01372-8

Yuan, J., Zhang, L., Li, J., Zhou, M., Zhang, X., & Wang, Y. (2019). Proteomic analysis of *Atractylodes lancea* infected by endophytic fungus *Gilmaniella* sp. AL12 reveals metabolic changes associated with growth promotion. *Frontiers in Plant Science, 10*, 1458. https://doi.org/10.3389/fpls.2019.01458

Zhai, M.-M., Qi, F.-M., Li, J., Jiang, C.-X., Hou, Y.-M., Cao, J.-K., Wu, Q.-X., & Zhao, Y.-X. (2016). The bioactive secondary metabolites from *Talaromyces* species. *Natural Products and Bioprospecting, 6*(1), 1–24. https://doi.org/10.1007/s13659-015-0086-3

Zhang, X., Li, S.-J., Li, J.-J., Liang, Z.-Z., & Zhao, C.-Q. (2018). Novel natural products from extremophilic fungi. *Marine Drugs, 16*(6), 194. https://doi.org/10.3390/md16060194

Zhong, G., Wang, Z.-J., Yan, F., Zhang, Y., & Huo, L. (2022). Recent advances in discovery, bioengineering, and bioactivity-evaluation of ribosomally synthesized and post-translationally modified peptides. *ACS Bio & Med Chem Au, 3*(1), 1–31. https://doi.org/10.1021/acsbiomedchemau.2c00062

Zhou, J., Zhang, H., Wu, Y., Li, X., & Liu, Y. (2022). Fungal secondary metabolites in food and pharmaceuticals in the era of multi-omics. *Critical Reviews in Food Science and Nutrition, 62*(23), 6391–6413. https://doi.org/10.1080/10408398.2021.1879730

Zhou, Y., Liao, L., Chen, S., & Ma, Y. (2021). Metabolomic insights into antifungal resistance and pathogenicity of *Candida* species. *Frontiers in Microbiology, 12*, 706100. https://doi.org/10.3389/fmicb.2021.706100

Fungal Metabolites in Nanoparticle and Nanomineral Production

Cheran Easwaran, Gokulakrishnan Moorthy, Muthukumar Govindan, Manimaran Gajendiran, Manish Kumar Dubey, and Sharmila Rahale Christopher

Abstract

Metabolites of the fungus have become a new path forward in formation of nanoparticles and nanominerals, they provide an environmentally friendly and sustainable substitute to traditional chemical practices of production of nanominerals. Nanoparticles and nanominerals have a repertoire of distinct characteristics because of their nanoscale size, which is applicable in many areas of application in medicine, agriculture as well as industry. Fungi serve as biofactories to make nanoparticles; using various metabolites including proteins, enzymes, and polysaccharides to minimize metal ions and stabilize the formed nanoparticles. These bio-systems have the ability to produce high precision and control of size, shape, and composition of metallic, oxide, and hybrid nanoparticles. The chapter describes several fungal species that produce different forms of nanoparticles such as copper, silver, gold, and magnetic, in terms of biosynthetic ability which is spectacular. Such nanoparticles produced by fungi have exclusive characteristics and can be used in various fields. They are promising in medicine in targeted drug delivery, antimicrobial treatment, and cancer diagnostics. Environmental applications in water purification, pollutant remediation, and sensing technologies are used in the environment. Fungal nanoparticles are used in industrial sectors to catalyze, as electronic components and developing advanced materials. Nevertheless, problems such as standardization of the production processes, knowledge of accurate biomolecular pathways, and large-scale production continue to be a challenge despite the great progress. More research within this field should be done in future to optimize the metabolic pathways, improve the stability of nanoparticles, and generate a more detailed understanding of the interactions of fungal nanomaterial.

Keywords

Fungal metabolites · Nanoparticle synthesis · Nanomineral production · Green synthesis · Biotechnology

C. Easwaran · S. R. Christopher (✉)
Centre for Agricultural Nanotechnology, Directorate of Natural Resource Management, Tamil Nadu Agricultural University, Coimbatore, India
e-mail: sharmila.r@tnau.ac.in

G. Moorthy (✉)
Indian Council of Agricultural Research—Indian Institute of Agricultural Biotechnology, Ranchi, India
e-mail: mrgokul95@gmail.com

M. Govindan
Department of Plant Pathology, Centre for Plant Protection Studies, Tamil Nadu Agricultural University, Coimbatore, India

M. Gajendiran
Department of Soil Science and Agricultural Chemistry, Directorate of Natural Resource Management, Tamil Nadu Agricultural University, Coimbatore, India

G. Moorthy
Indian Council of Agricultural Research—Indian Agricultural Research Institute, Pusa, New Delhi, Delhi, India

M. K. Dubey
Department of Biotechnology, University Centre for Research and Development (UCRD), Chandigarh University, Mohali, Punjab, India

1 Introduction to Nanoparticles and Nanominerals

Nanotechnology refers to the fast-growing area that investigates the materials and systems on the nanoscale, which is usually 1–100 nm. At this nanoscale, there are special physico-chemical properties of materials which differ greatly with those of the bulk. Nanoparticles (NPs), the building blocks of nanostructures, are precious to a great assortment of uses due to these characteristics, including, but not confined to, larger specific surface areas, lower melting points, peculiar optical and magnetic qualities

© The Author(s), under exclusive license to Springer Nature Switzerland AG 2026

A. Prakash and M. K. Dubey (eds.), *Advances in Fungal Secondary Metabolites: Bioactive Attributes and Biotechnological Potential of Industrial Interest*, Sustainable Economy and Ecotechnology, https://doi.org/10.1007/978-3-032-23507-7_14

(Horikoshi & Serpone, 2013; Sant, 2012). Nanotechnology has expanded due to phenomenal scientific instrumentation, which allows researchers to observe and control matter on an atomic and molecular scale (Mandal & Banerjee, 2020). This development has led to the emergence of new materials whose use and application in a number of areas such as biomedicine, environmental science, and information technology have been transformed. As an example, metallic NPs such as silver and gold have received much interest in the biomedical field (Mody et al., 2010).

They can be functionalized using various chemical groups to improve their applicability in the delivery of drugs and in diagnostic imaging (McNamara & Tofail, 2017). In addition to metallic NPs, a specialized subclass of NPs, nanomaterials (NMs), are important factors in biomineralization. These reactions entail the creation of minerals by living organisms and in most cases, these reactions begin with NP precursors. In the situation when the minerals crystals are prepared at the nanoscale (less than 100 nm), polymorphism, dominance of surface energy, and growth rate among other factors can be more controlled (Heiligtag & Niederberger, 2013). Not only do these distinct properties shed some light on how natural biomineralization works but also guide synthetic methods to mineralization which are valuable in material science in general (Navrotsky, 2004). Essentially, nanotechnology is one of the frontiers of innovation, which opens opportunities to create materials with unprecedented capabilities in dealing with urgent challenges in different areas.

NMs and NPs have become very crucial in many aspects such as industry and medicine. Their size dependent peculiarities make them very valuable in a broad production (Salata, 2004). Their unique structure and size enable these NMs to exhibit outstanding bulking, catalytic, plasticity, and exchangeability (Feng et al., 2024). In medicine and biology, NPs are applied in the delivery of drugs, vaccine creation, and diagnostic imaging improvement (Pelaz et al., 2017). They find their applications in pharmacy, cosmetics, and medicine as nanocarriers of therapeutic agents and in regenerative medicine (Gomes et al., 2021). Minerals, such as NMs are important to human, animal, and plant health by being important constituents in biochemical pathways and enzyme reactions (Soetan et al., 2010). According to geochemical and biogeochemical reactions on Earth, NMs may have chemical properties that depend on their size (Hochella et al., 2008). The sustained growth of NMs in a supercritical water provides the control of particle properties and possibility of industrial scalability, e.g., in case of talc, xonotlite, and tobermorite (Claverie et al., 2019). This size-dependence has an impact on their interactions with the hydrosphere, pedosphere, biosphere, and atmosphere. Nonetheless, NMs have the potential to be an environmental and health hazard such as phosphogypsum in the fertilizer industry containing complex NM assemblies, containing potentially toxic elements. The careful characterization of these NMs is necessary with the help of such sophisticated advanced electron microscopy tools [high-resolution transmission electron microscopy (HR-TEM) and field-emission scanning electron microscopy (FE-SEM)] to determine the effects on human health and establish effective interventions (Silva et al., 2010). Since nanomedicine continues to gain momentum, these materials are even more actively converted into feasible clinical products, which have new prospects in solving medical and health-related problems (Pelaz et al., 2017). Green chemistry and biotechnology are important in the synthesis of NPs with environmentally friendly alternatives to the traditional/conventional approaches. They can be described as corresponding to the twelve principles and future directions of green chemistry, which focus on the sustainability, energy efficiency, and low toxicity (Duan et al., 2015). Living organisms viz. algae, bacteria, fungi, lichens, yeast, and plants may be used to make NP thus requiring no aggressive chemicals and expensive processes. Such unification of biology and nanotechnology has brought about a considerable boost to the discipline of nanobiotechnology which employs biochemical and biophysical reactions in the production of NPs (Shah et al., 2015). The green synthetic practices produce NPs with a variety, high stability, and desired sizes, thus, making them suitable in a broad spectrum of applications, including the biomedical and diagnostic sectors (Patra et al., 2016). Recent efforts in the synthesis of green NPs have been directed at their potential as drug delivery systems in anticancer therapy and antimicrobial therapy. Computational modeling and in silico design have become useful in forecasting bioactivities and learning more about NP–target interactions, as a precursor to smart medicine and contemporary cancers therapy (Kazemi et al., 2023).

Fungal metabolites are finding more uses in NP and NM production, as they have clean, bio-compatible, non-hazardous, and environmentally sustainable properties. Fungi also have multiple benefits, among which are stabilizing the protein secretion, high yield of biomass, high levels of metal tolerance and simple growth/handling (Chauhan et al., 2022). The biological method offers NPs with the desirable size, morphology, and improved physicochemical traits at reduced levels of toxicity and environmental effects (Singh, 2020). Reducing agents secreted by fungi are enzymes and proteins which allow extracellular and intracellular synthesis of metal NPs using the metal salts (Siddiqi & Husen, 2016). It is also cost-effective and can be used in large-scale production which is an appeal to physical and chemical methods. NPs produced by fungi, and especially silver NPs, have demonstrated possible uses in biomedicine and industry, such as antimicrobial effects against pathogenic organisms (Castro-

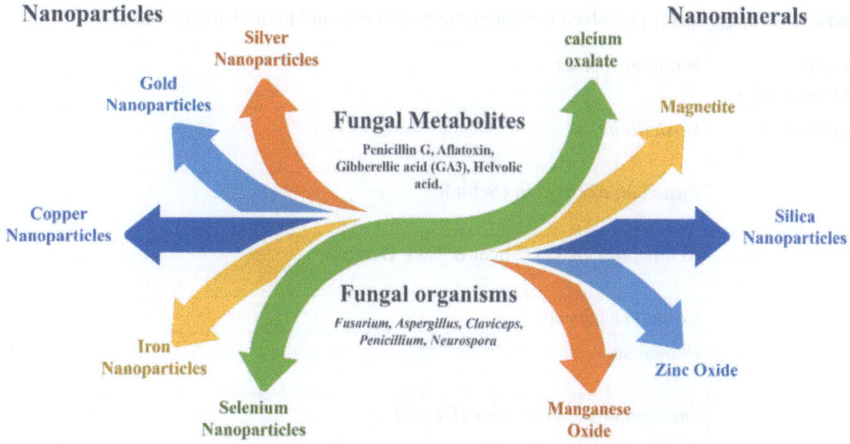

Fig. 1 Fungal metabolites in nanoparticle and nanomineral production. *Source* Authors

Longoria, 2016) (Fig. 1). With the growing prominence of green chemistry as the method of NP production is explained by the necessity to provide the alternatives that should be sustainable in the context of the energy crisis and environmental limitations. The main purpose of the chapter is to examine the activity of the metabolites on the fungi in relation to the NM synthesis and highlight the advantages over chemical processing. It is also used to answer the processes that occur in the fungal production of NPs, the effect of various fungal species on the NPs produced and the challenges associated with mass production of NPs.

2 Fungi as Biofactories for Nanoparticles Production

NPs synthesis using Fungi entails a complex biochemical pathway in which fungi are used as the source of biocatalysts as well as precursor molecules. Understanding these principles will help us to understand the processes of fungi mediated biosynthesis and develop improved synthesis regimes (Thakor et al., 2022). The ability of fungus to secrete extracellular enzymes and the innate capacity to mineralize intracellularly is a major characteristic of NP synthesis by fungi, which is essential to this process (Javed et al., 2024). Also, the versatility to a variety of environmental parameters available to fungi offers researchers the operational scalability and flexibility required to achieve scalable synthesis of NP. These unique properties enable researchers to experiment with new opportunities in NP production and this can result in new uses in various industries (Prasher et al., 2024). Table 1 describes the fungal species used in the biosynthesis of NPs.

2.1 Advantages of Using Fungi for Green Nanoparticle Production

Due to their enhanced metal tolerance and remarkable bioaccumulative potential, fungi become prominent players in the field of nanobiotechnology (Sastry et al., 2003). A clear advantage of NP synthesis is the ease of scale up, including by use of fermentation on thin solid substrate. The fungi are very effective producers of extracellular enzymes and hence they are an ideal target in industrial scale production of biocatalysts (Castro-Longoria et al., 2012). To this, other supplementary merits that favor the fungal mediated greener method of preparing the metallic NP include economic viability and convenience of biomass usage. More so, most fungal species have grown extremely fast making them easy to produce and maintain in the laboratory (Jain, 2007). In addition to this, majority of fungi have a high wall-binding capacity and favorable intracellular metal uptake capacity (Volesky & Holan, 1995). Also, besides these, fungi synthesize mesostructures, nanostructures, and metal NPs by intra- or extracellularly the intervention of the reducing enzymes and through the biomimetic mineralization (Scheme 2) (Ahmad et al., 2003; Durán et al., 2005).

The benefits of synthesizing nanomaterials using fungi as a biofactory are some of them.

- Enzymes highly effective in reducing various compounds, leading to the formation of NPs;
- Capability to withstand elevated levels of metals and metalloids;
- Capacity to produce substantial amounts of NPs, primarily outside the cell;
- Rapid NP synthesis;

Table 1 Fungal species involved in nanoparticle synthesis and their characteristics

Metal nanoparticles	Name of species	Size (nm)	References
Gold (Au)	*Verticillium* sp.	20	Mukherjee et al. (2001)
	Fusarium oxysporum (Schltdl.)	20–40	Mukherjee et al. (2001)
	Neurospora crassa (Shear & B.O. Dodge.)	32	Castro-Longoria et al. (2011)
	Colletotrichum sp.	20–40	Shankar et al. (2004)
	Phoma sp.	10–100	Soltani Nejad et al. (2022)
	Saccharomyces cerevisiae (Desm.)	15–20	Sen et al. (2011)
	Aspergillus niger (Tieghem.)	8.7 ± 6.6	Das et al. (2012)
	Aspergillus oryzae (Ahlb.)	6–37	
	Aspergillus tamari (Kita.)	3.5 ± 3	
	Penicillium ochrochloron (Biourge)	7.7 ± 4.3	
Silver (Ag)	*Aspergillus fumigatus* (Fresen.)	5–25	Bhainsa and D'Souza (2006)
	Hormoconis resinae (creosote fungus)	20–80 and 10–20	Varshney et al. (2009)
	WA 2315	23–105	Shaligram et al. (2009)
	Cladosporium cladosporioides (Fresen.)	10–100	Balaji et al. (2009)
	N. crassa (Shear & B.O. Dodge.)	11	Castro-Longoria et al. (2011)
	Purpureocillium lilacinum (Thom)	50	Khan et al. (2022)
	Phoma glomerata (Corda) (MTCC-2210)	19–65	Gade et al. (2014)
	Fusarium semitectum (Desm.)	10–60	Ingle et al. (2008)
	Aspergillus japonicus (Saito) PJ01	3.8 ± 1.1 and 9.1 ± 2.9	Li et al. (2021)
	Alternaria alternata (Fr.)	7.48–12.15	Kareem et al. (2019)
	Volvariella volvacea (Bull.)	15	Shaligram et al. (2009)
	A. niger (Tieghem.)	20	Gade et al. (2008)
Platinum (Pt)	*N. crassa* (Shear & B.O. Dodge.)	–	Sanghi and Verma (2009)
Copper (Cu)	*Trichoderma koningiopsis* (Samuels)	87.5	Salvadori et al. (2014)
	F. oxysporum (Schltdl.)	2–5	Hosseini et al. (2012)
	N. crassa (Shear & B.O. Dodge.)	10–20	Li and Gadd (2017b)
Magnesium (Mg)	*Trichoderma viride* (Persoon)	45.12–95.37	Alrabadi et al. (2017)
Palladium (Pd)	*Phanerochaete chrysosporium* (Burds.)	10–14	Tarver et al. (2019)
Selenium (Se) and tellurium (Te)	*Trichoderma harzianum* (Rifai), *Aureobasidium pullulans* (de Bary & Löwenthal), *Mortierella humilis* (Linnem. ex W. Gams), *P. glomerata* (Corda)	48–78	Liang et al. (2019)
Uranium (U)	*Geotrichum* sp.	50–100	Zhao et al. (2016)
Cadmium sulfide (CdS)	*Schizosaccharomyces pombe* (yeast)	1–1.5	Kowshik et al. (2002)
Zirconium (Zr/ ZrO_2)	*F. oxysporum*	7–8	Bansal et al. (2004)
Titanium dioxide (TiO_2)	*F. oxysporum*	6–13	Bansal et al. (2004)
	F. oxysporum	5–15	

(continued)

Table 1 (continued)

Metal nanoparticles	Name of species	Size (nm)	References
	T. harzianum	2–16	Jayaseelan et al. (2013)
Magnetite	*F. oxysporum* and *Verticillium* sp.	20–50	Bansal et al. (2005)
Zinc oxide (ZnO)	*Xylaria acuta*	34–55	Sumanth et al. (2020)
Cerium oxide (CeO_2)	*Humicola* sp.	12–20	Khan and Ahmad (2013)

Source Authors compilation

- Easy cultivation, straightforward NP processing, and scalability;
- Harmless to humans (when utilizing medicinal and edible fungi);
- Potential to create NPs with diverse therapeutic/bioactive properties (when employing medicinal fungi);
- Promising player in the production of active and cap compounds;
- The ability to synthesize NPs with advanced medicinal attributes using medicinal mushrooms.

2.2 Mechanisms of Fungal Metabolite-Mediated Nanoparticle Synthesis

The fungi produce NPs either intra- and extracellularly. Intracellular mycosynthesis involves the addition of metallic precursor to fungal mycelial culture, whereby it is taken across into the biomass. It entails extraction of NPs at the back by use of chemical, centrifugation, and filtration to disrupted and liberated biomass (Castro-Longoria et al., 2011; Molnár et al., 2018). On the other hand, extracellular mycosynthesis occurs when the metal precursor is incubated with an aqueous filtrate that is solely a combination of fungal biomolecules, therefore, the free NPs is generated in solution. This it is a more common procedure as it does not involve the disruption of the for-cell to get the NPs (Gudikandula et al., 2017; Silva et al., 2017). Intracellular NPs mycosynthesis has been suggested to occur in two steps (Fig. 2). When the precursors enter the fungal cells, they are subjected to cascade of transformation reactions, which are catalyzed by both enzymes and non-enzymes. Fungi utilize several different mechanisms to minimize and stabilize metal/metalloid ions both internally and externally, which leads to the generation of NPs with different shapes and properties. Within the fungal cells, certain enzymes are triggered in order to produce intracellular changes. As an example, enzymes such as reductases play a vital role in synthesis of AgNPs wherein silver ions are transformed to elemental silver. Enzymes such as hydrogenases have been detected to decrease selenium ions to an elemental state in the instance of selenium NPs. In the fungal environment, on the other hand, the extra cellular changes take place and are normally enabled by the released metabolites in the environment. The outcome of this exterior synthesis is the accumulation of NPs on the fungal cells or on the growth medium. Temperature, pH, and the components of the growth medium are also important variables that influence the extracellular synthesis, which, in turn, has significant effects on the size and morphology of the resulting NPs (Qin et al., 2020).

2.3 Role of Enzymes, Proteins, and Secondary Metabolites (SMs)

In response to their environmental contaminants, fungi generate NPs. Fungi reduce the ion toxicity by precipitating, immobilizing, transforming ion forms, co-precipitating, and binding to biological molecules (Bruins et al., 2000). The loss of ions leads to the precipitation of a metal in intra- or extracellular space in the form of nanomaterial (Das et al., 2012). Human beings use the above system in the synthesis of nanomaterials and bioremediation of ecosystems (Narayanan & Sakthivel, 2010). Fungal NP biosynthesis is a critical process in which the electron transfer step is caused by a donor molecule to the ion, thus, resulting in the ion precipitation as a NP. This electron transfer can be done by means of the activity of fungal enzymatic systems or fungal-derived molecules. The electron exchange may be promoted with the help of peptides, polysaccharides, vitamins, amino acids, alkaloids, enzymes, quinones, flavonoids, saponins, tannins, carboxylic acids, fatty acids, steroids, and other SMs (Keat et al., 2015; Moghaddam et al., 2015). Another group of enzymes that lower the toxic metal ions to NPs is involved in reducing toxic metal ions by some enzymes; they include, phytochelatin, a-NADPH-dependent nitrate reductase, and FAD-dependent glutathione reductase (Scott et al., 2008). Biosynthesis can be done either using direct contact between the ions and fungal biomass (Ahmad et al., 2002; Vahabi et al., 2011) or by

Fig. 2 Simplified representation of fungal mediated biogenic nanoparticle synthesis. *Source* Authors

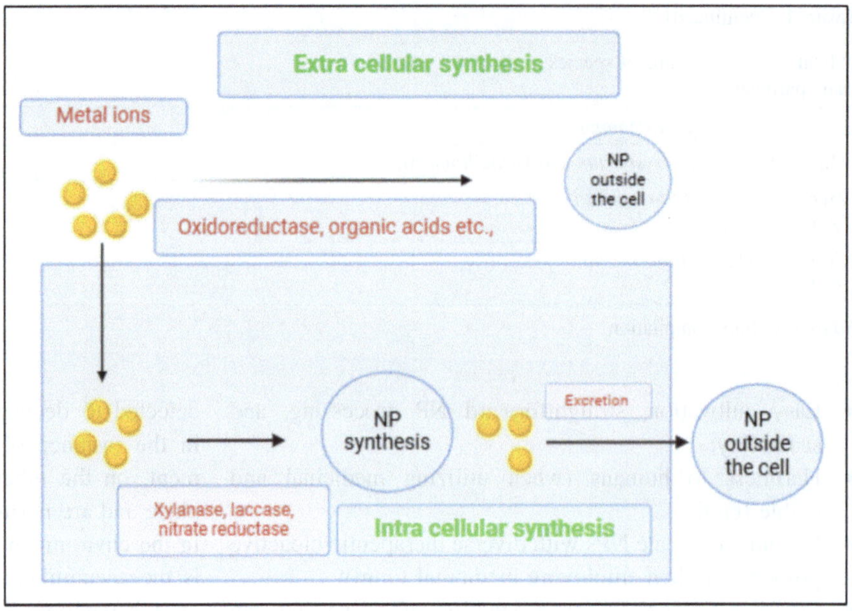

contact with extracts that are devoid of biomass (Shankar et al., 2004) such as enzymes and other molecules secreted by fungal (Mukherjee et al., 2001).

2.4 Fungal Biomolecules in Metal Reduction and Stabilization

Fungi incorporate the use of a protein-enzyme-molecule mix they release (exometabolites) and coat and stabilize NPs. This biological coating performs two very important functions: (1) to avoid clumps of NPs, and (2) to tailor the characteristics of the obtained nanomaterial. In the process of synthesis, the proteins synthesized by fungi have the effect of coating the NPs, preventing their undesirable proliferation. These biomolecules are also important in determining the particular physicochemical properties of NPs besides stabilization. Protein category as a coating also heavily influences the quantity of a surface charge, biocompatibility, and the interaction between synthesized NPs and other substances (Guilger-Casagrande et al., 2021; Javed et al., 2020).

Fungi have an outstanding capability of binding and internalizing metals and are therefore the best candidates in getting stable NPs (Dhillon et al., 2012). Such a synthesis usually yields NPs of high stability and good biocompatibility by fungi. Stable protein-coated AgNPs have been produced using a wide variety of fungi such as thermophilic *A. fumigatus* (Pandey et al., 2020), endophytic fungi such as *Catharanthus roseus* isolated in medicinal plants (Akther et al., 2019), and different strains of fungi (*Fusarium*) (Rai et al., 2021). Moreover, the xylotrophic basidiomycetes (*Ganoderma lucidum*, *Lentinus edodes*, *Grifola frondosa* and *Pleurotus ostreatus*) have been advantageous in producing stable Au, Ag, Se, and SiNPs because of the phenol oxidase (PO) activity (Vetchinkina et al., 2018). Finally, xylanases of *A. niger* and *Trichoderma longibrachiatum* enzymes have been used successfully to generate protein-coated and stabilized silver-gold NP alloys (Elegbede et al., 2019). Moreover, such fungi as *Penicillium janthinellum* DJP06 have been used to produce protein-capped and crystalline Au and silver NPs (Pareek et al., 2020). Other interesting ones are *A. niger* STA9 strain to synthesize CuNPs (Noor et al., 2020), *A. flavus* and *A. terreus* due to their ability to use their secondary protein structures to produce highly concentrated and stable silver NPs (Aboul-Nasr et al., 2021), *Penicillium* species due to their capacity to produce dispersed and stable Fe_3O_4 nanocomposites (Ding et al., 2015), and *F. oxysporum* due to their ability to produce terbium This wide range of examples highlights the great potential of fungi in the multifaceted-to-use biofactories in producing a wide spectrum of stable and functional NPs in a green and sustainable manner.

The protein capping is also an important process in the production of high biomineralization activity NPs. Research has identified proteins produced by *Pleurotus sajorcaju* to have the ability to deliver consistently and stably active antibacterial effects (Owaid, 2019). On the same note, fungal protein of *Coriolus versicolor* MUCL has demonstrated the capability of being successfully utilized in production of stable silver NPs in alkaline condition. This is brought about to a large extent by enhanced reducing power of such proteins. These proteins are reducing and capping agents during the synthesis process. The reducing capacity is also boosted by the alkaline environment, which makes the synthesized AgNPs to be stable. This bifunctional nature of proteins shows their importance in NP synthesis and gives credence to the fact that these stable AgNPs have a vast number of uses (Sanghi & Verma, 2009).

3 Common Types of Nanoparticles Synthesized by Fungal Metabolites

Fungi are also becoming good biological systems in producing NPs, as they have a capability of producing a wide range of metabolites, including SMs, proteins, and enzymes (Dhillon et al., 2012). These metabolites act as reductants and stabilizers in the production of NPs to regulate their shape, size, and surface properties. Their special characteristics of physicochemical and biological features of the fungi-produced NPs provide them with numerous fields of application in healthcare, medicine, food, agriculture, industry, and environmental pollution remediation (Saravanan et al., 2021). The syntheses of several types of NPs (Fig. 3) produced by the metabolites of fungi are discussed in more detail below based on the basic concepts and further applications.

3.1 Silver Nanoparticles (AgNPs)

The AgNPs are perhaps the most commonly investigated because of their excellent antimicrobial and anticancer properties (Sofi et al., 2022). The antimicrobial property of the silver NPs relates to disruption of microbial membranes, altered permeability, and production of ROS that prevent the growth of bacteria, fungi, and viruses, even including drug-resistant species. AgNPs cause apoptosis, but this time through an indirect pathway, disrupting the function of mitochondria, and producing oxidative stress, thereby making AgNPs selectively cytotoxic to cancer cells rather than to normal tissues (Sanati et al., 2022). The AgNPs can be produced by fungi using silver ions (Ag^+) reduced to metallic silver (Ag^0), with fungal enzymes and also the reducing agents in their form as phenolic that are also stabilizers. Among the common fungi, *A. flavus*, *F. oxysporum*, and *Penicillium* sp. are highly efficient producers of AgNPs (Guilger-Casagrande & Lima, 2019). Silver NPs have diverse applications in medicine and industry. In health care, they are used as antimicrobial coatings for medical devices, wound dressings, and as carriers in drug delivery systems (Saghazadeh et al., 2018). In applied environmental applications, AgNPs are used for water treatment or purification by effectively removing microbial contaminants. In addition, the textile industry uses AgNPs in fabrics for antimicrobial finishes to enhance hygiene and odor resistance.

A three-step process of bioreduction, nucleation, and stabilization of silver NPs using fungi is a non-toxic process that uses fungal enzymes (e.g., nitrate reductase) to reduce the silver ions and also the nucleation and stabilization of the particle by the fungal enzymes (e.g., reducing metabolites, e.g., phenolic compounds). In the process, nitrate (NO_3^-) is reduced by the activity of nitrate reductase to nitrite (NO_2^-) and the electrons are released to reduce Ag^+. Reduction is also helped by the reducing agents which encompass other metabolites in addition to the phenolic compounds. The primary reaction $Ag^+ + e^- \rightarrow Ag^0$ reduces silver ions to metallic silver atoms. It is then followed by nucleation, in which small aggregates of Ag^0 atoms nourish NP growth. The metabolites of fungi determine the size and shape of NPs at this stage. These clusters increase when the number of Ag^0 atoms condensed onto their surfaces increases. In this stage, NPs are covered by fungal proteins, polysaccharides, and SMs to prevent aggregation and oxidation. Hydroxyl (–OH), amine ($-NH_2$), and carboxyl (–COOH) functional groups of the metabolites bind to NPs which stabilize the colloidal state. The nitrate to nitrite reduction process produces electrons in nitrate reductase, which in turn reduces Ag + ions. Phenolic compounds give out electrons to facilitate the work of the enzymes. The chemical reaction decreases nitrate (NO_3^-) to nitrite (NO_2^-) and water and subsequently reduces Ag^+ to Ag^0. The metabolites generated by the fungi attach to the NP surface to inhibit agglomeration and enhance stability (Kannan et al., 2013).

$$(\mathrm{R - OH + AgNP \rightarrow R - O - AgNP}).$$

This reaction can be represented as:

$$\mathrm{Ag^+ + e^- \rightarrow Ag^0}$$

Simultaneously, the nitrate reductase reaction involves:

$$\mathrm{NO_3^- + 2H^+ + 2e^- \rightarrow NO_2^- + H_2O}$$

After the silver atoms are reduced, they combine to make seeds of NP growth in a process referred to as nucleation. Several silver atoms are combined as:

$$n\mathrm{Ag^0} \rightarrow \left(\mathrm{Ag}_n\right)^0$$

In the stabilization, the NPs are capped by fungal metabolites (proteins and polysaccharides) to prevent aggregation and oxidation. The interactions may be defined as:

$$\left(\mathrm{Ag}_n\right)^0 + \mathrm{Proteins/Polysaccharides} \rightarrow \mathrm{Capped\,AgNPs}$$

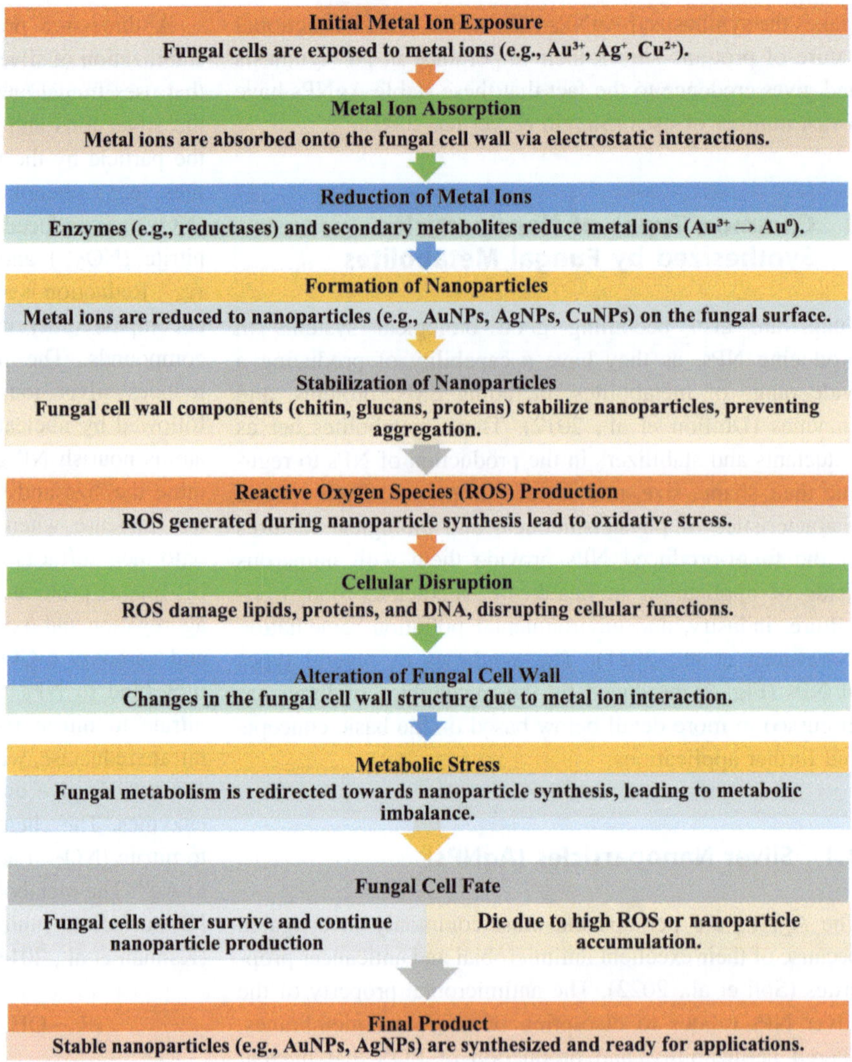

Fig. 3 Mechanism of fungal-mediated nanoparticle synthesis. *Source* Authors

It is an ideal approach in the medical, agricultural, and industrial applications of fungi-mediated stable, biocompatible, and ecologically friendly AgNPs. The enormous functional flexibility of regulated fungal-mediated AgNPs makes them promising for medical device coatings, water treatment/purification, and textiles. Due to the extraordinary antibacterial, optical, and high conductivity abilities of AgNPs, medicine and industry have been revolutionized. Antimicrobial AgNPs are used in wound dressings, surgical instruments, and medical equipment to reduce infections. Their optical properties are applied in disease diagnosis and customized drug delivery to enhance the effectiveness of treatment. AgNPs combat infections and heal wounds in dental and anti-inflammatory treatment (Zubair et al., 2023). AgNPs are also used in electronics where they enable miniaturization in printed circuit boards and touch screens via conductive inks. They provide antibacterial effects to clothes that enable smells to be avoided and high level of hygiene to garments and hospital trousers. AgNPs eliminate bacteria in water purification and increase the shelf life of food to prevent bacteria and skin diseases in cosmeceuticals (Agrawal et al., 2018). Moreover, they are the catalysts of chemical reactions and environmental technology and improve the efficiency of solar cells depending on their photonic and electrical characteristics. Sustainable procedures are required to optimize their potential due to the cytotoxicity and environmental impact of them.

3.2 Gold Nanoparticles (AuNPs)

Gold NPs are gold atomic structures between the ranges of the nanoscale 1–100 nm. AuNPs have certain optical, electrical, and catalytic characteristics due to the increase in the ratio of surface area to volume (SA:V), quantum and capacity to interact with light using localized surface plasmon resonance (LSPR) (Hassan et al., 2023). The mentioned properties precondition the application of AuNPs in a variety of areas such as catalysis, medicine, and diagnostics. Their ability to functionalize with biomolecules or ligands makes them even more versatile, thereby offering applications from green chemistry to targeted drug delivery.

3.2.1 Catalytic and Biomedical Applications

Reservoir properties of the AuNPs and high surface energies facilitate the AuNPs as highly active catalysts especially in redox reactions (Astruc, 2021). In catalysis, AuNPs function as active sites for the adsorption of reacting species; therefore, it is said to reduce activation energy and thus enhance reaction rates. Some of the applications of AuNPs include carbon monoxide oxidation and selective oxidation of alcohols which play a crucial role in the greener applications of chemical applications. The localized surface plasmon resonance of AuNPs, enhances catalytic activity by transferring energy to reactants under visible light, thus making reactions more energy-efficient.

3.2.2 Mechanism of Fungal AuNP Synthesis (Banerjee & Ravishankar Rai, 2018)

1. Gold Ion Uptake: Fungi sequester Au^{3+} ions from solutions via surface electrostatic interactions.
2. Reduction of Au^{3+} to Au^0: Intracellular enzymes, such as NADH-dependent reductases, transfer electrons to Au^{3+}, reducing it to Au^0. Reactive metabolites (e.g., flavonoids, and phenolics) act as additional reducing agents.
   - Reaction: $Au^{3+} + 3e^- \rightarrow Au^0$
3. Capping and Stabilization: Fungal proteins (e.g., extracellular enzymes) and SMs (e.g., alkaloids, and terpenoids) stabilize AuNPs, preventing aggregation, and controlling their size and shape.

AuNPs are highly valued in the biomedical field because of their biocompatibility and ability to be surface-functionalized. This means that functionalized AuNPs can be conjugated with drugs, targeting ligands, or imaging agents. This creates drug delivery, biosensors, and medical imaging uses. The AuNPs enhance the solubility, stability, and specificity of drugs. The specific delivery is by means of receptor-mediated endocytosis. As soon as environmental stimuli (change in pH or light) are detected within the cells, controlled drug release is enforced. AuNPs are also highly significant in diagnostics, and they can serve as contrast agents in modalities, including computed tomography (CT) and magnetic resonance imaging (MRI) (Luo et al., 2021). The ability to improve signals is essential in the use of biosensors to detect disease biomarkers and also in the lateral flow tests.

3.2.3 Fungal Synthesis of Gold Nanoparticles for Pharmaceutical Delivery

Fungal bioreduction is a highly environmental friendly route toward the synthesis of AuNPs. The cells of the fungi absorb the gold ions and eventually reduce them to elemental gold (AuO) through some enzyme-dependent reductases of NADH and metabolites such as phenolics and flavonoids. NPs are stabilized due to proteins and SMs that prevent aggregation and have specific dimensions and morphology as observed with fungi (El-Seedi et al., 2019). Extensively studied fungal species like *F. oxysporum* and *Trichoderma* sp. synthesize AuNPs with the desirable properties for biomedical applications. The biogenic AuNPs prepared through these fungi have remarkable efficiency in drug delivery applications. Functionalized fungal AuNPs can be bound specifically to the receptors expressed in diseased cells like folate receptors overexpressed in cancer (Adiga et al., 2025). This binding enhances the process of receptor-mediated endocytosis to make entry of AuNPs into the cells possible. In the acidic environment of tumors, these NPs would release their payloads of drugs in a regulated manner and hence reduce any side effects while enhancing therapy. With targeted drug delivery and the biocompatibility and cost-effectiveness offered by the AuNPs synthesized by fungi, this indeed presents a novel solution toward advanced drug delivery and nanomedicine.

3.3 Other Metal Nanoparticles

Among the MNPs, Zn, Fe, Cu, and Se NPs are highly significant due to their unique physico-chemical properties and versatility in various industrial applications (Shnoudeh et al., 2019). The synthesis of such NPs has been reported through physico-chemical and/or biological strategies. Each approach offers distinct advantages. Recent years have shown interest in the biological approach because it is eco-friendly and enables one to synthesize NPs of controlled size and morphology.

3.3.1 Zinc Nanoparticles (ZnNPs)

ZnNPs means nanoscale zinc metal or zinc oxide particles; they are between 1 and 100 nm. They have the highest surface area-to-volume ratio, excellent UV absorption, antimicrobial activity, and catalytic efficiency, which makes ZnNPs valuable for use in agriculture, cosmetics, coatings, and the environment (Kumar et al., 2021).

Synthesis of Zinc Nanoparticles

ZnNPs have been fabricated through various methods like physico-chemical, and biological ways. Among them, the popularity of green synthesis is growing greatly because it is non-toxic and eco-friendly in nature. In this process, plant extracts are very commonly used as reductors and stabilizers in order to reduce zinc salts into ZnNPs (El-Khateeb et al., 2019). A common reaction for green synthesis is the reduction of zinc nitrate [$Zn(NO_3)_2$] or zinc acetate [$Zn(CH_3COO)_2$], by phytochemicals derived from plants that are flavonoids and phenols (Virginia et al., 2024).

$$\text{Reaction: } Zn(NO_3)_2 + \text{Phytochemicals} \rightarrow \text{ZnNPs} + \text{Byproducts}$$

The plant-based phytochemicals serve as reducing agents for conversion of zinc ions (Zn^{2+}) into metallic zinc NPs and stabilize them against aggregation. ZnNPs find extensive applications in industries across the board due to their flexible properties (Fernandes et al., 2022). In cosmetics, it is a primary constituent of sunscreen, which offers excellent UV-blocking properties while ensuring an evenly spread application and greater skin protection. In coatings and paints, ZnNPs are greatly considered for preventing microbial growth in medical equipment, and food packaging, as well as high-touch surface hygiene. In water or wastewater treatment, ZnNPs are photocatalysts because they break down the harmful organic pollutants into harmless products exposed to light, bringing ecological sustainability (Alprol et al., 2023). ZnNPs are micronutrient fertilizers applied in agriculture. They enhance the bioavailability of zinc for activation of enzymes, synthesis of proteins, and production of chlorophyll, all leading to enhanced growth and crop productivity. Combined with the above applications and the eco-friendly methods of synthesis, ZnNPs have now become an integral material in innovative and sustainable solutions.

3.3.2 Iron Nanoparticles

FeNPs are nanoscale iron or its oxides. The size ranges from 1 to 100 nm. These materials have high reactivity, exhibit magnetic behavior, and also have large surface areas. They have huge versatility for this reason. The materials have been used to a huge extent in remediation, biomedical fields, and in energy storage systems because of their efficiency in catalysis, adsorption, and magnetic manipulation.

Synthesis of Iron Nanoparticles

FeNPs can be synthesized through various chemical, physical, and biological methods. Among these, chemical reduction methods and green synthesis are most commonly employed (Dhillon et al., 2012).

Chemical Reduction

A widely used approach involves reducing ferric chloride ($FeCl_3$) with sodium borohydride ($NaBH_4$) as a reducing agent. The reaction involves the reduction of Fe^{3+} ions to metallic iron, leading to the formation of FeNPs along with byproducts such as sodium chloride, hydrogen gas, and boric acid. Reaction as follows:

$$4FeCl_3 + 3NaBH_4 + 12H_2O \rightarrow 4\text{FeNPs} + 3NaCl + 12H_2 + 3B(OH)_3.$$

In this reaction, Fe^{3+} ions are reduced to metallic iron (Fe^0), forming FeNPs.

Green Synthesis

Biologically synthesized FeNPs use phytoextracts, bacteria, or fungi as reducing and capping agents, providing an eco-friendly alternative. Plant phytochemicals stabilize FeNPs and prevent oxidation during the synthesis process. FeNPs are very useful and can find applications in environmental and energy cleaning as well as medical applications. They adsorb and reduce heavy metals and organic compounds by transforming toxic chlorinated compounds into harmless hydrocarbons. Because of their super paramagnetic abilities, these NPs are important for drug targeting and improved MRI imaging applications. They also serve as catalysts in various industrial processes, for example in the Fenton-like reaction for wastewater treatment, whereby ferrous ions react with hydrogen peroxide to degrade pollutants (Chala, 2023). Moreover, Fe_3O_4NPs can increase the charge capacity and lifetime of lithium-ion batteries, therefore promoting improved storage and efficiency from a sustainability and energy point of view.

3.3.3 Copper Nanoparticles (CuNPs)

The attributes of CuNPs, such as electrical conduction, thermal stability, and antimicrobial characteristics, are valuable in many sectors. The unique surface plasmon resonance (SPR) and high surface area to volume (SA:V) ratio, which becomes more pronounced as the size of these NPs decreases from 1 to 100 nm, improve their performance in applications. Because of their high electrical conductive property and economically cheap cost when compared to gold and silver, CuNPs become important materials for electronics. They provide effective electrical connections in electronic devices using a conductive-integrated material in printed circuit boards (PCBs). For flexible and wearable electronics, CuNPs are being developed into conductive inks. Examples of next-generation technologies that will eventually require these inks to fabricate small, lightweight, flexible electric circuits are foldable screens, smart fabrics, and biosensors. By improving charge transfer and enabling material reduction, CuNPs, result in improvements in energy conversion performance in solar cells. These CuNPs can potentially inhibit microbial colonization in healthcare surfaces due to their antibacterial properties (Chatterjee et al., 2014). Copper ions generate ROS to destroy microbial membranes, thus producing an anticlose condition to bacteria. CuNPs can thus help to reduce infections by being integrated into coatings for hospital equipment, doorknobs, and fabrics. Antimicrobial CuNPs may also find application in water purification systems, where they will act on pathogens and pollutants. In areas where modern filtering methods are not available, this is a promise for drinking water safety in the future. Adding to that, the catalytic properties of CuNPs open the door for other industrial applications as fine chemical synthesis and organic pollutant degradation.

Synthesis of Copper Nanoparticles
CuNPs are typically synthesized using chemical reduction, electrochemical techniques, or green synthesis. A chemical method involves reducing copper sulfate ($CuSO_4$) with ascorbic acid as a reducing agent in the presence of a stabilizing agent:

$$CuSO_4 + 2C_6H_8O_6 \rightarrow CuNPs + C_6H_6O_6 + H_2SO_4$$

CuNPs exhibit excellent electrical conductivity, antimicrobial activity, and catalytic properties, making them valuable in electronics, antimicrobial coatings, and chemical synthesis. Their low cost compared to other noble metals further enhances their industrial potential.

3.3.4 Selenium Nanoparticles (SeNPs)

SeNPs are well recognized for their high bioavailability and their much potential therapeutic prospects in the pharmaceutical, food, and health industries (Ferro et al., 2021). Chemically, physically, or biologically synthesized SeNPs exhibit antioxidant, anticancer, and antibacterial properties, giving them great utility. The extent to which SeNPs fight the oxidative stress is still being actively investigated in the pharmaceutical industry because oxidative stress is linked to many diseases, including cancer, cardiovascular diseases, and neurological ailments. Anticancer activity is achieved by causing the production of ROS and modifying signaling pathways to stimulate apoptosis in cancer cells. These agents can preferentially kill the cancerous cells as a bonus, and reduce the harm to healthy cells. This is considered one of the very few prospects for next-generation therapies against cancer. SeNPs also enhance complementary anticancer drug activities. They are used in food applications as naturally occurring preservatives for killing foodborne pathogens (Yuan et al., 2023). They extend the shelf life of perishables by destroying bacterial cell membranes and inhibiting microbial enzymes. Antimicrobials in a SeNP film packaging material enhance food safety and quality throughout storage and transit. SeNPs are dietary supplements that remedy selenium deficiencies causing Keshan disease, infertility, and immune impairment. Selenium is a major micromineral that assists antioxidant enzymes like glutathione peroxidase to prevent cells from oxidative damage (Ndwandwe et al., 2021). Absorption and bioavailability of selenium supplements are much lower than for nano-formulation which offers adequate tissue delivery. Applications of nanomedicine include the use of SeNPs for wound healing, cardiovascular health, and immunological modulation. Clearly, their use as therapeutic agents and medicine carrier improves their potential.

Synthesis of Selenium Nanoparticles
SeNPs are synthesized through chemical, physical, and biological approaches. Biological methods often involve the reduction of selenite ions (SeO_3^{2-}) using microorganisms or plant extracts (Bisht et al., 2022). For example, *Bacillus* sp. can reduce selenite to elemental selenium:

$$SeO_3^{2-} + NADH \rightarrow SeNPs + NADa^{+}$$

SeNPs possess antioxidant property, anticancer property, and antimicrobial property. They are used in pharmaceuticals, food preservation, and dietary supplements, as they are important in human health. The unique properties of NPs such as zinc, iron, copper, and selenium have pioneered their industrial and biological applications. Among the most available sustainable NP synthesis technologies, green synthesis is worth mentioning. In agriculture, electronics, environmental clean-up, and medicine, these have proven their industrial use. New functionalizations have to be explored and optimized for new methods of nanomaterials or metal NPs; this can help beliefs in science and technology. Metal NPs could have diverse industrial applications. Nano-fertilizers improve soil fertility and reduce nutrient loss. CuNPs utilized in conductive inks for printed circuit board design are some examples. ZnNPs also aid in enhancing energy storage and solar cell efficiencies. Furthermore, NPs such as ZnNPs and SeNPs are used in skincare formulations as UV protectants and antioxidants.

3.4 Magnetic Nanoparticles (MNPs)

MNPs are a form of material in which one distinguishes sizes between 1 and 100 nm. These NPs have a magnetic nature, which is due to the composition they hold, usually iron oxides, i.e., magnetite (Fe_3O_4) or maghemite (γ-Fe_2O_3). Their characteristics include very high surface area to volume (SA:V) ratios, superparamagnetism, and the ability to tune surface chemistry. Such singular features account for their usefulness in several applications, especially in the biomedical field, imaging, diagnostics, and drug delivery.

3.4.1 Role of Fungi in Producing Magnetic Nanomaterials

Fungi have been used as effective biological factories for environmentally sustainable production of NPs with high magnetic value. Fungi offer a green and cost-effective alternative to chemical or physical means and minimize any kind of environmental impact. This involves the bioreduction of metal salts into NPs using fungi metabolites that can act as both stabilizing and reducing agents. The fungi reported to synthesize magnetic NPs include *A. niger*, *F. oxysporum*, and *Penicillium* sp. (Rai et al., 2021). This would typically be an enzymatic process involving enzymes like nitrate reductases and a pool of SMs like phenolic compounds and proteins that carry out the function by reducing metal precursors, such as Fe^{3+} or Fe^{2+}, into iron oxide NPs (Crichton, 2016). The biomolecules also aid in the stabilization of the NPs to prevent agglomeration and also to stabilize the size distribution in terms of size. The benefits of fungal-mediated synthesis are scalability, ability to control sizes, and synthesis of NPs with functionalized surfaces that could be vital in biomedical use. The secreted proteins in fungi, for example, could confer specific functional groups (–OH, –COOH, or $-NH_2$) on the NP surface, improving compatibility with the biological system; thus, making them suitable for targeted drug delivery or imaging techniques. Overall reaction would read as follows for synthesis through fungal metabolites:

$$Fe^{3+} + \text{Fungal Metabolites} \rightarrow Fe_3O_4\ \text{(Magnetite)} + \text{Stabilized Nanoparticles}$$

MNPs have excellent uses as agents for drug delivery and in medical imaging. They have superb paramagnetic characteristics, which are handy in MRI because they can increase contrasts between normal and sick tissues that can show important morphological changes in tumours and other abnormal growths. An external magnetic field modifies the relaxation time (or period) of surrounding water protons which in turn improves the picture resolution using MNPs. These functionalized MNPs modified with a biocompatible polymers coating [e.g., dextran or polyethylene glycol (PEG)] specifically target tumours hence making them safer and more effective as compared to medicines containing gadolinium. MNPs may be used for surface functionalization or encapsulation and magnetic field guidance to carry therapeutic chemicals or anticancer drugs to the destination site. Slow release of the drug at the location under environmental cues like pH or temperature helps in minimizing unwanted effects caused to the surrounding healthy tissue. Treatment by magnetic hyperthermia kills cancer cells by using MNPs. With these characteristics, they have enabled theragnosis delivering both therapy and real-time monitoring together. Add to this their specificity, biocompatibility, controlled release of drug, and multimodal functionality to use them in biomedicine; one can look at a way of avoiding multiple procedures and improved therapeutic outcomes.

4 Fungal Metabolites in Nanomineral Synthesis

Fungi have emerged as promising biological agents for the green synthesis of NPs, offering advantages such as high yields, low toxicity, and diverse metabolite production and act as sustainable alternatives to traditional chemical ways (Anjum et al., 2023; Loshchinina et al., 2022) (Fig. 4). Their potent enzymatic structures allow producing different types of metal NPs, such as gold, silver, and copper, with the intra- or extracellular process (Silva et al., 2016). The synthesis of NPs with varied physicochemical properties through fungi mediated synthesis enables the optimization of the synthesis conditions and fungal species (Anjum et al., 2023; Loshchinina et al., 2022). It has been used in the field of agriculture, medicine, and industry, especially in the management of plant diseases and crop enhancement (Anjum et al., 2023). The synthesis of NPs by fungi occurs in an intracellular and extracellular manner. The process of intracellular synthesis consists of metal ions uptake and reduction within the fungal cells through the help of proteins, polysaccharides, and enzymes. Instead, extracellular production is motivated by fungal secretions like proteins, peptides, and phenolic compounds into the surrounding medium, which are reducing and stabilizing agents (Xu et al., 2023). These metabolites are not only helpful in reducing metal ions but are also useful in stabilizing the NPs to improve their biocompatibility and functionality. The applications of these NPs are in different fields:

- Biomedicine: Silver and gold NPs exhibit antimicrobial and anticancer activities, making them suitable for medical applications (Syed et al., 2021).
- Agriculture: Fungal NPs are used in plant disease management and as environmentally friendly pesticides and fertilizers (Anjum et al., 2023).
- Industrial Applications: The diverse bioactivities of fungal metabolites, such as those from *F. oxysporum*, have potential uses in biotechnology and nanotechnology (Ibrahim et al., 2021).

4.1 Role of Fungi in Biomineralization

On Earth, fungi can flourish in nearly any type of environment, including many harsh ecosystems where other living things cannot. Fungi are important biosphere creatures that play important roles in symbioses, element cycle, plant pathogenicity, and the biodegradation of organic matter in both terrestrial and aquatic ecosystems. In terrestrial environments, several mycorrhizal and free-living filamentous fungus species carry out these functions. Bacteria and saprotrophic fungus are important organisms that mediate the breakdown of organic materials in the marine environment (Gadd, 2021). Compared to bacteria, fungi are more effective at breaking down particulate organic matter (POM) because their hyphae can penetrate and colonize solid substrates. Fungal metabolism has a role in all of the aforementioned processes, which can lead to the precipitation of biominerals like phosphates, metal carbonates, oxalates, oxides, and certain clay minerals. Numerous applications of this type of fungal biomineralization have been made, such as the immobilization of contaminating elements from terrestrial and aquatic habitats, the restoration of historic buildings and monuments, and the development of refined biomaterials (Bindschedler et al., 2016).

The mechanism by which organisms produce minerals is known as biomineralization, and it happens in all kingdoms of life, including archaea, bacteria, and eukarya. Plant colonization and the revegetation of contaminated ecosystems can be facilitated by the immobilization of toxic metals near plant roots resulting from biomineralization in the soil. Metal and mineral transformation is a major component of fungal activity wherever they exist, with biomineralization being a significant characteristic. Fungal biomineralization occurs widely in both natural and anthropogenic environments (Luo et al., 2023). The most prevalent fungal biominerals include carbonates, oxides, phosphates, sulphates, and sulfides, just like biominerals produced in other biological systems. Oxalates have a specific relationship with fungi; calcium oxalate is the best known example. In addition to calcium oxalate, various metal oxalates can form depending on the nature of substrate. The fungal mediated transformation of manganese oxides (MnO_2) leads to the precipitation of manganese oxalate dehydrate ($MnC_2O_4{\cdot}2H_2O$), also known as lindbergite. Hydrated copper oxalate, referred to as moolooite, is found in lichens that thrive on copper-enriched rocks, while magnesium oxalate dihydrate, or glushinskite, appears in lichens growing on manganese and serpentinite ore. The fungi that rot wood in the presence of copper-based preservatives, such as copper-chrome arsenate, can cause copper oxalate (CuC_2O_4) to develop, and in this case, may increase copper tolerance. Moreover, the formation of copper oxalate may occur in the process of fungal degradation of the copper metal. It is important to note that the copper oxalate stability may result in the formation of the protective crust on copper, which has been explored in conservation of historical copper products (Liu et al., 2021).

4.2 Fungal Synthesis of Nanocalcium Carbonates and Silica Nanoparticles

The creation of NPs via bio-technology, especially via fungi, has been an issue of much concern because of its environmental friendliness and affordability. This uses the natural mechanism of fungi to generate NPs including nanocalcium carbonates and silica NPs which find a wide variety of applications in various industries. Fungi use bioreduction processes with the help of biomolecules or enzymes to produce NPs (Fig. 5).

4.2.1 Nanocalcium Carbonates

Fungal biofilms and crusts often contain calcium carbonate, e.g., lichen thalli, especially on marble and limestone. Conversely, calcium carbonate variations can be readily

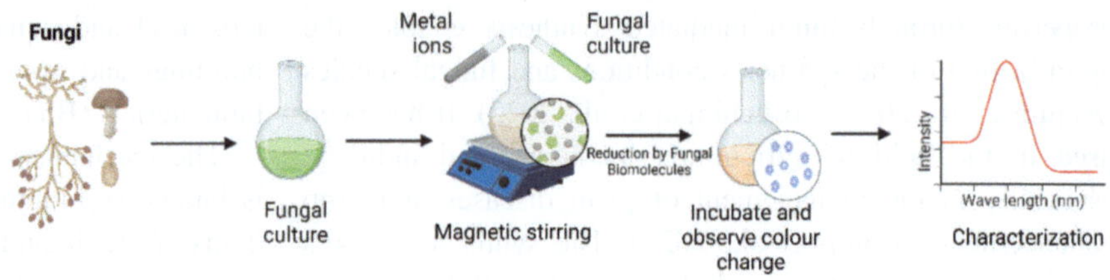

Fig. 4 Schematic representation of fungal mediated synthesis of nanoparticles. *Source* Authors

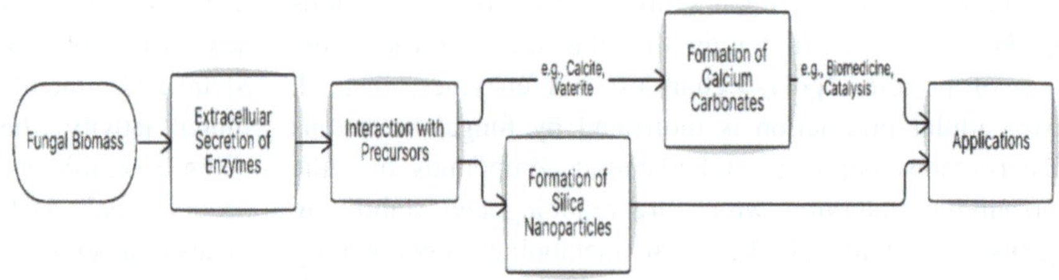

Fig. 5 Fungal synthesis of nanocalcium carbonates and silica nanoparticles. *Source* Authors

precipitated extracellularly like aragonite, viterite, and calcite in limestone-based microcosms or in related calcareous media. Biomineralization is a process used in synthesis of nanocalcium carbonates by fungi (Li et al., 2024). The urease-positive fungi, including *N. crassa*, can help make this possible by being able to precipitate calcium carbonate ($CaCO_3$) by their metabolic processes. When these fungi are grown in a medium which contains urea and calcium, they release urease which is an enzyme that breaks down urea to form ammonia and carbon dioxide. This reaction raises the pH of the immediate environment resulting in the calcium carbonate precipitation on the fungal mycelial filaments. This leads to the process of biomineralization of the crystals of calcite, which is a combination of calcium, carbon, and oxygen (Šebesta et al., 2022). One example of a fungus synthesizing nanocalcium carbonates is the fungus *N. crassa*. In one experiment, this fungus was cultured during 12 days at 25 °C in the medium with calcium and urea. The outcome of the process was wide-scale biomineralization of the fungal filaments and the energy dispersive X-ray examination revealed the presence of calcite crystals on the hyphae. It was demonstrated that more than 90% of calcium could be eliminated out of the media during the process of biomineralization, which indicated the efficiency of *N. crassa* in the production of nanocalcium carbonates (Li et al., 2014).

Nanocalcium carbonates produced by fungi could find some applications. It has found one important use in bioremediation and biorecovery, in which fungi such as *N. crassa* may be employed to isolate toxic or useful metals in solutions. It is also a process that contributes to the recycling of metals that are used again as well as in cleaning up the environment. Also, the biomineralization process may be implemented in the manufacturing of new electrochemical products because some metal carbonates produced by fungi can be utilized in the creation of advanced materials to use in electrochemically active applications (Li & Gadd, 2017a).

4.2.2 Silica Nanoparticles

Silica NPs produced within the framework of the green synthesis do not require the addition of hazardous chemicals to be capped contradicting the case of chemical synthesis. Silica, palladium, gold, and silver NPs have been produced through the usage of various fungus, including *Agaricus bisporus*, *F. oxysporum*, *Fusarium solani*, and *Penicillium crustosum* (Anjum et al., 2023). Fungal synthesis of silica NPs is a method that uses fungi as a biological agent to decrease and stabilize metal ions resulting in the formation of NPs. The biomolecules present in the fungi like enzymes, proteins, and other metabolites are the ones that facilitate this process as they are reducing and capping agents. Silica boasts of a number of advantages, including being cheap, safe, thermally stable, and being inert to chemical reactions.

Moreover, it is more effective in removing organic contaminants than other metals do. The use of fungus *F. oxysporum* is an example of fungal synthesis of silica NPs. This fungus has been used to produce silica NPs using wilted tomato plants. It entails characterization of the NPs with the help of the UV spectroscopy, FTIR, and SEM. The synthesized NPs have demonstrated a surface plasmon band at 280 nm, which means that they are synthesized (Marikani et al., 2016). Silica NP synthesized with fungal-based methods can be used in wide range of application because of their special characteristics like biocompatibility, stability, and large surface area. The biomedicine applications of them include biosensing and targeted drug delivery. Also, they are used in agriculture such as in the control of plant-based diseases and in the making of eco-friendly fertilizer and pesticides. Such NPs are also used in industrial sectors in different technological applications (Karande et al., 2021; Michael et al., 2022).

4.3 Applications in Bioremediation and Agriculture

NMs have appeared to be a bright instrument in such areas as bioremediation and agriculture because of their peculiarities and the possibility of improving the environmental sustainability. These applications take advantage of the small size and high reactivity of nanomaterials to solve the problems of soil and water management, crop production, and the environment.

4.3.1 Soil Reclamation

In the reclamation of the soil, NMs are being practical to solve the industrial and agricultural contamination. By increasing the effectiveness of bioremediation processes, they shorten the treatment time and cost as well as boosting microbial activity. They can be poor in effectiveness, however, due to the scope of the pollutants they are capable of (Nayak & Bhushan, 2021).

4.3.2 Water Quality Management

Nanomaterials are also used to treat water to eliminate pollutants including heavy metals and pesticides. Such methods as photocatalysis and nanobiosensors are being studied to identify and treat impure water bodies, rendering them safer to use in farms and by people (Spanos et al., 2021).

4.3.3 Crop Production and Protection

In the field of agriculture, nanotechnology aims at enhancing production and protection of crops by using nanofertilizers and nanopesticides. The products are nano-enabled to be more effective than the traditional products and deliver nutrients more effectively and provide better pest control with limited impacts on the environment.

4.3.4 Sustainable Agriculture

Nanomaterials also help in the sustainable farming process since they help to minimize the use of a large amount of agrochemicals. They facilitate accurate control of the inputs, which results into reduced cost of production and minimal environmental footprint. This is in line with the objectives of sustainable development because it facilitates efficient utilization of resources and reduction of wastages (Singh et al., 2020).

Nanomaterials-based applications in agriculture and bioremediation have certain advantages but the emerging issues about the environmental and health risks are worrisome. No one is confident of the long-term impacts of the exposure of nanomaterials on the environment and human health, so the research is required in the area of safety and the regulation of nanomaterials. Further detailed research is required on the fate and behavior of nanomaterials in soil and water systems. It is important to understand how they relate to soil constituents and plant systems to maximize on their utilization as well as how to reduce the risks they could cause. Further studies need to be aimed at the creation of safer and more efficient nanomaterials to be applied in agriculture and the environment (El-Ramady et al., 2023). There is a great potential in the use of these NMs in improving bioremediation and agriculture. Although they have many advantages, it is necessary to pay close attention to the environmental and health effects of their use so that its usage could be safe and sustainable.

5 Applications of Fungal Nanoparticles in Medicine

In the contemporary world nanobiotechnology methods that make use of the metals and microorganisms interaction in the creation of nanomaterials are trendy. The NPs synthesis biologically has therefore rendered nanotechnology to be termed as a Green Technology because the cost of synthesizing nanomaterials is significantly low compared to the chemical processes of synthesizing the NPs. Plants, fungi, algae, and bacteria are also used to produce the NPs, yet downstream processing costs, as well as low yields, limit their use in the industrial field (Prasad et al., 2016). Fungi is chosen to be used in the industry level to produce NPs because it can withstand high concentration of metals. Thus, a novel branch of fungi nanotechnology has been created and it is the Myconanotechnology (MNT). Fungal NPs are novel type of biological nanomaterials that have potential medical uses since it is synthesized in an environmentally friendly manner, and it is biocompatible and multifunctional. Such NPs are the derivatives of fungi and represent the next step of green nanotechnology and are applied in biomedical practice. They are applied in the antimicrobial effects and in the drug delivery systems, and in the diagnosis and treatment. As such, fungal NPs are a possible low cost/ecofriendly approach to the existing chemically synthesized NPs and can be applied in the resolution of issues like antimicrobial resistance, cancer treatment, and the development of effective approaches toward drug delivery (Fig. 6).

5.1 Nanoparticles for Drug Delivery and Targeted Therapy

Nanotechnology has transformed the health care system because it has allowed accurate, fast, and less invasive detection and treatment of infectious diseases. This development is fueled by superior technologies that employ nanos-

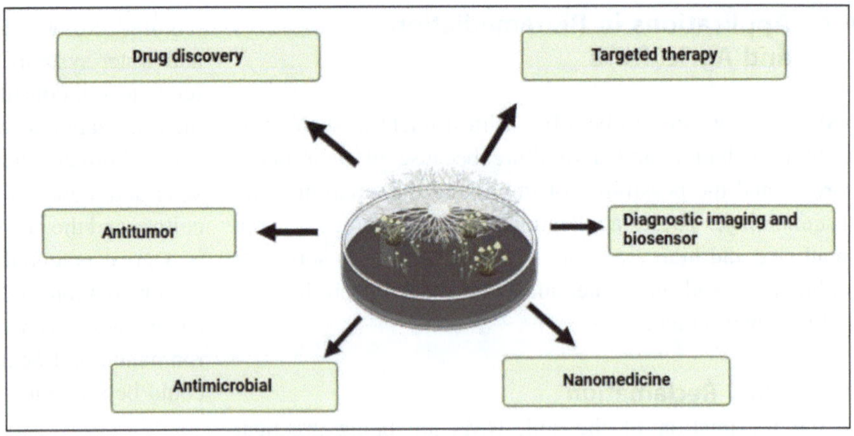

Fig. 6 Applications of fungal nanoparticles in medicine. *Source* Authors

tructures to develop desired delivery systems of drugs and diagnostics. As an example, polymeric NP and nanocapsules promote the delivery of therapeutic molecules (Fonte et al., 2016). In addition, antibiotic resistance is also being overcome successfully by the development of new antimicrobial NPs (Xie et al., 2014). Drugs developed by nanotechnology overcome the urgent necessity of developing stronger antimicrobial agents and allow releasing drugs into the specific cells and tissues with a minimum of dosage. Such innovations provide the healthcare systems with new opportunities to fight against multidrug-resistant pathogens and parasites, and the capability to cure cancer using highly toxic drugs with reduced strong side effects.

There are different categories of NPs that have demonstrated to have great potential in the fields of medicine. AgNPs with an excellent reputation as antimicrobial agents are applications in medical devices, dental implants, and cancer treatment (Wei et al., 2015). AuNPs are highly surface specific, and it allows them to bind to other drug molecules, DNA, and RNA (Furno et al., 2004). Other inorganic NPs, such as Al, Cd, Cu, Fe, Pa, Pt, Si, Ti, and Zn have been investigated as good drug delivery agents. Inorganic NP ceramics are also commonly utilized in drug delivery because of their large drug-carrying capacity, e.g., porous Si, Al, and Ti NPs. Silica nanoceramics are especially highly appropriate in the medical field due to its biocompatibility (Bamrungsap et al., 2012). Copper NPs are a cost-effective option that has less agglomeration and oxidation as stabilized and antimicrobial properties as well. With the help of human copper transporters, homeostasis difficulties related to these NPs are alleviated (Usman et al., 2013). The magnetic FeNPs provide a lot of potential in the rapid and accurate diagnosis of microbial infections (Mody et al., 2010). The common use of Zn and Cd in the quantum dots production is in optical applications (Kingsley et al., 2006). In non-medical areas, the noble metal NPs such as Au and Ag have extensive application in other fields of science such as electron microscopy, analysis, biosensors, electronics, and catalysis (Hiramatsu & Osterloh, 2004). A vast amount of research investment in nanotechnology, which is in billions of dollars, has been allocated to the use of NPs in life sciences and pharmacy (Bamrungsap et al., 2012). NP-based medications offer multiple advantages compared to conventional drugs, including improved treatment efficacy and reduced health risks. They hold great promise in treating various conditions, including cancer, infections, neurological diseases, cardiovascular issues, Alzheimer's, and ocular pathologies (Wei et al., 2015).

5.2 Antimicrobial Properties of Fungal Nanoparticles

NPs produced through a biological method have numerous uses in biomedicine. As an example, now-a-days mycologically synthesized silver NPs are utilized as a useful antibacterial agent. Silver AgNPs with high aspect ratio is able to interact with other particles, which induces their antimicrobial activity by increasing multi-folds. Special kind of clothings can be developed using silver or gold NPs that have been produced extracellularly by the *F. oxysporum.* This special cloths impregnate *F. oxysporum* generated silver or gold NPs which has antibacterial activity and is in application in a number of hospitals to help reduce infection caused by pathogenic bacteria including *Staphylococcus aureus* (Durán et al., 2007). The emerging technological developments in impregnation of AgNPs seems to inhibit proliferation of various drug resistant (MDR) strains of pathogenic organisms. The microorganisms will not develop resistance to silver as they normally develop resistance to the traditional/conventional narrow-spectrum antibiotics. In the medical field, AgNPs are applied in the construction of Ag-coated medical appliances, including nano-based lotions and gels (Rai et al., 2009). Due to their large biomedicine characteristics in fungal-based silver NPs can be produced either extracellularly or intracellularly within the fermentation (media). Other metallic NPs besides AgNPs include gold (Au), copper (Cu), selenium (Se), etc. are also prepared. It is also worth mentioning that the AuNPs possess widespread antimicrobial spectrum against high diversity of human and animal pathogens but are hardly applied in antimicrobial research due to expensive production process.

5.3 Antitumor Properties of Fungal Nanoparticles

Even unmodified biosynthesized NPs, without the delivery of a certain drug, have proven to have a lot of potential in antitumor treatments. Platinum NPs that are produced using *Saccharomyces boulardii* have demonstrated excellent potency against A431 (epidermoid carcinoma) and MCF-7 (breast cancer) cell lines (Borse et al., 2015). Correspondingly, AgNPs prepared through water extracts of the endophytic fungus *Cladosporium perangustum* were known to cause viability decrease in MCF-7 cells by elevating the actions of the cell death proteins caspase-3, 7 and 8 (Govindappa et al., 2020). Moreover, terbium oxide NPs produced with the help of *F. oxysporum* fungal biomass showed a dose-dependent cytotoxic reaction on MG-63 and Saos-2 cell lines as well as showed low levels of toxicity toward primary osteoblasts. These NPs also enhanced the generation of ROS, which eventually resulted in the death of cells (Iram et al., 2016). Other researchers found that the *Helminthosporium solani* fungi-derived AuNPs conjugated with doxorubicin, a popular anticancer drug, demonstrated greater cellular uptake and similar cytotoxicity as doxorubicin itself, in HEK293 cells (Kumar et al., 2008).

5.4 Fungal Nanoparticles in Diagnostic Imaging and Biosensors

The optical characteristics of metallic nanocrystals have recently been studied. With biosynthesis, metal NPs have been produced with a wide range of sizes, shapes, and optical characteristics such as high or low refractive indices, high transparency, new optical photoluminescence behavior, photonic crystals, and plasmon resonance (Iskandar, 2009). Nanophotonics is a branch of nanotechnology that studies the correlation between light and nanoparticles less than the wavelength of light resulting in such phenomena as localized surface plasmon resonance and size-dependent semiconductor bandgaps (Zhu et al., 2012). Indicatively, silver NPs produced through the *T. viride* showed photoluminescence emissions in a selected wavelength domain, showing the potential of their usage in labeling and imaging (Sarkar et al., 2010).

The other field of interest is optical radiation limiting which seeks to reduce the exposure of sensitive tissues to too much laser radiation during medical operations. This is through the use of dyes that absorb and dissipate excess radiations (Podgaetsky et al., 2004). Also, research on the biosynthesis of cadmium telluride quantum dots (CdTe QDs) with *S. cerevisiae* and *E. coli* showed size-dependent optical characteristics (Bao et al., 2010a, 2010b). Such protein-coated, water-soluble, and small CdTe QDs had a photoluminescence emission in the 488–551 nm range. In combination with folic acid, these QDs have been applied successfully to the in vitro imaging of cancer cells, and they showed a good biocompatibility (Bao et al., 2010a).

The emergence of the fungal nanobiotechnology has led to the design of new nanobiosensors that can be used to identify different biological species, such as antibodies, nucleic acids, infectious agents and metabolites. Such sensors have been effectively used in agriculture to identify fertilizers, herbicides, pesticides, pathogens, moisture levels, and soil pH (Prasad et al., 2014). The technology helps the farmers to maximize their crop management methods as it will dictate the best time to plant and when the crop is ready. Various researches have pointed out the possible use of fungal-based nanomaterials in agriculture. As an example, Elgorban et al. (2016) have shown the effectiveness of silver NPs synthesized by *A. versicolor* against such fungal pathogens as *Sclerotinia sclerotiorum* and *Botrytis cinerea* on strawberry plants. On the same note, Qian et al. (2013) also examined the antifungal properties of biologically produced NPs of *Epicoccum nigrum* against different fungal pathogens such as *Candida albicans, F. solani, Aspergillus* species.

Nanotechnology has tremendously advanced with the production of nanosensors especially wireless sensors as a result of the many benefits associated with them. Such sensors are able to efficiently watch diseases and development of crops, nutrient content, and environment. The newer nanosensors are able to detect trace levels of chemicals such as pesticides and herbicides in crops enhancing food safety and quality. These real-time monitoring devices when combined with nanofertilizers, nano pesticides and nano herbicides can go a long way into minimizing losses in crops and increasing agricultural output. As it was shown in a recent study, copper-doped montmorillonite could be used to detect the fungicide propineb in fresh and salt water symbols on-line with a high sensitivity (Abbaci et al., 2014).

5.5 Advances in Nanomedicine Using Fungal Nanoparticles

The introduction of other areas of nanotechnology such as biomedical nanotechnology has revolutionized the advancement of the fungal nanobiotechnology as well as broadening the delivery of diagnostic and therapeutic applications of drugs. Some of the diversities in the role of the fungus NPs are labeling; gene and drugs transferal; pathogen identifications; tissue making and engineering; tumor temperature increment; contrast enhancement by means of MRI and targeted drug release where nanomedicine is highly applied. The future of biosynthesized NPs has become a persisting field of study in drug delivery, gene therapy, cancer treatment, DNA analysis, antibacterial pharmaceuticals, separation science, biosensors, and MRI. Part of the study has been on the production of AuNPs using *Penicillium brevicompactum* (Mishra et al., 2011), the action of AgNPs on cancer cell lines (Jeyaraj et al., 2013), and biocompatible *T. viride* AgNPs with a diameter of between 5 and 40 nm (Fayaz et al., 2010). These NPs increase the antimicrobial effect of the ampicillin and kanamycin in the antibiotic. In this case, ampicillin had the greatest increase in activity as it was combined with AgNPs. Moreover, AgNPs of *F. oxysporum* exhibited antagonistic effects against *S. aureus* when they are exogenous. They are applicable in biomedical as well as textile industries (Duhan et al., 2017). Gajbhiye et al. (2009) demonstrated that *A. alternata* AgNPs increased the antifungal efficacy of fluconazole on *P. glomerata*, *Trichoderma* sp. and *C. albicans* but not on *P. herbarum* and *F. semitectum*. *Amylomyces rouxii* water extracts devoid of mycelia that had AgNPs were reported to be effective against pathogens like *Shigella dysenteriae*, *S. aureus*, *E. coli*, and *C. albicans* (Musarrat et al., 2010). Equally, the AuNPs produced on the surfaces of *Rhizopus oryzae* suppressed Gram-positive and Gram-negative bacteria (Das et al., 2009). In laser imaging, AgNPs of *T. viride* were notable with emission of 320–520 nm and overcame the issues of laser medicine by transfer of radiation blur to the tissue surface (Podgaetsky et al., 2004). Malaria vectors and disease therapies have also been reported to be fought using biosynthesized NPs produced by *Penicillium* spp. (Barabadi et al., 2019). Also, there is a study of metallic NPs obtained on the basis of the mushroom fruiting body, which has a variety of biomedical uses (Adebayo et al., 2021).

6 Environmental Applications of Fungal Nanoparticles

Fungal NPs are promising in terms of environmental issues. They enjoy special features, including a large surface area and biocompatibility, and thus they can be used in purifying water and remediation of soil. As an example, biomass-based biosorbents have the capability to absorb heavy metals and organic pollutants in wastewater, and this lowers the amount of environmental impact caused by industrial effluents. Also, those NPs may be employed to eliminate the persistent organic pollutants, including herbicides and pesticides, in waters and soils.

Moreover, fungal NPs can be used in the environmental monitoring. They can be programmed to identify certain pollutants or contaminants and thus early monitoring and swift action can be taken against environmental threats. These NPs will provide the solution to the current environmental problems in a sustainable and environmental-friendly way through utilizing the natural abilities of fungi. NP synthesis by means of fungi is a sustainable and environmentally friendly method of bioremediation. Metal ions can be reduced to NPs by fungi which have a variety of different metabolic abilities. As an example, fungi have the ability to reduce silver ions to AgNPs, which relationship has strong antimicrobial properties. These NPs may be employed to decontaminate water and soil which is polluted with heavy metals and organic pollutants. Also, NPs produced by fungi can increase the degradation of intractable organic pollutants both as catalysts or by producing ROS. The green synthesis technique offers a viable path to come up with environmentally sustainable and innovative technologies to bioremediate.

6.1 Bioremediation and Recovery of Heavy Metals

Bioremediation is an alternative technology that has been adopted to sustainably treat the environment of pollutants such as heavy metals, using microbes and plants. This is a more cost effective and efficient method of bio-approach compared to the conventional chemical, physical and enzymatic procedures. Fungal communities have great potential in bioremediation, especially biosorption, which is a promising field of study. Fungi may be utilized as absorption, amassing and alleviation of heavy metals in the environment (Verma & Kuila, 2019). Such fungi are able to decompose a very broad collection of environmental toxins, such as hydrocarbons, pesticides, heavy metals, and industrial chemicals. Fungal bioremediation has numerous advantages such as affordability, sustainability, adaptability, and friendliness to the environment, among others. Its applications are too many and they may be used to treat wastewater, remove oil spills, remediate contaminated soil. The bioremediation of fungi has colossal possibilities to resolve environmental complications and enhance a cleaner and healthier planet as its studies continue (Goutam et al., 2021) (Fig. 7; Table 2).

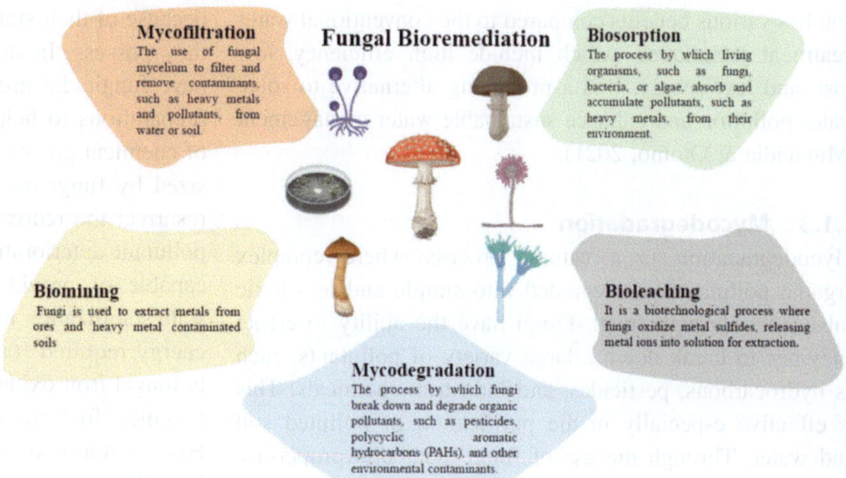

Fig. 7 Fungal bioremediation-harnessing the potential of fungi for a cleaner environment. *Source* Authors

Table 2 Fungal species involved in bioremediation by biomineralization of metals

Fungal species	Mechanism of fungal bioremediation	Targets	References
Mucor hiemalis (Wehmer)	Biomining	Ag, Al, Au, Cd, Co, Cr, Cu, Hg, Ni, Pb, Ti, U, and Zn	Hoque and Fritscher (2019)
A. niger (Tieghem)	Biomining	Zn, Ni, Mn, and Cu	Keshavarz et al. (2021)
	Bioaccumulation	Li	Sedlakova-Kadukova et al. (2020)
P. lilacinum (Thom)	Bioleaching	Cu, Al, Pb, Zn, and Sn	Xia et al. (2018)
Penicillium chrysogenum (Thom)	Bioaccumulation	Mn and Ag	Huerta-Rosas et al. (2020)
Penicillium simplicissimum (Oudem)	Biomining	Al	Shah et al. (2022)
A. bisporus (J.E. Lange) and *P. ostreatus* (Jacq.)	Mycofiltration	Pb, Cd, and Hg	Sarwar et al. (2023)
Lentinus squarrosulus (Mont.)	Mycofiltration	Pb, Cu, Cd, Cr, and Fe	Obayagbona et al. (2024)
Trametes versicolor (L.) and *P. ostreatus* (Jacq.)	Mycodegradation	Mn and Cu	Mallak et al. (2020)
Rhizopus arrhizus (A. Fisch.)	Biosorption	U, Zn, Cr, and Th	Tsezos and Volesky (1982)
Mucor rouxii	Biosorption	Cd, Ni, Pb, and Zn	Yan and Viraraghavan (2008)
S. cerevisiae (Desm.)	Biosorption	Cd, Co, Cr, Cu, Pb, and Zn	Farhan and Khadom (2015)
Candida tropicalis (Castell.)	Biosorption	Cd, Zn, Cr, and Zr	Bhattacharyya et al. (2024)
F. oxysporum (Schltdl.)	Biosorption	Pb and Co	Abdel-Galil et al. (2018)
T. viride (Persoon)	Biosorption	Ni, Cr, Pb, Cu, and Fe	Sujatha et al. (2013)
Gloeophyllum trabeum (Pers.)	Mycodegradation	Fe	Pozo et al. (2016)
Stropharia rugosoannulata (Farl.)	Mycodegradation	Cd and Ni	Yang et al. (2022)

Source Authors compilation

6.1.1 Fungal Biosorption

Fungal biosorption is a good environmental biotechnology being employed to exploit the capability of fungal biomass (living and dead) to efficiently eliminate pollutants that are considered detrimental to water. Pollution of the fungi is facilitated by the existence of different functional groups on the cell wall of fungi, including carboxyl, amino, and hydroxyl groups, which serve as the binding sites. The dead fungal biomass is usually desired in biosorption since it is stable and easy to handle. The advantages of this eco-friendly method are great efficiency, low cost, and it can be used on a large scale in wastewater treatment and ecological cleanup (Sintakindi & Ankamwar, 2021).

6.1.2 Mycofiltration

Mycofiltration is a new water treatment system, which makes use of the fungi powers in purifying the contaminated water. This sustainable strategy entails a flow of water through a system of fungal mycelium a microsystem of fungal threads. The mycelium is a biological filter and traps and decomposes a broad spectrum of contaminants, such as heavy metal, organic, and pathogens. Mycofiltration has various benefits compared to the conventional water treatment techniques, which include high efficiency, low cost, and low energy. It is a promising alternative to solve water pollution and enhance sustainable water management (Mnkandla & Otomo, 2021).

6.1.3 Mycodegradation

Mycodegradation is a natural process where complex organic pollutants are degraded into simple and less toxic substances through fungi. Fungi have the ability to create enzymes to break down a large variety of pollutants, such as hydrocarbons, pesticides, and industrial chemicals. This is effective especially in the purification of polluted soil and water. Through the use of fungi metabolic properties, mycodegradation can be considered an effective and affordable approach to environmental cleaning (Solanki et al., 2022).

6.1.4 Biomining

Biomining is a sustainable and clean technology of recovering valuable or precious metals out of low grade ores. In this process, microorganisms (except yeasts mostly bacteria and archaea) are used to oxidize the sulfides of metals and liberate the desired metals in a solution. Using the metabolic power of such microbes, biomining has a series of benefits as compared to conventional mining approaches, such as less energy usage, minimal ecological impact, and the capacity to recover metals in hitherto uneconomic ore deposits (Johnson, 2014).

6.1.5 Bioleaching

Bioleaching, is a biological clean-up process that involves the use of microorganisms to recover valuable metals in low grade ores. This is an environmentally friendly solution whereby special bacteria or archea are added to ore and oxidized metal sulfides are freed into a solution to extract the required metals. This is especially effective in mining other metals such as copper, gold, and uranium in an ore which is uneconomical to mine using conventional mining techniques. Bioleaching has a number of strengths, such as it consumes less energy, it has a lower environmental impact, and it is possible to extract precious metals out of waste materials (Dong et al., 2023a, 2023b).

7 Industrial Applications of Fungal Nanoparticles

NPs derived from fungi by the biological reduction of metal ions are applicable across possible industrial sectors because of their production process that is environmentally friendly; because of their stability, and in-built sophisticated qualities they possess. In the field of catalysis, these metallic NPs have fungicidal properties like those of AuNPs or AgNPs preparations to help increase the efficiency and selectivity of chemical processes. For example, the silver NPs, synthesized by fungi like *F. oxysporum* and *Aspergillus terreus*, resurrect the redox reactions which are usually needed for pollutant deterioration and fuel generation. These NPs are capable of providing active sites that support transition state stabilization in the reaction process while reducing the energy required for activation. Another area of application is fungal iron oxide NPs (Fe_3O_4) being used in wastewater treatment for Fenton-like processes to convert organic pollutants to non-toxic byproducts, thus being an environment-friendly water treatment system. Various applications of fungal derived nNPs are tabulated in Table 3.

In energy storage, fungal NPs improve battery and supercapacitor performance through increased charge capacity and longer cycle life. Anodes of lithium-ion batteries use fungal Fe_3O_4NPs, which provide high reversibility of intercalation and deintercalation of lithium ions and thereby improve battery life and efficiency. Similarly, manganese oxide NP synthesis by fungi is integrated into supercapacitors focusing on enhancing energy density and enabling shorter charging periods—a better capacity solution for energy storage devices.

Fantastic applications of fungal NPs have been integrated into electronics to enhance the conductivities and performance of various electronic parts. For instance, CuNPs produced by fungi such as *A. niger* are included in printed circuit boards (PCBs), solar cells, and flexible electronics. They further reduce resistivity in conductive pathways, thus making electronic parts more efficient and durable. Another widely accepted form of NPs is AgNPs for their commendably brilliant electrical conductivity and stability; thus, they are widely used in touch screens, other sensors, and electromagnetic shielding materials. Such qualities make them necessary for modern electronics.

Fungal NPs are added to materials science for modification of polymer properties, and coatings. The NPs of zinc oxide (ZnO) produced by fungi have been incorporated into polymer matrices to obtain improved mechanical properties, strength, and thermal resistance to give nanocomposites. These nanocomposites find application in packaging materials for improvement of UV resistance and durability. Silver and copper NPs are also used in making antimicrobial coating of medical devices, industrial tools, and high-touch surfaces to prevent microbial growth and maintain the necessary hygiene levels, thus making them applicable in health and safety.

Fungal NPs for food packaging and preservation serve to prolong life and ensure food safety. AgNPs are produced

Table 3 Applications of fungal mediated nanoparticles in various field

Application	Nanoparticle	Fungi/yeasts	References
Enhancement in antifungal activity	Ag	*A. alternata*	Gajbhiye et al. (2009)
Antimicrobial activity	Ag	*Aspergillus clavatus*	Saravanan et al. (2021)
	Ag	*Phytophthora infestans*	Thirumurugan et al. (2009)
	Ag	*P. ostreatus*	Devika et al. (2012)
	Au	*R. oryzae*	Das et al. (2009)
Antibacterial activity	Ag	*A. niger*	Gade et al. (2008)
	Ag	*Fusarium acuminatum*	Ingle et al. (2008)
	Ag	*Penicillium* sp.	Singh et al. (2014)
Wound healing activity	Ag	*A. niger*	Aguilar-Méndez et al. (2011)
Antifungal activity	Ag	*Colletotrichum gloeosporioides*	Durán et al. (2007)
	Ag	*Raffaelea* sp.	Kim et al. (2009)
Textile fabrics	Ag	*F. oxysporum*	Durán et al. (2007)
	Ag	*F. solani*	El-Rafie et al. (2010)
	Ag	*Lecanicillium lecanii*	Raja et al. (2011)
Antimicrobial properties against clinically relevant multidrug-resistant bacteria	Ag	*Macrophomina phaseolina*	Chowdhury et al. (2014)
Catalytic activity	Ag	*Penicillium oxalicum*	Du et al. (2015)
	Au	*A. japonicus AJP01*	Bhargava et al. (2015)
Antibacterial activity against MDR *E. coli* and *S. aureus*	Ag	*Penicillium* sp.	Singh et al. (2014)
Vegetable and fruit preservation	Ag	*T. viride*	Mohammed Fayaz et al. (2009)
Toxic to mosquito larvae	Au	*A. niger*	Soni and Prakash (2012)
Water hygiene management	Au	*R. oryzae*	Das et al. (2009)
Electric diode	CdS	*S. pombe*	Kowshik et al. (2002)
Live cell imaging and diagnostics	CdS	*F. oxysporum*	Kumar et al. (2007)

Source Authors compilation

by some fungi such as *Penicillium* sp., and incorporated into food packaging, which acts against microbes to prevent spoilage. Ag^{+} ions released from these NPs form ROS that damage microbial DNA and proteins as the food is stored. ZnO and TiO_2 NPs enhance the barrier properties of packaging materials, preventing oxidation and moisture ingress in perishable goods. Furthermore, SeNPs coatings are applied to fruits and vegetables and used on meat surfaces to keep food fresh through free radical neutralization and microbial contamination prevention. Biodegradable films laced with NPs of fungi are reported as environment-friendly alternatives to plastic packaging. They provide antimicrobial antioxidant properties for better preservation of food.

Fungal NPs have a number of advantages when used in industrial applications. The ability to synthesize it using non-toxic and renewable materials makes it green and thus determines its potential to be an environmentally friendly and economically favorable substitute to synthetic fabrics. Fungal systems have the benefit of being scaled to large-scale industrial production of NPs, so it is a more affordable and sustainable process than chemical processes. Fungal NP possess natural capping agents, including proteins and metabolites, which are biocompatible thus making them be applicable in many ways. They are also very stable in that, they last long and they are very durable in nearly all the industrial services.

8 Challenges and Future Directions in Fungal Nanoparticle Research

Fungal NPs can be utilized in several areas, such as medical, agricultural, and environmental remediation, to the greatest extent possible. Nevertheless, there are a number of obstacles to their extensive usage. Their complicated chemistry of synthesis is one of the key challenges. Actual control of the size, shape, and surface characteristics of the particles are critical determinants of performance and this consistency is a major challenge. Moreover, there is a lack of understanding of how fungal NPs exert their action, which means that their operation cannot be directed and advanced.

The other difficulty is that these NPs may be toxic. Although they demonstrate good antimicrobial and antifungal effects, their effects on human health and environment are yet to be investigated on a long-term basis. Thus, intensive toxicity testing is needed to make them safe to use. Besides, scalability of fungal NP production is an issue. The existing processes tend to be complicated and time-consuming, which can hardly be identified in terms of manufacturing large amounts to be used commercially.

In order to overcome these difficulties and capitalize on the potential of fungal NPs, the research of the future should be aimed at a few major sections. First, the standardization and reproducibility of synthesis processes is required to provide the regularity of quality and performance. Second, mechanisms of action of these NPs are critical to understanding to develop specific therapies and get the best out of them. Third, intense toxicity testing should be done to identify their safety profiles and determine the risks. Lastly, they should consider the production to come up with ways to create scalable and cost-effective production in order to implement it on a large-scale level. By overcoming these obstacles and expanding the research horizons, researchers can make the most of fungal NPs and transform different spheres.

9 Case Studies of Fungal Nanoparticle Production

Fungal NPs have demonstrated good usage in medicine especially in drug delivery, antimicrobial, antioxidant, and anticancer therapies. The following are some of the successful ones:

Case-1. *R. oryzae* was a fungal strain used as a biocatalyst in the production of MgO-NPs in environmentally friendly manner. This was optimized at the concentration of the precursor (4 mM), pH8, temperature (35 °C), and time of reaction (36 h). Techniques of characterization made sure that the resulting MgO-NPs were crystalline, homogeneous, and spherical in nature and had an average size of 20.38 ± 9.9 nm. These biogenic MgO-NPs had a significant antimicrobial activity (10.6 ± 0.4 mm) against *S. aureus*, *Bacillus subtilis*, *Pseudomonas aeruginosa*, *E. coli*, and *C. albicans* (200 mg/mL). They also had good effects in repelling *Culex pipiens* as larvae and adults with low concentration levels. Moreover, the MgO-NPs demonstrated significant 95.6 ± 1.6% decolorization of tannery effluents at 100 μg/100 mL dosage after 180 min that substantially reduced total suspended solids (TSS), total dissolved solids (TDS), biochemical oxygen demand (BOD), chemical oxygen demand (COD), and conductivity by 97.9%, 98.2%, 87.8%, 95.9%, and 97.3%, respectively, while adsorbing chromium ions by 98.2% (Hassan et al., 2021).

Case-2. Myco-synthesis of FeNPs was done using endophytic fungi *P. oxalicum*, which was isolated on *Tecomella undulata* and had an absorption band of 300 nm in the UV–Vis spectrum, and FTIR peaks of 3430.70–466.83 cm^{-1}. SEM analysis showed the existence of spherical particles with a diameter of 140 nm whereas AFM showed the width of the particles at 3.9 nm and height at 100 nm. The NPs successfully reduced methylene blue dye color by 99.17% in H_2O_2 in 6 h. The process emphasizes a green method of eliminating toxic effluents of textiles (Mathur et al., 2021).

Case-3. *Silver nanoparticles (POAgNPs) were prepared based on fungal endophyte P. oxalicum* in *Amoora rohituka* leaf leaves, and they appeared as sharp UV–Vis at 420 nm and functional groups via FTIR in reduction of silver salt. XRD indicated a high level of crystallinity with AFM, TEM and FESEM showing evenly spaced spherical NPs. POAgNPs exhibited antimicrobial activity, antioxidant activity against DPPH, superoxide, hydroxyl, and nitric oxide radicals with EC50 values of 9.034 ± 0.449, 56.378 ± 1.137, 34.094 ± 1.944, and 61.219 ± 0.69 μg/mL, respectively, and cytotoxic effects against breast cancer cell lines MDA-MB-231 (IC50: 20.080 ± 0.761 μg/mL) and MCF-7 (IC50: 40.038 ± 1.022 μg/mL). They were also found to induce apoptosis, prevent wound closure, change nuclear morphology, and elevate the concentration of Caspase-3, which is evidence of their potential as nanodrugs (Gupta et al., 2022).

Case-4. Fungal chitosan encapsulated *Gynura procumbens* mediated silver nanoparticles (FCs-GP-AgNPs) were prepared, evaluated using UV–vis, FTIR, XRD, TEM, and Zeta size analysis, and displayed sizes less than 100 nm and a zeta potential of − 32.9 mV (GP-AgNPs) and 53.6 mV (FCs-GP-AgNPs). Phenolic and flavonoid groups were confirmed by FTIR and crystal peaks of silver and chitosan were found by XRD. FCS-GP-AgNPs had a potent alpha-glucosidase (3.6 μg/mL) and alpha-amylase (7.5 μg/mL) inhibition, and *Bacillus cereus* (8.12 μg/mL), *S. aureus*

(4.08 μg/mL), *Listeria monocytogenes* (4.95 μg/mL), *E. coli* (8.25 μg/mL) and *Salmonella enterica* (4.12 μg/mL) MIC Biocompatibility assays of A549, LN229, and NIH3T3 cells were low in cytotoxicity, and FCS-GP-AgNPs appeared to be a good choice to apply in antibacterial and antidiabetic procedures (Sathiyaseelan et al., 2020).

The synthesis of fungal NPs has demonstrated successful usage in different industries because of their eco-friendly nature and other special qualities. The following are some of the outstanding cases.

9.1 Agriculture

An application of fungal NPs is in agriculture to manage plant diseases and as non-toxic pesticides and fertilizers, which increase crop production and limit environmental harm (Anjum et al., 2023). Fungi-based silver NPs have been useful in combating phytopathogens and have shown to have huge inhibition zones against a variety of plant pathogens (El-Moslamy et al., 2017).

9.2 Medicine

Fungi AgNPs have demonstrated potential in the medical field because of their antimicrobial, anticancer, and anti-HIV activity. They are applied to drug delivery systems and treatment of cancer (Zhao et al., 2018). Fungal NPs are also under consideration as a cure to fungal diseases in their invasive form, with it having a biocompatible and stable alternative to traditional medicine (Abd-Elsalam, 2023).

9.3 Environmental and Industrial Applications

The use of fungal NPs in environmental clean-up is explained by the fact that they can be used to degrade pollutants and cause less poisoning to the environment. The fungi-based synthesis of metallic nNPs such as silver and gold is a sustainable and economical process that can be used in industries, with such benefits as quick ion reduction and stabilization of NPs (Castro-Longoria et al., 2011).

9.4 Food Industry

Fungal NPs can be used in food systems to preserve food, where their antimicrobial properties increase the shelf life of food and provide food safety (Abd-Elsalam, 2023).

9.5 Comparative Studies on Fungal Versus Chemical Nanoparticle Synthesis

Comparative analyses of fungi and chemical NP synthesis show that there are certain advantages and disadvantages. Production using fungi is environmentally friendly, economical, and has the possibility of new NP structures. The chemical synthesis offers more control over the particle size and shape but utilizes toxic chemicals and uses a lot of energy. Synthesis Fungal synthesis is a prospective future of sustainable nanotechnology, whereas chemical synthesis is essential in the applications requiring a particular control over NP properties. The relative analysis of chemical and fungal synthesis is outlined in Table 4.

10 Conclusion

The application of the NPs that are produced by fungi on an industrial scale provides various lessons, especially in relation to the eco-friendly and sustainable production processes. These NPs are produced by means of fungi, reducing and stabilizing agents, which is a more environmentally friendly method compared to the traditional means of chemical production. The environmentally friendly process of synthesizing NPs through fungi involves use of low-energy processes that are cheap and do not use toxic substances. The process also uses biomolecules present in fungi, including enzymes, and proteins to reduce metal ions and stabilize NPs, which is why it can be described as a sustainable alternative to traditional approaches. By changing the synthesis conditions (temperature, pH, and precursor concentration) fungi can obtain a great variety of NPs with varying physicochemical properties. Such a versatility enables the creation of NPs with certain peculiarities in accordance with different industrial uses. The fungal NP can be used in agriculture in managing plant diseases and in the form of non-toxic pesticides, in medicine in antimicrobial as well as anticancer therapy, and in the environment in treatment of industrial effluents. Although fungal synthesis has potential, there are some hurdles like how best to optimize the synthesis environment as well as the mechanisms involved. The ongoing research aims at making the yield and stability better and the potential toxicity in the applications less possible. NPs synthesized by fungal metabolites are a sustainable and multifunctional mechanism of NP synthesis and have great potential in different industries. The greenness and versatility of the fungal production processes make them a good alternative to the conventional chemical procedure, yet more studies require optimization and understanding of the processes.

Table 4 Comparative studies on fungal versus chemical nanoparticle synthesis

Parameter	Fungal synthesis	Chemical synthesis	Comparative insights
Synthesis mechanism	Biological metabolic processes	Chemical reduction/precipitation	Fungal: enzyme-mediated, eco-friendly
Production cost	Low (€10–50/kg)	High (€100–500/kg)	80–90% cost reduction
Environmental impact	Minimal carbon footprint	Significant chemical waste	Fungal: sustainable approach
Processing temperature	25–37 °C	100–250 °C	Energy-efficient biosynthesis
Reaction time	5–15 days	2–6 h	Slower but more controlled
Nanoparticle morphology	Diverse, controlled shapes	Limited morphological control	Greater morphological diversity
Particle size range	10–100 nm	20–200 nm	Precise size modulation
Toxicity potential	Low biological toxicity	High chemical toxicity	Biocompatible alternative
Scalability	Moderate	High	Emerging large-scale potential
Purity levels	85–95%	90–99%	Comparable purification
Energy consumption	Low	High	Significant energy savings
Characterization techniques	SEM, TEM, XRD, FTIR	XRD, SEM, TEM	Comprehensive analysis methods
Functional group diversity	High	Moderate	Enhanced functional adaptability
Metal nanoparticle types	Ag, Au, ZnO, CuO	Ag, Au, Fe, Pt	Broader application spectrum

Source Authors compilation

References

Abbaci, A., Azzouz, N., & Bouznit, Y. (2014). A new copper doped montmorillonite modified carbon paste electrode for propineb detection. *Applied Clay Science, 90*, 130–134. https://doi.org/10.1016/j.clay.2013.12.036

Abdel-Galil, E. A., Aziz, O. A., Mostafa, A. Z., & Amin, M. (2018). Characterization and sorption behavior of some toxic metal ions on *Fusarium oxysporum* as biomass adsorbent. *Desalination and Water Treatment, 133*, 134–145. https://doi.org/10.5004/dwt.2018.23010

Abd-Elsalam, K. (2023). Special issue: Fungal nanotechnology 2. *Journal of Fungi, 9*. https://doi.org/10.3390/jof9050553

Aboul-Nasr, M. B., Mohamed, S. S., &Yasien, A. A. (2021). Fungus-mediated synthesis of silver nanoparticles using *Aspergillosis* causing fungi. *Journal of Environmental Studies, 24*(1), 42–52. https://doi.org/10.21608/jesj.2021.205306

Adebayo, E. A., Azeez, M. A., Alao, M. B., Oke, M. A., & Aina, D. A. (2021). Mushroom nanobiotechnology: Concepts, developments and potentials. In *Microbial nanobiotechnology: Principles and applications* (pp. 257–285). https://doi.org/10.1007/978-981-33-4777-9_9

Adiga, U., Rai, T., Desy, T. M., & Adiga, S. (2025). Role of gold nanoparticles as novel agents for cancer diagnosis and therapy. In *Metals in medicine* (pp. 393–435). Apple Academic Press.

Agrawal, S., Bhatt, M., Rai, S. K., Bhatt, A., Dangwal, P., & Agrawal, P. K. (2018). Silver nanoparticles and its potential applications: A review. *Journal of Pharmacognosy and Phytochemistry, 7*(2), 930–937.

Aguilar-Méndez, M. A., San Martín-Martínez, E., Ortega-Arroyo, L., Cobián-Portillo, G., & Sánchez-Espíndola, E. (2011). Synthesis and characterization of silver nanoparticles: Effect on phytopathogen Colletotrichum gloesporioides. *Journal of Nanoparticle Research, 13*(6), 2525–2532. https://doi.org/10.1007/s11051-010-0145-6

Ahmad, A., Mukherjee, P., Mandal, D., Senapati, S., Khan, M. I., Kumar, R., & Sastry, M. (2002). Enzyme mediated extracellular synthesis of CdS nanoparticles by the fungus, *Fusarium oxysporum. Journal of the American Chemical Society, 124*(41), 12108–12109. https://doi.org/10.1021/ja027296o

Ahmad, A., Senapati, S., Khan, M. I., Kumar, R., Ramani, R., Srinivas, V., & Sastry, M. (2003). Intracellular synthesis of gold nanoparticles by a novel alkalotolerantactinomycete. *Rhodococcus* species. *Nanotechnology, 14*(7), 824. https://doi.org/10.1088/0957-4484/14/7/323

Akther, T., Mathipi, V., Kumar, N. S., Davoodbasha, M., & Srinivasan, H. (2019). Fungal-mediated synthesis of pharmaceutically active silver nanoparticles and anticancer property against A549 cells through apoptosis. *Environmental Science and Pollution Research, 26*(13), 13649–13657. https://doi.org/10.1007/s11356-019-04718-w

Alprol, A. E., Mansour, A. T., El-Beltagi, H. S., & Ashour, M. (2023). Algal extracts for green synthesis of zinc oxide nanoparticles: promising approach for algae bioremediation. *Materials, 16*(7), 2819. https://doi.org/10.3390/ma16072819

Alrabadi, N. I., Thalij, K. M., Hussein, E. I., & Al-Trad, B. M. (2017). Antibacterial activity of ag and MgO nanoparticles synthesized by *Trichoderma viride. Journal of Applied Environmental and Biological Sciences, 7*(8), 94–101. https://doi.org/10.9734/bji/2017/29534

Anjum, S., Vyas, A., & Sofi, T. (2023). Fungi-mediated synthesis of nanoparticles: Characterization process and agricultural applications. *Journal of the Science of Food and Agriculture, 103*(10), 4727–4741. https://doi.org/10.1002/jsfa.12496

Astruc, D. (2021). On the roles of electron transfer in catalysis by nanoclusters and nanoparticles. *Chemistry–A European Journal, 27*(66), 16291–16308. https://doi.org/10.1002/chem.202102477

Balaji, D. S., Basavaraja, S., Deshpande, R., Mahesh, D. B., Prabhakar, B. K., & Venkataraman, A. (2009). Extracellular biosynthesis of functionalized silver nanoparticles by strains of Cladosporium cladosporioides fungus. *Colloids and Surfaces B: Biointerfaces, 68*(1), 88–92. https://doi.org/10.1016/j.colsurfb.2008.09.022

Bamrungsap, S., Zhao, Z., Chen, T., Wang, L., Li, C., Fu, T., & Tan, W. (2012). Nanotechnology in therapeutics: A focus on nanoparticles as a drug delivery system. *Nanomedicine, 7*(8), 1253–1271. https://doi.org/10.2217/nnm.12.87

Banerjee, K., & Ravishankar Rai, V. (2018). A review on mycosynthesis, mechanism, and characterization of silver and gold nanoparticles. *BioNanoScience, 8*(1), 17–31. https://doi.org/10.1007/s12668-017-0437-8

Bansal, V., Rautaray, D., Bharde, A. A., Ahire, K. C., Sanyal, A., Ahmad, A., & Sastry, M. (2005). Fungus-mediated biosynthesis of silica and titania particles. *Journal of Materials Chemistry, 15*, 2583–2589. https://doi.org/10.1039/B503008K

Bansal, V., Syed, A., Bhargava, S. K., Ahmad, A., & Sastry, M. (2007). Zirconia enrichment in zircon sand by selective fungus-mediated bioleaching of silica. *Langmuir, 23*(9), 4993–4998. https://doi.org/10.1021/la062535x

Bao, H., Hao, N., Yang, Y., & Zhao, D. (2010a). Biosynthesis of biocompatible cadmium telluride quantum dots using yeast cells. *Nano Research, 3*, 481–489. https://doi.org/10.1007/s12274-010-0008-6

Bao, H., Lu, Z., Cui, X., Qiao, Y., Guo, J., Anderson, J. M., & Li, C. M. (2010b). Extracellular microbial synthesis of biocompatible CdTe quantum dots. *Acta Biomaterialia, 6*(9), 3534–3541. https://doi.org/10.1016/j.actbio.2010.03.030

Barabadi, H., Tajani, B., Moradi, M., Damavandi Kamali, K., Meena, R., Honary, S., et al. (2019). *Penicillium* family as emerging nanofactory for biosynthesis of green nanomaterials: A journey into the world of microorganisms. *Journal of Cluster Science, 30*, 843–856. https://doi.org/10.1007/s10876-019-01554-3

Bhainsa, K. C., & D'souza, S. F. (2006). Extracellular biosynthesis of silver nanoparticles using the fungus Aspergillus fumigatus. *Colloids and Surfaces B: Biointerfaces, 47*(2), 160–164. https://doi.org/10.1016/j.colsurfb.2005.11.026

Bhargava, A., Jain, N., Gangopadhyay, S., & Panwar, J. (2015). Development of gold nanoparticle-fungal hybrid based heterogeneous interface for catalytic applications. *Process Biochemistry, 50*(8), 1293–1300. https://doi.org/10.1016/j.procbio.2015.04.012

Bhattacharyya, K., Bhattacharjee, N., Sen, D., Lai, T. K., Ghosh, A. K., Pal, R. R., & Ganguly, S. (2024). Unlocking Cd(II) biosorption potential of *Candida tropicalis* XTA 1874 for sustainable wastewater treatment. *Scientific Reports, 14*(1), 15690. https://doi.org/10.1038/s41598-024-66336-y

Bindschedler, S., Cailleau, G., & Verrecchia, E. (2016). Role of fungi in the biomineralization of calcite. *Minerals, 6*(2), 41. https://doi.org/10.3390/min6020041

Bisht, N., Phalswal, P., & Khanna, P. K. (2022). Selenium nanoparticles: A review on synthesis and biomedical applications. *Materials Advances, 3*(3), 1415–1431. https://doi.org/10.1039/D1MA00639H

Borse, V., Kaler, A., & Banerjee, U. C. (2015). Microbial synthesis of platinum nanoparticles and evaluation of their anticancer activity. *International Journal of Emerging Trends in Electrical and Electronics, 11*(2), 2320–9569. https://doi.org/10.13140/RG.2.1.3132.5283

Bruins, M. R., Kapil, S., & Oehme, F. W. (2000). Microbial resistance to metals in the environment. *Ecotoxicology and Environmental Safety, 45*(3), 198–207. https://doi.org/10.1006/eesa.1999.1860

Castro-Longoria, E. (2016). Fungal biosynthesis of nanoparticles, a cleaner alternative. https://doi.org/10.1007/978-3-319-42852-9_13

Castro-Longoria, E., Moreno-Velasquez, S. D., Vilchis-Nestor, A. R., Arenas-Berumen, E., & Avalos-Borja, M. (2012). Production of platinum nanoparticles and nanoaggregates using *Neurospora crassa. Journal of Microbiology and Biotechnology, 22*(7), 1000–1004. https://doi.org/10.4014/jmb.1110.10085

Castro-Longoria, E., Vilchis-Nestor, A. R., & Avalos-Borja, M. (2011). Biosynthesis of silver, gold and bimetallic nanoparticles using the filamentous fungus *Neurospora crassa. Colloids and Surfaces b: Biointerfaces, 83*(1), 42–48. https://doi.org/10.1016/j.colsurfb.2010.10.035

Chala, G. (2023). Review on green synthesis of iron-based nanoparticles for environmental applications. *Journal of Chemical Review, 5*(1), 1–14.

Chauhan, A., Anand, J., Parkash, V., & Rai, N. (2022). Biogenic synthesis: A sustainable approach for nanoparticles synthesis mediated by fungi. *Inorganic and Nano-Metal Chemistry, 53*, 460–473. https://doi.org/10.1080/24701556.2021.2025078

Chatterjee, A. K., Chakraborty, R., & Basu, T. (2014). Mechanism of antibacterial activity of copper nanoparticles. *Nanotechnology, 25*(13), 135101. https://doi.org/10.1088/0957-4484/25/13/135101

Chowdhury, S., Basu, A., Ray Chaudhuri, T., & Kundu, S. (2014). In-vitro characterization of the behaviour of Macrophomina phaseolina (Tassi) Goid at the rhizosphere and during early infection of roots of resistant and susceptible varieties of sesame. *European Journal of Plant Pathology, 138*(2), 361–375. https://doi.org/10.1007/s10658-013-0335-z

Claverie, M., Diez-Garcia, M., Martin, F., & Aymonier, C. (2019). Continuous synthesis of nanominerals in supercritical water. *Chemistry, 25*(23), 5814–5823. https://doi.org/10.1002/chem.201805435

Crichton, R. R. (2016). *Metal toxicity–an introduction.*

Das, S. K., Das, A. R., & Guha, A. K. (2009). Gold nanoparticles: Microbial synthesis and application in water hygiene management. *Langmuir, 25*(14), 8192–8199. https://doi.org/10.1021/la900585p

Das, S. K., Liang, J., Schmidt, M., Laffir, F., & Marsili, E. (2012). Biomineralization mechanism of gold by zygomycete fungi *Rhizopous oryzae. ACS Nano, 6*(7), 6165–6173. https://doi.org/10.1021/nn301502s

Devika, R., Elumalai, S., Manikandan, E., & Eswaramoorthy, D. (2012). Biosynthesis of silver nanoparticles using the fungus Pleurotus ostreatus and their antibacterial activity. *Open Access Science Report, 1*(12), 557. https://doi.org/10.4172/scientificreports.557

Dhillon, G. S., Brar, S. K., Kaur, S., & Verma, M. (2012). Green approach for nanoparticle biosynthesis by fungi: Current trends and applications. *Critical Reviews in Biotechnology, 32*(1), 49–73. https://doi.org/10.3109/07388551.2010.550568

Ding, C., Cheng, W., Sun, Y., & Wang, X. (2015). Novel fungus-Fe_3O_4 bio-nanocomposites as high performance adsorbents for the removal of radionuclides. *Journal of Hazardous Materials, 295*, 127–137. https://doi.org/10.1016/j.jhazmat.2015.04.032

Dong, Q., Chen, M., Zhang, Y., Song, P., Yang, H., Zhao, Y., & Zha, L. (2023a). Integrated physiologic and proteomic analysis of *Stropharia rugosoannulata* mycelia in response to Cd stress. *Journal of Hazardous Materials, 441*, Article 129877. https://doi.org/10.1016/j.jhazmat.2022.129877

Dong, Y., Zan, J., & Lin, H. (2023b). Bioleaching of heavy metals from metal tailings utilizing bacteria and fungi: Mechanisms, strengthen measures, and development prospect. *Journal of Environmental Management, 344*, Article 118511. https://doi.org/10.1016/j.jenvman.2023.118511

Du, L., Xu, Q., Huang, M., Xian, L., & Feng, J. X. (2015). Synthesis of small silver nanoparticles under light radiation by fungus Penicillium oxalicum and its application for the catalytic reduction of methylene blue. *Materials Chemistry and Physics, 160*, 40–47. https://doi.org/10.1016/j.matchemphys.2015.04.003

Duan, H., Wang, D., & Li, Y. (2015). Green chemistry for nanoparticle synthesis. *Chemical Society Reviews, 44*(16), 5778–5792. https://doi.org/10.1039/c4cs00363b

Duhan, J. S., Kumar, R., Kumar, N., Kaur, P., Nehra, K., & Duhan, S. (2017). Nanotechnology: The new perspective in precision agriculture. *Biotechnology Reports, 15*, 11–23. https://doi.org/10.1016/j.btre.2017.03.002

Durán, N., Marcato, P. D., Alves, O. L., De Souza, G. I., & Esposito, E. (2005). Mechanistic aspects of biosynthesis of silver nanoparticles by several *Fusarium oxysporum* strains. *Journal of Nanobiotechnology, 3*, 1–7. https://doi.org/10.1186/1477-3155-3-8

Durán, N., Marcato, P. D., De Souza, G. I., Alves, O. L., & Esposito, E. (2007). Antibacterial effect of silver nanoparticles produced by fungal process on textile fabrics and their effluent treatment. *Journal of Biomedical Nanotechnology, 3*(2), 203–208. https://doi.org/10.1166/jbn.2007.022

Elegbede, J. A., Lateef, A., Azeez, M. A., Asafa, T. B., Yekeen, T. A., Oladipo, I. C., et al. (2019). Silver-gold alloy nanoparticles biofabricated by fungal xylanases exhibited potent biomedical and catalytic activities. *Biotechnology Progress, 35*(5), e2829. https://doi.org/10.1002/btpr.2829

Elgorban, A. M., Aref, S. M., Seham, S. M., Elhindi, K. M., Bahkali, A. H., Sayed, S. R., & Manal, M. A. (2016). Extracellular synthesis of silver nanoparticles using *Aspergillus versicolor* and evaluation of their activity on plant pathogenic fungi. *Mycosphere, 7*(6), 844–852. https://doi.org/10.5943/mycosphere/7/6/15

El-Khateeb, A. Y., Hamed, E., Ibrahim, F. Y., & Hamed, S. E. (2019). Eco-friendly synthesis of selenium and zinc nanoparticles with biocompatible sargassum latifolium algae extract in preservation of edible oils. *Journal of Food and Dairy Sciences, 10*(5), 141–146.

El-Moslamy, S., Elkady, M., Rezk, A., & Abdel-Fattah, Y. (2017). Applying Taguchi design and large-scale strategy for mycosynthesis of nano-silver from endophytic *Trichoderma harzianum* SYA.F4 and its application against phytopathogens. *Scientific Reports, 7*. https://doi.org/10.1038/srep45297

El-Rafie, M. H., Mohamed, A. A., Shaheen, T. I., & Hebeish, A. (2010). Antimicrobial effect of silver nanoparticles produced by fungal process on cotton fabrics. *Carbohydrate Polymers, 80*(3), 779–782. https://doi.org/10.1016/j.carbpol.2009.12.028

El-Ramady, H., Abdalla, N., Sari, D., Ferroudj, A., Muthu, A., Prokisch, J., Fawzy, Z., Brevik, E., & Solberg, S. (2023). Nanofarming: Promising solutions for the future of the global agricultural industry. *Agronomy*. https://doi.org/10.3390/agronomy13061600

El-Seedi, H. R., El-Shabasy, R. M., Khalifa, S. A., Saeed, A., Shah, A., Shah, R., ... & Guo, W. (2019). Metal nanoparticles fabricated by green chemistry using natural extracts: Biosynthesis, mechanisms, and applications. *RSC Advances, 9*(42), 24539–24559. https://doi.org/10.1039/C9RA02225B

Farhan, S. N., & Khadom, A. A. (2015). Biosorption of heavy metals from aqueous solutions by *Saccharomyces cerevisiae*. *International Journal of Industrial Chemistry, 6*, 119–130. https://doi.org/10.1007/s40090-015-0038-8

Fayaz, A. M., Balaji, K., Girilal, M., Yadav, R., Kalaichelvan, P. T., & Venketesan, R. (2010). Biogenic synthesis of silver nanoparticles and their synergistic effect with antibiotics: A study against gram-positive and gram-negative bacteria. *Nanomedicine: Nanotechnology, Biology and Medicine, 6*(1), 103–109. https://doi.org/10.1016/j.nano.2009.04.006

Feng, F., Zhang, Y., Zhang, X., Mu, B., Qu, W., & Wang, P. (2024). Natural nano-minerals (NNMs): Conception, classification and their biomedical composites. *ACS Omega, 9*(16), 17760–17783. https://doi.org/10.1021/acsomega.4c00674

Fernandes, M., Padrão, J., Ribeiro, A. I., Fernandes, R. D., Melro, L., Nicolau, T., ... & Zille, A. (2022). Polysaccharides and metal nanoparticles for functional textiles: A review. *Nanomaterials, 12*(6), 1006. https://doi.org/10.3390/nano12061006

Ferro, C., Florindo, H. F., & Santos, H. A. (2021). Selenium nanoparticles for biomedical applications: From development and characterization to therapeutics. *Advanced Healthcare Materials, 10*(16), 2100598. https://doi.org/10.1002/adhm.202100598

Fonte, P., Reis, S., & Sarmento, B. (2016). Facts and evidences on the lyophilization of polymeric nanoparticles for drug delivery. *Journal of Controlled Release, 225*, 75–86. https://doi.org/10.1016/j.jconrel.2016.01.034

Furno, F., Morley, K. S., Wong, B., Sharp, B. L., Arnold, P. L., Howdle, S. M., et al. (2004). Silver nanoparticles and polymeric medical devices: A new approach to prevention of infection. *Journal of Antimicrobial Chemotherapy, 54*(6), 1019–1024. https://doi.org/10.1093/jac/dkh478

Gadd, G. M. (2021). Fungal biomineralization. *Current Biology, 31*(24), R1557–R1563. https://doi.org/10.1016/j.cub.2021.10.041

Gade, A. K., Bonde, P., Ingle, A. P., Marcato, P. D., Duran, N., & Rai, M. K. (2008). Exploitation of *Aspergillus niger* for synthesis of silver nanoparticles. *Journal of Biobased Materials and Bioenergy, 2*(3), 243–247. https://doi.org/10.1166/jbmb.2008.401

Gade, A., Gaikwad, S., Duran, N., & Rai, M. (2014). Green synthesis of silver nanoparticles by Phoma glomerata. *Micron, 59*, 52–59. https://doi.org/10.1016/j.micron.2013.12.005

Gajbhiye, M., Kesharwani, J., Ingle, A., Gade, A., & Rai, M. (2009). Fungus-mediated synthesis of silver nanoparticles and their activity against pathogenic fungi in combination with fluconazole. *Nanomedicine: Nanotechnology, Biology and Medicine, 5*(4), 382–386. https://doi.org/10.1016/j.nano.2009.06.005

Gomes, C. S., Santos, D., & Amaral, M. H. (2021). Nanominerals and nanomaterials utilized in pharmacy and therapeutics. https://doi.org/10.1007/978-3-030-65706-2_10

Goutam, J., Sharma, J., Singh, R., & Sharma, D. (2021). Fungal-mediated bioremediation of heavy metal–polluted environment. *Microbial Rejuvenation of Polluted Environment:, 2*, 51–76. https://doi.org/10.1007/978-981-15-7455-9_3

Govindappa, M., Lavanya, M., Aishwarya, P., Pai, K., Lunked, P., Hemashekhar, B., et al. (2020). Synthesis and characterization of endophytic fungi, *Cladosporium perangustum* mediated silver nanoparticles and their antioxidant, anticancer and nano-toxicological study. *BioNanoScience, 10*, 928–941. https://doi.org/10.1007/s12668-020-00719-z

Gudikandula, K., Vadapally, P., & Charya, M. S. (2017). Biogenic synthesis of silver nanoparticles from white rot fungi: Their characterization and antibacterial studies. *OpenNano, 2*, 64–78. https://doi.org/10.1016/j.onano.2017.07.002

Guilger-Casagrande, M., & Lima, R. D. (2019). Synthesis of silver nanoparticles mediated by fungi: a review. *Frontiers in Bioengineering and Biotechnology, 7*, 287. https://doi.org/10.3389/fbioe.2019.00287

Guilger-Casagrande, M., Germano-Costa, T., Bilesky-José, N., Pasquoto-Stigliani, T., Carvalho, L., Fraceto, L. F., & De Lima, R. (2021). Influence of the capping of biogenic silver nanoparticles on their toxicity and mechanism of action towards *Sclerotinia sclerotiorum*. *Journal of Nanobiotechnology, 19*, 1–18. https://doi.org/10.1186/s12951-021-00797-5

Gupta, P., Rai, N., Verma, A., Saikia, D., Singh, S. P., Kumar, R., et al. (2022). Green-based approach to synthesize silver nanoparticles using the fungal endophyte *Penicillium oxalicum* and their antimicrobial, antioxidant, and in vitro anticancer potential. *ACS Omega, 7*(50), 46653–46673. https://doi.org/10.1021/acsomega.2c05605

Hassan, S., Fouda, A., Saied, E., Farag, M., Eid, A., Barghoth, M., Awad, M., Hamza, M., & Awad, M. (2021). *Rhizopus oryzae*-mediated green synthesis of magnesium oxide nanoparticles (MgO-NPs): A promising tool for antimicrobial, mosquitocidal action, and tanning effluent treatment. *Journal of Fungi, 7*. https://doi.org/10.3390/jof7050372

Hassan, A. S., & Rasheed, Z. S. (2023). The biological activity and surface plasmon resonance effect of gold and silver nanoparticles. *Egyptian Journal of Botany, 63*(3), 683–691.

Heiligtag, F., & Niederberger, M. (2013). The fascinating world of nanoparticle research. *Materials Today, 16*, 262–271. https://doi.org/10.1016/J.MATTOD.2013.07.004

Hiramatsu, H., & Osterloh, F. E. (2004). A simple large-scale synthesis of nearly monodisperse gold and silver nanoparticles with adjustable sizes and with exchangeable surfactants. *Chemistry of Materials, 16*(13), 2509–2511. https://doi.org/10.1021/cm049532v

Hochella, M. F., Lower, S. K., Maurice, P. A., Penn, R. L., Sahai, N., Sparks, D. L., & Twining, B. S. (2008). Nanominerals, mineral nanoparticles, and earth systems. *Science, 319*, 1631–1635. https://doi.org/10.1126/science.1141134

Hoque, E., & Fritscher, J. (2019). Multimetal bioremediation and biomining by a combination of new aquatic strains of *Mucor hiemalis*. *Scientific Reports, 9*(1), 10318. https://doi.org/10.1038/s41598-019-46560-7

Horikoshi, S., & Serpone, N. (2013). Introduction to nanoparticles (pp. 1–24). https://doi.org/10.1002/9783527648122.CH1

Hosseini, M. R., Schaffie, M., Pazouki, M., Darezereshki, E., & Ranjbar, M. (2012). Biologically synthesized copper sulfide nanoparticles: Production and characterization. *Materials Science in Semiconductor Processing, 15*(2), 222–225. https://doi.org/10.1016/j.mssp.2012.03.012

Huerta-Rosas, B., Cano-Rodríguez, I., Gamiño-Arroyo, Z., Gómez-Castro, F. I., Carrillo-Pedroza, F. R., Romo-Rodríguez, P., & Gutiérrez-Corona, J. F. (2020). Aerobic processes for bioleaching manganese and silver using microorganisms indigenous to mine tailings. *World Journal of Microbiology and Biotechnology, 36*, 1–16. https://doi.org/10.1007/s11274-020-02902-6

Ibrahim, S., Sirwi, A., Eid, B., Mohamed, S., & Mohamed, G. (2021). Bright side of *Fusarium oxysporum*: Secondary metabolites bioactivities and industrial relevance in biotechnology and nanotechnology. *Journal of Fungi, 7*. https://doi.org/10.3390/jof7110943

Ingle, A., Gade, A., Pierrat, S., Sonnichsen, C., & Rai, M. (2008). Mycosynthesis of silver nanoparticles using the fungus Fusarium acuminatum and its activity against some human pathogenic bacteria. *Current Nanoscience, 4*(2), 141–144. https://doi.org/10.2174/157341308784340804

Iram, S., Khan, S., Ansary, A. A., Arshad, M., Siddiqui, S., Ahmad, E., et al. (2016). Biogenic terbium oxide nanoparticles as the vanguard against osteosarcoma. *Spectrochimica Acta Part A: Molecular and Biomolecular Spectroscopy, 168*, 123–131. https://doi.org/10.1016/j.saa.2016.05.053

Iskandar, F. (2009). Nanoparticle processing for optical applications—A review. *Advanced Powder Technology, 20*(4), 283–292. https://doi.org/10.1016/j.apt.2009.07.001

Jain, K. K. (2007). Applications of nanobiotechnology in clinical diagnostics. *Clinical Chemistry, 53*(11), 2002–2009. https://doi.org/10.1373/clinchem.2007.090795

Javed, M. R., Bilal, M., Ijaz, A., Seemab, R., Abbasi, A. Z., Farooq, K., et al. (2024). Mycogenesis of nanoparticles and their potential applications. In *Nanofungicides* (pp. 103–121). Elsevier. https://doi.org/10.1016/B978-0-323-95305-4.00006-6

Javed, R., Zia, M., Naz, S., Aisida, S. O., Ain, N. U., & Ao, Q. (2020). Role of capping agents in the application of nanoparticles in biomedicine and environmental remediation: Recent trends and future prospects. *Journal of Nanobiotechnology, 18*, 1–15. https://doi.org/10.1186/s12951-020-00704-4

Jayaseelan, C., Rahuman, A. A., Roopan, S. M., Kirthi, A. V., Venkatesan, J., Kim, S. K., & Siva, C. (2013). Biological approach to synthesize TiO_2 nanoparticles using *Aeromonas hydrophila* and its antibacterial activity. *Spectrochimica Acta Part a: Molecular and Biomolecular Spectroscopy, 107*, 82–89. https://doi.org/10.1016/j.saa.2012.12.083

Jeyaraj, M., Sathishkumar, G., Sivanandhan, G., MubarakAli, D., Rajesh, M., Arun, R., et al. (2013). Biogenic silver nanoparticles for cancer treatment: An experimental report. *Colloids and Surfaces b: Biointerfaces, 106*, 86–92. https://doi.org/10.1016/j.colsurfb.2013.01.027

Johnson, D. B. (2014). Biomining—Biotechnologies for extracting and recovering metals from ores and waste materials. *Current Opinion in Biotechnology, 30*, 24–31. https://doi.org/10.1016/j.copbio.2014.04.008

Kannan, R. R. R., Arumugam, R., Ramya, D., Manivannan, K., & Anantharaman, P. (2013). Green synthesis of silver nanoparticles using marine macroalga Chaetomorpha linum. *Applied Nanoscience, 3*(3), 229–233. https://doi.org/10.1007/s13204-012-0125-5

Karande, S., Jadhav, S., Garud, H., Kalantre, V., Burungale, S., & Patil, P. (2021). Green and sustainable synthesis of silica nanoparticles. *Nanotechnology for Environmental Engineering, 6*. https://doi.org/10.1007/s41204-021-00124-1

Kareem, S., Familola, O. T., Oloyede, A. R., & Dare, E. O. (2019). Microbial synthesis and characterization of silver nanoparticles using *Alternaria alternata. Applied Environmental Research, 41*(1), 1–7. https://digital.car.chula.ac.th/aer/vol41/iss1/1

Kazemi, S., Hosseingholian, A., Gohari, S. D., Feirahi, F., Moammeri, F., Mesbahian, G., Moghaddam, Z. S., & Ren, Q. (2023). Recent advances in green synthesized nanoparticles: From production to application. *Materials Today Sustainability*. https://doi.org/10.1016/j.mtsust.2023.100500

Keat, C. L., Aziz, A., Eid, A. M., & Elmarzugi, N. A. (2015). Biosynthesis of nanoparticles and silver nanoparticles. *Bioresources and Bioprocessing, 2*, 1–11. https://doi.org/10.1186/s40643-015-0076-2

Keshavarz, S., Faraji, F., Rashchi, F., & Mokmeli, M. (2021). Bioleaching of manganese from a low-grade pyrolusite ore using *Aspergillus niger*: Process optimization and kinetic studies. *Journal of Environmental Management, 285*, Article 112153. https://doi.org/10.1016/j.jenvman.2021.112153

Khan, S. A., & Ahmad, A. (2013). Fungus mediated synthesis of biomedically important cerium oxide nanoparticles. *Materials Research Bulletin, 48*(10), 4134–4138. https://doi.org/10.1016/j.materresbull.2013.06.038

Khan, M., Khan, A. U., Rafatullah, M., Alam, M., Bogdanchikova, N., & Garibo, D. (2022). Search for effective approaches to fight microorganisms causing high losses in agriculture: Application of P. lilacinum metabolites and mycosynthesised silver nanoparticles. *Biomolecules, 12*(2), 174. https://doi.org/10.3390/biom12020174

Kingsley, J. D., Dou, H., Morehead, J., Rabinow, B., Gendelman, H. E., & Destache, C. J. (2006). Nanotechnology: A focus on nanoparticles as a drug delivery system. *Journal of Neuroimmune Pharmacology, 1*, 340–350. https://doi.org/10.1007/s11481-006-9032-4

Kim, S. W., Kim, K. S., Lamsal, K., Kim, Y. J., Kim, S. B., Jung, M., ... & Lee, Y. S. (2009). An in vitro study of the antifungal effect of silver nanoparticles on oak wilt pathogen Raffaelea sp. *Journal of Microbiology and Biotechnology, 19*(8), 760–764.

Kowshik, M., Deshmukh, N., Vogel, W., Urban, J., Kulkarni, S. K., & Paknikar, K. M. (2002). Microbial synthesis of semiconductor CdS nanoparticles, their characterization, and their use in the fabrication of an ideal diode. *Biotechnology and Bioengineering, 78*(5), 583–588. https://doi.org/10.1002/bit.10233

Kumar, S. A., Ansary, A. A., Ahmad, A., & Khan, M. I. (2007). Extracellular biosynthesis of CdSe quantum dots by the fungus, Fusarium oxysporum. *Journal of Biomedical Nanotechnology, 3*(2), 190–194. https://doi.org/10.1166/jbn.2007.027

Kumar, S. A., Peter, Y. A., & Nadeau, J. L. (2008). Facile biosynthesis, separation and conjugation of gold nanoparticles to doxorubicin. *Nanotechnology, 19*(49), Article 495101. https://doi.org/10.1088/0957-4484/19/49/495101

Kumar, A., Singh, I. K., Mishra, R., Singh, A., Ramawat, N., & Singh, A. (2021). The role of zinc oxide nanoparticles in plants: A critical appraisal. In *Nanomaterial biointeractions at the cellular, organismal and system levels* (pp. 249–267). Cham: Springer. https://doi.org/10.1007/978-3-030-65792-5_10

Li, Q., Csetenyi, L., & Gadd, G. (2014). Biomineralization of metal carbonates by *Neurospora crassa. Environmental Science & Technology, 48*(24), 14409–14416. https://doi.org/10.1021/es5042546

Li, Q., & Gadd, G. (2017a). Fungal nanoscale metal carbonates and production of electrochemical materials. *Microbial Biotechnology, 10*, 1131–1136. https://doi.org/10.1111/1751-7915.12765

Li, Q., & Gadd, G. M. (2017b). Biosynthesis of copper carbonate nanoparticles by ureolytic fungi. *Applied Microbiology and Biotechnology, 101*(19), 7397–7407. https://doi.org/10.1007/s00253-017-8451-x

Li, P. J., Pan, J. J., Tao, L. J., Li, X., Su, D. L., Shan, Y., & Li, H. Y. (2021). Green synthesis of silver nanoparticles by extracellular extracts from aspergillus japonicus PJ01. *Molecules, 26*(15), 4479. https://doi.org/10.3390/molecules26154479

Li, Q., Zhang, M., Wei, B., Lan, W., Wang, Q., Chen, C., et al. (2024). Fungal biomineralization of toxic metals accelerates organic pollutant removal. *Current Biology, 34*(10), 2077–2084. https://doi.org/10.1016/j.cub.2024.04.005

Liang, X., Perez, M. A. M. J., Nwoko, K. C., Egbers, P., Feldmann, J., Csetenyi, L., & Gadd, G. M. (2019). Fungal formation of selenium and tellurium nanoparticles. *Applied Microbiology and Biotechnology, 103*, 7241–7259. https://doi.org/10.1007/s00253-019-09995-6

Liu, F., Shah, D. S., & Gadd, G. M. (2021). Role of protein in fungal biomineralization of copper carbonate nanoparticles. *Current Biology, 31*(2), 358–368. https://doi.org/10.1016/j.cub.2020.10.044

Loshchinina, E. A., Vetchinkina, E., & Kupryashina, M. A. (2022). Diversity of biogenic nanoparticles obtained by the fungi-mediated synthesis: A review. *Biomimetics, 8*. https://doi.org/10.3390/biomimetics8010001

Luo, Z., Li, F., Hu, C., Yin, L., Li, D., Ji, C., ... & Luo, K. (2021). Highly dispersed silver nanoparticles for performance-enhanced lithium oxygen batteries. *Journal of Materials Science & Technology, 73*, 171–177. https://doi.org/10.1016/j.jmst.2020.07.039

Luo, M., Li, Z., Su, M., Gadd, G. M., Yin, Z., Benton, M. J., et al. (2023). Fungal-induced fossil biomineralization. *Current Biology, 33*(12), 2417–2424. https://doi.org/10.1016/j.cub.2023.04.067

Mallak, A. M., Lakzian, A., Khodaverdi, E., Haghnia, G. H., & Mahmoudi, S. (2020). Effect of *Pleurotus ostreatus* and *Trametes versicolor* on triclosan biodegradation and activity of laccase and manganese peroxidase enzymes. *Microbial Pathogenesis, 149*, Article 104473. https://doi.org/10.1016/j.micpath.2020.104473

Mandal, A., & Banerjee, E. (2020). Introduction to nanoscience, nanotechnology and nanoparticles (pp. 1–39). https://doi.org/10.1007/978-981-15-5274-8_1

Marikani, K., Sangareswari, K., Suganya, P., Ganesan, R., & Rajarathinam, K. (2016). Biobased approach for the synthesis, characterization, optimization and application of silica nanoparticles by fungus *Fusarium oxysporum*. *Pharmaceutical and Biological Evaluations, 2*, 223–233.

Mathur, P., Saini, S., Paul, E., Sharma, C., & Mehtani, P. (2021). Endophytic fungi mediated synthesis of iron nanoparticles: Characterization and application in methylene blue decolorization. *Current Research in Green and Sustainable Chemistry, 4*, Article 100053. https://doi.org/10.1016/j.crgsc.2020.100053

McNamara, K., & Tofail, S. (2017). Nanoparticles in biomedical applications. *Advances in Physics: X, 2*, 54–88. https://doi.org/10.1080/23746149.2016.1254570

Michael, A., Singh, A., Roy, A., & Islam, M. R. (2022). [Retracted] Fungal- and algal-derived synthesis of various nanoparticles and their applications. *Bioinorganic Chemistry and Applications, 2022*(1), 3142674. https://doi.org/10.1155/2022/3142674

Mishra, A., Tripathy, S. K., Wahab, R., Jeong, S. H., Hwang, I., Yang, Y. B., et al. (2011). Microbial synthesis of gold nanoparticles using the fungus *Penicillium brevicompactum* and their cytotoxic effects against mouse mayo blast cancer C_2C_{12} cells. *Applied Microbiology and Biotechnology, 92*, 617–630. https://doi.org/10.1007/s00253-011-3556-0

Mnkandla, S. M., & Otomo, P. V. (2021). Effectiveness of mycofiltration for removal of contaminants from water: A systematic review protocol. *Environmental Evidence, 10*, 1–8. https://doi.org/10.1186/s13750-021-00232-0

Mody, V. V., Siwale, R., Singh, A., & Mody, H. R. (2010). Introduction to metallic nanoparticles. *Journal of Pharmacy and Bioallied Sciences, 2*(4), 282–289. https://doi.org/10.4103/0975-7406.72127

Moghaddam, A., Namvar, F., Moniri, M., Tahir, P., Azizi, S., & Mohamad, R. (2015). Nanoparticles biosynthesized by fungi and yeast: A review of their preparation, properties, and medical applications. *Molecules, 20*(9), 16540–16565. https://doi.org/10.3390/molecules200916540

Mohammed Fayaz, A., Balaji, K., Girilal, M., Kalaichelvan, P. T., & Venkatesan, R. (2009). Mycobased synthesis of silver nanoparticles and their incorporation into sodium alginate films for vegetable and fruit preservation. *Journal of Agricultural and Food Chemistry, 57*(14), 6246–6252. https://doi.org/10.1021/jf900337h

Molnár, Z., Bódai, V., Szakacs, G., Erdélyi, B., Fogarassy, Z., Sáfrán, G., et al. (2018). Green synthesis of gold nanoparticles by thermophilic filamentous fungi. *Scientific Reports, 8*(1), 3943. https://doi.org/10.1038/s41598-018-22112-3

Mukherjee, P., Ahmad, A., Mandal, D., Senapati, S., Sainkar, S. R., Khan, M. I., Parishcha, R., Ajaykumar, P. V., Alam, M., Kumar, R., & Sastry, M. (2001). Fungus-mediated synthesis of silver nanoparticles and their immobilization in the mycelial matrix: A novel biological approach to nanoparticle synthesis. *Nano Letters, 1*, 515–519. https://doi.org/10.1021/nl0155274

Musarrat, J., Dwivedi, S., Singh, B. R., Al-Khedhairy, A. A., Azam, A., & Naqvi, A. (2010). Production of antimicrobial silver nanoparticles in water extracts of the fungus *Amylomyces rouxii* strain KSU-09. *Bioresource Technology, 101*(22), 8772–8776. https://doi.org/10.1016/j.biortech.2010.06.065

Narayanan, K. B., & Sakthivel, N. (2010). Biological synthesis of metal nanoparticles by microbes. *Advances in Colloid and Interface Science, 156*(1–2), 1–13. https://doi.org/10.1016/j.cis.2010.02.001

Navrotsky, A. (2004). Energetic clues to pathways to biomineralization: Precursors, clusters, and nanoparticles. *Proceedings of the National Academy of Sciences of the United States of America, 101*(33), 12096–12101. https://doi.org/10.1073/PNAS.0404778101

Nayak, A., & Bhushan, B. (2021). Nano-bioremediation technologies for potential application in soil reclamation (pp. 510–529). https://doi.org/10.4018/978-1-7998-7062-3.CH020.

Ndwandwe, B. K., Malinga, S. P., Kayitesi, E., & Dlamini, B. C. (2021). Advances in green synthesis of selenium nanoparticles and their application in food packaging. *International Journal of Food Science and Technology, 56*(6), 2640–2650. https://doi.org/10.1111/ijfs.14916

Noor, S., Shah, Z., Javed, A., Ali, A., Hussain, S. B., Zafar, S., et al. (2020). A fungal based synthesis method for copper nanoparticles with the determination of anticancer, antidiabetic and antibacterial activities. *Journal of Microbiol Methods, 174*, Article 105966. https://doi.org/10.1016/j.mimet.2020.105966

Obayagbona, O. N., Dunkwu-Okafor, A., & Odigie, O. (2024). Mycofiltration of urban derived raw stormwater using *Lentinus squarrosulus*. *Bio-Research, 22*(2), 2336–2341. https://doi.org/10.4314/br.v22i2.2

Owaid, M. N. (2019). Green synthesis of silver nanoparticles by *Pleurotus* (oyster mushroom) and their bioactivity. *Environmental Nanotechnology, Monitoring & Management, 12*, Article 100256. https://doi.org/10.1016/j.enmm.2019.100256

Pandey, S., De Klerk, C., Kim, J., Kang, M., & Fosso-Kankeu, E. (2020). Eco friendly approach for synthesis, characterization and biological activities of milk protein stabilized silver nanoparticles. *Polymers, 12*(6), 1418. https://doi.org/10.3390/polym12061418

Pareek, V., Bhargava, A., & Panwar, J. (2020). Biomimetic approach for multifarious synthesis of nanoparticles using metal tolerant fungi: A mechanistic perspective. *Materials Science and Engineering: B, 262*, Article 114771. https://doi.org/10.1016/j.mseb.2020.114771

Patra, J. K., Das, G., & Baek, K. (2016). Towards a greener environment: Synthesis and applications of green nanoparticles. *Pakistan Journal of Agricultural Sciences, 53*, 345–354. https://doi.org/10.21162/PAKJAS/16.3027

Pelaz, B., Alexiou, C., Alvarez-Puebla, R. A., Alves, F., Andrews, A. M., Ashraf, S., et al. (2017). Diverse applications of nanomedicine. *ACS Nano, 11*(3), 2313–2381. https://doi.org/10.1021/acsnano.6b06040

Podgaetsky, V. M., Tat'yana, N. K., Tereshchenko, S. A., Reznichenko, A. V., & Selishchev, S. V. (2004, March). Laser-limiting materials for medical use. In *Smart medical and biomedical sensor technology* (Vol. 5261, pp. 183–191). SPIE. https://doi.org/10.1117/12.521641

Pozo, C., Melín, V., Elissetche, J. P., Contreras, D., Freer, J., & Rodríguez, J. (2016). Translocation of transition metals during the degradation of *Pinus radiata* by *Gloeophyllum trabeum* on the forest soil. *Journal of the Chilean Chemical Society, 61*(3), 3152–3156. https://doi.org/10.4067/S0717-97072016000300030

Prasad, R., Kumar, V., & Prasad, K. S. (2014). Nanotechnology in sustainable agriculture: Present concerns and future aspects. *African Journal of Biotechnology, 13*(6), 705–713. https://doi.org/10.5897/AJBX2013.13554

Prasad, R., Pandey, R., & Barman, I. (2016). Engineering tailored nanoparticles with microbes: Quo vadis. *Wiley Interdisciplinary Reviews: Nanomedicine and Nanobiotechnology, 8*(2), 316–330. https://doi.org/10.1002/wnan.1363

Prasher, P., Sharma, M., & Kumar, T. (2024). Bacteriogenic metal nanoparticles: Novel green fungicides. In *Nanofungicides* (pp. 85–102). Elsevier. https://doi.org/10.1016/B978-0-323-95305-4.00005-4

Qian, Y., Yu, H., He, D., Yang, H., Wang, W., Wan, X., & Wang, L. (2013). Biosynthesis of silver nanoparticles by the endophytic fungus *Epicoccum nigrum* and their activity against pathogenic fungi. *Bioprocess and Biosystems Engineering, 36*, 1613–1619. https://doi.org/10.1007/s00449-013-0937-z

Qin, W., Wang, C. Y., Ma, Y. X., Shen, M. J., Li, J., Jiao, K., et al. (2020). Microbe-mediated extracellular and intracellular mineralization: Environmental, industrial, and biotechnological applications. *Advanced Materials, 32*(22), 1907833. https://doi.org/10.1002/adma.201907833

Rai, M., Yadav, A., & Gade, A. (2009). Silver nanoparticles as a new generation of antimicrobials. *Biotechnology Advances, 27*(1), 76–83. https://doi.org/10.1016/j.biotechadv.2008.09.002

Rai, M., Bonde, S., Golinska, P., Trzcińska-Wencel, J., Gade, A., Abd-Elsalam, K. A., et al. (2021). *Fusarium* as a novel fungus for the synthesis of nanoparticles: Mechanism and applications. *Journal of Fungi, 7*(2), 139. https://doi.org/10.3390/jof7020139

Raja, K., Namasivayam, S., & Avimanyu, T. (2011). Silver nanoparticle synthesis from lecanicillium lecanii and evalutionary treatment on cotton fabrics by measuring their improved antibacterial activity with antibiotics against Staphylococcus aureus (ATCC 29213) and E. coli (ATCC 25922) strains. *International Journal of Pharmacy and Pharmaceutical Sciences, 4*(3).

Saghazadeh, S., Rinoldi, C., Schot, M., Kashaf, S. S., Sharifi, F., Jalilian, E., ... & Khademhosseini, A. (2018). Drug delivery systems and materials for wound healing applications. *Advanced Drug Delivery Reviews, 127*, 138–166. https://doi.org/10.1016/j.addr.2018.04.008

Salata, O. (2004). Applications of nanoparticles in biology and medicine. *Journal of Nanobiotechnology*. https://doi.org/10.1186/1477-3155-2-3

Salvadori, M. R., Ando, R. A., Oller do Nascimento, C. A., & Corrêa, B. (2014). Intracellular biosynthesis and removal of copper nanoparticles by dead biomass of yeast isolated from the wastewater of a mine in the Brazilian Amazonia. *PLoS One, 9*(1), e87968. https://doi.org/10.1371/journal.pone.0087968

Sanati, M., Ahmadi, R., & Torabi, P. (2022). The cytotoxic effects of green synthesized $AgCl_2$ nanoparticles on colon cancer (SW480) cells. *Journal of Biological Studies, 5*(1), 155–158. https://doi.org/10.62400/jbs.v5i1.6401

Sanghi, R., & Verma, P. (2009). Biomimetic synthesis and characterization of protein capped silver nanoparticles. *Bioresource Technology, 100*(1), 501–504. https://doi.org/10.1016/j.biortech.2008.05.048

Sant, S. (2012). Nanoparticles: From theory to applications. *Materials and Manufacturing Processes, 27*, 1462–1463. https://doi.org/10.1080/10426914.2012.663137

Saravanan, S. (2021). Functional, structural and morphological property of green synthesized silver nanoparticles using azadirachta indica leaf extract. *Materials Today: Proceedings, 47*, 1815–1818. https://doi.org/10.1016/j.matpr.2021.03.227

Sarkar, R., Kumbhakar, P., & Mitra, A. K. (2010). Green synthesis of silver nanoparticles and its optical properties. *Digest Journal of Nanomaterials and Biostructures, 5*(2), 491–496.

Sarwar, A., Nayyar, B. G., Irshad, H., Anwar, P., Olihk, N., & Ajmal, M. (2023). Mycofiltration of heavy metals (Pb, Cd, Hg) from aqueous solution by living biomass of two mushrooms *Pleurotus ostreatus* and *Agaricus bisporus* as biosorbents. *Journal of Water Chemistry and Technology, 45*(6), 599–606. https://doi.org/10.3103/S1063455X23060097

Sastry, M., Ahmad, A., Khan, M. I., & Kumar, R. (2003). Biosynthesis of metal nanoparticles using fungi and actinomycete. *Current Science*, 162–170.

Sathiyaseelan, A., Saravanakumar, K., Mariadoss, A. V. A., & Wang, M. H. (2020). Biocompatible fungal chitosan encapsulated phytogenic silver nanoparticles enhanced antidiabetic, antioxidant and antibacterial activity. *International Journal of Biological Macromolecules, 153*, 63–71. https://doi.org/10.1016/j.ijbiomac.2020.02.291

Scott, D., Toney, M., & Muzikár, M. (2008). Harnessing the mechanism of glutathione reductase for synthesis of active site bound metallic nanoparticles and electrical connection to electrodes. *Journal of the American Chemical Society, 130*, 865–874. https://doi.org/10.1021/ja074660g

Šebesta, M., Vojtková, H., Cyprichová, V., Ingle, A., Urík, M., & Kolenčík, M. (2022). Mycosynthesis of metal-containing nanoparticles—Fungal metal resistance and mechanisms of synthesis. *International Journal of Molecular Sciences, 23*. https://doi.org/10.3390/ijms232214084

Sedlakova-Kadukova, J., Marcincakova, R., Luptakova, A., Vojtko, M., Fujda, M., & Pristas, P. (2020). Comparison of three different bioleaching systems for Li recovery from lepidolite. *Scientific Reports, 10*(1), 14594. https://doi.org/10.1038/s41598-020-71596-5

Sen, K., Sinha, P., & Lahiri, S. (2011). Time dependent formation of gold nanoparticles in yeast cells: A comparative study. *Biochemical Engineering Journal, 55*(1), 1–6. https://doi.org/10.1016/j.bej.2011.02.014

Shah, M., Fawcett, D., Sharma, S., Tripathy, S. K., & Poinern, G. E. (2015). Green synthesis of metallic nanoparticles via biological entities. *Materials, 8*, 7278–7308. https://doi.org/10.3390/ma8115377

Shah, S. S., Palmieri, M. C., Sponchiado, S. R. P., & Bevilaqua, D. (2022). A sustainable approach on biomining of low-grade bauxite by *P. simplicissimum* using molasses medium. *Brazilian Journal of Microbiology, 53*(2), 831–843. https://doi.org/10.1007/s42770-022-00683-8

Shaligram, N. S., Bule, M., Bhambure, R., Singhal, R. S., Singh, S. K., Szakacs, G., & Pandey, A. (2009). Biosynthesis of silver nanoparticles using aqueous extract from the compactin producing fungal strain. *Process Biochemistry, 44*(8), 939–943. https://doi.org/10.1016/j.procbio.2009.04.009

Shankar, S. S., Ahmad, A., Pasricha, R., Khan, M. I., Kumar, R., & Sastry, M. (2004). Immobilization of biogenic gold nanoparticles in thermally evaporated fatty acid and amine thin films. *Journal of Colloid and Interface Science, 274*, 69–75. https://doi.org/10.1016/j.jcis.2003.12.011

Shnoudeh, A. J., Hamad, I., Abdo, R. W., Qadumii, L., Jaber, A. Y., Surchi, H. S., & Alkelany, S. Z. (2019). Synthesis, characterization, and applications of metal nanoparticles. In *Biomaterials and bionanotechnology* (pp. 527–612). Academic Press. https://doi.org/10.1016/B978-0-12-814427-5.00015-9

Siddiqi, K. S., & Husen, A. (2016). Fabrication of metal nanoparticles from fungi and metal salts: Scope and application. *Nanoscale Research Letters, 11*. https://doi.org/10.1186/s11671-016-1311-2

Silva, L. F., Hower, J. C., Izquierdo, M. T., & Querol, X. (2010). Complex nanominerals and ultrafine particles assemblages in phosphogypsum of the fertilizer industry and implications on human exposure. *The Science of the Total Environment, 408*(21), 5117–5122. https://doi.org/10.1016/j.scitotenv.2010.07.023

Silva, L. P., Bonatto, C. C., & Polez, V. L. (2016). Green synthesis of metal nanoparticles by fungi: Current trends and challenges. https://doi.org/10.1007/978-3-319-42990-8_4

Silva, L. P. C., Oliveira, J. P., Keijok, W. J., da Silva, A. R., Aguiar, A. R., Guimarães, M. C. C., & Braga, F. R. (2017). Extracellular biosynthesis of silver nanoparticles using the cell-free filtrate of nematophagous fungus *Duddingtonia flagrans*. *International Journal of Nanomedicine, 12*, 6373. https://doi.org/10.2147/IJN.S137703

Singh, R., Handa, R., & Manchanda, G. (2020). Nanoparticles in sustainable agriculture: An emerging opportunity. *Journal of Controlled Release: Official Journal of the Controlled Release Society*. https://doi.org/10.1016/j.jconrel.2020.10.051

Singh, S. (2020). Nanoparticles from fungi: A novel approach toward eco-friendly drug designing. *Modern Approaches in Drug Designing*. https://doi.org/10.31031/madd.2020.03.000558

Singh, D., Rathod, V., Ninganagouda, S., Hiremath, J., Singh, A. K., & Mathew, J. (2014). Optimization and characterization of silver nanoparticle by endophytic fungi Penicillium sp. isolated from Curcuma longa (turmeric) and application studies against MDR E. coli and S. aureus. *Bioinorganic Chemistry and Applications, 2014*(1), 408021. https://doi.org/10.1155/2014/408021

Sintakindi, A., & Ankamwar, B. (2021). Fungal biosorption as an alternative for the treatment of dyes in waste waters: A review. *Environmental Technology Reviews, 10*(1), 26–43. https://doi.org/10.1080/21622515.2020.1869322

Soetan, K. O., Olaiya, C. O., & Oyewole, O. E. (2010). The importance of mineral elements for humans, domestic animals and plants: A review. *African Journal of Food Science, 4*, 200–222. https://api.semanticscholar.org/CorpusID:46633650

Sofi, M. A., Nanda, A., Sofi, M. A., Sheikh, T., & Rather, G. A. (2022). In vitro evaluation of antimicrobial and anticancer potential of Artemisia absinthium growing in Kashmir Himalayas. *Journal of Applied & Natural Science, 14*(3). https://doi.org/10.31018/jans.v14i3.3619

Solanki, S., Sinha, S., & Singh, R. (2022). Myco-degradation of microplastics: An account of identified pathways and analytical methods for their determination. *Biodegradation, 33*(6), 529–556. https://doi.org/10.1007/s10532-022-10001-6

Soltani Nejad, M., Samandari Najafabadi, N., Aghighi, S., Pakina, E., & Zargar, M. (2022). Evaluation of Phoma sp. biomass as an endophytic fungus for synthesis of extracellular gold nanoparticles with antibacterial and antifungal properties. *Molecules, 27*(4), 1181. https://doi.org/10.3390/molecules27041181

Soni, N., & Prakash, S. (2012). Efficacy of fungus mediated silver and gold nanoparticles against Aedes aegypti larvae. *Parasitology Research, 110*(1), 175–184. https://doi.org/10.1007/s00436-011-2467-4

Spanos, A., Athanasiou, K., Ioannou, A., Fotopoulos, V., & Krasia-Christoforou, T. (2021). Functionalized magnetic nanomaterials in agricultural applications. *Nanomaterials, 11*. https://doi.org/10.3390/nano11113106

Sujatha, P., Kalarani, V., & Kumar, B. N. (2013). Effective biosorption of nickel(II) from aqueous solutions using *Trichoderma viride*. *Journal of Chemistry, 2013*(1), Article 716098. https://doi.org/10.1155/2013/716098

Sumanth, B., Lakshmeesha, T. R., Ansari, M. A., Alzohairy, M. A., Udayashankar, A. C., Shobha, B., ... & Almatroudi, A. (2020). Mycogenic synthesis of extracellular zinc oxide nanoparticles from Xylaria acuta and its nanoantibiotic potential. *International Journal of Nanomedicine*, 8519–8536. https://doi.org/10.2147/IJN.S271743

Syed, A., Saedi, M., Bahkali, A., Elgorgan, A., Kharat, M., Pai, K., Pichtel, J., & Ahmad, A. (2021). αAu_2S nanoparticles: Fungal-mediated synthesis, structural characterization and bioassay. *Green Chemistry Letters and Reviews, 15*, 61–70. https://doi.org/10.1080/17518253.2021.1999509

Tarver, S., Gray, D., Loponov, K., Das, D. B., Sun, T., & Sotenko, M. (2019). Biomineralization of Pd nanoparticles using *Phanerochaete chrysosporium* as a sustainable approach to turn platinum group metals (PGMs) wastes into catalysts. *International Biodeterioration & Biodegradation, 143*, Article 104724. https://doi.org/10.1016/j.ibiod.2019.104724

Thakor, R., Mistry, H., Patel, H., Jhala, D., Parmar, N., & Bariya, H. (2022). Biogenic synthesis of silver nanoparticles mediated by the consortium comprising the marine fungal filtrates of *Penicillium oxalicum* and *Fusarium hainanense* along with their antimicrobial, antioxidant, larvicidal and anticancer potency. *Journal of Applied Microbiology, 133*(2), 857–869. https://doi.org/10.1111/jam.15611

Thirumurugan, G., Shaheedha, S. M., & Dhanaraju, M. D. (2009). In vitro evaluation of antibacterial activity of silver nanoparticles synthesised by using Phytophthora infestans. *International Journal of ChemTech Research, 1*(3), 714.

Tsezos, M., & Volesky, B. (1982). The mechanism of uranium biosorption by *Rhizopus arrhizus*. *Biotechnology and Bioengineering, 24*(2), 385–401. https://doi.org/10.1002/bit.260240211

Usman, M. S., Zowalaty, M. E. E., Shameli, K., Zainuddin, N., Salama, M., & Ibrahim, N. A. (2013). Synthesis, characterization, and antimicrobial properties of copper nanoparticles. *International Journal of Nanomedicine*, 4467–4479. https://doi.org/10.2147/ijn.s50837

Vahabi, K., Mansoori, G. A., & Karimi, S. (2011). Biosynthesis of silver nanoparticles by fungus *Trichoderma reesei* (a route for large-scale production of AgNPs). *Insciences Journal, 1*(1), 65–79. https://doi.org/10.5640/insc.010165

Varshney, R., Mishra, A. N., Bhadauria, S., & Gaur, M. S. (2009). A novel microbial route to synthesize silver nanoparticles using fungus Hormoconis resinae. *Digest Journal of Nanomaterials and Biostructures, 4*(2), 349–355.

Verma, S., & Kuila, A. (2019). Bioremediation of heavy metals by microbial process. *Environmental Technology & Innovation, 14*, Article 100369. https://doi.org/10.1016/j.eti.2019.100369

Vetchinkina, E., Loshchinina, E., Kupryashina, M., Burov, A., Pylaev, T., & Nikitina, V. (2018). Green synthesis of nanoparticles with extracellular and intracellular extracts of basidiomycetes. *PeerJ, 6*, Article e5237. https://doi.org/10.7717/peerj.5237

Virginia, F., Cathrine, L., Leenus, P., Pratheema, P., & Philo, A. J. (2024). A Bioinspired green synthesis of ZnO nanoparticles using the ethanolic root extract of Coleus forskohlii: characterization and evaluation of its synergistic antimicrobial and anticancer activity: Nanochemistry. *Moroccan Journal of Chemistry, 12*(4), 1683–1706. https://doi.org/10.48317/IMIST.PRSM/morjchem-v12i4.50730

Volesky, B., & Holan, Z. R. (1995). Biosorption of heavy metals. *Biotechnology Progress, 11*(3), 235–250. https://doi.org/10.1021/bp00033a001

Wei, L., Lu, J., Xu, H., Patel, A., Chen, Z. S., & Chen, G. (2015). Silver nanoparticles: Synthesis, properties, and therapeutic applications. *Drug Discovery Today, 20*(5), 595–601. https://doi.org/10.1016/j.drudis.2014.11.014

Xia, M., Bao, P., Liu, A., Wang, M., Shen, L., Yu, R., & Zeng, W. (2018). Bioleaching of low-grade waste printed circuit boards by mixed fungal culture and its community structure analysis. *Resources, Conservation and Recycling, 136*, 267–275. https://doi.org/10.1016/j.resconrec.2018.05.001

Xie, S., Tao, Y., Pan, Y., Qu, W., Cheng, G., Huang, L., et al. (2014). Biodegradable nanoparticles for intracellular delivery of antimicrobial agents. *Journal of Controlled Release, 187*, 101–117. https://doi.org/10.1016/j.jconrel.2014.05.034

Xu, F., Li, Y., Zhao, X., Liu, G., Pang, B., Liao, N., Li, H., & Shi, J. (2023). Diversity of fungus-mediated synthesis of gold nanoparticles: Properties, mechanisms, challenges, and solving methods. *Critical Reviews in Biotechnology,* 1–17. https://doi.org/10.1080/07388551.2023.2225131

Yan, G., & Viraraghavan, T. (2008). Mechanism of biosorption of heavy metals by *Mucor rouxii. Engineering in Life Sciences, 8*(4), 363–371. https://doi.org/10.1002/elsc.200820237

Yang, Y., Meng, G., Ni, S., Zhang, H., & Dong, C. (2022). Genomic analysis of Strophoria rugosoannulata reveals its nutritional strategy and application potential in bioremediation. *Journal of Fungi, 8*(2), 162. https://doi.org/10.3390/jof8020162

Yuan, Q., Xiao, R., Afolabi, M., Bomma, M., & Xiao, Z. (2023). Evaluation of antibacterial activity of selenium nanoparticles against food-borne pathogens. *Microorganisms, 11*(6), 1519. https://doi.org/10.3390/microorganisms11061519

Zhao, C., Li, X., Ding, C., Liao, J., Du, L., Yang, J., ... & Sun, Q. (2016). Characterization of uranium bioaccumulation on a fungal isolate Geotrichum sp. dwc-1 as investigated by FTIR, TEM and XPS. *Journal of Radioanalytical and Nuclear Chemistry, 310*(1), 165–175. https://doi.org/10.1007/s10967-016-4797-2

Zhao, X., Zhou, L., Rajoka, M., Yan, L., Jiang, C., Shao, D., Zhu, J., Shi, J., Huang, Q., Yang, H., & Jin, M. (2018). Fungal silver nanoparticles: Synthesis, application and challenges. *Critical Reviews in Biotechnology, 38*, 817–835. https://doi.org/10.1080/07388551.2017.1414141

Zhu, T., Cloutier, S. G., Ivanov, I., Knappenberger, K. L., Robel, I., & Zhang, F. (2012). Nanocrystals for electronic and optoelectronic applications. *Journal of Nanomaterials, 2012*, 8. https://doi.org/10.1155/2012/392742

Zubair, M., Rafique, M. S., Khalid, A., Yaqub, T., Shahid, M. F., Alomar, S. Y., & Shar, M. A. (2023). The fabrication of gold–silver bimetallic colloids by microplasma: a worthwhile strategy for counteracting the surface activity of avian influenza virus. *Crystals, 13*(2), 340. https://doi.org/10.3390/cryst13020340

MIX
Papier aus verantwortungsvollen Quellen
Paper from responsible sources
FSC® C105338
FSC
www.fsc.org

If you have any concerns about our products,
you can contact us on
ProductSafety@springernature.com

In case Publisher is established outside the EU,
the EU authorized representative is:
Springer Nature Customer Service Center GmbH
Europaplatz 3, 69115 Heidelberg, Germany

Printed by Libri Plureos GmbH
in Hamburg, Germany